Innovations in Green Nanoscience and Nanotechnology

This book discusses how greener synthetic pathways are amenable and productive for the synthesis of novel nanomaterials. It furthers the integration of advances in green nanoscience and nanotechnology, including pathways dedicated to the design, development, and fabrication of a range of products and devices. Topics such as green nanotechnology for advanced energy systems, sustainable delivery systems, medicine, agri-nanotechnology for sustainable agriculture, nanotechnology in crop protection, and nanotechnology for soil conservation are included.

Features

- Provides a holistic view of green nanotechnology and its applications
- Places an emphasis on synthesis, characterization, and applications of green nanomaterials
- Discusses the development of innovative green synthetic pathways to produce novel biomaterials
- Includes characterization tools used in the material synthesis via green synthetic pathways
- Advocates green nanotechnology solutions for sustainability and energy

This book is aimed at researchers and professionals in nanotechnology, green chemistry, and chemical engineering.

Emerging Materials and Technologies

Series Editor: Boris I. Kharissov

The *Emerging Materials and Technologies* series is devoted to highlighting publications centered on emerging advanced materials and novel technologies. Attention is paid to those newly discovered or applied materials with potential to solve pressing societal problems and improve quality of life, corresponding to environmental protection, medicine, communications, energy, transportation, advanced manufacturing, and related areas.

The series takes into account that, under present strong demands for energy, material, and cost savings, as well as heavy contamination problems and worldwide pandemic conditions, the area of emerging materials and related scalable technologies is a highly interdisciplinary field, with the need for researchers, professionals, and academics across the spectrum of engineering and technological disciplines. The main objective of this book series is to attract more attention to these materials and technologies and invite conversation among the international R&D community.

Emerging Nanomaterials for Catalysis and Sensor Applications
Edited by Anitha Varghese and Gurumurthy Hegde

Advanced Materials for a Sustainable Environment: Development Strategies and Applications
Edited by Naveen Kumar and Peter Ramashadi Makgwane

Nanomaterials from Renewable Resources for Emerging Applications
Edited by Sandeep S. Ahankari, Amar K. Mohanty, and Manjusri Misra

Multifunctional Polymeric Foams: Advancements and Innovative Approaches
Edited by Soney C George and Resmi B. P.

Nanotechnology Platforms for Antiviral Challenges: Fundamentals, Applications and Advances
Edited by Soney C George and Ann Rose Abraham

Carbon-Based Conductive Polymer Composites: Processing, Properties, and Applications in Flexible Strain Sensors
Dong Xiang

Nanocarbons: Preparation, Assessments, and Applications
Ashwini P. Alegaonkar and Prashant S. Alegaonkar

Emerging Applications of Carbon Nanotubes and Graphene
Edited by Bhanu Pratap Singh and Kiran M. Subhedar

Micro to Quantum Supercapacitor Devices: Fundamentals and Applications
Abha Misra

Application of Numerical Methods in Civil Engineering Problems
M.S.H. Al-Furjan, M. Rabani Bidgoli, Reza Kolahchi, A. Farrokhian, and M.R. Bayati

For more information about this series, please visit: www.routledge.com/Emerging-Materials-and-Technologies/book-series/CRCEMT

Innovations in Green Nanoscience and Nanotechnology

Synthesis, Characterization, and Applications

Edited by
Shrikaant Kulkarni

CRC Press
Taylor & Francis Group
Boca Raton London New York

CRC Press is an imprint of the
Taylor & Francis Group, an **informa** business

First edition published 2023
by CRC Press
6000 Broken Sound Parkway NW, Suite 300, Boca Raton, FL 33487-2742

and by CRC Press
4 Park Square, Milton Park, Abingdon, Oxon, OX14 4RN

CRC Press is an imprint of Taylor & Francis Group, LLC

Library of Congress Cataloging-in-Publication Data
Names: Kulkarni, Shrikaant, editor.
Title: Innovations in green nanoscience and nanotechnology : synthesis,
characterization, and applications / edited by Shrikaant Kulkarni.
Description: First edition. | Boca Raton : CRC Press, 2023. |
Series: Emerging materials and technologies | Includes bibliographical references and index. |
Identifiers: LCCN 2022033774 (print) | LCCN 2022033775 (ebook) |
ISBN 9781032333281 (hbk) | ISBN 9781032333298 (pbk) |
ISBN 9781003319153 (ebk)
Subjects: LCSH: Nanotechnology. | Green chemistry.
Classification: LCC T174.7 .I488 2023 (print) | LCC T174.7 (ebook) |
DDC 620/.5–dc23/eng/20221006
LC record available at https://lccn.loc.gov/2022033774
LC ebook record available at https://lccn.loc.gov/2022033775

ISBN: 978-1-032-33328-1 (hbk)
ISBN: 978-1-032-33329-8 (pbk)
ISBN: 978-1-003-31915-3 (ebk)

DOI: 10.1201/9781003319153

Typeset in Times
by codeMantra

Contents

SECTION I Green Nanomaterials: Synthesis and Characterization

SECTION II Green Nanomaterials: Applications in Bio-Medicine, Drug Delivery, Energy, Sensing and Other

SECTION III   Green Nanomaterials: Case Studies

Preface

Sustainable development has become a buzz word in the world today. It has driven the inclusive development of the coming generations by means of judiciously using available natural resources among humans and the ecosystems wherein they sustain. Over the years, we have witnessed a significant loss of ecology due to the overuse of resources bestowed upon us by nature apart from the application of synthetically derived materials on a large scale. This vexed context asks for right and sustainable measures such as the use of eco-friendly materials for the cause of preservation of the sanctity of ecology. Therefore, new vistas have to be opened for the cause of the advancement of novel, green, and sustainable materials in tandem with the applications so as to sustain the ecology and planet earth such that earth remains a planet worth living for generations to come. Indeed, the attainment of sustainable development goals has emerged as a movement at global level that covers the design and adoption of new synthesis routes and modifications of the existing pathways with the help of green and environmentally friendly materials and protocols. These approaches showcase the development of capabilities for the cause of conservation and sustenance of resources for providing better service to the generations of humanity.

Adoption, application, and advancement in green and sustainable nanomaterials is one such measure that covers the synthesis and fabrication of nanomaterials, development of process technologies, and optimization of processes, other than testing, studying their functional behavior, stability, reproducibility, durability, etc. Further, future trends in these materials will largely focus on the sustenance in R&D initiatives. This book presents advancements in the synthesis, characterization, and applications of some state-of-the-art nanomaterials prepared by using green and sustainable pathways in a plethora of domains such as agriculture, environmental sustainability, and biomedicine. This book will also help nurture future directions in green and sustainable nanoscience and nanotechnology.

Sustenance in the novel nanomaterials can be accomplished with the application of new synthesis routes or by maneuvering existing materials having unique properties. These materials are used in the making of equipment, instruments, and gadgets with innovative qualities and find place in a range of application areas such as biomedicine, textiles, building, and paper. A host of current nanomaterials are known for their novelty and have been produced over the years as bio-nanomaterials, ceramics, polymeric materials, etc. The origin of many of these materials is in nature, such as plants, animals, ores, etc. extracted from plants and obtained in different geometrical morphologies which are amenable to prepare composites from various materials in tune with the specific applications. Many of them are prepared synthetically with given geometries and size for fulfilling the needs of society. The major purpose of this volume is to take a review of the advancements in the design of nanomaterials over the time that are eco-friendly and sustainable.

Keeping in view the addition to the current knowledge base by these innovative materials obtained by converting conceptual understanding into reality, and to bring about mass-scale production of nanomaterials, this book consists of three sections with each one comprising some chapters throwing light on one or the other aspect. The first chapter provides an overview of the use of graphene as a thermoelectric nanomaterial for the production and conversion of green energy. The second chapter deals with environmentally benign applications, particularly in devices used for energy harvesting and storage applications. The third chapter dwells upon advances in high volumetric carbon nanotubes in storing electrochemical energy. The fourth chapter encompasses details about nanomaterial-embedded membranes as filtering media for getting rid of heavy metal pollutants. Chapter 5 refers to the use of green nanotechnology for reinventing the phytomedicine concept for using it to advantage. Chapter 6 is an extensive overview of how green nanomaterials are looked upon as boon in the frontier area of biomedical engineering. Chapter 7 is all about trends that are developed in the green synthesis of carbon-based nanostructures. Chapter 8 extensively discusses the applications of biodegradable polymeric materials in the field of biomedical science and technology. Chapter 9 gives an account of the role of green nanomaterials for bringing in sustainability in agriculture at large. Chapter 10 presents the results obtained in the form of silica nanoparticles derived from agricultural wastes as a renewable source to be used as a novel catalyst support material. Chapter 11 lists the futuristic applications of green and sustainable nanomaterials in the remediation of contaminants in the environment and thereby in cleaning the environment.

Overall, this volume can certainly be used as an excellent reference book by academicians, research scholars, practitioners, industrialists, and the scientific community who are interested in the synthesis and applications of advanced, novel, and sustainable nanomaterials. The authors are well-known academicians, material engineers, and researchers. I express my sincere gratitude to the contributors of all the chapters for their timely and overwhelming support in taking this volume to its logical end. Finally, I express my sincere thanks to CRC Press for accepting this book for publication.

Editor

Shrikaant Kulkarni, PhD, is an Adjunct Professor in the Science and Technology Department at Vishwakarma University, Pune, India. Dr. Kulkarni has been an academician and researcher for 39 years. He has delivered invited lectures and conducted sessions at national and international conferences as well as faculty development programs. He was a Professor in the Department of Civil Engineering at Padm. Dr. V. B. Kolte College of Engineering, Malkapur (M.S.), India. He teaches engineering chemistry, green chemistry, nanotechnology, analytical chemistry, catalysis, chemical engineering materials, industrial organization, and management, to name a few. He has published over 100 research papers in national and international journals and conferences. He has authored 36 book chapters in CRC Press, Springer Nature, AAP Press, Elsevier, Wiley, and IET books. He has edited five books in green engineering and green nanoscience and nanotechnology published by Apple Academic Press/CRC Press. Another three books are in the offing. He has authored four textbooks and worked as a resource person for many national and international events. Dr. Kulkarni earned MSc, MPhil, and PhD degrees in chemistry in addition to master's degrees in economics, business management, and political science. He has expertise in the fields of material science, green chemistry and engineering, analytical chemistry, and green nanoscience and nanotechnology.

Contributors

Jatinder Singh Aulakh
Department of Chemistry
Punjabi University
Patiala, India

Om M. Bagade
Department of Pharmaceutics
Vishwakarma University School of Pharmacy
Pune, India

Akash Balakrishnan
Department of Chemical Engineering
National Institute of Technology Rourkela
Rourkela, India

Sachin Chavan
Bharati Vidyapeeth Deemed to be University
Pune, India

Yogesh Chendake
Department of Chemical Engineering
College of Engineering
Bharati Vidyapeeth Deemed to be University
Pune, India

Mahendra Chinthala
Department of Chemical Engineering
National Institute of Technology Rourkela
Rourkela, India

Sonali Dhamal
Bharati Vidyapeeth Deemed to be University
Pune, India

Priyanka E. Doke-Bagade
Department of Pharmaceutics
D. Y. Patil International University School of
    Pharmacy
Pune, India

Deepti Goyal
Department of Applied Chemistry
School of Vocational Studies and Applied
    Sciences
Gautam Buddha University
Greater Noida, India

Shikha Gulati
Department of Chemistry
University of Delhi
Delhi, India

Rutuja Gumathannavar
Symbiosis Center for Nanoscience and
    Nanotechnology (SCNN)
Symbiosis International University
Pune, India

Heena
Department of Chemistry
GSSDGS Khalsa College
Patiala, India

Nidhi Jain
Department of Engineering Science
Bharati Vidyapeeth's College of Engineering
Pune, India

Kalpana Joshi
Sinhgad College of Engineering
Pune, India

Shweta Kashid
Sinhgad College of Engineering
Pune, India

Stuti Katara
Department of Pure and Applied Chemistry
University of Kota
Kota, India

Mona Kejariwal
Department of Botany
R.D. and S.H. National College of Arts and
    Commerce and S.W.A. Science College
Mumbai, India

Ajit R. Kulkarni
Department of Metallurgical Engineering and
    Materials Science
Indian Institute of Technology Bombay
Mumbai, India

Anand Kulkarni
Department of Chemical Engineering
College of Engineering
Bharati Vidyapeeth Deemed to be University
Pune, India

Atul Kulkarni
Symbiosis Center for Nanoscience and
    Nanotechnology
Symbiosis International (Deemed University)
Pune, India

Kavita Kulkarni
Department of Chemical Engineering
College of Engineering
Bharati Vidyapeeth Deemed to be University
Pune, India

Sanjay Kumar
Department of Chemistry
University of Delhi
Delhi, India

Sunil Kumar
Special Centre for Nano Sciences and AIRF
Jawaharlal Nehru University
New Delhi, India

Vimal Kumar
Special Centre for Nano Sciences and AIRF
Jawaharlal Nehru University
New Delhi, India

Sakshi Kabra Malpani
Save the Water™
Plantation, Florida

Rajesh Kumar Meena
Department of Chemistry
Kalindi College
Delhi, India

Prem Pandey
Symbiosis Center for Nanoscience and
    Nanotechnology
Symbiosis International (Deemed University)
Pune, India

Tulika Prasad
Special Centre for Nano
    Sciences and AIRF
Jawaharlal Nehru University
New Delhi, India

Anam Rais
Special Centre for Nano
    Sciences and AIRF
Jawaharlal Nehru University
New Delhi, India

Neha Saini
Symbiosis Center for Nanoscience and
    Nanotechnology
Symbiosis International
    (Deemed University)
Pune, India

Nidhi Sapre
Symbiosis Center for
    Nanoscience and
    Nanotechnology (SCNN)
Symbiosis International University
Pune, India

Sayoni Sarkar
Centre for Research in Nanotechnology and
    Science
Indian Institute of Technology Bombay
Mumbai, India

Suvidha Sehrawat
Department of Chemistry
Chandigarh University
Mohali, India

Varinder Singh
Department of Chemistry
RIMT University
Mandi Gobindgarh, India

Itika Varshney
Special Centre for Nano
    Sciences and AIRF
Jawaharlal Nehru University
New Delhi, India

Swapnali Walake
Symbiosis Center for Nanoscience and
    Nanotechnology (SCNN)
Symbiosis International University
Pune, India

Shashwati Wankar
Symbiosis Center for Nanoscience and
    Nanotechnology (SCNN)
Symbiosis International University
Pune, India

Section I

Green Nanomaterials

Synthesis and Characterization

1 Green Synthesis, Characterization and Applications of Quantum Dots

Varinder Singh
RIMT University

Heena
GSSDGS Khalsa College

Jatinder Singh Aulakh
Punjabi University

CONTENTS

1.1 INTRODUCTION

The development of the nanostructured material "quantum dots" at a quick pace in the domain of nanoscience has attracted the attention of various researchers as they covered a gap between bulk materials and molecular levels. Because of their distinct structural features, they show remarkable size-dependent functions with diverse applications, especially in the area of biology and electronics [1,2]. Nanoparticles with particle size smaller than 100 nm are categorized as zero-dimensional (termed as quantum dots (QDs)), one-dimensional (quantum wires) and two-dimensional (quantum wells). QDs are semiconductor crystals with ultra-small particles of size lying within the range 1.5–10 nm and were first discovered by Alexei Ekimov in 1981. After that, Efros in 1982 introduced the quantum effects based on the size of these crystals that can change their electronic and optical properties [3]. Further, Louis Brus, in 2008, was awarded in the field of nanotechnology the Kavli Prize for his notable contribution to developing these semiconductor crystals. Now, it has been almost more than two decades since their introduction in nanoscience, and the rate of their utility is increasing day by day. Because of their unique, advanced properties and exceptional features, today,

DOI: 10.1201/9781003319153-2

QDs being the most researched materials are currently used for various applications in energy generation, catalysis, bio-imaging, drug delivery, gene therapy and biotechnology.

QDs are known for exhibiting exclusive optical properties that depend upon changes in band gap energy [4]. This energy is required for the migration of electron from ground state to another higher electronic state that results in electron and hole pair formation, known as an exciton. After that, subsequently, recombination of electron and hole leads to emission of radiation in the form of a fluorescent photon. It is found that the energy needed to produce exciton increases because the size of QDs lies under the Bohr first excitation radius, and this effect is called quantum confinement. So, when the size of QDs decreases, the energy of the emitting photon rises, resulting in a greater band gap energy, and thus, different optical characteristics may be seen by changing the particle size. Due to their small size, QDs feature high photostability, high quantum yield, high resistance to photobleaching and size-controlled photoemissions, which altogether make these particles potential candidates for many applications in several fields of technology [5].

The literature revealed that the research on nanomaterials such as QDs has increased over the last few years in both theoretical and experimental ways. However, despite the effective use of nanocrystals as multifunctional materials for scientific and industrial purposes, their production and applications are associated with the use of expensive and toxic chemicals and by-products. Thus, because of these side effects and risks, their application in the medical field is limited. Therefore, it is an urgent need and a great challenge for researchers to develop green routes for their synthesis to increase the sustainability of the environment in order to lessen the environmental risk and effective cost related to their production. This chapter briefly discusses QDs' structure, characteristics, green synthesis methods, characterization techniques and applications.

1.2   STRUCTURE OF QUANTUM DOTS

The structure of QDs consists of an inorganic core and other layers named as shell and capping ligands layers, which all collectively resemble an earth-like structure. The inorganic core is the basis of QDs that control all the optical and fundamental semiconducting properties. InP and CdSe are two commonly used core materials. CdSe possesses high efficiency and quantum yield but are toxic, whereas InP is more eco-friendly and safe. The surface of the core is further encapsulated with another inorganic layer called a shell that has a larger band gap than the material of the core itself [6]. QDs having a shell band gap higher than the core, e.g., CdSe/ZnS, exhibit exceptional stability along with quantum yield.

On the other hand, when the shell material has a narrower band gap than the core, as is the case with CdS/HgS, the stability is reduced. Thus, the shell of QDs works as a barrier between the surrounding environment and the core. The main component of QDs is the third layer of capping agents that can be introduced by a selective synthetic method depending on the application in which they will be used [7–9]. In organic capped QDs, carboxylic acid, amines, phosphenes and mercaptans are the mainly employed ligands to develop suitable nanomaterials for biomedical applications. As the charged heads of ligands have a key role in nanomaterial's solubility, fundamental properties and applications, the selection of capping ligands has a significant role in designing nanomaterials. Hence, the composition and nature of core, shell and capping ligands are all responsible for the physicochemical properties of QDs (Figure 1.1).

Concerning different types of nanomaterials, the core part (CdSe, InP, GaN, CdTe and PbSe) mainly constitutes atoms of the groups II-IV, III-V and IV-VI in the periodic table, whereas the shell of QDs usually consists of ZnS and SiO_2. Despite the remarkable applications of QDs made up of such materials, they are associated with undesirable toxic effects on cells and tissues of living organisms. These harmful effects are due to the leaching of cadmium ions from the core; therefore, silicon, carbon and InP QDs are some new-generation materials that can overcome these limitations. ZnS QDs are being used as a photocatalyst for the removal of pollutants as another environmentally friendly approach. The development of such new-generation QDs is an urgent need.

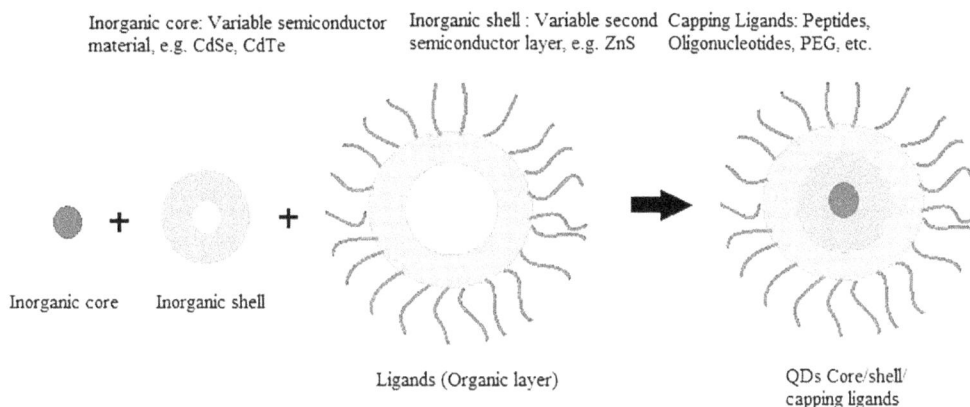

Inorganic core: Variable semiconductor material, e.g. CdSe, CdTe

Inorganic shell : Variable second semiconductor layer, e.g. ZnS

Capping Ligands: Peptides, Oligonucleotides, PEG, etc.

Inorganic core   Inorganic shell

Ligands (Organic layer)

QDs Core/shell/ capping ligands

FIGURE 1.1   Basic structure and components of QDs.

1.3   GREEN SYNTHETIC METHODS FOR QUANTUM DOTS

After the discovery of quantum dots, the primary concern of researchers was developing synthetic routes to produce highly pure QDs with good yields. Therefore, during this era, different top-down and bottom-up techniques involving the physical, chemical and biological pathways were employed and developed to synthesize nanocrystals exhibiting controlled shapes with particular size distribution [10–12]. Using the method apt for synthesis, the diameter of crystals can be varied from nanometers to micrometers and size distribution of particles can be controlled up to 2% as the applications of QDs predominantly depend upon these parameters [13]. However, despite the diverse applications of QDs, their synthetic methods are mainly associated with hazardous chemicals and toxic metal precursors under non-green and high energy consumption conditions. Nowadays, the production of nanomaterials using non-toxic, biosafe and eco-friendly reagents under low energy consumption reaction conditions termed as "green synthesis" is an emerging area that can provide us environmental and economic benefits.

Generally, two synthetic routes are being used, while others are their modified forms exploiting different chemistry with minor changes in reaction time, temperature, coordinating solvents and metal precursors, which fabricate the nanocrystals with specific optical properties and stability. The first method that is commonly used is the organometallic synthesis in a colloidal medium that was given by Murrays et al. [14]. This bottom-up approach consists of three steps: nucleation, growth of particles and, finally, termination, which requires a typical procedure using coordinating solvents with high boiling points (such as hexadecylamine, trioctylphosphine and tri-n-octyl phosphine oxide) with metal precursors under refluxing conditions at high temperatures and inert environment. The metal and dried degassed coordinating solvent undergo nucleation at high temperature around 300°C along with successive growth of particles where temperature, solvent and reaction time are significant parameters that control the size of QDs. Variation in these parameters affects the size of crystals, which further results in a change in emission wavelength; with decreasing size, a shift in red color to blue can be seen. In the typical mechanism of the process, the tail of the solvent consisting of groups such as oxides and amines binds with the surface of the particles, whereas alkyl chain parts pose toward the bulk. Hence, this phenomenon regulates the growth and nucleation processes. QDs produced by this method are hydrophobic and cannot be dissolved in polar and aqueous mediums. Therefore, further transformations are required to make them hydrophilic and their practical usage in polar mediums possible, which can be done by employing a number of methods. "Cap exchange" is one of the methods where bifunctional molecules with two polar heads replace the layer of the coordinating solvent [15–17]. Bifunctional molecules contain an acidic group on one side, mainly thiol, and hydrophilic parts such as amines or carboxylic acids on the other side. The addition of silica shells to QDs is another route for surface modification where usually the source of silica is alkoxysilanes used during the process of polycondensation [18]. Similarly, introducing amphiphilic molecules such as phospholipids encapsulates QDs to produce biocompatible nanoparticles [19]. Encapsulation of QDs into solid-lipid nanoparticles consisting of lipids with increased physical and chemical stability, on the other hand, is also an effective route.

In conclusion, organometallic synthesis is helpful to prepare a variety of QDs such as PbS, CdTe and CdS with higher quantum yield and significant performance in different fields. Still, this method is associated with the use of non-green conditions and chemicals. The way to alter this method to a green synthesis method is to opt for stable, inexpensive and non-toxic precursors such as $Cd(Ac)_2$ [20], $CdCl_2$ [21] and CdO [22] instead of volatile and highly toxic precursors such as $Cd(Me)_2$. The use of environmentally friendly benign solvents such as castor [23], olive oil [24], oleic acid [25], paraffin and glycerol is also a remarkable step toward employing green synthesis that can avoid costly, highly toxic and air-sensitive chemicals such as HDA and TOPO.

Another widely used method to synthesize QDs is aqueous synthesis in a colloidal medium. It is more environmentally friendly than organometallic synthesis because it requires no solubilization step, with biocompatible solvents under mild energy reaction conditions [26]. The method is extensively applicable to fabricate all types of QDs such as $AgInS_2$-ZnS, CdS and ZnSe. Metal precursors such as $CdCl_2$, $Cd(NO_3)_2$ and $Cd(CHCOO)_2$ used in this method can easily solubilize in water and bind with stabilizing reagents such as glutathione, L-cysteine, thioglycolic acid and 3-mercaptopropionic acid [27,28], where the size of particles and emission wavelength depend upon reaction conditions just like the organometallic synthesis. In this synthesis route, after adding the metal precursor, the reaction mixture generally refluxes at a low temperature around 100°C for the growth of crystals, which is relatively low compared to the organometallic synthesis. Some reports in the literature have shown the growth phenomenon at room temperature or freezing temperature. Because of the hydrophilic surface of QDs produced by this method, the hydrodynamic diameter of these QDs is very small in comparison with those synthesized by the organometallic route we discussed above. Also, this route is proved to be less expensive and more biosafe and can be considered the best approach in green technology. To grow highly monodispersed, stable and high-quality QDs, a method in which ultrasonic and microwaves interact with an aqueous solution of metal precursors and dissociate the precursors and water molecules to promote the growth of particles has been used, where the temperature of the reaction required is relatively lower than the other methods [29,30]. Similarly, the hydrothermal process [31,32] to crystallize the inorganic salts using aqueous medium while controlling pressure and temperature has been used to synthesize QDs where the change in temperature and pressure affects the size and shape of QDs. Glucose [33], ammonium citrate [34,35] and chitosan [36] are used as organic precursors to synthesize carbon QDs applying these methods.

A new, effective and sustainable approach to synthesize QDs such as carbon, CdSe, CdS and CdTe using microorganisms such as *Saccharomyces cerevisiae*, *Gluconacetobacter xylinus*, *Fusarium oxysporum*, *Bacillus megaterium* and *Escherichia coli* has been employed [37–39]. The synthesis of nanomaterials using plants and microorganisms has been developing and generating highly fluorescent QDs.

1.4 CHARACTERIZATION OF THE SYNTHESIZED GREEN QUANTUM DOTS

There are myriads of techniques used for the characterization of quantum dots, including structure elucidation, establishment of purity of sample, functional group determination, morphology analysis, size of the formed particle based on zeta potential and visible analysis via UV-visible spectroscopy and special arrangements. The details of the techniques generally used for the characterization of QDs are as follows:

UV-visible spectroscopy: This approach compares the amount of distinct wavelengths of visible light absorbed or transmitted through the material under examination with blank sample. For the absorption to take place, a sufficient amount of energy is used to excite a molecule from its ground state. In an aqueous suspension, the UV-visible spectroscopy has been proved to be the best characterization technique used to detect QDs. GQDs and CDs are efficient photon harvesters in the short wavelength region due to the (C=C bond) π-π* transition. In the UV range (260–320 nm), these dots showed the greatest optical adsorption, along with a tail extending into the visible region. For instance, the qualitative study of the green synthesis of QDs of zinc ion was associated with visual observations made by Singh et al. They observed the color transformation from brown to yellow as soon as an extract of leaf is added to zinc acetate solution in order to prepare its QDs [40]. The mechanism proceeds in three steps: First, the reduction of zinc ion occurs in a leaf extract due to the presence of flavonoids and alkaloids; second, the reduced zinc ion reacts with the diffused oxygen content to form zinc oxide nanoparticles, and finally, they attain the proper shape by uniting together and form stable QDs. Here, the UV-visible spectroscopy has emerged as an imperative tool for the characterization of colored solutions. Mahalakshmi et al. showed the optical absorption of the synthesized cadmium tellurium quantum dots. Its absorption spectra exhibited a blueshift from bulk materials to nanomaterial confinement [41]. Anooj et al. synthesized green QDs via the use of graphene from *Rosa gallica* petal extract and characterized them primarily by UV spectroscopy [42].

1.4.1 SCANNING ELECTRON MICROSCOPY (SEM)

Scanning electron microscopy has been used worldwide in numerous disciplines. It is an effective tool for the analysis of organic and inorganic materials ranging from nanometer to micrometer scale. It offers very precise high-magnification images of a wide variety of materials and depicts their morphology. The device has an inbuilt high-pressure system that has the capacity to uphold any number of samples. Generally, a thermal source is used for the emission of electrons. By this, an image is produced by point depending upon the scan coils which ponders the movement of an electron beam from discrete locations to a straight line. The morphology of a material depends entirely on the presence of stabilizers or the reducing power of reducing agents. Bakshi et al. observed that reducing agents with low power allow lipid molecules to be adsorbed on nanoparticle surfaces for an extended period of time [43]. Zaheer et al. synthesized green QDs and characterized them by SEM images the films embedded to silver nanoparticles [44]. Conventional optical and scanning electron microscopes in the range of 23–245 nm were used to study the nanoparticle films. In addition to scanning electron microscopy (SEM), transmission electron microscopy (TEM), atomic force microscopy and scanning tunneling microscopy are used to measure the particle size.

The quantum dots surface chemistry plays an imperative role according to Lees et al. because it contributes toward the hydrodynamic diameter, so it's an essential to find out the method to know this parameter. Photoelectron spectroscopy, NMR spectroscopy and Rutherford's backscattering experiment have all been explained to date. Using pomelo peel aqueous extract, Mahalakshmi et al. synthesized cadmium tellurium quantum dots and analyzed them using various methods. The findings of photoelectron spectroscopy indicated that the binding energy of Cd^+ ions was measured at the core level, with metallic Cd possessing binding energies of 413–406 eV [41].

1.4.2 X-RAY DIFFRACTION (XRD)

Carbon-based QDs and related information such as crystal structure, particle size and phase purity are characterized using the XRD technique. Furthermore, the crystalline phases of carbon-based QDs were determined using this method [45]. Thambiraj et al. presented a green synthesis of fluorescent CQDs by using sugarcane bagasse and exfoliation, along with chemical oxidation. At

$2\Theta = 11.4°, 20.6°, 22.8°, 42.3°$ and $45.7°$, XRD distinctive peaks have been found. As a consequence, the primary peak of the CQDs was determined to lie at $2\Theta = 20.6°$ and $42.3°$, which is assumed to be due to the graphite carbon presence [46]. For the process of diffraction to occur, each atom's electrons scatter light in a uniform manner. The atoms in a crystal are arranged in a periodic array with long-range order allowing diffraction to occur. Scattering results in the diffraction process containing information of atomic arrangement within the crystal. Chen et al. synthesized graphene quantum dots using starch as a unique precursor in a hydrothermal process. The structure was created utilizing starch by a hydrothermal reaction, as shown by the XRD results of graphene quantum dots, which revealed a broader peak at 24 compared to the (0 0 2), which corresponds to JCPDS card number 75-0444 [47].

1.4.3 Transmission Electron Microscopy (TEM)

Electron microscopy is the domain of chemical analysis of quantum dots at high spatial resolution. The beam of electrons fired in this technique is not limited to imaging. Many electrons experience inelastic scattering processes as they travel through the specimen, resulting in energy loss [41,47]. Relaxation processes can result in the emission of X-ray quanta, Auger electrons or light, among other things. Chemical reactions are made possible by inelastic processes. The electron's penetrating capacity is low, and it is absorbed by the thick specimen. As a result, the specimen thickness should be limited to a few hundred angstroms (1 Å equals one ten-billionth of a meter). In the high-voltage electron microscope, however, significantly thicker samples are sometimes employed.

A high-voltage electron beam is emitted from a tungsten filament (cathode) by electrical heating; the electron beam's shaft is attracted toward an anode (magnetic lenses) and passes through an aperture in TEM. The beam passes through the electromagnetic condenser, objective, intermediate and projector lenses after traversing the aperture. The electron beam was focused and modulated by a goniometer as it passed through a very thin specimen loaded on a grid inserted in the route [48]. The part of the beam that was scattered and transmitted through the objective aperture was projected on the fluorescence screen by the projector lens after being adjusted by intermediate lenses. The image is examined by the use of an optical binocular.

1.5 APPLICATIONS

Because of their unique properties, including high thermal and mechanical conductance, biocompatibility, non-toxicity, water solubility and simple derivatization, these materials have received a lot of attention from researchers. Having all these unique characteristics, QDs have widespread use in energy, catalysis, bio-imaging, drug delivery, gene therapy and biotechnological applications [49–53]. They exhibit astounding attributes toward electrochemical biosensing because of their remarkable solubility, strong inertness and chemical reactivity, large surface area, cheap cost, adaptability and the capacity to change surface chemistry significantly involving nanostructured form [54].

1.5.1 Luminescent Properties

The term luminescence was coined in 1888 by Eilhard Wiedemann. It is the spontaneous emission of light by materials. It is often known as cold light caused by chemical reactions. Many of the semiconductor compounds with direct band gap are suitable for light emission. In general, the value of energy difference should be negative, which means that it lies usually above the ground state level having localized carriers to be activated to states with higher energies [55]. A novel green method was established by Hoan et al. for the synthesis of quantum dots, a hydrothermal method from lemon peel showing optoelectronic, imaging and luminescent probing of vanadium and molybdenum ions [56]. These lemon-derived carbon dots as luminescent probes are green, eco-friendly, less expensive and selective in nature for the detection of ions. These also act as a promising material

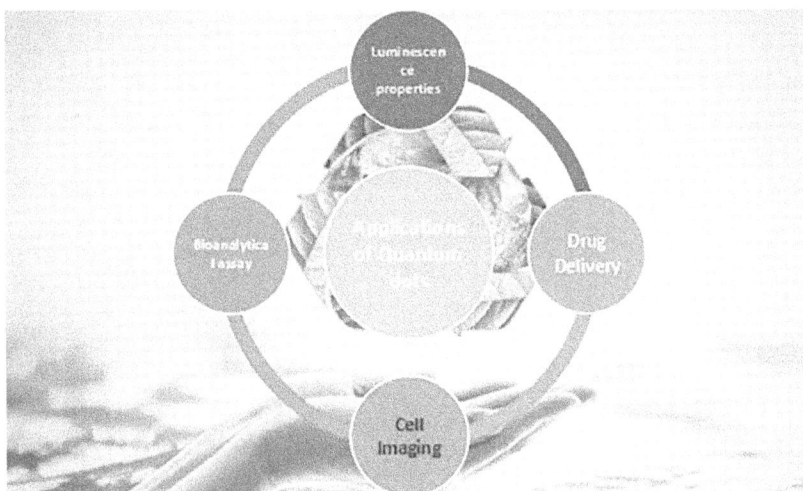

FIGURE 1.2 Applications of quantum dots.

for application in biomedical and clinical sciences, especially in diagnosis, bioanalytical assays and biosensors. Currently, carbon dots could be used as a luminescent or bio-imaging probe and in light-emitting diodes due to their strong luminescent properties [57]. These light-emitting diodes have a myriad of applications in medical instrumentation, barcode readers, fiber-optic communication, laptops, digital cameras, traffic signals and tower lights. The luminescent quantum dots have also been employed in recent research to examine metal ions such as ferric ions, cuprous ions, mercurous ions and beryllium ions [58–62].

For energy-efficient lighting, QDs have extensively been studied in recent years. LEDs, or light-emitting diodes, have made a significant contribution (Figure 1.2) to applied science in contemporary revolutionary period. Alivisatos et al. demonstrated the first quantum dot exhibiting lighting effects. In comparison with previous conventional diodes, these newly developed diodes offer an abundant advantage in terms of cost, production efficiency, stability and purity [63].

Due to their unique properties such as solubility in different solvents, availability of functionalization sites, low cost, large surface area and versatility, they exhibit an outstanding role in electrochemical biosensing. Electrochemiluminescence (a type of chemiluminescence) is a process in which electrochemical processes anticipate light-emitting reactions [53]. The multidox species green quantum dots exhibit a myriad of applications including electron transfer, biosensors and molecular switches due to their large specific surface area and edge sites [64,65]. Campuzano et al. reported the electrochemical biosensing phenomenon to illustrate the applications of QDs that can be used as electrode surface modifiers or single tags [54]. Another method was reported by Bano et al., who made fluorescent green quantum dots from tamarind leaves via a hydrothermal method, which can be used for sensing, bio-imaging, disease diagnosis and control. Yro et al. processed the *Escherichia coli* bacteria fluoresced with carbon dots used for the hydrothermal treatment of pineapples and calamansi wastes [66]. Novel graphene/riboflavin-modified dots as a sensor for the determination of persulfate were reported by Roushani et al. in 2014. The reaction was carried out on an amperometer with a limit of detection $0.2\,\mu m$ and sensitivity of $4.7\,nA/\mu M$. The electrode used for the determination was modified with wide pH range (1–10) and had confined surface characteristics [67]. Table 1.1 represents the different materials synthesized and used for application as a biosensor.

1.5.2 DRUG DELIVERY

QDs have been shown to be the most effective for tracking nanocarriers in drug delivery and monitoring drug release. Generally, most of the biological molecules emit light in blue green spectra, and

TABLE 1.1

Electrochemical Methods Using QDs as a Biosensors

S. No.	Electrochemical Method	Analyte	LOD	Material Used	Reference
1	Pyrolytic graphite electrode with differential pulse voltammetry	DNA sequence	50 nM	Green quantum dots-modified electrode coupled with DNA	[65]
2	Hydrothermal method via cyclic voltammetry	Glucose	1.25 μM	Modified carbon ceramic electrode	[68]
3	Pyrolysis differential pulse voltammetry	Hydrogen peroxide	0.375 μM	Hemoglobin immobilized on chitosan nanocomposite	[69]
4	Microwave-assisted hydrothermal cyclic voltammetry	Caffeic acid	0.095 μM	Nanocomposite MoS_2 on screen-printed electrode	[70]
5	Hydrothermal pyrolysis amperometry	Hydrogen peroxide	5 nM	Palladium–silver graphene oxide doped with hollow nanospheres	[71]
6	Hydrothermal synthesis from multiwalled carbon nanotubes with differential pulse voltammetry	HCV	0.0125 fg/mL	Silver nanoparticles and green quantum dots	[72]
7	Hydrothermal synthesis from multiwalled carbon nanotubes with differential pulse voltammetry	TNT	0.00025 pM	TNT aptamer	[73]

in combination with carbon dots, they help in the shifting of luminescence spectra from red region and near-infrared region. In biomedical research, the most attracting feature of QDs includes high fluorescence [74]. The QDs have a tunable size with fluorescence emission from visible to infrared region emitting different colors. Dots smaller in size usually have large band gaps, which leads to blueshift, and vice versa, dots larger in size brings redshift with smaller band gaps. It is the most important application in drug delivery and biosensing because it permits simultaneous excitation of multicolor quantum dots from a single light source.[75]. These are used in drug delivery and photodynamic therapy due to the presence of stable fluorophore and more lifetime stability [76]. The quantum dot molecules easily get attached via electrostatic, adsorption and covalent linkage with biomolecules without altering their physical essence of biomolecules [77,78].

Green quantum dots are safe and efficiently channelize the release of drug molecules, especially those that are water-insoluble. The specific unique applications include sensing, tracing probe, antioxidation and neurodegenerative disease treatment. Recently, the world has been facing the threat from coronavirus, which rapidly mutates. Green carbon dots exhibit a good prospect in antiviral application and can be effectively employed to deal with the problem. Devi et al. found that quantum dots produced from the aloe vera extract have a good antimicrobial activity against *Escherichia coli* (*E. coli*) and *Staphylococcus aureus* [79]. The fluorescence feature is also useful in evaluating the efficacy of drug loading and tracking the drug delivery. Eco-friendly green quantum dots have been prepared by *Daucus carota* subspecies for drug delivery through hydrogen bonding that breaks in an extracellular microenvironment to release the right amount of drug. This proves to give a high efficacy toward cancer cell membrane facilitating the internalization of dots by suitable bacteria.

1.5.3 Cell Imaging

The green quantum dots has been found as good candidates in cell imaging and drug administration after penetrating culture cells and being examined under a fluorescence microscope. The most imperative feature of them includes biomedical and clinical applications due to their low toxicity.

Toxicity is the major parameter in detecting drug delivery and cell imaging. Ramezani made carbon quantum dots by a hydrothermal method to penetrate HT-29 cells in a small time period and observed them with the help of fluorescence spectroscopic technique having emission wavelength (at 518 nm) along with excitation wavelength (at 488 nm) [80]. Results demonstrate that the calculated $IC_{50}\%$ value for HT-29 cells was as low as 924.25 µg/mL and can be used in cell imaging. Similarly, the cytotoxicity of natural carbon quantum dots was examined for hepatoblastoma cells using the methylthiazolyldiphenyl-tetrazolium bromide assay [81]. Cell viabilities were tested within the stipulated time period, and a low toxicity was found, which is an essential factor for cell imaging. In another experiment, Kumari et al. used Dulbecco's modified Eagle's medium containing roughly 10% fetal bovine serum to culture the cells. The cells were incubated with quantum dots for about 8 hours for drug delivery and use as a cell imaging source [82].

1.6 CONCLUSIONS

The structure, characteristics, green production methods, application and performance of QDs are all explored in this chapter. Nanotechnology provides the possibility of finding solutions to global concerns that touch society at all levels. If we can make it green, though, discoveries, advancements and collaborative efforts will have a genuine impact on everyone's life. It means that researchers must stop being potential and produce a comprehensive eco-friendly procedure using environmentally friendly technologies, a role that is now in their hands and brains. There seems to be no doubt that researchers will create a viable and cost-effective method for producing QDs in the near future, omitting the need of quotes in the word "green" to make it completely green. Actually, the colloidal synthesis of quantum dots with microorganism-mediated or natural materials considers being sustainable, which avoids the usage of hazardous chemicals. Moreover, ongoing beneficial cooperation spanning a variety of fields, including biotech, photonics, nanoscience, toxicology, physics and engineering, may establish QD development as a milestone in this global revolution of nanostructures. QDs are the greatest candidate for this type of research because of their high stability and extreme luminance.

REFERENCES

1. S. Kargozar, and M. Mozafari, Nanotechnology and nanomedicine: Start small, think big. *Materials Today: Proceedings*, 5 (2018) 15492–15500.
2. X.T. Zheng, A. Ananthanarayanan, K.Q. Luo, P. Chen, Glowing graphene quantum dots and carbon dots: Properties, syntheses, and biological applications. *Small (Weinheim ander Bergstrasse, Germany)*, 14 (2015) 1620–1636.
3. A.L. Efros, Interband light absorption in a semiconductor sphere. *Fiz Tekh Poluprovodn (Leningrad)*, 7 (1982) 1209–1214.
4. P. Harrison, A. Valavanis, *Quantum Wells, Wires and Dots*. Theoretical and Computational Physics of Semiconductor Nanostructures, John Wiley & Sons, New York (2016).
5. M. Sun, X. Ma, X. Chen, Y. Sun, X. Cui, Y. Lin, A nanocomposite of carbon quantum dots and TiO2 nanotube arrays: Enhancing photoelectrochemical and photocatalytic properties. *RSC Advances*, 4 (2014) 1120–1127.
6. V. G. Reshma, P.V. Mohanan, Quantum dots: Applications and safety consequences. *Journal of Luminescence*, 205 (2019) 287–298.
7. V.L. Colvin, A.N. Goldstein, A.P. Alivisatos, Semiconductor nanocrystals covalently bound to metal-surfaces with self-assembled monolayers. *Journal of the American Chemical Society*, 114 (1992) 5221–5230.
8. B.O. Dabbousi, C.B. Murray, M.F. Rubner, M.G. Bawendi, Langmuir-Blodgett manipulation of size-selected CdSe nanocrystallites. *Chemistry of Materials*, 6 (1994) 216–219.
9. C.B. Murray, C.R. Kagan, M.G. Bawendi, Self-organization of CdSe nanocrystallites into 3- dimensional quantum-dot superlattices. *Science*, 270 (1995) 1335–1338.
10. S.B. Brichkin, V.F. Razumov, Colloidal quantum dots: Synthesis, properties and applications. *Russian Chemical Reviews*, 85 (2016) 1297.

11. M. A. Faramarzi, A. Sadighi, Insights into biogenic and chemical production of inorganic nanomaterials and nanostructures. *Advances in Colloid and Interface Science*, 189 (2013) 1–20.

12. S. Kargozar, S. Ramakrishna, M. Mozafari, Chemistry of biomaterials: Future prospects. *Current Opinion in Biomedical Engineering*, 10 (2019) 181–190.

13. S. Santra, K. Wang, R. Tapec, W. Tan, Development of novel dye-doped silica nanoparticles for biomarker application. *Journal of Biomedical Optics*, 2 (2001) 160–166.

14. C.B. Murray, D. Norris, M.G. Bawendi, Synthesis and Characterization of Nearly Monodisperse CdE (E= S, Se, Te) Semiconductor Nanocrystallites. *Journal of the American Chemical Society*, 4 (1993) 8706–8715.

15. P.A.S. Jorge, M.A. Martin, T. Trindade, J.L. Santos, F. Farahi, Optical fiber sensing using quantum dots. *Sensors*, 7 (2007) 3489–3534.

16. H.Q. Wang, H.L. Zhang, X.Q. Li, J.H. Wang, Z.L. Huang, Y.D. Zhao, Solubilization and bioconjugation of QDs and their application in cell imaging. *Journal of Biomedical Materials Research, Part A*, 86A (2008) 833–841.

17. J. Wang, J. Xu, M.D. Goodman, Y. Chen, M. Cai, J. Shinar, Z.Q. Lin, A simple biphasic route to water soluble dithiocarbamate functionalized quantum dots. *Journal of Materials Chemistry*, 18 (2008) 3270–3274.

18. R. Koole, M.M. Van Schooneveld, J. Hilhorst, C.D. Donega, D.C.T. Hart, A. Van Blaaderen, D. Vanmaekelbergh, A. Meijerink, On the incorporation mechanism of hydrophobic quantum dots in silica spheres by a reverse microemulsion method. *Chemistry of Materials*, 20 (2008) 2503–2512.

19. W. Liu, Z.K. He, J.G. Liang, Y.L. Zhu, H.B. Xu, X.L. Yang, Preparation and characterization of novel fluorescent nanocomposite particles: CdSe/ZnS core-shell quantum dots loaded solid lipid nanoparticles. *Journal of Biomedical Materials Research Part A*, 84A (2008) 1018–1025.

20. L. Qu, Z.A. Peng, X. Peng, Alternative routes toward high quality CdSe nanocrystals. *Nano Letters*, 1 (2001) 333–337.

21. C. Ge, M. Xu, J. Liu, J. Lei, H. Ju, Facile synthesis and application of highly luminescent CdTe quantum dots with an electrogenerated precursor. *Chemical Communications*, 4 (2008) 450–452.

22. Z.A. Peng, X. Peng, Formation of high-quality CdTe, CdSe, and CdS nanocrystals using CdO as precursor. *Journal of the American Chemical Society*, 123 (2001) 183–184.

23. J.W. Kyobe, E.B. Mubofu, Y.M.M. Makame, S. Mlowe, N. Revaprasadu, CdSe quantum dots capped with naturally occurring biobased oils. *New Journal of Chemistry*, 39 (2015) 7251–7259.

24. S. Sapra, A.L. Rogach, J. Feldmann, Phosphine-free synthesis of monodisperse CdSe nanocrystals in olive oil. *Journal of Materials Chemistry*, 16 (2006) 3391–3395.

25. R.B. Vasiliev, S.G. Dorofeev, D.N. Dirin, D.A. Belov, T.A. Kuznetsova, Synthesis and optical properties of PbSe and CdSe colloidal quantum dots capped with oleic acid. *Mendeleev Communications*, 14 (2004) 169–171.

26. D. Wang, T. Xie, Y. Li, Nanocrystals: Solution-based synthesis and applications as nanocatalysts. *Nano Research*, 2 (2009) 30–46.

27. A.L. Rogach, T. Franzl, T.A. Klar, V. Feldmann, N. Gaponik, V. Lesnyak, A. Shavel, A. Eychmüller, Y.P. Rakovich, J.F. Donegan, Aqueous Synthesis of Thiol-Capped CdTe Nanocrystals: State-of-the-Art. *The Journal of Physical Chemistry C*, 111 (2007) 14628–14637.

28. X. Jiang, J. Shao, B.Q. Li, Ceiling temperature and photo-thermal sensitivity of aqueous MSA-CdTe quantum dots thermometers. *Applied Surface Science*, 394 (2017) 554–561.

29. H.F. Qian, L. Li, J.C. Ren, One-step and rapid synthesis of high quality alloyed quantum dots (CdSe-CdS) in aqueous phase by microwave irradiation with controllable temperature. *Materials Research Bulletin*, 40 (2005) 1726–1736.

30. J.J. Zhu, Y. Koltypin, A. Gedanken, General sonochemical method far the preparation of nanophasic selenides: Synthesis of ZnSe nanoparticles. *Chemistry of Materials*, 12 (2000) 73–78.

31. L.C. Wang, L.Y. Chen, T. Luo, Y.T. Qian, A hydrothermal method to prepare the spherical ZnS and flower-like CdS microcrystallites. *Materials Letters*, 60 (2006) 3627–3630.

32. H.Q. Yang, W.Y. Yin, H. Zhao, R.L. Yang, Y.Z. Song A complexant-assisted hydrothermal procedure for growing well-dispersed InP nanocrystals. *Journal of Physics and Chemistry of Solids*, 69 (2008) 1017–1022.

33. H. Zhu, X. Wang, Y. Li, Z. Wang, F. Yang, X. Yang, Microwave synthesis of fluorescent carbon nanoparticles with electro-chemiluminescence properties. *Chemical Communications*, 34 (2009) 5118–5120.

34. A.B. Bourlinos, A. Stassinopoulos, D. Anglos, R. Zboril, M. Karakassides, E.P. Giannelis, Surface functionalized carbogenic quantum dots. *Small*, 4 (2008) 455–458.

35. Z. Yang, M. Xu, Y. Liu, F. He, F. Gao, Y. Su, H. Wei, Y. Zhang, Nitrogen-doped, carbon-rich, highly photoluminescent carbon dots from ammonium citrate. *Nanoscale*, 6 (2014) 1890–1895.

36. X. Liu, J. Pang, F. Xu, X. Zhang, Simple approach to synthesize amino-functionalized carbon dots by carbonization of chitosan. *Scientific Reports*, 6 (2016) 31100.

37. K.B. Narayanan, N. Sakthivel, Biological synthesis of metal nanoparticles by microbes. *Advances in Colloid and Interface Science*, 156 (2010) 1–13.

38. A. Prakash, S. Sharma, N. Ahmad, A. Ghosh, P. Sinha, Bacteria mediated extracellular synthesis of metallic nanoparticles. *International Research Journal of Biotechnology*, 1 (2010) 71–79.

39. R.Y. Sweeney, C. Mao, X. Gao, J.L. Burt, A.M. Belcher, G. Georgiou, B.L. Iverson, Bacterial biosynthesis of cadmium sulfide nanocrystals. *Chemistry & Biology*, 11 (2004) 1553–1559.

40. A.K. Singh, P. Pal, V. Gupta, T.P. Yadav, V. Gupta, S.P. Singh, Green synthesis, characterization and antimicrobial activity of zinc oxide quantum dots using Eclipta alba. *Materials Chemistry and Sciences*, 203 (2018) 40–48.

41. A. Mahalakhsmi, G. Baskar, Green synthesis and characterization of cadmium-tellurium quantum dots using pomelo peel aqueous extract. *Journal of Electronic Materials*, 48 (2019) 5975–5979.

42. E.S. Anooj, P.K. Praseetha, Green synthesis and characterization of graphene quantum dots from rosa gallica petal extract. *Plant Archives*, 20 (2020) 6151–6155.

43. M.S. Bakshi, F. Possmayer, N.O. Petersen, Role of different phospholipids in the synthesis of pearl-necklace-type gold-silver bimetallic nanoparticles as bioconjugate materials. *The Journal of Physical Chemistry C*, 111 (2007) 14113.

44. Z. Zoya, Rafiuddin, Silver nanoparticles to self-assembled films: Green synthesis and characterization. *Colloids and Surfaces B: Biointerfaces*, 90 (2012) 48–52.

45. P. Zuo, X. Lu, Z. Sun, Y. Guo, Y.A. He, A review on syntheses, properties, characterization and bioanalytical applications of fluorescent carbon dots. *Microchimica Acta*, 183 (2016) 519–542.

46. S. Thambiraj, R. Shankaran, Green synthesis of highly fluorescent carbon quantum dots from sugarcane bagasse pulp. *Applications of Surface Science*, 390 (2016) 435–443.

47. W. Chen, D. Li, L. Tian, W. Xiang, T. Wang, W. Hu, Z. Dai, Synthesis of graphene quantum dots from natural polymer starch for cell imaging. *Green Chemistry*, 20 (2018) 4438.

48. K.S. Subramanian, K. Raja, M. Kannan, *Fundamentals and Applications of Nanotechnology*. DAYA Publishing House, New Delhi (2018).

49. S.N. Baker, G.A. Baker, Luminescent carbon nanodots: Emergent nanolights. *Angewandte Chemie International Edition*, 49 (2010) 6726–6744.

50. L. Cao, X. Wang, M.J. Meziani, F. Lu, H. Wang, P.G. Luo, Y. Lin, B.A. Harruff, L.M. Veca, D. Murray, Carbon dots for multiphoton bioimaging. *Journal of American Chemical Society*, 129 (2007) 11318–11319.

51. J. Shen, Y. Zhu, X. Yang, C. Li, Graphene quantum dots: Emergent nanolights for bioimaging, sensors, catalysis and photovoltaic devices. *Chemical Communications*, 48 (2012) 3686–3699.

52. Q. Wang, X. Huang, Y. Long, X. Wang, H. Zhang, R. Zhu, L. Liang, P. Teng, H. Zheng, Hollow luminescent carbon dots for drug delivery. *Carbon*, 59 (2013) 192–199.

53. C. Frigerio, D.S. Ribeiro, S.S. Rodrigues, V.L. Abreu, J.A. Barbosa, J.A. Prior, K.L. Marques, J.L. Santos, Application of quantum dots as analytical tools in automated chemical analysis: A review. *Analytica Chimica Acta*, 735 (2012) 9–22.

54. S. Campuzano, P. Yáñez-Sedeño, J.M. Pingarrón, Carbon dots and graphene quantum dots in electrochemical biosensing. *Nanomaterials (Basel)*, 9 (2019) 634.

55. W. Bao, Z.C. Su, C.C. Zheng, J. Ning, S. Xu, Carrier localization effects in InGaN/GaN Multiple-quantum-wells LED nanowires: Luminescence quantum efficiency improvement and "Negative" thermal activation energy. *Scientific Reports*, 6 (2016) 34545.

56. B.T. Hoana, T.T. Thanha, P.D. Tamc, N.N. Trungd, S. Choe, V.H. Pham, A green luminescence of lemon derived carbon quantum dots and their applications for sensing of V^{5+} ions. *Materials Science & Engineering B*, 251 (2019) 114455.

57. L. Tang, R. Ji, X. Cao, J. Lin, H. Jiang, X. Li, K.S. Teng, C.M. Luk, S. Zeng, J. Hao, S.P. Lau, Deep ultraviolet photoluminescence of water-soluble self-passivated graphene quantum dots. *ACS Nano*, 6 (2012) 5102–5110.

58. X. Li, S. Zhang, S.A. Lulinich, Y. Liu, H. Zeng, Engineering surface states of carbon dots to achieve controllable luminescence for solid-luminescent composites and sensitive Be^{2+} detection. *Scientific Reports*, 4 (2014) 4976.

59. W. Lu, X. Qin, S. Liu, G. Chang, Y. Zhang, Y. Luo, A.M. Asiri, A.O. Al-Youbi, X. Sun, Economical, green synthesis of fluorescent carbon nanoparticles and their use as probes for sensitive and selective detection of mercury (II) ions. *Analytical Chemistry*, 84 (2012) 5351–5357.

60. Q. Xu, J. Zhao, Y. Liu, P. Pu, X. Wag, Y. Chen, C. Go, J. Chen, H. Zhou, Enhancing the luminescence of carbon dots by doping nitrogen element and its application in the detection of Fe (III). *Journal of Material Sciences*, 50 (2015) 2571–2576.

61. Y. Guo, L. Yang, W. Li, X. Wang, Y. Shang, B. Li, Carbon dots doped with nitrogen and sulfur and loaded with copper(II) as a "turn-on" fluorescent probe for cystein, glutathione and homocysteine, *Microchimica Acta*, 183 (2016) 1409–1416.

62. M. Yang, W. Kong, H. Li, J. Liu, H. Huang, Y. Liu, Z. Kang, Fluorescent carbon dots for sensitive determination and intracellular imaging of zinc (II) ion. *Microchimica Acta*, 182 (2015) 2443–2450.

63. W. KiBae, J. Kwak, J.W. Park, K. Char, C. Lee, S. Lee, Highly efficient green-light- emitting diodes based on CdSe@ZnS quantum dots with a chemical-composition gradient. *Advanced Materials*, 21 (2009) 1690–1694.

64. A. Ambrosi, C.K. Chua, A. Bonanni, M. Pumera, Electrochemistry of graphene and related materials. *Chemical Reviews*, 114 (2014) 7150–7188.

65. J. Zhao, G. Chen, L. Zhu, G. Li, Graphene quantum dots-based platform for the fabrication of electrochemical biosensors. *Electrochemical Communications*, 13 (2011) 31–33.

66. D.Y. Pan, G.M.O. Quaichon, R.A.T. Cruz, C.S. Emolaga, M.C.O. Que, E.R. Magdaluyo, B.A. Basilia, Hydrothermal synthesis of carbon quantum dots from biowaste for bio-imaging. *AIP Conference Proceedings*, 20 (2019) 83.

67. M. Roushani, Z. Abdi, Novel electrochemical sensor based on graphene quantum dots/riboflavin nanocomposite for the detection of persulfate. *Sensors and Actuators B: Chemical*, 201 (2014) 503–510.

68. H. Razmi, R. Mohammad-Rezaei, Graphene quantum dots as a new substrate for immobilization and direct electrochemistry of glucose oxidase: Application to sensitive glucose determination. *Biosensors and Bioelectronics*, 41 (2013) 498–504.

69. R. Mohammad-Rezaei, H. Razmi, Preparation and characterization of hemoglobin immobilized on graphene quantum dots-chitosan nanocomposite as a sensitive and stable hydrogen peroxide biosensor. *Sensor Letters*, 14 (2016) 685–691.

70. I. Vasilescu, S.A.V. Eremia, M. Kusko, A. Radoi, E. Vasile, G.L. Radu, Molybdenum disulphide and graphene quantum dots as electrode modifiers for laccase biosensor. *Biosensors and Bioelectronics*, 75 (2016) 232–237.

71. J. Xi, C. Xie, Y. Zhang, L. Wang, J. Xiao, X. Duan, J. Ren, F. Xiao, S. Wang, Pd Nanoparticles Decorated N-Doped Graphene Quantum Dots@N-Doped carbon hollow nanospheres with high electrochemical sensing performance in cancer detection. *ACS Applied Materials and Interfaces*, 8 (2016) 22563–22573.

72. A. Valipour, M. Roushani, Using silver nanoparticle and thiol graphene quantum dots nanocomposite as a substratum to load antibody for detection of hepatitis C virus core antigen: Electrochemical oxidation of riboflavin was used as redox probe. *Biosensors and Bioelectronics*, 89 (2017) 946–951.

73. F. Shahdost-fard, M. Roushani, Designing an ultra-sensitive aptasensor based on an AgNPs/thiol-GQD nanocomposite for TNT detection at femtomolar levels using the electrochemical oxidation of Rutin as a redox probe. *Biosensors and Bioelectronics*, 87 (2017) 724–731.

74. X. Liu, Y. Luo, Surface modifications technology of quantum dots based biosensors and their medical applications. *Chinese Journal of Analytical Chemistry*, 42 (2014) 1061–1069.

75. A. Hoshino, K. Fujioka, T. Oku, M. Suga, Y.F. Sasaki, T. Ohta, M. Yasuhara, K. Suzuki, K. Yamamoto, physicochemical properties and cellular toxicity of nanocrystal quantum dots depend on their surface modification. *Nano Letters*, 4 (2004) 2163–2169.

76. J. Li, J.J. Zhu, Quantum dots for fluorescent biosensing and bio-imaging applications. *Analyst*, 138 (2013) 2506–2515.

77. K.E. Sapsford, T. Pons, I.L. Medintz, H. Mattoussi, Biosensing with luminescent semiconductor quantum dots. *Sensor*, 6 (2006) 925–953.

78. T.Q. Vu, W.Y. Lam, E.W. Hatch, D.S. Lidke, Quantum dots for quantitative imaging: From single molecules to tissue. *Cell Tissue Research*, 360 (2015) 71–86.

79. P. Devi, A. Thakur, S.K. Bhardwaj, S. Saini, P. Rajput, P. Kumar, Metal ion sensing and light activated antimicrobial activity of Aloe vera derived carbon dots. *Journal of Material Science: Materials in Electronics*, 29 (2018) 17254–17261.

80. Z. Ramezani, M. Qorbanpour, N. Rahbar, Green synthesis of carbon quantum dots using quince fruit (Cydonia oblonga) powder as carbon precursor: Application in cell imaging and As3+ determination. *Colloids and Surfaces A: Physicochemical and Engineering Aspects*, 549 (2018) 58–62.

81. M. Lu, Y. Duan, Y. Song, J. Tan, L. Zhou, Green preparation of versatile nitrogen-doped carbon quantum dots from watermelon juice for cell imaging, detection of Fe^{3+}ions and cysteine, and optical thermometry. *Journal of Molecular Liquids*, 269 (2018) 766–774.

82. A. Kumari, A. Kumar, S.K. Sahu, S. Kumar, Synthesis of green fluorescent carbon quantum dots using waste polyolefins residue for Cu^{2+} ion sensing and live cell imaging. *Sensors and Actuators B: Chemical*, 254 (2017) 197–215.

2 Biopolymer-Based Nanomaterials
Synthesis, Characterization, and Applications

Shweta Kashid and Kalpana Joshi
Sinhgad College of Engineering

CONTENTS

DOI: 10.1201/9781003319153-3

LIST OF ABBREVIATIONS

AA-LDHs:	Alginate-lactate dehydrogenases
AgNPs:	Silver nanoparticles
Alginate-g-PNIPAAm:	Alginate-g-poly(N-isopropylacrylamide)
Alginate-PAMAM:	Alginate-poly(amidoamine)
APA:	Alginate/poly-L-lysine
AuNPs:	Gold nanoparticles
BNC:	Bacterial nanocellulose
BSA:	Bovine serum albumin
CHAP:	Cholesterol-modified amino-pullulan
CHCP:	Cholesterol-modified carboxyethyl pullulan
ChNPs:	Chitosan nanoparticles
CHP:	Cholesterol-modified pullulans
CNCs:	Cellulose nanocrystals
CNFs:	Cellulose nanofibers
CRT:	Cathode-ray tube
CS:	Chitosan
CSHD-AgNCs:	Chitosan cross-linked silver nanocomposites
DASNPs:	Drug carriers/di-aldehyde starch nanoparticles
DEAMEA:	Di-ethylaminoethyl methacrylate
DEAP:	3-(Diethylamino)propylamine
DEX:	Dexamethasone
DLS:	Dynamic light scattering
DMF:	Dimethylformamide
DOX:	Doxorubicin
DPA:	Dithiodipropionic acid
ESG:	Enzymatically synthesized glycogen
FA:	Folic acid
FDDDSs:	Fast-dissolving drug delivery systems
FTIR:	Fourier transform infrared
g/L:	Grams per liter
GG:	Gellan gum
GN-ZnO:	Gelatin-incorporated zinc oxide nanoparticles
HA:	Hyaluronic acid
HES-paclitaxel:	Hydroxyethyl starch
HMP:	Hexametaphosphate
HPMC:	Hydroxypropyl methylcellulose
ICDD:	International Centre for Diffraction Data
INPs:	Iron nanoparticles
JCPDS:	Joint Committee on Powder Diffraction Standards
JFP:	Jellyfish polysaccharide

kDa:	Kilodalton
KHz:	Kilohertz
kv:	Kilovolt
LNFs:	Lysozyme nanofibers
MBA:	Mulberry anthocyanins
mg:	Milligram
MION:	Mono-crystalline iron oxide nanoparticles
ml:	Milliliter
mM:	Millimole
MMT:	Montmorillonite clay
MP:	Methylprednisolone
MPA:	Microscopic polyangiitis
MRI:	Magnetic resonance imaging
MSCs:	Mesenchymal stem cells
MWCNTs:	Multi-walled carbon nanotubes
nHAP:	Nano-silica or hydroxyapatite
nm:	Nanometer
NMR:	Nuclear magnetic resonance
NOE:	Nuclear Overhauser effect
NPs:	Nanoparticles
NRs:	Nanorods
NTs:	Nanotubes
ODDMAC:	(Octadecyldimethyl-(3-triethoxy silylpropyl)ammonium chloride
PDI:	Polydispersity index
PdNPs:	Palladium nanoparticles
PEG:	Polyethylene glycol
PEGMA:	Polyethylene glycol monomethacrylate
PL:	Pullalan
PLA:	Polylactic acid
PNIPAM:	Poly(N-isopropylacrylamide)
PPN:	Phthalyl pullulan nanoparticles
Pt:	Platinum
PVA:	Polyvinyl alcohol
RCSPs:	Rana chensinensis skin peptides
RH:	Relative humidity
rpm:	Revolutions per minute
SEM:	Scanning electron microscopy
SERS:	Surface-enhanced Raman scattering
SETC:	Starch Encapsulated Triphala Churna
SiRNA:	Small interfering ribonucleic acid
SNPs:	Starch nanoparticles
SPLA50:	Secretory phospholipase
St-DPA:	Starch-dithiodipropionic acid
TC:	Triphala Churna
TEM:	Transmission electron microscopy
TEMPO:	2,2,6,6-Tetramethyl-1-piperidinyloxy
TFA:	Trifluoroacetic acid
TiO$_2$NPs:	Titanium dioxide nanoparticles
TOCNs:	TEMPO-treated cellulose nanofibrils
TPP:	Tripolyphosphate
US FDA:	Food and Drug Administration

UV/V:	Ultraviolet/visible
XG:	Xanthan gum
XRD:	X-ray diffraction
ZnNPs:	Zinc nanoparticles
μm:	Micrometer
5-Fu:	5-Fluorouracil

2.1 INTRODUCTION

Biopolymers are polymers derived from bio-based materials and are biodegradable and biocompatible. Biopolymers consist of covalently bonded monomeric units, which are synthesized by living organisms. These are mainly classified into three types based on repeating units of monomers present in a chemical structure as proteins, polynucleotides and polysaccharides. The class of polysaccharides is found to be superior compared to the proteins and polynucleotides due to their thermal stability and structural diversity, providing a lot of potential in different applications. The broad classification on the basis of their natural origin is shown in Figure 2.1.

Among hundreds of identified polysaccharides, starch, cellulose, chitin, chitosan, alginate, gellan, pullulan, and xanthan are some of the important examples that have material properties in a wide range of medical and industrial applications. Based on functional and molecular weight variations, polysaccharides are available in two different forms, viz. mono- and multifunctional polysaccharides. Monofunctional type contains hydroxyl groups only, whereas multifunctional type consists of amine, hydroxyl, or carboxylic acid groups. Polysaccharides also have certain characteristics such as higher chirality, varied water solubility (miscible to immiscible), and eco-friendliness, which can be further utilized in the preparation of nanomaterials and nanocomposites [1].

Polysaccharide-based green materials are available in various types such as aerogels, sponges, films, hydrogels, fibers, food packaging, and membranes, which have a wide scope in different industries such as pharmaceuticals, food, wastewater treatment, cosmetics, biomedical, and biosensors [2–7]. The concerns about the depletion of fossil fuels lead to increased demand in search of other fuel sources. Polysaccharides can be considered as a promising biomass for the production of biofuels (e.g., cellulose) because of their availability and economic feasibility [8]. As a result of their abundance, eco-friendliness, and cost-effectiveness, biopolymeric polysaccharides are an important ingredient in the synthesis of a wide range of products.

This chapter provides an overview on various polysaccharide-based nanomaterials such as cellulose, starch, chitin, chitosan, alginate, pullulan, gellan, and xanthan with their synthesis, characterization, and applications.

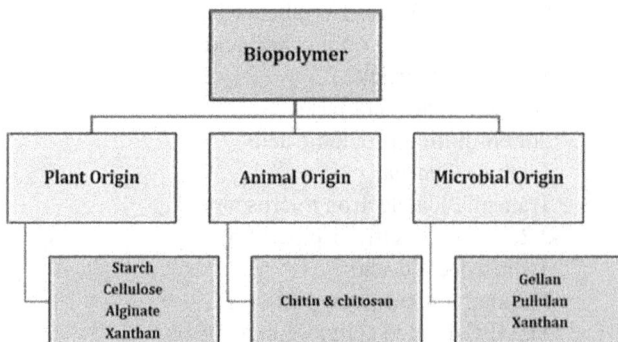

FIGURE 2.1 Classification of natural source biopolymers

2.2 BIOPOLYMERIC NANOMATERIALS

2.2.1 CELLULOSE

Cellulose, present in cotton, hemp, jute, and flax, is the primary source of polymers and a fundamental structural element of plant cell walls with an worldwide annual yield of 1.5×10^{12} tons [9]. It consists of a linear chain of units of D-anhydroglucopyranose linked by $\beta(1 \rightarrow 4)$ glycosidic linkage (Figure 2.2) resulting in bundles of microfibrils and nanocrystals that can be cleaved by mineral acids. Due to the formed hydrogen bonds from its hydroxyl groups, cellulose has a high degree of crystallinity and stereo-regularity [10]. Moreover, some bacteria and algae are also capable of synthesizing cellulose [10,11].

Naturally available cellulose has two forms: cellulose triclinic form Iα and monoclinic Iβ unit cell, which possess the same configuration, but differ in crystal structure [12]. Cellulose from plant cell walls is rich in Iβ, whereas that from algal and bacterial cell walls contains Iβ [13,14]. Celluloses Iα and Iβ have the identical molecular structure and O_3-H-O_5 having intra-chain hydrogen bonding, while O_2-H-O_6 inter-chain hydrogen bonding is different. Furthermore, each monoclinic unit cell includes two chains of cellulose Iβ, but the triclinic unit cell contains only one chain of cellulose Iα [15,16]. It is a versatile, biodegradable, and non-toxic material used in the production of valuable products such as cellulose-based threads, films, and their derivatives used in various applications [17,18].

Cellulose nanofibers (CNFs), bacterial nanocellulose (BNC), and cellulose nanocrystals (CNCs) are three types of cellulosic nanoparticles that can be prepared through acid hydrolysis, enzymatic treatment, or mechanical disintegration [11]. CNCs are long rod-shaped particles with varying width of 3–5 nm and length of 50–500 nm, which are synthesized by acid hydrolysis to eliminate all amorphous areas from cellulose fibers, leaving only crystalline cellulose [19]. This process results in highly crystalline (54%–88%) materials. However, CNFs contain both amorphous and crystalline regions in the nanostructure. CNF nanostructures are long and flexible with width between 5 and 25 nm and length between 500 and 3,000 nm and are synthesized from pulp fibers by mechanical refining. This method is initiated by pre-treatment with chemicals because the attractive binding forces between fibrils lead to the release of them by reducing binding effect [11]. The difference in sizes of CNCs and CNFs alters the fabrication process. Solvent casting or filtration through a thin membrane can be used to make CNFs, whereas CNCs are produced by casting with solvent. Although cellulose of plant origin possesses excellent features, it limits the usage of plant-derived CNFs facing difficulties during extraction. The bacterial CNFs overcome some of these limitations because of the extracellular product that makes extraction much easier. The films prepared from these materials have high transparency [20] and high specific strength and also act as a good oxygen barrier. The characteristics of these materials may get affected with humidity, which can be altered by mixing nanocellulose with other components in various proportions [20]. CNFs have applications in food packaging, bioimaging [21], and electronics [22], in making nanocomposites.

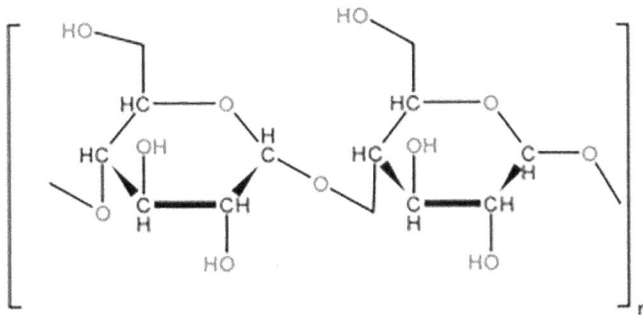

FIGURE 2.2 Cellulose Structure

2.2.1.1 Synthesis of Cellulose Nanoparticles

Various pre-treatments and synthesis strategies are available in the literature to minimize synthesis costs and improve characteristics of cellulosic materials. While producing CNFs, the properties of CNFs are altered using different techniques such as reduction of the cellulose chain length using enzymatic pre-treatment [23], or introducing electrostatic charges, TEMPO-mediated oxidation [24], phosphorylation, and carboxymethylation [25]. They are synthesized using the following methods.

2.2.1.1.1 Acid Hydrolysis

This method is commonly used for the synthesis of CNCs from organic and mineral acids of various forms and concentrations, such as phosphoric acid (H_3PO_4) [26], citric acid ($C_6H_8O_7$) [27], hydrochloric acid (HCl) [28], sulfuric acid (H_2SO_4) and oxalic acid ($C_2H_2O_4$) etc. [29]. The acid concentration varies depending on the system. The acid hydrolyzes the amorphous portions of cellulose fibril aggregates at low concentrations and reacts with cellulose at high concentrations to introduce new functionality. This method is utilized for the synthesis of CNCs from different cellulosic materials such as softwood kraft pulp, cotton, sisal, coconut husk, rice straw, and banana waste [30]. Cellulosic source material is hydrolyzed using H_2SO_4/HCl/H_3PO_4 at a temperature of 50°C for 3–6 hours under continuous mechanical stirring. The suspension obtained is neutralized with distilled water and centrifuged for 30 minutes at 3,000 rpm. After that, the solution is sonicated for 10 minutes in an ice bath. In acid hydrolysis treatment, the length of CNC is determined by cellulose microfibrils, temperature, and the time required for the process [26].

2.2.1.1.2 Enzymatic Hydrolysis

Cellulosic fibers are pre-treated and washed with the surfactant buffer Tween 20 for an hour at a 25°C. The acid content varies depending on the system. The fibers are rinsed three times with deionized water, and the moisture is removed with overnight slow heating in an oven at 95°C. Pre-treated fibers are hydrolyzed using a cellulase enzyme. This enzymatic hydrolysis is carried out with 5 g/L fibers and with 2.5% concentration of cellulose enzyme at 48°C for 180 hours using acetate buffer 0.1 M (pH 5). The enzyme is inactivated first, and then hydrolysis is terminated by increasing reaction temperature to 80°C for 30 minutes. The suspension is rinsed twice with deionized water and then centrifuged for 15 minutes at 5,000 rpm. The hydrolyzed suspension is sonicated further for 15 minutes, and the top turbid part containing nanoparticles is separated from the bottom solid part. The sonication cycle is repeated twice, and a clear top layer can be obtained [31].

2.2.1.1.3 Chemo-mechanical Treatment

CNFs are sometimes synthesized with mechanical treatment, which maintains the non-crystalline parts and the length of the fibrils. In this treatment, the fibers are dipped in deionized water and homogenized for 15–20 minutes at 8–10,000 rpm. The suspension is passed through a microfluidizer to prepare cellulose nanofibers, or its regeneration and electrospinning of the cellulose polymer melt is performed. The TEMPO oxidation and microfluidizer for the preparation of cellulose nanofibers are utilized for introducing charged groups to the surface. The mechanically pre-treated fibers are added to a solution of equal amount of 0.000125 wt% sodium bromide and 2,2,6,6-tetramethyl-1-piperidinyloxy (TEMPO). Further, 0.01 wt% sodium hypochlorite is added, which initiates the reaction by maintaining pH at 10.5 using sodium hydroxide. The reaction is carried out till there is no more requirement of addition of sodium hydroxide. At 25°C, the pH of the solution is adjusted by adding HCl dropwise. The solution mixture is then passed through a microfluidizer. The precipitates are removed by passing cellulose nanofibers through a dialysis membrane tube [24].

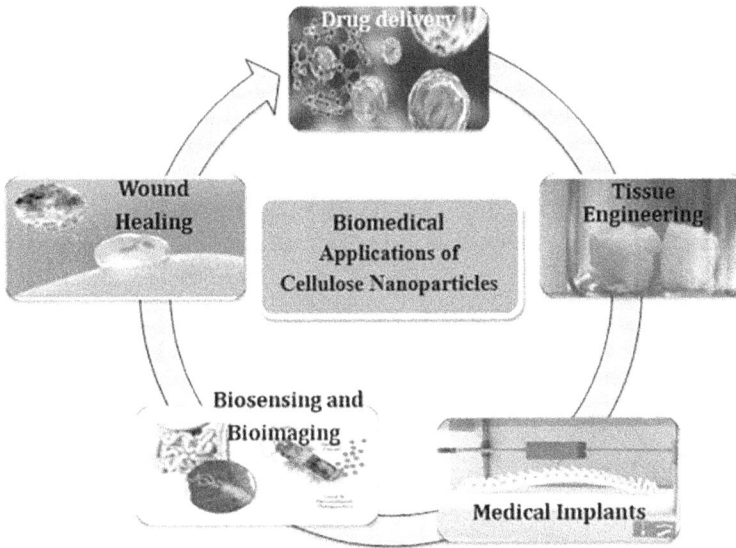

FIGURE 2.3 Applications of cellulose nanoparticles in Bio-medical field

2.2.1.2 Applications of Cellulose Nanomaterials

2.2.1.2.1 Application in Biomedical Field

Nanocellulose offers multiple applications in bioengineering field. Because of excellent biological and physical characteristics, especially surface chemistry, low toxicity, and biocompatibility, it is applicable to biomedical industry, specifically in drug delivery, tissue engineering, medical implants, biosensors, diagnostics, etc., which is depicted in Figure 2.3 [32].

2.2.1.2.2 Drug Delivery

CNFs, BNC, and CNCs are used as a vehicle for drug delivery with their varying properties. Hydrogels, which have a double-membrane structure and are biocompatible in nature, are prepared from CNCs. These hydrogels have the ability to release one drug quickly and the other slowly. CNCs along with cyclodextrin loaded with curcumin are used for anticancer drug delivery. CNCs are also used in drug delivery for bone cell-related diseases [33]. CNFs are used in drug formulations and help to improve encapsulation efficiency. They are also used as film and foam formers, as well as stabilizers for long-lasting sustained drug release. CNF aerogels are used as carriers for oral controlled drug delivery systems. CNF hydrogels are used for the delivery of small molecules and proteins [34]. BNC as a matrix is used for controlled protein delivery [35].

2.2.1.2.3 Tissue Engineering

Nanocellulose is considered to be suitable for tissue engineering support and creates various shapes such as membrane-like structures, electrospun fibers, and hydrogels. When combined with other biomaterials, nanocellulose provides a platform for supporting and promoting cellular activity for tissue regeneration and repair. For example, a group of hyaluronic acid $(C_{14}H_{21}NO_{11})_n$ and chitosan on a matrix is used in the bone tissue engineering [36].

2.2.1.2.4 Medical Implants

BNC is used for replacing damaged blood vessels. BNC is also used in the field of tissue replacement. Nanocellulose is used for *in situ* softening cortical implants. One of the examples is brain-computer interface used to connect the central nervous system to the rest of the world, which holds great potential for restoring brain functions in paralyzed people and helps to improve human health [37].

2.2.1.2.5   Biosensing and Bioimaging

Biosensors are getting much attention due to their capability to measure important parameters, especially in medical diagnostics. CNCs are used in biosensing. For example, AgNPs and bacterial cellulose nanocomposites are used as binding compounds in surface-enhanced Raman scattering (SERS) for the recognition of amino acids, 2,2-dithiodipyridine, and thiosalicylic acid. CNCs labeled with pH-responsive fluorescent dyes are used for pH sensing. Fluorescently labeled CNCs are used to develop emergent diagnostic tools. Folic acid-conjugated CNCs are developed for cancer diagnosis. A dye named Alexa Fluor is attached on the surface of CNCs and used in bioimaging applications in case of living animals [38].

2.2.1.2.6   Wound Healing

Bacterial cellulose is used in the treatment of wounds. BC membranes have excellent wound-healing capability with elasticity and conformability, liquid-holding capacity, and mechanical material strength. Several wound dressing products such as XCell®, Bioprocess®, and Biofill® are based on BC [39].

2.2.1.2.7   Application in Paper and Pulp Industries

Green nano-additives are used to improve paper performance and paper functionalization with unique properties. The strength of paper is primarily determined by the strength of the fibers, the length of the paper, and the number of hydrogen bonds in the matrix. When nanocellulose is added to pulp, it is distributed in the gaps between the fibers, which helps to increase the adhesion between the fibers and fills the gaps in the paper improving the paper's strength [40].

2.2.1.2.8   Application in Food Packaging

A wide research is being conducted to incorporate nanocellulose into the food packaging. In food packaging, it is essential to use eco-friendly packaging, i.e., renewable and biodegradable, and it should maintain the food stability and food quality. To achieve this, bio-based materials such as nanocellulose are used. Nanocellulose has usually been applied as reinforcing phases, or as matrices in a variety of materials. CNCs are used with bio-nanocomposites for food packaging. For example, gelatin-CNC-glycerol film-forming dispersions help to improve modulus, elongation, transparency of the films, and water vapor barriers. CNCs are hydrophilic in nature, but can be utilized as water-resistant hydrophilic matrices such as carrageenan, alginate, soya isolate, and chitosan. It is used to improve the thermal sustainability of some polymers. CNFs are used for reinforcement purpose. For example, CNFs are used to make PLA sheets by mixing CNFs with water and PLA acetone. In amylopectin films, CNFs serve as a plasticizer [41].

Hence, due to unique properties of cellulose nanomaterials, such as strength, rheological properties, and reactivity, they are widely used in many applications.

2.2.2   Starch

Starch is a naturally occurring biopolymer and is part of our diet every day. It is found in roots of plants and cereals such as corn, rice, wheat, and potato. Plant starch is typically converted into insoluble powders [42]. It is made up of an amylopectin branched chain molecule and an amylose linear chain molecule. The granular form of these two starch components is assembled in the size range of 1–100 μm and is shown in Figure 2.4 [43]. Although starch has many applications in biomedical and food industries, due to some drawbacks such as retrogradation, limited digestibility, poor solubility, and poor functional properties, starch nanoparticles (SNPs) are gaining more attention because of their improved structure qualities and wide applications [44].

2.2.2.1   Starch Nanoparticles (SNPs)

SNPs are considered as the promising biomaterials due to their availability, biodegradability, and compatibility in food, pharmaceuticals, and biomedical applications [45].

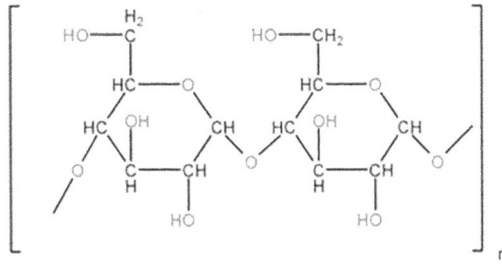

FIGURE 2.4 Chemical Structure of Starch

SNPs can be synthesized using various techniques such as hydrolysis of acid and alkali and enzymes. Furthermore, extrusion, sonication, or precipitation by co-solvent, chemical, and enzymatic treatment can also be used for synthesis. SNPs also have the potential to be utilized in food quality indicators (nanoencapsulation), in drug transporters, and in the reinforcement of biodegradable and non-biodegradable polymeric matrices [46].

2.2.2.2 Synthesis of Starch Nanoparticles

2.2.2.2.1 Acid or Alkali Hydrolysis

The synthesis of starch nanoparticles is commonly achieved using mild acid or alkali hydrolysis. When treated with mild acids and alkalis, the amorphous regions erode, allowing the resistant crystalline moieties to form. In one of the studies [47], the preparation of nanoscale starch granules (30–80 nm) has been reported, in which waxy corn starch is treated with HCl under controlled conditions in a submerged reactor. Mudasir Ahmed et al. [44] gave a mild alkali treatment to the starch extracted from lotus stem, horse chestnut and water chestnut, and lotus stem. In this treatment, starch solution (1.5%) is heated at 80°C in the 0.1M solution of NaOH with continuous stirring for 30 minutes. The starch slurry obtained is then ultrasonicated at 40 KHz for 30 minutes with 5-minute intervals to avoid overheating. Further precipitation is carried out by the dropwise addition of ethanol in 1:2 ratio. After centrifugation (15 minutes, 8,000 rpm), the solution is collected, lyophilized, and stored in the form of powder with the average particle size of 400–600 nm.

2.2.2.2.2 Acetylated SNPs by Precipitation

The acetylated SNPs are synthesized using the nanoprecipitation technique [48–50]. The acetylation of starch is carried out using acid hydrolysis and sudden quenching of reaction mixture using the method proposed by Murray [51]. Further, in 20 mL of acetone, 100 mg of starch acetate is dissolved with the dropwise addition of distilled water to the polymeric solution. The resulting suspension is vigorously stirred, and acetone is vaporized and centrifuged (8,000 rpm, 20 minutes). The precipitate is dried in at 50°C for 24 hours. This technique is used to obtain the oxidized SNPs.

2.2.2.2.3 Electrospinning

Electrospinning is a one-step, simple, and accurate technique for producing nanofibers with desired morphology from both natural and synthetic polymers [52]. In this method, an electrical potential is applied between two ends. To overcome the tension at the surface of a polymeric drop, a droplet of polymeric solution is held at the pinpoint of the capillary tube and an electric field (16–20 kv) is applied to control the jet of solution [53]. Various factors such as viscosity, conductivity, surface tension, and electric charge can affect the electrospinning process. Electrospinning method can result in uniform and high-quality nanofibers of size range 1–10 nm.

2.2.2.2.4 Ultrasonication

Ultrasonication is performed to study the physical properties of starch suspensions. In this technique, high-frequency pulses in the range of 20–40 KHz are used, which creates rapid vibrations in the liquid suspension resulting in the collapse of bubbles causing cavitation. Ultrasonication can result in the distortion of the crystalline regions and result in the swelling of the treated starch [54]. The waxy corn starch is ultrasonicated at the high frequency of 24 kHz at a low temperature of 8–10°C; it gives SNPs in the size range of 20–80 nm. It is observed that with the increase in sonication time, fine particles of 20–100 nm size are obtained [55].

2.2.2.2.5 Cross-linking and Homogenization

The chemical technique of cross-linking is used to make nanoparticles and nanospheres. The side chains are opened and dispersed by acids, enzymes, and alkalis. The hydrophilic nature of starch can form aggregates in both water and oil emulsions [56]. The stable molecules are preserved by the emulsions in dispersed phase, but the unstable small mini-emulsions can be effectively dispersed by using pressure homogenizers. The purified nanoparticles obtained after homogenization have shown particle size in the range of 50–250 nm [57].

2.2.2.3 Applications of SNPs

Depending on the starch origin, different shapes and sizes of starch nanocrystals can be obtained. Nanomaterials such as SNPs, nanogels, and nanodendrites can be obtained from starch. Due to some poor mechanical properties, it is also added with various materials to increase their strength and elasticity.

2.2.2.3.1 SNPs in Food

Starch films can be used for packaging of food materials in industries. Due to the lack of mechanical properties, they are used with some additives in food industries. For example, the addition of glycerol improves their thermostability. Thermoplastic starch with silver nanoparticles or with talc nanoparticles is used as biofilms in food packaging [58].

All over the world, almost 80% of the starch includes corn starch. Some biocomposite and nanocomposite films made up of corn starch and various biopolymers such as cellulose and chitosan are used with varying thickness in packaging food materials. When gelatin is treated with corn starch, it leads to biocomposite films of 50 μm particle size that can then be used in the edible food packaging [59]. Starch also has good compatibility with proteins. Cooled corn starch and protein pastes improve the quality of starch containing dishes [60]. The hydrogel beds prepared by corn starch and alginate are used to protect and deliver antioxidants such as yerba mate in food products [61]. The antimicrobial films (Figure 2.5) are very much protective against infective agents in food packaging.

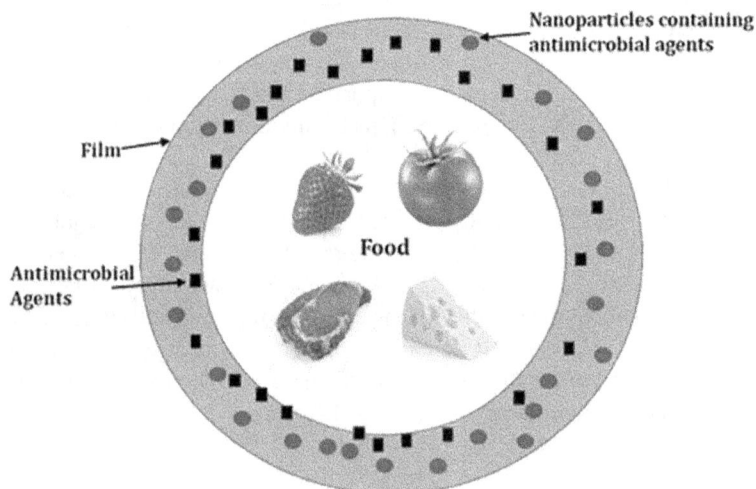

FIGURE 2.5 Edible films loaded with antimicrobial nano-carriers in packaged foods [59].

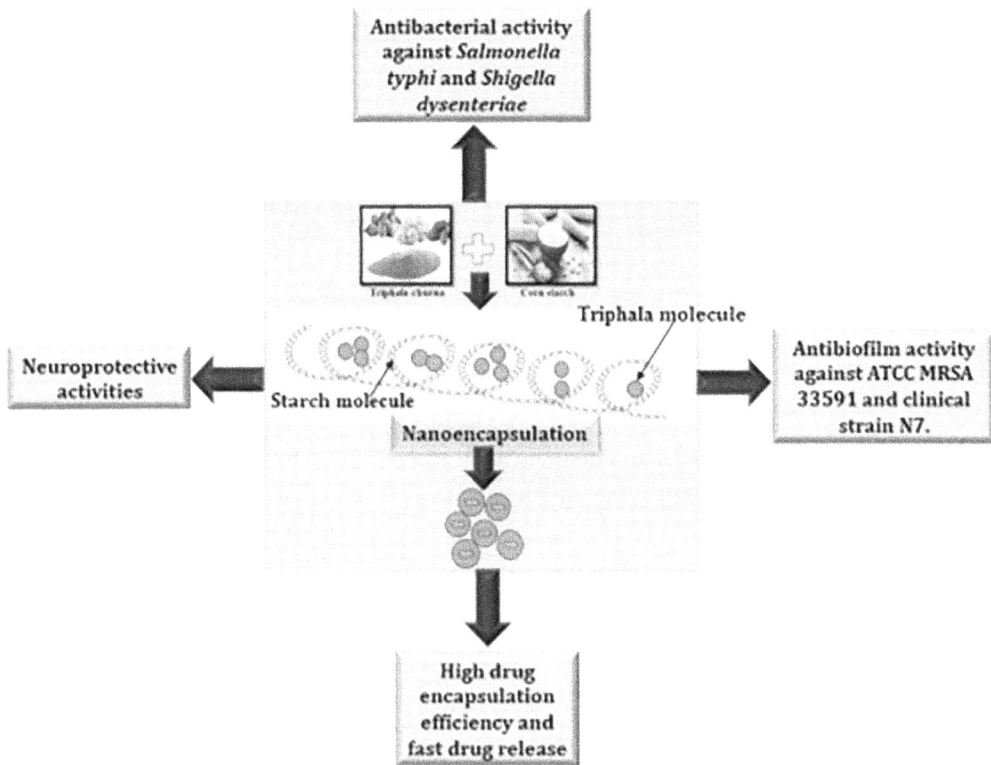

FIGURE 2.6 Effects of Encapsulation of Triphala Churna into starch polymer [63].

2.2.2.3.2 Pharmaceutical Applications

The corn starch is widely used in the pharmaceutical industry. Biocomposite films used in pharmaceutical packaging are prepared by combining corn starch with chitosan as copolymer [62]. A tri-doshic rasayana, *Triphala Churna* (TC), rich in citric acid and phenols, has a balanced restorative impact on three constitutional elements that control human existence, according to Ayurveda. Encapsulation of TC into starch polymer was done to enhance the solubility and efficiency of it. This Starch Encapsulated *Triphala Churna* (SETC) exhibits high levels of free radical scavenging and acetylcholinesterase inhibitory action. It also shows potent antimicrobial and anti-biofilm activity against *Salmonella typhi* and against ATCC MRSA 33591 and clinical strain N7, respectively. Thus, the nanoprecipitation of corn starch leads to the formation of *Triphala Churna*. The various effects of encapsulation of TC using starch are highlighted in Figure 2.6 [63].

2.2.2.3.3 Drug Delivery

It is very important to deliver drugs in the right way without causing any harm to the human body. One has to be more careful about its complete solubility during drug release so that it doesn't lead to any toxicity. A variety of starch-based nanomaterials used in drug delivery are shown in Figure 2.7.

There are mainly two approaches for the production of starch-based SNPs, which include "top-down" and "bottom-up" approaches. Various preparation methods such as acid hydrolysis, ultrasonication, homogenization, crushing, and gamma irradiation are employed to form SNPs. These pH-dependent SNPs show varied response at different values of pH. Poly(methacrylic acid)-polysorbate 80-grafted SNPs show swelling in the pH range of 7.4–4.0. SNPs containing $NaHCO_3$ at pH 5 have shown CO_2 gas release, which is used to control drug release. Nanoparticles have

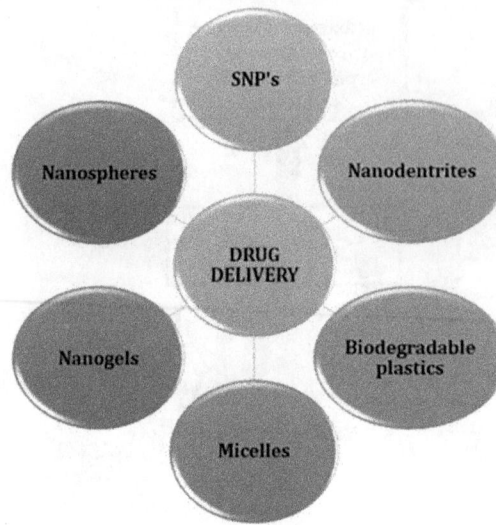

FIGURE 2.7   Forms of Starch-based nanomaterials in drug delivery

a wide scope of research in anticancer drug delivery treatments. In 2012, Xiao et al. reported a drug which carries 5-fluorouracil anticancerous drug carriers/di-aldehyde starch nanoparticles (DASNPs) and can be significantly utilized as a model medicine in enhanced breast cancer cell inhibition (MCF-7) [64].

The starch nanospheres can be prepared by various methods such as reverse microemulsion, precipitation, self-assembly, and sacrificial template. These nanospheres have good biocompatibility and biodegradability and are available at low cost; hence, they are used as a drug carrier. For example, HES-paclitaxel NPs are used to control drug release. The temperature-responsive nano-sized hydrogels (nanogels) such as PEG/DEAP, Fe_3O_4-g-(PNIPAAm-co-PMA), PNIPAM-g-(starch-MNPs), and spiropyran-modified pullulan nanogels can be used in drug delivery systems. Micelles are core-shell nanocarriers made by using different kinds of modified starch such as DOSA starch, OSA starch, and corn starch. Starch-g-P micelles (DEAEMA-co-PEGMA), mPEG-St-DCA micelles, mPEG-St-SeSex micelles, HES=DOX micelles, starch-g-PEG-LA micelles, and mPEG-St-DPA micelles are used to control drug release. Nowadays, the starch-based biodegradable plastics are created to enhance drug delivery and avoid post-drug release treatments reducing toxicity [65]. A chemically modified enzymatically synthesized glycogen (ESG) nanodendrite is made for SiRNA delivery [65].

2.2.2.3.4   Biomedical Applications

Biomedical applications include the formation of materials for bone regeneration treatments in the field of tissue engineering [65]. Nanocomposites of starch containing carbon nanotubes (MWCNTs) are used to make bone regeneration and in tissue engineering scaffolds [66]. The nanocomposite tubes have the ability to stimulate bone formation, and they show antimicrobial properties [67]. Natural materials such as starch are always a better choice and can be preferred in tissue engineering scaffolds. The agglomerated beads of 100-micrometer-sized particles combined with corn starch and gelatin are used to control the structure and function of engineered tissue [67]. Different corticosteroids such as secretory phospholipase (SPLA50), dexamethasone (DEX), methylprednisolone (MP), and microscopic polyangiitis (MPA) are prepared with modified starch. Corticosteroids have widespread use in the medical field as they show anti-inflammatory activities and can be used in treating infections, such as acute respiratory distress, and degenerative diseases, and as immune suppressors in organ transplantation [68].

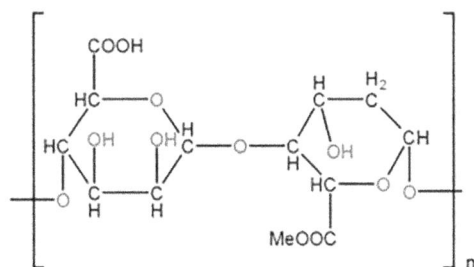

FIGURE 2.8 Chemical Structure of Alginate

2.2.2.3.5 Industrial Applications of SNPs

Industrially, corn starch can be used as a cleansing agent, as a fabric stiffener, as a fast absorbent polymer in water management systems, and also in organic pesticides. It also prevents mildew from ruining books and documents in storage [69].

2.2.3 ALGINATE

Edible seaweeds have remarkable potential as functional food, and they can act against human non-communicable diseases such as diabetes, auto-immune diseases, cardiovascular diseases, and cancer type II. One of the main components of edible brown seaweed is alginate (Figure 2.8), an anionic polysaccharide that is made up of two components: mannuronic and guluronic acids [70]. Along with this, it contains other bioactive components such as peptides, polyunsaturated fatty acids, polyphenols, vitamins, and minerals [70,71]. Although commercially available alginate is extracted from brown seaweed, bacterial biosynthesis of alginate can also provide defined chemical structures and physical properties compared to seaweed-derived alginate. Biocompatibility, minimal toxicity, and mild gelation caused by the addition of divalent cations are all inherent features of alginate [72].

2.2.3.1 Alginate Nanoparticles

Alginate nanoparticles are rapidly evolving with their applications in biomedical, pharma, and food industries. The gelling property of an alginate is frequently used in drug delivery, where alginate can be used as a biodegradable nanocarrier material for carrying out targeted drug delivery. Alginate nanoparticles are synthesized using a variety of techniques to achieve more rational, coherent, efficient, and cost-effective properties. Recent techniques for producing alginate-based nanoparticles include emulsification/gelation complexation, electrospraying, layer-by-layer deposition, electrospinning, and spray drying. The method selection for synthesis depends on the nature of encapsulants and the required characteristics nanoparticles should possess. In the following section, the commonly used synthesis techniques are discussed.

2.2.3.2 Synthesis of Alginate-Based Nanoparticles

2.2.3.2.1 Emulsification/Gelation

When compared to nozzle-based techniques, this is a simple and cost-effective technique. Emulsification disperses droplets from oil phase to nanospheres [73]. This process is carried out in two steps: preparation of alginate-in-oil emulsion and internal or external gelation by an ionic or covalent cross-linker. Internal gelation occurs when cations are released from the inner core of alginate emulsion droplets [74]. External gelation occurs when a cross-linker, such as calcium chloride, diffuses and reacts with the acid group of guluronic acid. Before emulsification, the salt of $CaCO_3$ is dissolved with alginate to initiate gelation by improving the solubility of calcium ions with a decrease in its pH from 7 to 6.5. Calcium ions also start to migrate from the external to internal surface of droplets. Internal gelation results in symmetrical, porous, and less denser nanospheres [75,76].

2.2.3.2.2 Coating Techniques of Alginate Nanocarriers

2.2.3.2.2.1 *Layer-by-layer Technique*   It is a coating technique in which film-wise coating is done by applying micro- or nanometric thick layers sequentially to form the core-shell. Alternate cationic and anionic layers are arranged, which creates electrostatic interactions between oppositely charged polyelectrolytes. Furthermore, polymeric carriers allow drug to be embedded between the layers, promoting encapsulation efficiency and controlled drug release [77]. In this technique, various parameters such as medium pH, temperature, polyelectrolyte solution salt concentration, saturation adsorption time, and the matrix or carrier material used have a significant impact on the final coat [78].

2.2.3.2.2.2 *Spray Drying*   Spray drying is a technique that involves atomizing liquid droplets and drying them with a hot gas to produce micro- and nano-sized particles [79]. In this process, the moisture from sprayed wet droplets is removed. Spray drying consists of the following steps: (1) emulsification or dispersion of drug in the solvent, (2) atomization in droplets using nozzle, (3) drying of drops with hot gas, and (4) product collection [80]. The spray drying process is based on removing moisture content from wet droplets in a hot environment. The drying gas is introduced from the top of the chamber through an air dispenser. The feed solution is atomized with a constant flow rate and at a certain temperature in the hot gas chamber. The wet drops formed are dehydrated using evaporation. The dry particles are collected in an electrostatic chamber, which separates surface-charged particles and deflects them using an electric field if any defects persist. An electrostatic particle-collecting chamber consisting of a round stainless steel tube linked to a high-voltage source (anode) and a grounded star electrode (cathode) is fixed inside the tube. Finally, the exhaust gas is carried away through a small filter, which retains the free particles from the gas flow [81].

2.2.3.2.2.3 *Electrospray*   The electrospray technique is a single-step approach for controlling the particle size with less solvent impurities. It can be used to create micro- and nanoparticles [82]. This method is commonly used to encapsulate bioactive macromolecules such as cells, nucleic acids, and proteins [39,83]. In this method, using a syringe pump, a liquid is gently pushed via a thin metal needle at a steady flow rate. The needle is then subjected to a high voltage to accelerate the liquid away from it, thereby overcoming the surface tension of droplets. These droplets are turned into fine micro- or nano-sized spray at the tip, forming a cone [84]. The size of the droplets can be controlled using a variety of processing parameters such as voltage supply, size of the needle, flow rate, and distance from jet to surface. The droplet size can also vary with material concentration, type of cross-linker, and surfactant. Although it is a time-consuming technique, it has the benefit of yielding better nanoparticles [65].

2.2.3.2.2.4 *Electrospinning*   Electrospinning technique is used to produce ultrafine nanofibers from polymeric solution or melt. Nanofibers are utilized in tissue engineering, regenerative medicine, and drug transport systems [85,86]. An electrospinning setup includes a specialized pump with a high voltage and a collection chamber at a specific length. Electrospinning is carried out with a high polymer concentration and a more stable jet [87].

2.2.3.3 Applications of Alginate-Based Nanomaterials

2.2.3.3.1 Biomedical Applications

Along with synthetic polymers, natural biopolymers also have widespread applications in the biomedical field. Alginate is considered safe for biomedical applications because it is utilized in food preparation for human consumption. Some other detailed applications are given below.

2.2.3.3.2 Wound Dressing

The materials used in wound dressing must be antibacterial and should have good vascularization. The cloth dressing can lead to bacterial infection, and it also sticks to the wound causing disruption in the wound healing process. So, natural polymers are used in the dressing of wounds. Alginate is biocompatible and hydrophilic in nature; hence, it is effectively used in wound dressing. Kaltostat® alginate dressing is a primary dressing applied on acute and chronic wounds. It is made up of calcium alginate or sodium alginate gel absorbent fiber matrices in contact with a fluid. It can control minor bleeding from wound [88,89]. The nanofibers are prepared from alginates of two different molecular weights. Low molecular weight alginates are employed as scaffolds, while high molecular weight alginates are used to cover wounds due to their mechanical strength [90]. Nowadays, alginate composites are used instead of natural alginates for wound dressing because naive alginates are unable to reduce bacterial infections. The additives used for composites must have the properties such as hemostasis, antibacterial activity, high cell viability, and enhanced clotting of blood and cell adhesion. The honey-loaded alginate/PVA nanofibrous membrane shows antibacterial and antioxidant properties and is used for wound dressing [91].

Alginate with gelatin, nHAP (nano-silica or hydroxyapatite), and *Rana chensinensis* of skin peptides (RCSPs), which are extracts from *Rana chensinensis* as discarded skin forms different types of composites and shows various effects in wound healing, is shown in Figure 2.9.

2.2.3.3.3 Drug Delivery

This system aims at supplying drugs to specific organs or tissues in the body with the maximum possible effectiveness. Among alginate nanomaterials, hydrogels are very much effective due to their gentle gelation process and can be used for encapsulation of cells and release of drug on a continuous basis. Recently, biocompatible double-membrane hydrogels have been prepared using anionic alginate and cationic cellulose nanocrystals (CNCs). These are specially made for complex drug deliveries [33].

The word "cancer" is just enough to express the fear about it. But nowadays, treating cancer is possible. Treatments such as chemotherapy, radiation therapy, and the combination of both are used to treat cancer. Sometimes, these treatments can cause systemic toxicity and create damage to the

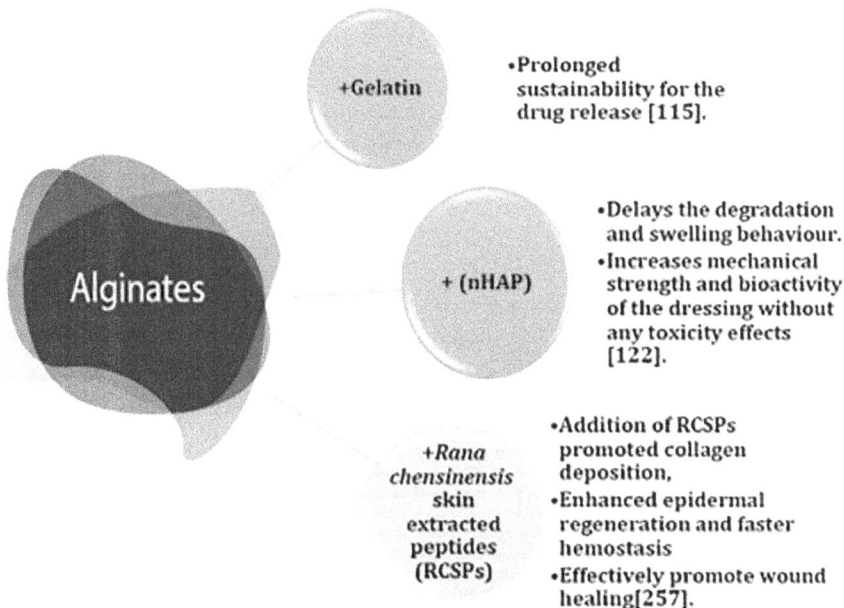

Alginates

+Gelatin
- Prolonged sustainability for the drug release [115].

+ (nHAP)
- Delays the degradation and swelling behaviour.
- Increases mechanical strength and bioactivity of the dressing without any toxicity effects [122].

+Rana chensinensis skin extracted peptides (RCSPs)
- Addition of RCSPs promoted collagen deposition,
- Enhanced epidermal regeneration and faster hemostasis
- Effectively promote wound healing[257].

FIGURE 2.9 Applications of Alginate composites in wound dressing [92–94].

healthy tissues or cells. To reduce this toxicity, biopolymers such as alginate are used. Injectable hydrogels made up of alginate-g-poly(N-isopropylacrylamide) (alginate-g-PNIPAAm) are used in the treatment of prostate cancer (AT3B-1) [95]. The colon cancer can also be treated with alginate-cyclodextrin nanogels, which enhance the chemotherapeutic efficacy by pressure-controlled drug release [95].

Among the cancer-related deaths, breast cancer has the highest death rate in women. To control this, a variety of drugs have been made in which alginate plays an important role. Magnetic alginate nanoparticles are used in the treatment of breast cancer (MDA-MB-231) to develop a Cur (curcumin) drug [96]. The keratin-alginate nanogel composite shows good antitumor results and lesser side reactions against breast cancer (B16 and 4T1) [97]. The use of alginate nanogel platforms in making the drug to cure breast cancer (MCF-7) leads to the inhibition of tumor growth and also reduces side effects and improves the life of cancer patients [98]. Hybrid alginate-PAMAM (G5) nanogels are advantageous in case of moderate tumor growth for sustained release, higher efficacy, and reduced level of toxicity against breast cancer (MCF-7) [99].

2.2.3.3.4 Tissue Engineering

Tissue engineering is just like a magic in the bioengineering. Regeneration and creation of tissues, cells, and organs can be done with tissue engineering tools. The growth of these tissues and organs should be proper and less toxic. Alginates have properties such as biodegradability and biocompatibility, so they have wide applications in tissue engineering for different organs.

Magnetic resonance spectroscopy is an advanced method used in the medical and tissue engineering field. For the evaluation of TC-tet cells, mono-crystalline iron oxide nanoparticles (MION) are used as a contrast NMR agent. The alginate APA/alginate/poly-L-lysine/microbeads are used for encapsulation of the βTC-tet cells. The location of the cells within the APA beads can be determined [100].

The variety of properties make alginate always a better choice in biomedical and tissue engineering field. For example, the acid dissolution by Pt nanoparticles chitosan bubbles causes limitations in the applications due to narrow pH conditions; hence, when Pt nanoparticles combined with alginate bubbles are developed, they have shown applicability in wide pH range. These composite bubbles are termed as "smart vehicles" for biomedical imaging and targeted medicine transport [101].

2.2.3.3.5 Applications in Food Industry

Due to the increasing demand for processed and packaged foods, food industries need to provide safe and qualitative food to mankind. Hence, the food industries are more focusing on the techniques to improve processing and packaging. Various materials are used for this process. Natural biopolymers such as alginate are used in food industry, which enhances the qualitative production of food materials. For more effective results, alginate is treated with some other materials to form nano-biocomposites.

Recently, nano-biocomposite fibers made up of zein/sodium/TiO$_2$NPs-alginate have been used in food packaging, which show resistance to microorganisms *E. coli and S. aureus* and have good biocompatibility [102]. The edible food coatings are mostly used to store food for a long time, as shown in Figure 2.10. The nanocoatings of alginate/lysozyme nanolaminates are produced by the layer-by-layer methodology to maintain the freshness of food items such as "coalho" cheese during long-term storage. These coatings have increased the shelf life of "coalho" cheese [103]. The AA-LDHs were blended into a biocompatible alginate matrix to create an edible coating, which was developed on freshly plucked strawberries using a coat, which increases the shelf life of the strawberries with best preservation [104].

2.2.3.3.6 Applications in Textile Industry

Currently, the textile industry needs the antimicrobial finishing process based on biopolymers. Different types of biopolymers are used to develop antimicrobial finishing in the textile industry.

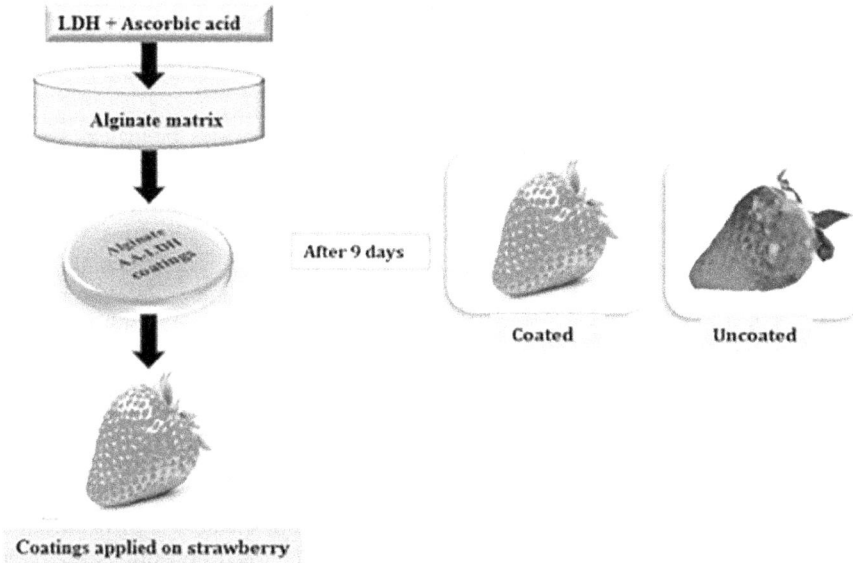

FIGURE 2.10   Edible food coating made up of alginate-AA-LDH composite [104].

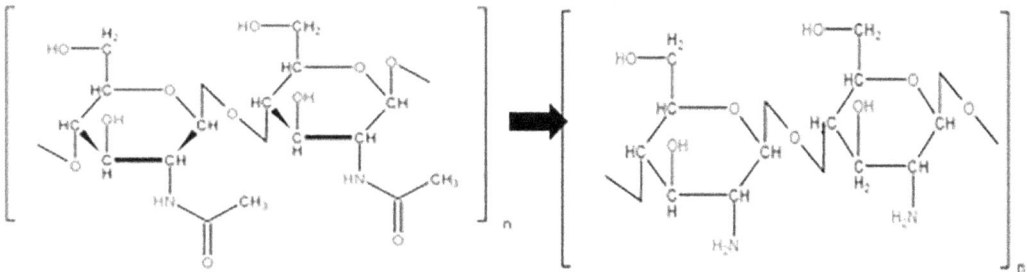

FIGURE 2.11   Chemical structure of chitin and chitosan.

For example, the calcium alginate/silver chloride (Ca/AgCl) nanocomposites possess an excellent flame-retardant and antibacterial activity, so they can be used in decoration and textile industry [105].

2.2.4   CHITIN AND CHITOSAN

These are the most easily available polysaccharides from cell walls/shells of crabs, insects, and fungi. Chitin is commercially produced from the shell waste of crabs, shrimps, and krill using deproteinizing and demineralizing processes to remove the minerals and proteins that form the composite structure of the shells along with chitin. The dried shell waste consists of about 15–25% of chitin [106].

The chemical structure of chitin (Figure 2.11) is a glycan $(1{\rightarrow}4)$ β-linkage containing 2-deoxy-2-acetamido-d-glucose. It is highly insoluble in organic solvents with high inter- and intra-molecular H-H bonding in its crystals. The acetamide groups in chitin are hydrolyzed to form primary amino groups, and the deacetylated chitosan derivative obtained is highly soluble in aqueous acidic solutions, making it suitable for fibrous production. The enzymatic biodegradation of chitosan by lyso-zyme gives non-toxic products. This ease in processability and characteristics of chitosan can be uti-lized in various biomedical applications [107]. In recent years, chitosan has been gaining importance in different types of pharmaceutical, food, environmental, and biomedical applications because of its mucoadhesive and antibacterial properties, biodegradability, and biocompatibility [108,109].

2.2.4.1 Chitosan Nanoparticles (ChNPs)

ChNPs are considered as a carrier material of nanoparticles in various biological systems [110]. Chitosan nanoparticles' sizes and surface charges are considered important during cellular uptake and biodistribution *in vivo*. NPs with slight negative charges and less pH have better accumulation in tumor cells. For the synthesis of ChNPs, various methods are used, such as ionic gelation, reverse micellar extraction, and emulsification; furthermore, nanofibers can be prepared by electrospinning technique. Ionotropic gelation is the most widely used in ChNPs' preparation due to many benefits in the drug delivery system, but clinically, there are some barriers such as unstable blood circulation, toxicity, and ineffective oral bioavailability, which need to be resolved to check its clinical potential [111].

2.2.4.2 Synthesis of ChNPs

2.2.4.2.1 Ionic Gelation Method

Ionic gelation is dependent on the electrostatic interactions of chitosan's amino-sugar monomeric units with polyanions with opposite charges, such as tripolyphosphate (TPP) or hexametaphosphate (HMP), or dextran sulfate [112–114]. These negatively charged polyanions are multivalent, non-toxic, and capable of forming gels through ionic interactions. Their capacity to form gels is dependent on the density of charge present on both ions. The effect of different kinds of factors such as concentration, pH, composition ratio, and the type of mixing on chitosan-TPP nanoparticles preparation has been studied and reported by Nasti et al. [115]. It is found that these factors were unaffected by TPP concentration.

Another simple method for the synthesis of magnetic chitosan-Fe_3O_4 nanoparticles is cross-linking with TPP, using NaOH precipitation and oxidation in aqueous HCl with the formation of nanoparticles containing chitosan and $Fe(OH)_2$. The particles obtained are in the size range of 50–100 nm [116].

2.2.4.2.2 Reverse Micellar Method

This method is used to prepare ultrafine polymeric nanoparticles. The surfactant is dissolved in an organic solvent to make reverse micelles. An aqueous chitosan solution is added by mixing it continuously to avoid any precipitate formation. The solution obtained is optically transparent in emulsive form. A small amount of water is added in low molar chitosan solution to control the particle size and particle distribution [117].

Mansouri et al. studied and used the reverse micellar method to create bovine serum albumin (BSA)-loaded chitosan nanoparticles with size ranging from 143 to 428 nm. This methodology is also used in the preparation of magnetic chitosan nanoparticles using Span 80 emulsifier and glutaraldehyde as a binding or cross-linking reagent [118].

2.2.4.2.3 Electrospinning

Chitosan nanofibers are made by electrospinning. Several factors influence it, including operational voltage, flow rate, temperature, and pressure. The applied voltage decides the degree of electrostatic interaction causing a polymeric jet. There is an increase in fiber size due to increased voltage [86,119]. Chitosan amino groups produce salts with trifluoroacetic acid (TFA) in this process, which can break down stiff connections between molecules. The transformation of spherical beads into networks of fibers is maintained under controlled conditions of electrospinning, and homogenous fibers of average size of 330 nm can be generated [120].

2.2.4.3 Applications of ChNPs

Chitosan (CS) is a deacetylated form of chitin. Chitosan is a cationic charge and has the potential to form gels or act as a cross-linking agent. Due to these properties, chitosan is used in combination with nanotechnology, which leads to the formation of chitosan nanoparticles (ChNPs) [53].

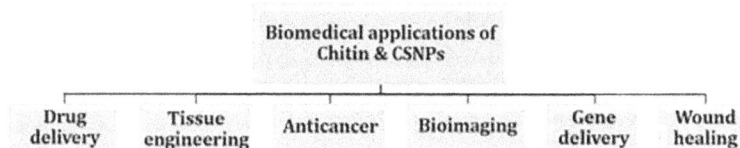

FIGURE 2.12 Biomedical applications of Chitin and CSNPs.

2.2.4.3.1 Applications in Biomedical Field

Biomedical applications of ChNPs are summarized in Figure 2.12, which include drug delivery, vaccine delivery, wound healing, and use as an antibacterial agent [121].

2.2.4.3.2 Drug Delivery

These nanoparticles have great significance in the application of drug delivery. Chitin nanocrystals from acid hydrolysis treatment are sharp, stable molecules and can be utilized as a promising agent in targeted drug delivery [122]. The drug release from chitosan NPs occurs in three ways: release via erosion, diffusion, and swelling. Hydrophilic drugs, peptides, proteins, and nucleic acids are encapsulated using ChNPs [123]. They have the ability to overcome barriers in biological media to protect from the degradation of molecules and controlled release of drug to a targeted site. To deliver peptides and proteins, chitosan-based nanoparticles can be used in mucous membrane systems such as the oral, nasal, and pulmonary routes [124].

2.2.4.3.3 Tissue Engineering

Tissue engineering deals with the repair, augmentation, and control of a specific tissue's or organ's function [65]. Chitin-based materials are utilized in various forms such as films, scaffolds including microcapsules, porous and fibrous structures, and hydrogels. These materials have favourable biological properties for tissue regeneration of the skeletal system. They can also be combined with bioactive factors that promote osteogenesis [125]. Chitosan nanoparticles are utilized to promote permeability through transmucosal membrane, which enhances the flow through the cellular channels of nanoparticles rearranging the structure of firmly bound compounds using binding properties. Chitosan nanoparticles along with electrically conductive gold hydrogels are used for cardiac tissue engineering. CS gold nanoparticles were combined with mesenchymal stem cells (MSCs) and cultured for 14 days without electrical stimulations in this technique. CS-gold nanoparticles (GNP) aided in the MSC viability, metabolism, migration, and proliferation, as well as in the formation of uniform cellular constructs. This technique helped to enhance the properties of myocardial constructs [126].

2.2.4.3.4 Cancer Diagnosis

Chitin- and chitosan-derived nanoparticles are small in size, which helps in anticancer activity. The synthesis of chitosan oligosaccharide-arachidic acid conjugate is successfully achieved and used in developing self-assembled nanoparticles for doxorubicin delivery. Folic acid combined with carboxymethyl chitosan and synchronized to manganese-doped zinc sulfide nanoparticles is used in cancer cell targeting, controlled drug delivery, and bioimaging [127].

2.2.4.3.5 Gene Therapy and Bioimaging

Chitosan has the excellent characteristics required for a perfect gene delivery system. Amino groups on the chitosan and phosphorous groups on the DNA combine to generate stable chitosan-DNA nanoparticles with enhanced genetic characteristics. Chitosan is also used extensively in bioimaging; for example, the combination of Fe_3O_4 imaging compound into structured nanoparticles for magnetic resonance imaging (MRI) is used to improve targeted tumor imaging.

2.2.4.3.6 Wound Healing

Chitosan nanoparticles produced in different formulations such as sponges of chitosan, CS composite scaffolds, scaffolds loaded with drug, and immobilized scaffolds are used in various wound healing applications [128].

2.2.4.3.7 Applications in Food Packaging

ChNPs are used as an encapsulating agent in encapsulation technology. For example, ChNPs obtained from the emulsion-electrostatic interaction method with the help of tripolyphosphate (TPP) microparticles reduce the formation of H_2O_2 in krill oil stored for 14 days at 45°C. Clove essential oil encapsulated into ChNPs resulted in a lower degradation of volatile compounds as well as improved *in vitro* antifungal activity. The storage period of food can also be enhanced by the application of chitosan and its derivatives. ChNPs can exert control over the release of various bioactive components, which help to enhance the antimicrobial properties of food packaging materials. For example, Beluga fish fillets treated with edible coating, which has ChNPs and is loaded with fennel essential oil along with modified atmospheric packaging, lead to a lower value of peroxide. ChNPs are added to improve the film forming of the materials [129].

2.2.4.3.8 Applications in Agriculture

Chitosan plays an important role in the growth and development of many plants. It has the ability to increase plant chlorophyll content and uptake of nutrient. Chitosan nanoparticles are able to cause a significant impact on coffee seedlings by enhancing the rate of pigmentation, photosynthesis and uptake of nutrient. Nanochitosan-NPK fertilizers are used to increase the growth performance and yield of wheat crops. To control the proliferation on insect larvae, microspheres of chitosan and cashew tree gum are used. Nanochitosan biosensors containing paramagnetic Fe_3O_4 are used to detect and eliminate heavy metals [128]. Chitosan-based nanoparticles are also used in controlling salinity stress. For example, chitosan-polyvinyl alcohol hydrogels, if applied to tomato plants under salt stress, enhanced the plant growth and gene expression necessary for detoxification. Figure 2.13 illustrates the applications of chitosan in agriculture [130].

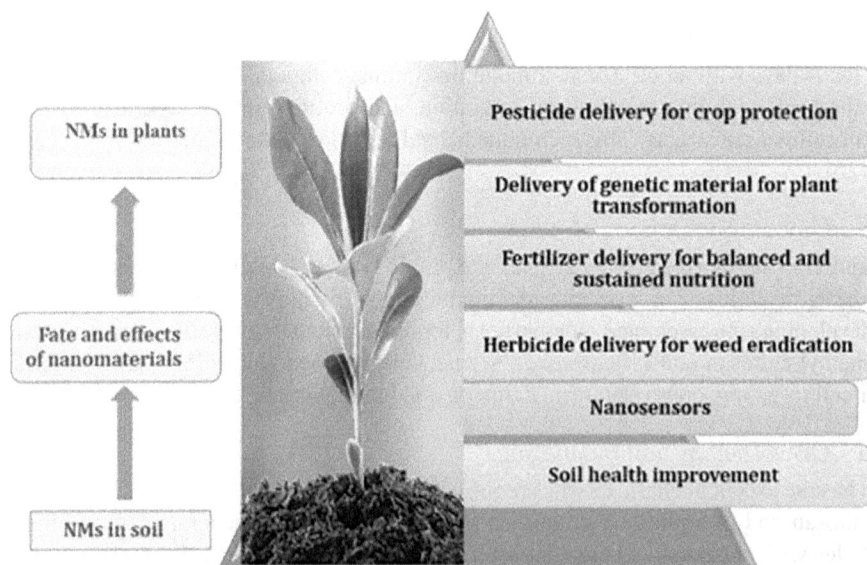

FIGURE 2.13   Application of CSNPs in agriculture [130].

2.2.4.3.9 Applications in Wastewater Treatment

Water pollution is the most serious concern in today's world, owing to the inadequacy of traditional water treatment methods. To treat water, the prominent groups present in chitosan, amino groups, and hydroxyl groups are used as an absorbent and they help to remove pesticides and metal pollutants. Chitosan nanofibers are used as an adsorbent due to their high porosity and surface area per unit weight and are tested to remove Pb(II) and Cu (II). ChNPs-coated 4-micron membrane have been used for drinking water purification. They held good bacterial growth compared to non-coated membranes. It is found that chitosan combined with zinc oxide nanoparticle composites was able to get rid of 99% of the unwanted pigments from the textile effluent [130].

Chitin- and chitosan-derived nanoparticles are gaining demand in various disciplines due to their non-toxicity and resistance to bacteria.

2.2.5 PULLULAN

Pullulan is an extracellular bacterial polysaccharide material synthesized by strains of *Aureobasidium pullulans*, a polymorphic fungus [131,132]. Various substrates, such as starch, bakery waste, and agro-industrial leftovers, are used to produce this biopolymer. During its life cycle, pullulan produced by the fungus has a long branched septate hypha and numerous chlamydospores. Its chemical formula is $(C_6H_{10}O_5)_n$, and its structural composition consists of maltotriose with α-1,6-repeated units and an α-1,4 glycosidic linkage as depicted in Figure 2.14. The presence of α-linkages offers water solubility, fiber flexibility, and film-forming capacity similar to petrol-derived plastics [133]. It also possesses several properties such as hydrophobic, non-toxic, non-immunogenic, and anti-carcinogenic properties [134,135]. Thus, it can be widely applied in the biomedical, pharmaceutical, food, and cosmetic applications.

2.2.5.1 Pullulan Nanoparticles

Similar to chitosan, pullulan is also one of the polysaccharides that have been reported in the application of drug delivery. The nanoparticles of pullulan have better cell adherence [136,137] and matrix-forming capacity that can be utilized in drug and gene nanocarriers [138,139]. These carriers utilize hydrophobic components of pullulan, e.g., cholesteryl-pullulan, which has the ability to form nanoparticles by self-aggregation. It also has the tendency to form stable complexes with both hydrophilic and hydrophobic drugs [140]. There are some hydrophilic synthesized derivatives of pullulan nanoparticles such as carboxylated, phosphorylated, aminated, and sulfated NPs that can be produced using a mild technique of polyelectrolyte complexation. Another type of prebiotics, phthalyl pullulan nanoparticles are synthesized using the simple technique of self-assembly of hydrophobically modified hydrophilic polymers [141].

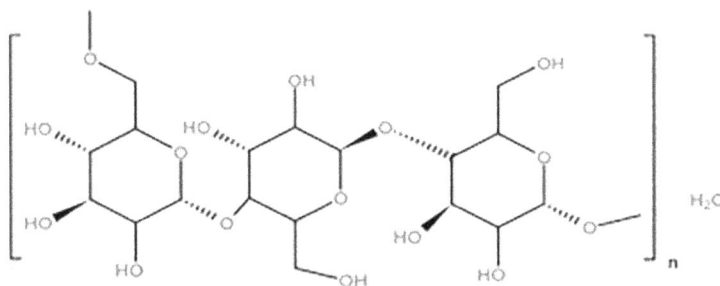

FIGURE 2.14 Chemical Structure of Pullulan

2.2.5.2   Synthesis of PNPs

2.2.5.2.1   *Polyelectrolyte Complexation*

The synthesized derivatives of pullulan are prepared using this technique. The hydrophilic conditions are maintained in a simple mixture of solutions. Rapid formation of nanoparticles takes place within 10 minutes [142].

2.2.5.2.2   *Self-assembly Process*

Pullulan nanoparticles such as cholesterol-modified pullulan (CHP) NPs, cholesterol-modified carboxyethyl pullulan (CHCP), cholesterol-modified amino-pullulan (CHAP), and phthalyl pullulan nanoparticles are synthesized using this method. One gram of pullulan is solubilized in 10 ml dimethyl formamide (DMF), and 0.1% dimethyl-amino-pyridine residue per pullulan sugar molecule was added as a catalyst in the solution. For phthalyl pullulan nanoparticles (PPN), phthalic anhydride is then added in different ratios (phthalic anhydride:pullulan) such as 6:1, 9:1, and 12:1 to produce PPN with different degrees of phthalic group substitution. This reaction is carried out at 54°C for 48 hours under nitrogen. The nanoparticles are dialyzed in DMF to eliminate unreacted solution. The unreacted pullulan from the prepared NPs is removed using ultracentrifugation. Finally, particles are freeze-dried and stored at −20°C [143].

2.2.5.3   Applications of PNPs

2.2.5.3.1   *Biomedical Applications of Pullulan*

Pullulan in the form of nanoparticles and nanogels can be employed as drug carriers, reducing adverse therapeutic effects while boosting activity and stability [144]. It can be useful in oral drug delivery because it dissolves quickly in water. The current research reported that JFP/pullulan nanofibers can be utilized as a replacement material in fast-dispersing drug delivery systems (FDDDSs) [145]. Pullulan nanoparticles with folic acid (FA) have dual efficacy and are effectively used in liver cancer therapy and other anticancer treatments (Figure 2.15) [83]. In studies on nasal vaccination, pullulan-based nanoparticles adhere to the nasal epithelium [146]. *Lactobacillus plantarum* treated with phthalyl pullulan nanoparticles (PPN) is synthesized and classified to create prebiotic for *Lactobacillus plantarum*. These polymeric nanoparticles can have a major impact on probiotics production, which are effective against various bacteria when used as prebiotics [141].

2.2.5.3.2   *Food and Food Packaging*

Aureobasidium pullulans generates a water-soluble food-grade polymer called pullulan made up of starch and sugar as a carbon source. Pullulan is a non-hygroscopic polymer with high intermolecular mechanical strength that can be converted into nanoparticles, nanofibers, and films [138].

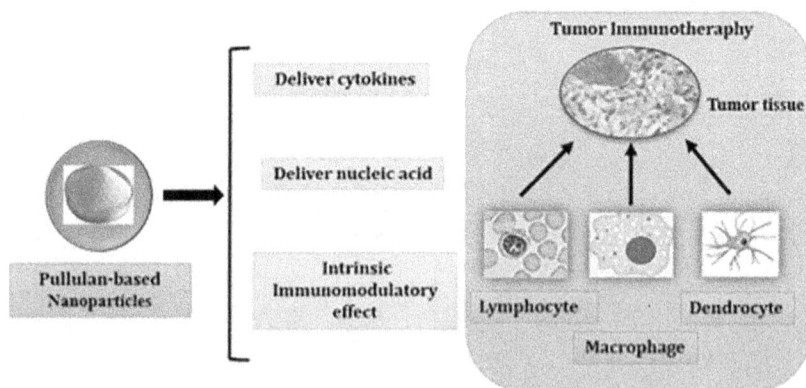

FIGURE 2.15   Pullulan based nanomaterials in Cancer therapy [83].

TABLE 2.1

Pullulan-Based Nanomaterials Used in Active Packaging for Food Preservation

Pullulan-Based Nanomaterials	Type of Food Items	Effective against Bacterial Strains
Pullulan with silver nanoparticles	Edible *food* items, especially meat	*S. aureus* and *L. monocytogenes* [149]
Pullulan (PL) and lysozyme nanofibers (LNFs)	Fresh *fruit preservation*	*Staphylococcus aureus*, i.e., lysozyme-resistant bacteria [148,149]
Pullulan with nanoparticles/ essential oils	Poultry meat products	*S. aureus*, *L. monocytogenes*, *S. typhimurium*, and *E. coli* [149]
Pullulan with zinc nanoparticles	Turkey meat	*S. aureus*, *L. monocytogenes*, *E. coli*, and *S. typhimurium* [149]

Pullulan (PL), TEMPO-treated cellulose nanofibrils (TOCNs), and montmorillonite clay (MMT) nanocomposite films can be used as green and biodegradable food packaging and protection films, and the details are summarized in Table 2.1 [147]. Pullulan (PL) and lysozyme nanofibers (LNFs) were used to create homogeneous, shiny, and translucent nanocomposite films and show antibacterial activity against *Staphylococcus aureus*; that is, lysozyme-resistant bacteria are used in eco-friendly edible films for active packaging [148].

2.2.5.3.3   Other Applications

Pullulan is a tasteless, white, highly water-soluble, and non-toxic powder. Due to these properties, it can be used as nanofibrous materials. Electrospun pullulan and polyvinyl alcohol (PVA) nanofiber matrix are used in UV-resistant packaging and dressing materials. PVA and montmorillonite were incorporated into pullulan fibers to increase electrospun nanofibrous matrix strength and thermal properties. Tannic acid (TA)/chitosan (CS)/pullulan (PL) composite nanomembranes are cross-linked together to become more stable in water for possible wound dressing applications [145]. The gellan-pullulan composite nanogel was found to have superior adsorption capability to any other material, so it can be utilized for cationic color adsorption [83].

2.2.6   GELLAN

Gellan gum (GG) is a linear exopolysaccharide and possesses high molecular weight, and it is commercially synthesized by microbial fermentation of the Proteobacteria *Sphingomonas paucimobilis* [150]. The composition of gellan gum comprises a repetitive chain of 1,3-β-D-glucose, 1,4-α-L-rhamnose, and 1,4-β-D-glucuronic acid in the percentage proportion of 60:20:20. Additionally, it contains a considerable amount of non-polysaccharide materials, which can be easily separated by filtration and centrifugation [151]. Acetate and glycated functionalities are also present in gellan gum. It is commercially available in various gels with trade names Gelrite™ (acylated) and Kelcogel™ (deacylated) [152–155].

The gellan gum is thermo-responsive, biocompatible, and non-toxic, but has poor mechanical strength; it can be added with other materials to form soft gels with improved elasticity [156]. Biomaterials such as gellan gum, alginate, and xanthan gum can be utilized in artificial dressings because they can enhance healing rate and skin formation. Gellan gum can be combined with different nanoparticles that enhance the material properties and its performance [157–159].

2.2.6.1   Gellan Gum Nanoparticles

Nanocomposites are considered as the second generation of nanotechnology, and gellan gum nanoparticles are also part of these composites. It can be in the form of homo- or hetero-nanomaterial structures, which can be used in various functions. The gellan gum combined with metal oxide nanomaterials such as titanium dioxide, gold, and silver are used in different forms, e.g., nanoparticles

(NPs), nanorods (NRs), and nanotubes (NTs). These nanocomposites possess good mechanical and biological properties; thus, they can be widely applied in wound dressing, tissue engineering scaffolds, and food packaging [160–162].

2.2.6.2   Synthesis of Gellan Nanocomposites

2.2.6.2.1   Gellan Gum TiO₂ Biofilms

Gellan gum solution is prepared by dissolving 1g GG in deionized water (100 ml) with constant mixing at 70°C and for 2 hours. 50% (w/w %) glycerol, 5mM calcium chloride, and 1% TiO_2 is mixed for 2.5 hours. The mixture is transferred to a mold and is dried for 20 hours at 50°C to form mold-shaped films. Further films formed are preconditioned in a desiccator for 2 days prior to testing at 27°C with 50% relative humidity (RH) [161].

2.2.6.2.2   Gellan Gum-Reduced Au/Ag NPs

An aqueous solution of $AuCl_4$ 100mM in 100 ml is heated and reduced in 0.02% (w/v) aqueous solution of GG (100 ml). The distribution of AuNPs is obtained at an absorbance of 520nm; furthermore, this solution is dialyzed using dialysis tubing of 12kDa cut-off size for 24 hours to remove other impurities. The concentration of gellan gum gold nanoparticles is determined using atomic absorption spectrometry [163].

In case of GG silver nanoparticles (AgNPs), gellan gum is mixed with NaOH solution with a constant stirring at 95°C till gellan gum gets completely dissolved into it. One milliliter of 10mM $AgNO_3$ is added to dissolved solution gradually till the solution turns colorless to yellow. A change in color represents the formation of AgNPs. The maximum absorbance is checked at 420nm using UV/Vis spectroscopy [164].

2.2.6.3   Applications of Gellan Gum Nanocomposites

Gellan gum is thermo-sensitive, biocompatible, non-toxic, and ductile, and it can withstand heat and acid stress during the manufacturing process. These versatile properties are beneficial in many applications in pharmaceuticals, biomedical, food, and other industries and are summarized in Table 2.2 [165].

2.2.6.3.1   Applications in Drug Delivery

Gellan gum is used in the form of tablets, gel pads, micro- and nanoparticles, films, and beads in biological systems. In recent years, gellan gum has gained its importance in developing nano-formulations to enable effective targeted delivery [166].

Gellan-polyvinyl alcohol (PVA) nanofibrous scaffolds are biodegradable, and thus, they are used to induce skin tissue regeneration in a temporary substrate or biomedical graft [167]. Gellan gum nanohydrogels with anti-inflammatory formulations and anticancer compounds are used in cancer treatment as part of combination therapy; additionally, anticancer compounds and entrapped gold nanoparticles within a gel matrix effectively treat cancer cells [168,169]. Self-assembled gellan-based nanohydrogels and chitosan/GG resveratrol-loaded nanofibers have great potential as drug and prednisolone delivery carriers [170]. Nanocomposites loaded with rifampicin (RF), pyrazinamide (PZA), and maleate, silk sericin-gellan gum-chitosan (MA-SS-GG-CS) are utilized to solve the problems related to TB (tuberculosis) treatment [171]. The system consisting of vancomycin-containing nanoparticles and vancomycin embedded in a matrix of gellan gum has shown antimicrobial activity against *Staphylococcus spp.* while also being compatible with osteoblast cells. This system can also be applied as an intrabone anti-biotherapy for osteomyelitis [172].

2.2.6.3.2   Applications in Composite Biofilms

Silver nanocomposite hydrogels based on gellan gum have a high antibacterial activity. This property is used to create antibacterial coatings in implantable devices and medicinal materials

TABLE 2.2

Gellan Gum-Based Nanomaterials Properties and Applications in Various Fields

Gellan Gum-Based Nanomaterials	Product	Applications
Gellan gum-based silver nanocomposite hydrogels	Antibacterial coatings	Implantable devices and medicinal materials, in antibacterial vaccines to control bacterial infections [177]
TiO$_2$-NPs (gellan gum/titanium dioxide nanoparticles)	Biofilm	Wound healing [173]
Gellan gum methacrylate and laponite nanohydrogel	Wound dressing material	The treatment of burn wounds [178]
The GC–SiO$_2$ (gellan gum-sodium carboxymethyl cellulose (GC) silicon dioxide (SiO$_2$))	Antibacterial films	Against gram-positive and gram-negative pathogens [175]
Cellulose nanocrystals and gellan gum	Edible coating	Covering mushrooms, providing surface protection, and extending product shelf life [165]
Gellan gum-capped silver nanoparticles	Colorimetric hydrogen sulfide (H$_2$S) sensor	To monitor meat spoilage in intelligent food packaging [167]
The GG+TiO$_2$-NRs (titanium dioxide nanorods incorporating gellan gum)	Films	Antibacterial packaging films [168]
Gelatin (GN) with ZnO nanoparticles as the top layer (GN-ZnO) and gellan gum (GG) with mulberry anthocyanins (MBA)	Bilayer colorimetric films	Monitoring fish spoilage in smart packaging [169]

to control bacterial infections in antibacterial vaccines [173]. Gellan gum embedded with TiO$_2$-NPs (gellan gum/titanium dioxide nanoparticles) biofilms show good antibacterial properties, so they are used in wound healing; likewise, gellan gum methacrylate and laponite nanohydrogels are also used in wound dressing materials for the treatment of burn wounds [174,175]. The GC-SiO$_2$ (gellan gum-sodium carboxymethyl cellulose (GC) silicon dioxide (SiO$_2$)) films demonstrated a substantial antimicrobial activity against pathogenic bacteria, while the GC-SiO$_{2-5}$ with [176] ODDMAC (octadecyldimethyl-(3-triethoxy silylpropyl)ammonium chloride) nanocomposites well dispersed and demonstrated with enhanced antibacterial activity against 5–6 different pathogens [177].

2.2.6.3.3 Food Processing and Packaging

Gellan gum is used in a variety of food applications as a thickener, binder, and stabilizer. Due to its high tensile strength, excellent gas barrier properties, and moisture resistance, it's also used in the production of edible films [178].

The surface of the *Agaricus bisporus* mushrooms coated with cellulose nanocrystals and gellan gum-based edible coating is applied for covering mushrooms, providing surface protection, and extending product shelf life [179]. To monitor meat spoilage in real time for intelligent food packaging as shown in Figure 2.16, a colorimetric sensor of hydrogen sulfide (H$_2$S) based on gellan gum-capped silver nanoparticles was established. This strategy provided a simple, but useful platform for intelligent food packaging that was non-destructive, robust, cost-effective, and user-friendly [164]. The GG+TiO$_2$-NRs films were found to have a good antibacterial activity against pathogenic strains such as *S. aureus*, *Streptococcus*, *E. coli*, and *P. aeruginosa* and could be used as antibacterial packaging films [174]. Gelatin (GN) with ZnO nanoparticles as the top layer (GN-ZnO) and gellan gum (GG) with mulberry anthocyanins (MBA) as the bottom layer were combined to form bilayer colorimetric films for monitoring fish spoilage in smart packaging [180].

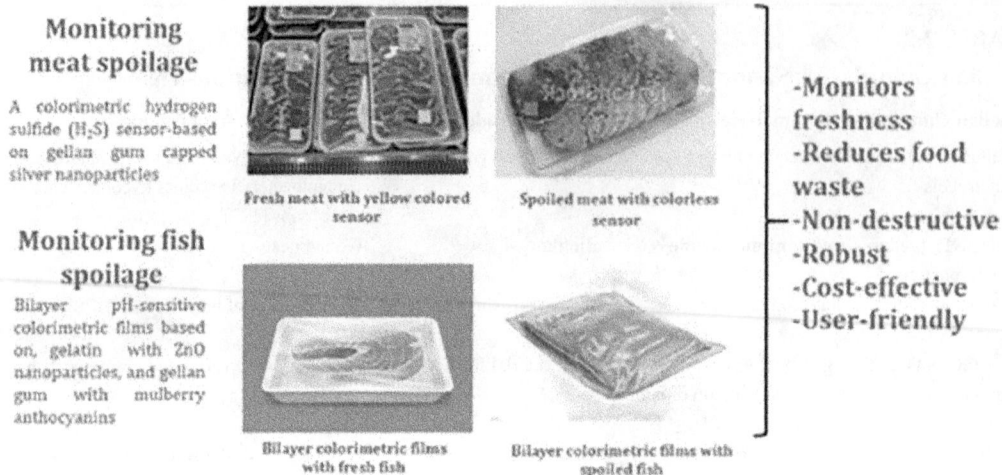

Monitoring meat spoilage

A colorimetric hydrogen sulfide (H₂S) sensor-based on gellan gum capped silver nanoparticles

Fresh meat with yellow colored sensor

Spoiled meat with colorless sensor

Monitoring fish spoilage

Bilayer pH-sensitive colorimetric films based on, gelatin with ZnO nanoparticles, and gellan gum with mulberry anthocyanins

Bilayer colorimetric films with fresh fish

Bilayer colorimetric films with spoiled fish

-Monitors freshness
-Reduces food waste
-Non-destructive
-Robust
-Cost-effective
-User-friendly

FIGURE 2.16 Gallen gum in Intelligent food packaging

2.2.6.3.4 *Other Applications*

Gellan gum combined with TiO_2 nanoparticles hydrogels are effective at washing, disinfecting, and preserving contaminated samples with *Penicillium chrysogenum* and *Cladosporium cladosporioides* [181]. The combination of gellan gum hydrogel and deacidified compound or titanium dioxide nanoparticles was studied to clean and decolor paper infected by *Aspergillus versicolor* [182].

2.2.7 XANTHAN GUM

Xanthan gum has gained importance as additives in food from the last decade after FDA approval [184]. It is a high molecular weight, polyanionic, extracellular polysaccharide obtained from gram-negative bacteria *Xanthomonas campestris* [185,186]. The chemical composition of xanthan gum comprises β-1,4-D-glucopyranose glucan as background with a pendant trisaccharide branched chain, containing mannose (β-1,4), glucuronic acid (β-1,2), and mannose residues at the terminal. The alternate glucose residue in the periphery is attached with the branches of α-1,3 linkages, and terminal mannose is substituted with pyruvate residue and acetylated at C-6 [187,188]. The chemical structure of XG is illustrated in Figure 2.17. The non-toxic and biocompatible characteristics of xanthan gum make them beneficial in food, pharma, and biomedical industries [74,189].

2.2.7.1 Xanthan Gum Nanoparticles

It has a huge market demand for the expansion of green synthesis of xanthan-embedded nanoparticles which can act as an ideal stabilizer for metallic nanoparticles improving its absorptivity and transport, such as palladium (PdNPs) [190], gold (AuNPs) [191], iron (INPs) [192], zinc (ZnNPs) [193], silver (AgNPs) [194], and silicon dioxide [195,196]. ZnNPs exhibit promising antibacterial, anti-biofilm and antivirulence properties that can be potentially used in food industries.

2.2.7.2 Synthesis of Xanthan Nanoparticles

2.2.7.2.1 *Xanthan Gum Gold Nanoparticles*

Xanthan gum-stabilized gold particles (XG-AuNPs) can be synthesized by mixing equal volume of hydrogen tetrachloroaurate(III) hydrate (HAuCl₄.3H₂O) 0.25 mM with xanthan gum XG (1.5 mg/ml as reported [197]). 50 mM ascorbic acid is added dropwise with constant stirring into a prepared solution at 25°C. The development of xanthan-stabilized gold nanoparticles is indicated when the color changes from dark yellow to red (XG-AuNPs). After centrifuging the mixture for 15 minutes

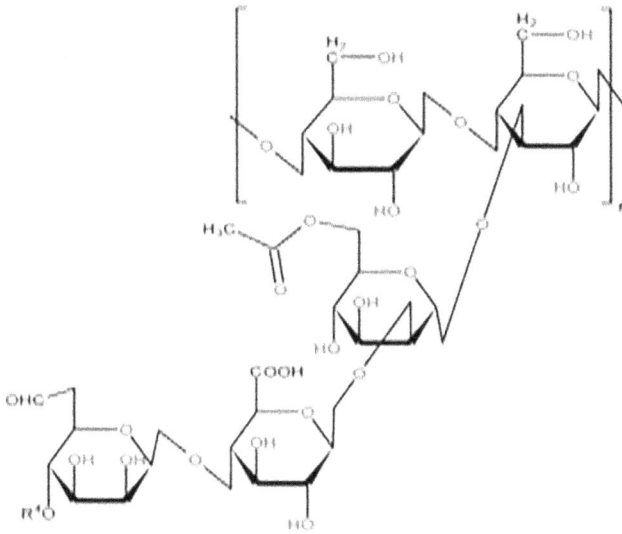

FIGURE 2.17 Chemical structure of xanthan gum

at 15,000 rpm, the upper phase containing excess ascorbic acid is removed [117,198] and the pellet at the bottom is re-suspended and mixed in distilled water.

2.2.7.2.2 Xanthan Gum Silver Nanoparticles

One milligram of xanthan gum is dissolved in ultrapure water (100 mL) with steady stirring to achieve a 1 percent (w/v) solution for the production of XGAgNps. To achieve final concentrations of 1, 2, and 3 mM, $AgNO_3$ is added. The reaction mixtures are then mixed regularly and incubated in the dark for 12, 24, 36, and 48 hours at 60°C, 80°C, and 100°C temperature, respectively [194].

2.2.7.2.3 Xanthan-Based Iron Nanoparticles (INPs)

X. campestris culture is grown in an appropriate medium as reported [193]. The pellet is obtained after centrifugation, and it is dissolved further in deionized water and mixed with vortexing. The cell suspension is sonicated for 1 minute to isolate INPs from cell surface without disrupting microbial cells [199]. At the end, cells were harvested by mild centrifugation.

2.2.7.2.4 XG Palladium Nanoparticles (PdNPs)

To form an H_2PdCl_4 solution, 0.0356 g of $PdCl_2$ is mixed in 100 ml of HCl (0.41 mM). An aliquot of equivolume solution is mixed with a 0.2% XG in a boiling tube. The reaction is carried out in an autoclave at 120°C temperature for 10 minutes [192].

2.2.7.2.5 Xanthan-Based ZnO Nanoparticles

The nanoparticles are combined using chemical co-precipitation [200]. In a three-corked round-bottom flask, a 100 mL solution of 0.5 M Zn $(NO_3)_2.6H_2O$ is mixed with constant stirring at 1,000 rpm for 25 minutes to get a homogeneous solution. A xanthan gum solution of 2.3% is prepared in distilled water with mixing at 50°C for 2 hours to get a completely dissolved solution. 0.1 M NaOH is added dropwise in a homogeneous suspension to reduce xanthan gum into the bulk of Zn^{2+} solution. The mixture is constantly stirred at 50°C, and the change is monitored by collecting samples after specific time intervals and verified using UV-Vis spectroscopy. After 8 hours, a white precipitate is observed. Further, it is separated by centrifugation. The product is washed with deionized water several times for the removal of other impurities and dried in an oven for 120 minutes at 60°C [193].

2.2.7.3 Applications of Xanthan-Based Nanoparticles

Xanthan is used as a hydrogel, dispersion stabilizer, gelling agent, viscosity enhancer, etc. It also shows excellent water solubility, biocompatibility, and biodegradability; hence, the xanthan-based nanomaterials are used in different areas such as food industry, drug delivery, and tissue engineering [201].

2.2.7.3.1 Applications in Food Industry

The use of protective and appropriate packaging is essential for increasing the shelf life and quality of food products. Xanthan gum was used in food industry for the first time by US FDA in 1969 [202]. Due to its unique behavior, it has various applications in food industry. The nanocomposites films of hydroxypropyl methylcellulose (HPMC) and xanthan gum incorporated with Ag have better solubility and good antibacterial and thermal properties and can be used in food packaging [203].

Nowadays, everyone wants to take less efforts in cooking the food, even chopping the vegetables, so now it has become easier to get high-quality chopped fresh vegetables and fruits. For the quality preservation of this fresh-cut packages, xanthan gum/pectin/sodium alginate edible films have been made [204]. Gellan gum (GG)- and xanthan gum (XG)-based nanocomposite antibacterial films containing zinc oxide nanoparticles (ZnONPs) are made for food packaging [205].

2.2.7.3.2 Applications in Drug Delivery

Xanthan gum (XG) is a polyanionic polysaccharide with a high molecular weight. It is non-toxic and non-irritating and possesses good bioadhesive properties, which can be used in drug delivery. A carrier comprised of XG thiolated with L-cysteine is used to treat sialorrhea. Xanthan gum thiolation is required to increase the adhesion on buccal mucosa rather than native XG. This carrier works efficiently resulting in the reduction in the salivary flow of patients with sialorrhea [206]. XG is widely used along with other particles to provide better targeted drug deliveries. The XG-stabilized gold nanoparticles are prepared for high drug loading (Figure 2.18). In this, XG is used as a stabilizing

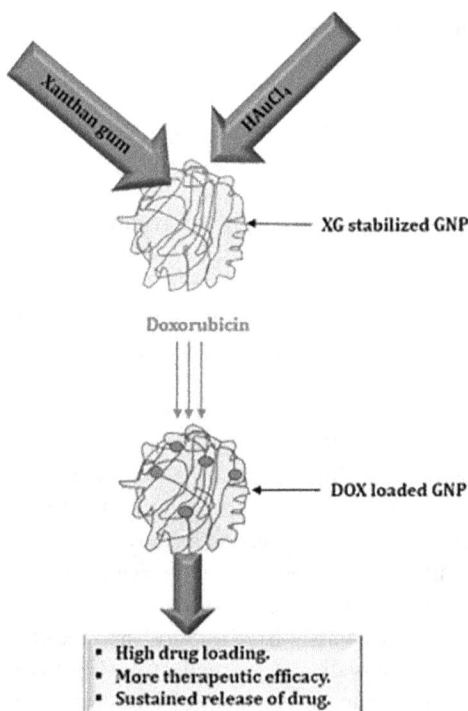

FIGURE 2.18 Synthesis of xanthan gum stabilized gold nanoparticles for drug delivery [197].

and reducing agent. The nanoparticles prepared have good biocompatibility and non-toxicity [197]. Nowadays, topical drug delivery also includes the use of XG as a gelling base. In 2009, Bhaskar et al. produced lipid nanoparticles structures of flurbiprofen for transdermal drug transport system. These nanocarriers were embedded into XG gel. The designed gels have improved permeation and bioavailability and sustained release for 24 hours [207]. One of the most sensitive parts of a body is brain. It's difficult to deliver the drug in brain without any harm and disruption. The intranasal route is one of the effective ways for drug delivery into brain. Carbamazepine mucoadhesive nanoemulgel was prepared with 0.1% of XG as an anionic mucoadhesive polymeric substance. This was done for targeting the brain via the olfactory mucosa [208].

2.2.7.3.3 *Tissue Engineering Applications*

Xanthan gum can be used in tissue engineering as a natural biopolymer to create a variety of XG-based nanomaterials. BNC hydrogels reinforced with CNCs were made from xanthan gum (XG) solutions, and chitosan (CS) has cytocompatibility and the ability to release chemotherapeutic agents, indicating their suitability for use in tissue engineering and drug delivery applications [209]. Recently, the gum-based nanoparticles are used for the production of nanofibers. In 2018, Shekarforoushet et al. formulated XG-CS nanofibers for the delivery of hydrophobic compounds [210].

2.3 CHARACTERIZATION OF BIOPOLYMERIC NANOMATERIALS

The characterization of biopolymer-based nanomaterials include various techniques depending on the size, form, chemical composition, molecular weight, purity, solvent medium, and nature of reducing and stabilizing agents. These materials has special importance when treated as nanoparticles in different areas such as biomedical applications, pharmaceuticals, artificial tissue regeneration, biosensors, agriculture, and food. The most widely used techniques are Fourier transform infrared (FTIR) spectroscopy, X-ray diffraction (XRD), transmission electron microscopy (TEM), scanning electron microscopy (SEM), dynamic light scattering (DLS), nuclear magnetic resonance (NMR) spectroscopy, UV-visible spectroscopy, briefly discussed in this section. Table 2.3 provides an overview of characterization techniques utilized for various biopolymeric nanomaterials with specifications.

2.3.1 SCANNING ELECTRON MICROSCOPY (SEM)

SEM studies play a key role in analyzing the nanoparticles of biopolymers and nanocomposites based on the structure, morphology, size, shape, surface modification, etc. [231].

In this technique, an electron microscope of scans the sample surface through a high-energy beam of electrons [232]. A high-voltage (20 kV) electron beam is accelerated and passed through an aperture using electromagnetic lenses, and the electron beam is narrowed. Further, the beam uses scan coils to scan the specimen's surface. The images are created after SEM signals from the area of beam and specimen interactions are produced [198]. However, sometimes, before working with SEMs, it is essential to cover a sample by an additional thin layer of gold sputter of 10 nm to obtain a good-quality SEM image [233]. SEM is used to take images with areas of width 1 cm to 5 μm (magnification range 20X–30,000X, spatial resolution 50–100 nm) [183].

SEM imaging is often used to detect surface texture, homogeneous coats, and the difference between various components in biopolymeric or composite materials [231]. SEM also provides information on the samples' purity. It also determines the position of the secondary and tertiary nanostructures and the degree of aggregation. The nanoparticles distribution in the sample is also determined by SEM [232]. SEM analyzes the texture of surface in case of dextran-based hydrogels [215], nanogels [234], cellulose-gellan gum nanocomposites [177], chitosan nanoparticles [211], alginate nanoparticles [212], and starch nanoparticles [213].

TABLE 2.3

Characterization Techniques Used for Biopolymer-Based Nanomaterials

Characterization Techniques	Particle Size	Specifications	Biopolymeric NPs Applications
Scanning electron microscopy (SEM)	< 1 nm	Morphological analysis	Surface morphology of chitosan nanoparticle [211] Morphology of alginate nanoparticles [212] Starch nanoparticles shape analysis [213] Alignment of cellulose nanocrystals [214] Dextran-based nanogel morphology [215]
X-ray diffraction (XRD)	1 nm	Determination of size, shape, structure, and crystallinity of nanomaterials	Starch nanoparticles crystallinity [69] Crystalline nature of nanoparticles of chitosan cross-linked silver nanocomposites (CSHD-AgNCs) [216] Gellan gum-capped silver nanoparticles analysis [217] Nanoparticles of sodium alginate nanocomposites [218] Cellulose nanocrystals physical analysis [219]
Transmission electron microscopy (TEM)	<100 nm	Estimation of the thickness, grain composition, size, and morphology of nanoparticles	Morphology of alginate nanoparticles [212] Cellulose nanofibers-particle dispersion [214] Chitosan electrospun nanofiber analysis [220] Dextran nanoparticle size estimation [221] Starch nanoparticles (SNPs) study [222]
Nuclear magnetic resonance (NMR) spectroscopy analysis	1–100 nm	Detailed description on the structure, responses, and environment dynamics is provided.	Starch nanocomposites structure study [69] Nanoparticle analysis in cellulose nanocomposites [223] Dextran nanohydrogels [224]
UV-visible spectroscopy	UV-visible regions of 200–800 nm	Determines the physiochemical characterization of nanomaterials, which includes size, concentration, and aggregation of molecules	Confirmation of chitosan nanoparticle formation [212] Cellulose nanomaterials suspension properties [225] Alginate/silver nanoparticle physical properties [226] Gellan gum nanoparticles study [217] Starch and pullulan nanoparticle physical analysis [227] Xanthan conformation-based silver nanoparticle structure morphology [194] Gold nanoparticles stabilized in dextran absorbance study [221] Chitosan nanoparticle size determination [211]
Fourier transform infrared (FTIR) spectroscopy	20°A–1 μm	Transmittance and relative crystallinity	Synthesis of the chitosan nanoparticles [211] Starch nanoparticles study [69] Cellulose nanoparticle crystallinity [228] Alginate nanoparticle functional group analysis [218]
Dynamic light scattering (DLS)	Below 5 nm to several microns	Determines the size distribution of particles and polymers in suspension	Measurement of average particle size and polydispersity index (PDI) of chitosan nanoparticles [211] Size of alginate nanoparticles [212] Monitoring of a xanthan gum nanogel [229] Characterization of cellulose nanocrystals [230]

2.3.2 X-Ray Diffraction (XRD)

This method gives an estimation of the size, shape, structural changes, deformations, and crystallinity of the nanomaterials and operates in the fine range of 0.2–10 nm. The interference of X-rays with crystalline sample is the basis for X-ray diffraction. When X-rays from a CRT (cathode-ray tube) are filtered and collimated, they are directed at a sample. The constructive interference is based on Bragg's law, which gives the correlation between incident radiation and angle of diffraction with lattice spacing. The lattice spacing is important in case of powdered compounds and their structure orientation [235].

X-ray diffraction is mostly used for the characterization of the nanoparticles such as starch [69] and also to study the crystalline nature of nanoparticles of chitosan cross-linked silver nanocomposites (CSHD-AgNCs) [216]. It has the advantages of performing samples in the dried powdered form colloidal solutions, which represents average statistical and volume values. By comparing the sample peaks with standard reference peaks from the databases, the particle composition can be estimated. As XRD peaks are too wide for particles of size (3 nm) and are not suitable for the amorphous materials, an analysis of gellan gum-capped silver nanoparticles [217], study of the nanoparticles of sodium alginate nanocomposites [218], and cellulose nanocrystals physical analysis [219] are mainly done by using XRD.

2.3.3 Transmission Electron Microscopy (TEM)

This technique interprets and characterizes morphological and compositional features of different nanomaterials [236]. TEM is a much more sophisticated technique than SEM as it generates a clear image with higher resolution in particle size <1 nm [237].

TEM is based on the electron-electron interaction. In this, electrons are impinged on the sample of very less thickness (<100). These electrons show different transformations leading to the electron-electron interaction due to interaction with the sample. Then a series of electromagnetic lenses are used to focus these electrons and to form images based on the density of unscattered electrons, such as phase with contrast, amplitude and its contrast, and shadow images with percentage darkness [238].

The synthesis of the biopolymeric nanomaterials requires the analysis of the particles, their size, shape, behavior, etc. TEM is used in studying the dispersion pattern of the particles on cellulose nanofibers [214]. Additional magnification allows for the confirmation of metal nanoparticle adhesion to the fiber surface as well as the calculation of the average particle size [239]. It also includes photos of electrospun polyethylene oxide (PEO)-chitosan fibers with a core-shell configuration [220]. Furthermore, TEM helps in identifying various concentric and eccentric structures to determine the core and shell diameters of fiber components. The characteristics of gold nanoparticles in dextran solution [221] and starch nanoparticles (SNPs) structure can be determined by TEM [222].

2.3.4 Nuclear Magnetic Resonance Spectroscopy (NMR)

Nuclear magnetic spectroscopy is a useful tool for analyzing the chemical structure of a wide range of organisms. It's an essential tool for biopolymers [240]. NMR analysis reveals important structural details such as functional groups, covalent linkages, and non-covalent interactions such as stereochemistry [241].

It is a phenomenon that occurs when the nuclei from atoms are exposed to a second magnetic oscillating field [242]. This spectroscopy technique relies on the particle interaction and electromagnetic radiation from the surface. This method can detect the local magnetic fields around the atomic nuclei of each individual substance.

When sensitive radio receivers detect the activation of particle nuclei into nuclear magnetic resonance with radio waves, the NMR signal is obtained [243]. It estimates the time taken by the nuclei to regain its lowest possible energy, providing precise data on chemical structure, response time, and molecular interaction [136].

NMR spectroscopy can be used to obtain chemical characterization of solutions as well as solid states [244]. The ability to study biochemical processes directly in living organisms represents a major benefit of NMR spectroscopy [245]. With several revolutionary milestones such as Fourier transform (FT) NMR spectroscopy, two-dimensional (2D) NMR method, and nuclear Overhauser effect (NOE) spectroscopy, NMR spectroscopy has blossomed as a major characterization tool in organic chemistry, analytical chemistry, and structural biology [246]. Some studies on the use of NMR for structural characterization of bio-nanocomposites have been reported, such as starch nanocomposites [69] and cellulose nanocomposites [223]. It is also used in the characterization of dextran nanohydrogels [224].

2.3.5 UV-Visible Spectroscopy

Ultraviolet-visible spectroscopy is a basic technique used to determine the physiochemical description of nanomaterials, which includes size concentration and aggregation of molecules [247].

The basis of spectroscopy is absorption of ultraviolet and visible light in the electromagnetic spectrum. It makes use of light in the visible and neighboring wavelength ranges (UV and infrared (NIR)). The beam of light passes through the sample, and the remaining light is analyzed in a detector within the wavelength range of 800–200 nm. When the light passes through the sample, molecules absorb light according to the wavelength spectrum depending upon chemical bonds and structure [248]. This process generates absorption spectra specific to the material composition [235]. Spectroscopy can determine the characteristics of nanomaterials such as suspension properties of cellulose nanoparticles [225], alginate/silver nanoparticle physical properties [226], gellan gum nanoparticles study [217], starch and pullulan nanoparticles physical analysis [227], xanthan conformation-based silver nanoparticle structure morphology [194], gold nanoparticles stabilized in dextran absorbance study [221], chitosan nanoparticle size determination [211], and nanoparticle formation [212]. The surface resonance passing through metallic nanoparticles such as Au, Ag, and Cu is characterized, which examines the size and shape of nanomaterials. This can be further applied in biosensors, biochemical catalysis, surface-enhanced scattering studies, etc.

2.3.6 Fourier Transform Infrared (FTIR) Spectroscopy

Fourier transform spectroscopy is a mathematical transformation-based investigation of the generated signal and is very sensitive and rapid compared to other spectroscopy techniques. The infrared light from the light source is passed through an interferometer with a certain optical path in FTIR analysis. The light from the infrared source is routed through the interferometer and distributed by a beam splitter. The output signal is called interferogram, which records the signal intensity [228].

The resultant signal intensity is then subjected to Fourier transformation to obtain the spectral information. Hence,

$$S(v) = \int_{-\infty}^{+\infty} I(x)e^{i2\pi vx} \, dx = F-1\big[I(x)\big] \tag{2.1}$$

is referred to as the inverse Fourier transform.

$$I(v) = \int_{-\infty}^{+\infty} S(v)e^{i2\pi vx} dv = F[s(v)] \tag{2.2}$$

is called the Fourier transform.

Thus, equation (2.1) converts $I(x)$, as a function of path difference x into a spectrum $S(v)$, where v is frequency.

It is the used to identify and characterize the biopolymeric nanomaterials [249]. The FTIR spectroscopy is specially used to estimate the intensity and wavelength of the absorbed radiation. Nanomaterials based on biopolymers such as starch [69], cellulose [225] and alginate [218] use FTIR technique to study about their nanostructures. The study about the synthesis of the chitosan nanoparticles can also be done with the help of FTIR spectroscopy [211]. Due to the use of interferometers, the FTIR technique is known to be different from other techniques. Michelson interferometer is mostly used in FTIR spectroscopy [249].

2.3.7 DYNAMIC LIGHT SCATTERING (DLS)

Dynamic light scattering (DLS) measures the particle size from the scattered light in a solution. This determines the nanoparticle size distribution [250]. In this technique, particle fluctuations are measured in the form of scattered light through their motion. This approach provides particle size data such as hydrodynamic diameter as well as the polydispersity index of the polymeric solution [251].

When a light of beam travels through a colloidal solution, particles are distributed. A monochromatic light illuminates the particles in DLS. The pace at which the intensity of scattered light fluctuates is regulated by the particle size. The coefficient of diffusion can be determined with the time taken by particle with respect to its intensity fluctuations, which further determines hydraulic size of the particle [250].

DLS (dynamic light scattering) is a non-invasive, label-free technique for the characterization of nanometer- to micrometer-sized particles [252]. The advantages of this technique are its speed, non-destructive nature, requirement of small sample size, and the ability to provide quick results [113]. Dynamic light scattering determines the size of nanoparticles in suspension along with their stability with respect to time at various pH and temperature conditions [157]. It also investigates particle-particle interactions with other species, including polymers, DNA, and biomarkers [250]. It is used for calculating the average particle size, polydispersity index (PDI) of chitosan [211], alginate nanoparticles [212], and the length (L) of rod-like cellulose crystals [230], and for analyzing a xanthan gum nanogel [229].

2.4 SUMMARY

The biopolymer-based nanomaterials are versatile components and can be considered as sustainable solutions to various future challenges in various sectors such as agriculture, food packaging, biomedical engineering, and waste disposal and recycle. The inherent properties of biopolymeric nanomaterials and their abundance make them a choice for biodegradable and biocompatible medical implants, tissue regeneration, wound dressings, and targeted drug delivery systems. These biomedical applications of green nanomaterials require a thorough understanding of the complex nano-biological molecular-level interactions on the cell surfaces, and their size- and shape-dependent reactivity. A better understanding of the specific handling, usage, and information about toxicity limits of biomaterials in various systems would aid in developing novel strategies for their further applications. The mechanical properties of polysaccharides can be transformed by using special additive materials and reinforcement, and they can be synthesized as discussed in this chapter with reported references. Biopolymer-based nanomaterials can be a versatile "green" alternative to synthetic nanomaterials.

2.5 KEYWORDS

Alginate
Biopolymers
Biocomposites

Biopolymeric nanomaterials
Cellulose
Chitin and chitosan
Gellan
Polysaccharides
Pullulan
Starch
Xanthan

REFERENCES

1. Dassanayake, S., Acharya, S. and Abidi, N. 2019. Biopolymer-based materials from polysaccharides: Properties, processing, characterization and sorption applications. In Advanced Sorption Process Applications. IntechOpen. doi: 10.5772/intechopen.80898.
2. Hu, Y. and Abidi, N. 2016. Distinct chiral nematic self-assembling behavior caused by different size-unified cellulose nanocrystals via a multistage separation. *Langmuir.* **32**: 9863–9872.
3. Hu, Y. and Catchmark, J. M. 2011. In vitro biodegradability and mechanical properties of bioabsorbable bacterial cellulose incorporating cellulases. *Acta Biomaterialia.* **7**: 2835–2845.
4. Hu, Y., Catchmark, J. M. and Vogler, E. A. 2013. Factors impacting the formation of sphere-like bacterial cellulose particles and their biocompatibility for human osteoblast growth. *Biomacromolecules.* **14**: 3444–3452.
5. Hu, Y., Catchmark, J. M., Zhu, Y., Abidi, N., Zhou, X., Wang, J. and Liang, N. 2014. Engineering of porous bacterial cellulose toward human fibroblasts ingrowth for tissue engineering. *Journal of Materials Research.* **29**: 2682–2693.
6. Hu, Y., Li, S., Jackson, T., Moussa, H. and Abidi, N. 2016. Preparation, characterization, and cationic functionalization of cellulose-based aerogels for wastewater clarification. *Journal of Materials.* **2016**: 3186589.
7. Wang, S., Lu, A. and Zhang, L. 2016. Recent advances in regenerated cellulose materials. *Progress in Polymer Science.* **53**: 169–206.
8. Popp, J., Lakner, Z., Harangi-Rákos, M. and Fári, M. 2014. The effect of bioenergy expansion: Food, energy, and environment. *Renewable and Sustainable Energy Reviews.* **32**: 559–578.
9. Chauhan, Y. P., Sapkal, R. S. and Sapkal, V. S. 2009. Microcrystalline cellulose from cotton rags (waste from garment and hosiery industries. *International Journal of Chemical Sciences.* **7**: 681–688.
10. Klemm, D., Heublein, B., Fink, H.-P. and Bohn, A. 2005. Cellulose: Fascinating biopolymer and sustainable raw material. *Angewandte Chemie International Edition.* **44**: 3358–3393.
11. Moon, R. J., Martini, A., Nairn, J., Simonsen, J. and Youngblood, J. 2011. Cellulose nanomaterials review: Structure, properties and nanocomposites. *Chemical Society Reviews* **40**: 3941–3994.
12. Saxena, I. M. and Brown Jr., R. M., 2005. Cellulose biosynthesis: Current views and evolving concepts. *Annals of Botany.* **96**: 9–21.
13. Festucci-Buselli, R. A. and Joshi, C. P. 2007. Structure, organization, and functions of cellulose synthase complexes in higher plants. *Brazilian Journal of Plant Physiology.* **19**: 1–13.
14. Gümüşkaya, E., Usta, M. and Kirci, H. 2003. The effects of various pulping conditions on crystalline structure of cellulose in cotton linters. *Polymer Degradation and Stability.* **81**: 559–564.
15. Jarvis, M. 2003. Cellulose stacks up. *Nature.* **426**: 611–612.
16. Šturcová, A., His, I., Apperley, D. C., Sugiyama, J. and Jarvis, M. C. 2004. Structural details of crystalline cellulose from higher plants. *Biomacromolecules.* **5**: 1333–1339.
17. Biswas, A., Saha, B. C., Lawton, J. W., Shogren, R. L. and Willett, J. L. 2006. Process for obtaining cellulose acetate from agricultural by-products. *Carbohydrate Polymers.* **64**: 134–137.
18. Peng, B. L., Dhar, N., Liu, H. L. and Tam, K. C. 2011. Chemistry and applications of nanocrystalline cellulose and its derivatives: A nanotechnology perspective. *The Canadian Journal of Chemical Engineering.* **89**: 1191–1206.
19. Rånby, B. G. 1951. Fibrous macromolecular systems. Cellulose and muscle. The colloidal properties of cellulose micelles. *Discussions of the Faraday Society* **11**: 158–164.
20. Aulin, C., Salazar-Alvarez, G. and Lindström, T. 2012. High strength, flexible and transparent nanofibrillated cellulose–nanoclay biohybrid films with tunable oxygen and water vapor permeability. *Nanoscale.* **4**: 6622–6628.

21. Arora, A. and Padua, G. W. 2010. Review: Nanocomposites in food packaging. *Journal of Food Science.* **75**: R43–R49.
22. Jung, Y. H., Chang, T.-H., Zhang, H., Yao, C., Zheng, Q., Yang, V. W., Mi, H., Kim, M., Cho, S. J., Park, D.-W., Jiang, H., Lee, J., Qiu, Y., Zhou, W., Cai, Z., Gong, S. and Ma, Z. 2015. High-performance green flexible electronics based on biodegradable cellulose nanofibril paper. *Nature Communications.* **6**: 7170.
23. Henriksson, M. and Berglund, L. A. 2007. Structure and properties of cellulose nanocomposite films containing melamine formaldehyde. *Journal of Applied Polymer Science.* **106**: 2817–2824.
24. Saito, T., Hirota, M., Tamura, N., Kimura, S., Fukuzumi, H., Heux, L. and Isogai, A. 2009. Individualization of nano-sized plant cellulose fibrils by direct surface carboxylation using TEMPO catalyst under neutral conditions. *Biomacromolecules.* **10**: 1992–1996.
25. Wågberg, L., Decher, G., Norgren, M., Lindström, T., Ankerfors, M. and Axnäs, K. 2008. The build-up of polyelectrolyte multilayers of microfibrillated cellulose and cationic polyelectrolytes. *Langmuir.* **24**: 784–795.
26. Rosa, M. F., Medeiros, E. S., Malmonge, J. A., Gregorski, K. S., Wood, D. F., Mattoso, L. H. C., Glenn, G., Orts, W. J. and Imam, S. H. 2010. Cellulose nanowhiskers from coconut husk fibers: Effect of preparation conditions on their thermal and morphological behavior. *Carbohydrate Polymers.* **81**: 83–92.
27. Brunel, F., Véron, L., David, L., Domard, A. and Delair, T. 2008. A novel synthesis of chitosan nanoparticles in reverse emulsion. *Langmuir.* **24**: 11370–11377.
28. Spinella, S., Maiorana, A., Qian, Q., Dawson, N. J., Hepworth, V., McCallum, S. A., Ganesh, M., Singer, K. D. and Gross, R. A. 2016. Concurrent cellulose hydrolysis and esterification to prepare a surface-modified cellulose nanocrystal decorated with carboxylic acid moieties. *ACS Sustainable Chemistry & Engineering.* **4**: 1538–1550.
29. Chen, L., Zhu, J. Y., Baez, C., Kitin, P. and Elder, T. 2016. Highly thermal-stable and functional cellulose nanocrystals and nanofibrils produced using fully recyclable organic acids. *Green Chemistry.* **18**: 3835–3843.
30. Azeh, Y., Olatunji, G. and Adekola, F. 2017. Synthesis and characterization of cellulose nanoparticles and its derivatives using a combination of spectro-analytical techniques research article open access. *International Journal of Nanotechnology in Medicine & Engineering.* **2**: 65–94.
31. Tumkur, P., Ronur Praful, T., Lamani, B., Nazario, N., Prabhakaran, K., Hall, J. and Ramesh, T. 2018. Enzymatic synthesis, characterization and biocompatibility studies of cellulose nanoparticles from cotton fibers. *Advances in Natural Sciences: Nanoscience and Nanotechnology.* **2**: 1–4.
32. Gorgieva, S. and Trček, J. 2019. Bacterial cellulose: Production, modification and perspectives in biomedical applications. *Nanomaterials.* **9**: 1352.
33. Lin, N., Gèze, A., Wouessidjewe, D., Huang, J. and Dufresne, A. 2016. Biocompatible double-membrane hydrogels from cationic cellulose nanocrystals and anionic alginate as complexing drugs codelivery. *ACS Applied Materials & Interfaces.* **8**: 6880–6889.
34. Löbmann, K. and Svagan, A. J. 2017. Cellulose nanofibers as excipient for the delivery of poorly soluble drugs. *International Journal of Pharmaceutics.* **533**: 285–297.
35. Liyaskina, E., Revin, V., Paramonova, E., Nazarkina, M., Pestov, N., Revina, N. and Kolesnikova, S. 2017. Nanomaterials from bacterial cellulose for antimicrobial wound dressing. *Journal of Physics: Conference Series.* **784**: 012034.
36. Favi, P. M., Benson, R. S., Neilsen, N. R., Hammonds, R. L., Bates, C. C., Stephens, C. P. and Dhar, M. S. 2013. Cell proliferation, viability, and in vitro differentiation of equine mesenchymal stem cells seeded on bacterial cellulose hydrogel scaffolds. *Materials Science and Engineering: C.* **33**: 1935–1944.
37. Fink, H., Hong, J., Drotz, K., Risberg, B., Sanchez, J. and Sellborn, A. 2011. An in vitro study of blood compatibility of vascular grafts made of bacterial cellulose in comparison with conventionally-used graft materials. *Journal of Biomedical Materials Research Part A.* **97A**: 52–58.
38. Tortorella, S., Locatelli, E., Maturi, M., Sambri, L. and Comes Franchini, M. 2020. Surface-modified nanocellulose for application in biomedical engineering and nanomedicine: A review. *International Journal of Nanomedicine.* **15**: 9909–9937.
39. Hua, K. Nanocellulose for Biomedical Applications: Modification, Characterisation and Biocompatibility Studies. http://urn.kb.se/resolve?urn=urn:nbn:se:uu:diva-267301.
40. Mohieldin, S., Zainudin, E. S., Tahir, P. M. and Mohamed, A. Z. 2011. Nanotechnology in pulp and paper industries: A review. *Key Engineering Materials.* **471–472**: 251–256.
41. Azeredo, H., Rosa, M. and Mattoso, L. 2017. Nanocellulose in bio-based food packaging applications. *Industrial Crops and Products.* **97**: 664–671.
42. Le Corre, D., Bras, J. and Dufresne, A. 2010. Starch nanoparticles: A review. *Biomacromolecules.* **11**: 1139–1153.

43. Pérez, S. and Bertoft, E. 2010. The molecular structures of starch components and their contribution to the architecture of starch granules: A comprehensive review. *Starch - Stärke*. **62**: 389–420.

44. De, B., Medeiros, S., Souza, M., Pinheiro, A., Bourbon, A., Cerqueira, M., Vicente, A. and Carneiro-da-Cunha, M. 2013. Physical characterisation of an alginate/lysozyme nano-laminate coating and its evaluation on "Coalho" cheese shelf life. *Food and Bioprocess Technology*. **7**: 1088.

45. He, X., Deng, H. and Hwang, H. 2019. The current application of nanotechnology in food and agriculture. *Journal of Food and Drug Analysis*. **27**: 1–21.

46. Le Corre, D. and Angellier-Coussy, H. 2014. Preparation and application of starch nanoparticles for nanocomposites: A review. *Reactive and Functional Polymers*. **85**: 97–120.

47. Šárka, E. and Dvoracek, V. 2017. Waxy starch as a perspective raw material (a review). *Food Hydrocolloids*. **69**: 402–409.

48. Tan, Y., Xu, K., Li, L., Liu, C., Song, C. and Wang, P. 2009. Fabrication of size-controlled starch-based nanospheres by nanoprecipitation. *ACS Applied Materials & Interfaces*. **1**: 956–959.

49. Teodoro, A. P., Mali, S., Romero, N. and de Carvalho, G. M. 2015. Cassava starch films containing acetylated starch nanoparticles as reinforcement: Physical and mechanical characterization. *Carbohydrate Polymers*. **126**: 9–16.

50. Venkatesan, J., Anil, S., Kim, S.-K. and Shim, M. S. 2016. Seaweed polysaccharide-based nanoparticles: Preparation and applications for drug delivery. *Polymers*. **8**: 30.

51. Murray J. D. 1945. Acetylation process and production. U.S. Patent. 376–378.

52. Bhardwaj, N. and Kundu, S. C. 2010. Electrospinning: A fascinating fiber fabrication technique. *Biotechnology Advances*. **28**: 325–347.

53. Crini, N., Lichtfouse, E., Torri, G. and Grégorio, C. 2019. Fundamentals and applications of chitosan. *Sustainable Agriculture Reviews* **35**: 49–123.

54. Czechowska-Biskup, R., Rokita, B., Lotfy, S., Ulański, P. and Rosiak, J. 2005. Degradation of chitosan and starch by 360 kHz ultrasound. *Carbohydrate Polymers*. **60**: 175–184.

55. Režek Jambrak, A., Mason, T., Lelas, V. and Kresic, G. 2010. Ultrasonic effect on physicochemical and functional properties of α-lactalbumin. *Lwt - Food Science and Technology*. **43**: 254–262.

56. Fahmy, H. M., Salah Eldin, R. E., Abu Serea, E. S., Gomaa, N. M., AboElmagd, G. M., Salem, S. A., Elsayed, Z. A., Edrees, A., Shams-Eldin, E. and Shalan, A. E. 2020. Advances in nanotechnology and antibacterial properties of biodegradable food packaging materials. *RSC Advances*. **10**: 20467–20484.

57. Shi, A., Li, D., Wang, L., Li, B. and Adhikari, B. 2011. Preparation of starch-based nanoparticles through high-pressure homogenization and miniemulsion cross-linking: Influence of various process parameters on particle size and stability. *Carbohydrate Polymers*. **83**: 1604–1610.

58. Yong, S. X. M., Song, C. P. and Choo, W. S. 2021. Impact of high-pressure homogenization on the extractability and stability of phytochemicals. *Frontiers in Sustainable Food Systems*. **4**: 294.

59. Wang, K., Wang, W., Ye, R., Xiao, J., Liu, Y., Ding, J., Zhang, S. and Liu, A. 2017. Mechanical and barrier properties of maize starch–gelatin composite films: Effects of amylose content. *Journal of the Science of Food and Agriculture*. **97**: 3613–3622.

60. Ji, N., Qiu, C., Xu, Y., Xiong, L. and Sun, Q. 2017. Differences in rheological behavior between normal and waxy corn starches modified by dry heating with hydrocolloids. *Starch - Stärke*. **69**: 1600332.

61. López Córdoba, A., Deladino, L. and Martino, M. 2013. Effect of starch filler on calcium-alginate hydrogels loaded with yerba mate antioxidants. *Carbohydrate Polymers*. **95**: 315–323.

62. Ren, L., Yan, X., Zhou, J., Tong, J. and Su, X. 2017. Influence of chitosan concentration on mechanical and barrier properties of corn starch/chitosan films. *International Journal of Biological Macromolecules*. **105**: 1636–1643.

63. Nallasamy, P., Ramalingam, T., Nooruddin, T., Shanmuganathan, R., Arivalagan, P. and Natarajan, S. 2020. Polyherbal drug loaded starch nanoparticles as promising drug delivery system: Antimicrobial, antibiofilm and neuroprotective studies. *Process Biochemistry*. **92**: 355–364.

64. Xiao, S., Liu, X., Tong, C., Zhao, L., Liu, X., Zhou, A. and Cao, Y. 2012. Dialdehyde starch nanoparticles as antitumor drug delivery system: An in vitro, in vivo, and immunohistological evaluation. *Chinese Science Bulletin*. **57**: 3226–3232.

65. Torres, F. and Arce, D. 2015. Starch-based nanocomposites for biomedical applications, pp. 73–94. In.: *Biodegradable Polymeric Nanocomposites*. doi: 10.1201/b19314-5.

66. Famá, L. M., Pettarin, V., Goyanes, S. N. and Bernal, C. R. 2011. Starch/multi-walled carbon nanotubes composites with improved mechanical properties. *Carbohydrate Polymers*. **83**: 1226–1231.

67. Harrison, B. S. and Atala, A. 2007. Carbon nanotube applications for tissue engineering. *Biomaterials*. **28**: 344–353.

68. Silva, G. A., Costa, F. J., Neves, N. M., Coutinho, O. P., Dias, A. C. P. and Reis, R. L. 2005. Entrapment ability and release profile of corticosteroids from starch-based microparticles. *Journal of Biomedical Materials Research Part A.* **73A**: 234–243.

69. Palanisamy, C. P., Cui, B., Zhang, H., Jayaraman, S. and Kodiveri Muthukaliannan, G. 2020. A comprehensive review on corn starch-based nanomaterials: Properties, simulations, and applications. *Polymers.* **12**: 2161.

70. Tanna, B. and Mishra, A. 2019. Nutraceutical potential of seaweed polysaccharides: Structure, bioactivity, safety, and toxicity. *Comprehensive Reviews in Food Science and Food Safety.* **18**: 817–831.

71. Bera, H. and Saha, S. 2021. *Biopolymer-Based Nanomaterials in Drug Delivery and Biomedical Applications.* London: Academic Press.

72. Gombotz, W. R. and Wee, S. 1998. Protein release from alginate matrices. *Advanced Drug Delivery Reviews.* **31**: 267–285.

73. Paques, J. P., Sagis, L. M. C., van Rijn, C. J. M. and van der Linden, E. 2014. Nanospheres of alginate prepared through w/o emulsification and internal gelation with nanoparticles of $CaCO_3$. *Food Hydrocolloids.* **40**: 182–188.

74. Qais, F. A., Samreen and Ahmad, I. 2019. Green synthesis of metal nanoparticles: Characterization and their antibacterial efficacy, pp. 635–680. In: *Antibacterial Drug Discovery to Combat MDR: Natural Compounds, Nanotechnology and Novel Synthetic Sources*, (I. Ahmad, S. Ahmad and K. P. Rumbaugh, eds.). Springer, Singapore.

75. Baek, S., Joo, S. and Toborek, M. 2019. Treatment of antibiotic-resistant bacteria by encapsulation of ZnO nanoparticles in an alginate biopolymer: Insights into treatment mechanisms. *Journal of Hazardous Materials.* **373**: 122–130.

76. Scolari, I. R., Páez, P. L., Musri, M. M., Petiti, J. P., Torres, A. and Granero, G. E. 2020. Rifampicin loaded in alginate/chitosan nanoparticles as a promising pulmonary carrier against Staphylococcus aureus. *Drug Delivery and Translational Research.* **10**: 1403–1417.

77. Jardim, K. V., Palomec-Garfias, A. F., Andrade, B. Y. G., Chaker, J. A., Báo, S. N., Márquez-Beltrán, C., Moya, S. E., Parize, A. L. and Sousa, M. H. 2018. Novel magneto-responsive nanoplatforms based on $MnFe_2O_4$ nanoparticles layer-by-layer functionalized with chitosan and sodium alginate for magnetic controlled release of curcumin. *Materials Science and Engineering: C.* **92**: 184–195.

78. Ye, S., Wang, C., Liu, X. and Tong, Z. 2005. Multilayer nanocapsules of polysaccharide chitosan and alginate through layer-by-layer assembly directly on PS nanoparticles for release. *Journal of Biomaterials Science.* Polymer edition. **16**: 909–23.

79. Poozesh, S. and Bilgili, E. 2019. Scale-up of pharmaceutical spray drying using scale-up rules: A review. *International Journal of Pharmaceutics.* **562**: 271–292.

80. Arpagaus, C., Collenberg, A., Rütti, D., Assadpour, E. and Jafari, S. M. 2018. Nano spray drying for encapsulation of pharmaceuticals. *International Journal of Pharmaceutics.* **546**: 194–214.

81. Ziaee, A., Albadarin, A. B., Padrela, L., Femmer, T., O'Reilly, E. and Walker, G. 2019. Spray drying of pharmaceuticals and biopharmaceuticals: Critical parameters and experimental process optimization approaches. *European Journal of Pharmaceutical Sciences.* **127**: 300–318.

82. Yaghoobi, N., Faridi Majidi, R., Faramarzi, M. A., Baharifar, H. and Amani, A. 2017. Preparation, optimization and activity evaluation of PLGA/streptokinase nanoparticles using electrospray. *Advanced Pharmaceutical Bulletin* **7**: 131–139.

83. Huang, L., Chaurasiya, B., Wu, D., Wang, H., Du, Y., Tu, J., Webster, T. J. and Sun, C. 2018. Versatile redox-sensitive pullulan nanoparticles for enhanced liver targeting and efficient cancer therapy. *Nanomedicine: Nanotechnology, Biology and Medicine.* **14**: 1005–1017.

84. Rahmam, S., Naim, M. N., Ng, E., Mokhtar, M. N. and Bakar, N. F. A. 2016. Encapsulation of bioactive compound from extracted jasmine flower using β-Cyclodextrin via electrospray. *IOP Conference Series: Earth and Environmental Science.* **36**: 012054.

85. Correia, C. R. 2019. Cell encapsulation in liquified compartments: Protocol optimization and challenges. *PLoS ONE.* **14**: e0218045.

86. Jiang, H., Zhao, P. and Zhu, K. 2007. Fabrication and characterization of zein-based nanofibrous scaffolds by an electrospinning method. *Macromolecular Bioscience.* **7**: 517–525.

87. Li, W.-J. and Tuan, R. S. 2009. Fabrication and application of nanofibrous scaffolds in tissue engineering. *Current Protocols in Cell Biology.* **42**: 25.2.1–25.2.12.

88. Parani, M., Lokhande, G., Singh, A. and Gaharwar, A. K. 2016. Engineered nanomaterials for infection control and healing acute and chronic wounds. *ACS Applied Materials & Interfaces.* **8**: 10049–10069.

89. Szekalska, M., Puciłowska, A., Szymańska, E., Ciosek, P. and Winnicka, K. 2016. Alginate: Current use and future perspectives in pharmaceutical and biomedical applications. *International Journal of Polymer Science.* **2016**: 7697031.

90. Bonino, C. A., Krebs, M. D., Saquing, C. D., Jeong, S. I., Shearer, K. L., Alsberg, E. and Khan, S. A. 2011. Electrospinning alginate-based nanofibers: From blends to crosslinked low molecular weight alginate-only systems. *Carbohydrate Polymers.* **85**: 111–119.

91. Tang, Y., Lan, X., Liang, C., Zhong, Z., Xie, R., Zhou, Y., Miao, X., Wang, H. and Wang, W. 2019. Honey loaded alginate/PVA nanofibrous membrane as potential bioactive wound dressing. *Carbohydrate Polymers.* **219**: 113–120.

92. Alves, N. M. and Mano, J. F. 2008. Chitosan derivatives obtained by chemical modifications for biomedical and environmental applications. *International Journal of Biological Macromolecules.* **43**: 401–414.

93. Li, R., Cheng, Z., Wen, R., Zhao, X., Yu, X., Sun, L., Zhang, Y., Han, Z., Yuan, Y. and Kang, L. 2018. Novel SA@Ca2+/RCSPs core–shell structure nanofibers by electrospinning for wound dressings. *RSC Advances.* **8**: 15558–15566.

94. Zhang, X., Huang, C., Zhao, Y. and Jin, X. 2017. Preparation and characterization of nanoparticle reinforced alginate fibers with high porosity for potential wound dressing application. *RSC Advances.* **7**: 39349–39358.

95. Hosseinifar, T., Sheybani, S., Abdouss, M., Hassani Najafabadi, S. A. and Shafiee Ardestani, M. 2018. Pressure responsive nanogel base on Alginate-Cyclodextrin with enhanced apoptosis mechanism for colon cancer delivery. *Journal of Biomedical Materials Research Part A.* **106**: 349–359.

96. Song, W., Su, X., Gregory, D. A., Li, W., Cai, Z. and Zhao, X. 2018. Magnetic Alginate/Chitosan nanoparticles for targeted delivery of curcumin into human breast cancer cells. *Nanomaterials.* **8**: 907.

97. Sun, Z., Yi, Z., Zhang, H., Ma, X., Su, W., Sun, X. and Li, X. 2017. Bio-responsive alginate-keratin composite nanogels with enhanced drug loading efficiency for cancer therapy. *Carbohydrate Polymers.* **175**: 159–169.

98. Mirrahimi, M., Abed, Z., Beik, J., Shiri, I., Dezfuli, A. S., Mahabadi, V., Kamrava, S. K., Ghaznavi, H. and Shakeri-Zadeh, A. 2019. A thermo⁻ responsive alginate nanogel platform co⁻loaded with gold nanoparticles and cisplatin for combined cancer chemo⁻ photothermal therapy. *Pharmacological Research.* **143**: 178–185.

99. Matai, I. and Gopinath, P. 2016. Chemically cross-linked hybrid nanogels of alginate and PAMAM dendrimers as efficient anticancer drug delivery vehicles. *ACS Biomaterials Science & Engineering.* **2**: 213–223.

100. Constantinidis, I., Grant, S., Simpson, N., Oca-Cossio, J., Sweeney, C., Mao, H., Blackband, S. and Sambanis, A. 2009. Use of magnetic nanoparticles to monitor alginate-encapsulated βTC-tet cells. *Magnetic Resonance in Medicine.* **61**: 282–90. Official journal of the Society of Magnetic Resonance in Medicine/Society of Magnetic Resonance in Medicine.

101. Yang, C.-H., Wang, W.-T., Grumezescu, A. M., Huang, K.-S. and Lin, Y.-S. 2014. One-step synthesis of platinum nanoparticles loaded in alginate bubbles. *Nanoscale Research Letters.* **9**: 277.

102. Omerović, N., Djisalov, M., Živojević, K., Mladenović, M., Vunduk, J., Milenković, I., Knežević, N. Ž., Gadjanski, I. and Vidić, J. 2021. Antimicrobial nanoparticles and biodegradable polymer composites for active food packaging applications. *Comprehensive Reviews in Food Science and Food Safety.* **20**: 2428–2454.

103. Ahmad, M., Mudgil, P., Gani, A., Hamed, F., Masoodi, F. A. and Maqsood, S. 2019. Nano-encapsulation of catechin in starch nanoparticles: Characterization, release behavior and bioactivity retention during simulated in-vitro digestion. *Food Chemistry.* **270**: 95–104.

104. Madhusha, C., Munaweera, I., Karunaratne, V. and Kottegoda, N. 2020. Facile mechanochemical approach to synthesizing edible food preservation coatings based on alginate/ascorbic acid-layered double hydroxide bio-nanohybrids. *Journal of Agricultural and Food Chemistry* **68**: 8962–8975.

105. Zhang, X., Zhang, Q., Xue, Y., Wang, Y., Zhou, X., Li, Z. and Li, Q. 2021. Simple and green synthesis of calcium alginate/AgCl nanocomposites with low-smoke flame-retardant and antimicrobial properties. *Cellulose.* **28**: 5151–5167.

106. Qin, Y. 2016. 5- Applications of advanced technologies in the development of functional medical textile materials, pp. 55–70. In: *Medical Textile Materials,* (Y. Qin, eds.). Woodhead Publishing, Sawston.

107. Rohani, A., Shakeri, M. and Bashari, A. 2019. Recent advances in application of chitosan and its derivatives in functional finishing of textiles, pp. 107–133. In: *The Impact and Prospects of Green Chemistry for Textile Technology.* doi: 10.1016/B978-0-08-102491-1.00005-8.

108. Kurczewska, J., Pecyna, P., Ratajczak, M., Gajęcka, M. and Schroeder, G. 2017. Halloysite nanotubes as carriers of vancomycin in alginate-based wound dressing. *Saudi Pharmaceutical Journal.* **25**: 911–920.

109. Riva, R., Ragelle, H., des Rieux, A., Duhem, N. and Jérôme, C., Préat, V. 2011. Chitosan and chitosan derivatives in drug delivery and tissue engineering. In: *Chitosan for Biomaterials II*, (R. Jayakumar, M. Prabaharan and R. Muzzarelli, eds.). Advances in Polymer Science. Springer, Heidelberg. doi:10.1007/12_2011_137.

110. Almaaytah, A., Qaoud, M. T., Khalil Mohammed, G., Abualhaijaa, A., Knappe, D., Hoffmann, R. and Al-Balas, Q. 2018. Antimicrobial and antibiofilm Activity of UP-5, an ultrashort antimicrobial peptide designed using only arginine and biphenylalanine. *Pharmaceuticals.* **11**: 3.

111. Honary, S. and Zahir, F. 2013. Effect of zeta potential on the properties of nano-drug delivery systems - A review (Part 1). *Tropical Journal of Pharmaceutical Research.* **12**. doi: 10.4314/tjpr.v12i2.19.

112. Katas, H., Raja, M. A. G. and Lam, K. L. 2013. Development of chitosan nanoparticles as a stable drug delivery system for protein/siRNA. *International Journal of Biomaterials.* **2013**: 146320.

113. Kiilll, C. P., da Silva Barud, H., Santagneli, S. H., Ribeiro, S. J. L., Silva, A. M., Tercjak, A., Gutierrez, J., Pironi, A. M. and Gremião, M. P. D. 2017. Synthesis and factorial design applied to a novel chitosan/sodium polyphosphate nanoparticles via ionotropic gelation as an RGD delivery system. *Carbohydrate Polymers.* **157**: 1695–1702.

114. Rassu, G., Porcu, E. P., Fancello, S., Obinu, A., Senes, N., Galleri, G., Migheli, R., Gavini, E. and Giunchedi, P. 2019. Intranasal delivery of genistein-loaded nanoparticles as a potential preventive system against neurodegenerative disorders. *Pharmaceutics.* **11**: 8.

115. Nasti, A., Zaki, N. M., de Leonardis, P., Ungphaiboon, S., Sansongsak, P., Rimoli, M. G. and Tirelli, N. 2009. Chitosan/TPP and Chitosan/TPP-hyaluronic acid nanoparticles: Systematic optimisation of the preparative process and preliminary biological evaluation. *Pharmaceutical Research.* **26**: 1918–1930.

116. Wu, Y., Wang, Y., Luo, G. and Dai, Y. 2009. In situ preparation of magnetic Fe_3O_4-chitosan nanoparticles for lipase immobilization by cross-linking and oxidation in aqueous solution. *Bioresource Technology.* **100**: 3459–3464.

117. Araki, J., Wada, M., Kuga, S. and Okano, T. 1998. Flow properties of microcrystalline cellulose suspension prepared by acid treatment of native cellulose. *Colloids and Surfaces A: Physicochemical and Engineering Aspects.* **142**: 75–82.

118. Fang, H., Huang, J., Ding, L., Li, M. and Chen, Z. 2009. Preparation of magnetic chitosan nanoparticles and immobilization of laccase. *Journal of Wuhan University of Technology-Materials Science Edition* **24**: 42–47.

119. Yao, C., Li, X. and Song, T. 2007. Electrospinning and crosslinking of zein nanofiber mats. *Journal of Applied Polymer Science.* **103**: 380–385.

120. Ohkawa, K., Cha, D., Kim, H., Nishida, A. and Yamamoto, H. 2004. Electrospinning of Chitosan. *Macromolecular Rapid Communications.* **25**: 1600–1605.

121. De Jong, W. H. 2008. Drug delivery and nanoparticles: Applications and hazards. *Int J Nanomedicine.* **3**: 133–149.

122. Ou, X., Zheng, J., Zhao, X. and Liu, M. 2018. Chemically cross-linked chitin nanocrystal scaffolds for drug delivery. *ACS Applied Nano Materials.* **1**: 6790.

123. Laskar, K. and Rauf, A. 2017. Chitosan based nanoparticles towards biomedical applications. *Journal of Nanomedicine Research.* **5**: 1–4.

124. Jhaveri, J., Raichura, Z., Khan, T., Momin, M. and Omri, A. 2021. Chitosan nanoparticles-insight into properties, functionalization and applications in drug delivery and theranostics. *Molecules.* **26**: 272.

125. Ahmed, S., Annu, Ali, A. and Sheikh, J. 2018. A review on chitosan centred scaffolds and their applications in tissue engineering. *International Journal of Biological Macromolecules.* **116**: 849–862.

126. Yang, T.-L. 2011. Chitin-based materials in tissue engineering: Applications in soft tissue and epithelial organ. *International Journal of Molecular Sciences.* **12**: 1936–1963.

127. Ghadi, A., Mahjoub, S., Tabandeh, F. and Talebnia, F. 2014. Synthesis and optimization of chitosan nanoparticles: Potential applications in nanomedicine and biomedical engineering. *Caspian Journal of Internal Medicine.* **5**: 156–161.

128. Ahmed, S. and Ikram, S. 2016. Chitosan based scaffolds and their applications in wound healing. *Achievements in the Life Sciences.* **10**: 27–37.

129. Singh, A., Mittal, A. and Benjakul, S. 2021. Chitosan nanoparticles: Preparation, food applications and health benefits. *Science Asia.* **47**: 1–10.

130. Bandara, S., Du, H., Carson, L., Bradford, D. and Kommalapati, R. 2020. Agricultural and biomedical applications of chitosan-based nanomaterials. *Nanomaterials.* **10**: 1903.

131. Shukla, A., Mehta, K., Parmar, J., Pandya, J. and Saraf, M. 2019. Depicting the exemplary knowledge of microbial exopolysaccharides in a nutshell. *European Polymer Journal*. **119**: 298–310.
132. Singhal, R. S. and Kulkarni, P. R. 1999. Production of food additives by fermentation, pp. 144–200. In: *Biotechnology: Food Fermentation*, (V. K. Joshi, ed.). Asia Tech Publishers Inc., New Delhi.
133. Mhd Haniffa, M. A., Ching, Y. C., Abdullah, L. C., Poh, S. C. and Chuah, C. H. 2016. Review of bionanocomposite coating films and their applications. *Polymers*. **8**: 246.
134. Leathers, T. D. 2003. Biotechnological production and applications of pullulan. *Applied Microbiology and Biotechnology*. **62**: 468–473.
135. Li, R., Tomasula, P., De Sousa, A. M., Liu, S.-C., Tunick, M., Liu, K. and Liu, L. 2017. Electrospinning pullulan fibers from salt solutions. *Polymers*. **9**: 32.
136. Autissier, A., Letourneur, D. and Le Visage, C. 2007. Pullulan-based hydrogel for smooth muscle cell culture. *Journal of Biomedical Materials Research Part A*. **82A**: 336–342.
137. Na, K., Shin, D., Yun, K., Park, K.-H. and Lee, K. C. 2003. Conjugation of heparin into carboxylated pullulan derivatives as an extracellular matrix for endothelial cell culture. *Biotechnology Letters*. **25**: 381–385.
138. Cheng, K.-C., Demirci, A. and Catchmark, J. M. 2011. Pullulan: Biosynthesis, production, and applications. *Applied Microbiology and Biotechnology*. **92**: 29–44.
139. Ramesan, R. and Sharma, C. 2007. Pullulan as a promising biomaterial for biomedical applications: A perspective. *Trends in Biomaterials and Artificial Organs*. **20**: 111–116.
140. Akiyoshi, K., Kobayashi, S., Shichibe, S., Mix, D., Baudys, M., Kim, S. and Sunamoto, J. 1998. Self-assembled hydrogel nanoparticle of cholesterol-bearing pullulan as a carrier of protein drugs: Complexation and stabilization of insulin. *Journal of Controlled Release*. **54**: 313–320.
141. Hong, L., Kim, W.-S., Lee, S.-M., Kang, S.-K., Choi, Y.-J. and Cho, C.-S. 2019. Pullulan nanoparticles as prebiotics enhance the antibacterial properties of lactobacillus plantarum through the induction of mild stress in probiotics. *Frontiers in Microbiology*. **10**: 142.
142. Dionísio, M., Cordeiro, C., Remuñán-López, C., Seijo, B., Rosa da Costa, A. M. and Grenha, A. 2013. Pullulan-based nanoparticles as carriers for transmucosal protein delivery. *European Journal of Pharmaceutical Sciences*. **50**: 102–113.
143. Na, K., Bum Lee, T., Park, K.-H., Shin, E.-K., Lee, Y.-B. and Choi, H.-K. 2003. Self-assembled nanoparticles of hydrophobically-modified polysaccharide bearing vitamin H as a targeted anti-cancer drug delivery system. *European Journal of Pharmaceutical Sciences*. **18**: 165–173.
144. Hriday, B., Hossain, C. M. and Saha, S. 2021. *Biopolymer-Based Nanomaterials in Drug Delivery and Biomedical Applications*. Elsevier, Amsterdam.
145. Ponrasu, T., Chen, B.-H., Chou, T.-H., Wu, J.-J. and Cheng, Y.-S. 2021. Fast dissolving electrospun nanofibers fabricated from jelly fig polysaccharide/pullulan for drug delivery applications. *Polymers (Basel)*. **13**: 241.
146. Nochi, T., Yuki, Y., Takahashi, H., Sawada, S., Mejima, M., Kohda, T., Harada, N., Kong, I. G., Sato, A., Kataoka, N., Tokuhara, D., Kurokawa, S., Takahashi, Y., Tsukada, H., Kozaki, S., Akiyoshi, K. and Kiyono, H. 2010. Nanogel antigenic protein-delivery system for adjuvant-free intranasal vaccines. *Nature Materials*. **9**: 572–578.
147. Yeasmin, S., Yeum, J. H. and Yang, S. B. 2020. Fabrication and characterization of pullulan-based nanocomposites reinforced with montmorillonite and tempo cellulose nanofibril. *Carbohydrate Polymers*. **240**: 116307.
148. Silva, N. H. C. S., Vilela, C., Almeida, A. and Marrucho, I. M. 2019. Pullulan-based nanocomposite films for functional food packaging: Exploiting lysozyme nanofibers as antibacterial and antioxidant reinforcing additives. *Food Hydrocolloids*. **77**: 921–930.
149. Khalaf, H. H., Sharoba, A. M., El-Tanahi, H. H. and Morsy, M. K. 2013. Stability of antimicrobial activity of pullulan edible films incorporated with nanoparticles and essential oils and their impact on turkey deli meat quality. *Journal of Food and Dairy Sciences*. **4**: 557–573.
150. Maji, K., Dasgupta, S., Pramanik, K. and Bissoyi, A. 2016. Preparation and evaluation of gelatin-chitosan-nanobioglass 3D porous scaffold for bone tissue engineering. *International Journal of Biomaterials*. **2016**: 9825659.
151. Osmałek, T., Froelich, A. and Tasarek, S. 2014. Application of gellan gum in pharmacy and medicine. *International Journal of Pharmaceutics*. **466**: 328–340.
152. Chakraborty, S., Jana, S., Gandhi, A., Sen, K. K., Zhiang, W. and Kokare, C. 2014. Gellan gum microspheres containing a novel α-amylase from marine Nocardiopsis sp. strain B2 for immobilization. *International Journal of Biological Macromolecules*. **70**: 292–299.
153. Mahdi, M. H., Conway, B. R. and Smith, A. M. 2015. Development of mucoadhesive sprayable gellan gum fluid gels. *International Journal of Pharmaceutics*. **488**: 12–19.

154. Rosas-Flores, W., Ramos-Ramírez, E. G. and Salazar-Montoya, J. A. 2013. Microencapsulation of Lactobacillus helveticus and Lactobacillus delbrueckii using alginate and gellan gum. *Carbohydrate Polymers*. **98**: 1011–1017.

155. Salunke, S. R. and Patil, S. B. 2016. Ion activated in situ gel of gellan gum containing salbutamol sulphate for nasal administration. *International Journal of Biological Macromolecules*. **87**: 41–47.

156. Morris, E. R. 2012. Gelation of gellan – A review. *Food Hydrocolloids*. **28**: 373–411.

157. Chen, H., Lan, G., Ran, L., Xiao, Y., Yu, K., Lu, B., Dai, F., Wu, D. and Lu, F. 2018. A novel wound dressing based on a Konjac glucomannan/silver nanoparticle composite sponge effectively kills bacteria and accelerates wound healing. *Carbohydrate Polymers*. **183**: 70–80.

158. Ding, L., Shan, X., Zhao, X., Zha, H., Chen, X., Wang, J., Cai, C., Wang, X., Li, G., Hao, J. and Yu, G. 2017. Spongy bilayer dressing composed of chitosan–Ag nanoparticles and chitosan–Bletilla striata polysaccharide for wound healing applications. *Carbohydrate Polymers*. **157**: 1538–1547.

159. Haider, A., Haider, S., Kang, I.-K., Kumar, A., Kummara, M. R., Kamal, T. and Han, S. S. 2018. A novel use of cellulose based filter paper containing silver nanoparticles for its potential application as wound dressing agent. *International Journal of Biological Macromolecules*. **108**: 455–461.

160. Liu, J., Tang, J., Wan, J., Wu, C., Graham, B., Kerr, P. G. and Wu, Y. 2019. Functional sustainability of periphytic biofilms in organic matter and Cu^{2+} removal during prolonged exposure to TiO_2 nanoparticles. *Journal of Hazardous Materials*. **370**: 4–12.

161. Razali, M. H., Amin, K. A. M. and Ismail, N. A. 2020. Titanium dioxide nanotubes incorporated gellan gum bio-nanocomposite film for wound healing: Effect of TiO_2 nanotubes concentration. *International Journal of Biological Macromolecules*. **153**: 1117–1135.

162. Zhu, N., Wang, S., Tang, C., Duan, P., Yao, L., Tang, J., Wong, P. K., An, T., Dionysiou, D. D. and Wu, Y. 2019. Protection mechanisms of periphytic biofilm to photocatalytic nanoparticle exposure. *Environmental Science & Technology*. **53**: 1585–1594.

163. Dhar, S., Mali, V., Bodhankar, S., Shiras, A., Prasad, B. L. V. and Pokharkar, V. 2011. Biocompatible gellan gum-reduced gold nanoparticles: Cellular uptake and subacute oral toxicity studies. *Journal of Applied Toxicology*. **31**: 411–420.

164. Zhai, X., Li, Z., Shi, J., Huang, X., Sun, Z., Zhang, D., Zou, X., Sun, Y., Zhang, J., Holmes, M., Gong, Y., Povey, M. and Wang, S. 2019. A colorimetric hydrogen sulfide sensor based on gellan gum-silver nanoparticles bionanocomposite for monitoring of meat spoilage in intelligent packaging. *Food Chemistry*. **290**: 135–143.

165. Muthukumar, T., Song, J. E. and Khang, G. 2019. Biological role of gellan gum in improving scaffold drug delivery, cell adhesion properties for tissue engineering applications. *Molecules*. **24**: 4514.

166. Pereira, L. 2017. *Therapeutic and Nutritional Uses of Algae*, 1st edn. CRC Press, Boca Raton, FL.

167. Vashisth, P., Nikhil, K., Roy, P., Pruthi, P. A., Singh, R. P. and Pruthi, V. 2016. A novel gellan-PVA nanofibrous scaffold for skin tissue regeneration: Fabrication and characterization. *Carbohydrate Polymers*. **136**: 851–859.

168. D'Arrigo, G., Navarro, G., Di Meo, C., Matricardi, P. and Torchilin, V. 2014. Gellan gum nanohydrogel containing anti-inflammatory and anti-cancer drugs: A multi-drug delivery system for a combination therapy in cancer treatment. *European Journal of Pharmaceutics and Biopharmaceutics*. **87**: 208–216.

169. Kudaibergenov, S. 2019. Gellan gum immobilized anticancer drugs and gold nanoparticles in nanomedicine. *Academic Journal of Polymer Science*. **2**. Doi: 10.19080/AJOP.2019.02.555588.

170. D'Arrigo, G., Di Meo, C., Gaucci, E., Chichiarelli, S., Coviello, T., Capitani, D., Alhaique, F. and Matricardi, P. 2012. Self-assembled gellan-based nanohydrogels as a tool for prednisolone delivery. *Soft Matter*. **8**: 11557–11564.

171. Mehnath, S., Ayisha Sithika, M. A., Arjama, M., Rajan, M., Amarnath Praphakar, R. and Jeyaraj, M. 2019. Sericin-chitosan doped maleate gellan gum nanocomposites for effective cell damage in Mycobacterium tuberculosis. *International Journal of Biological Macromolecules*. **122**: 174–184.

172. Posadowska, U., Brzychczy-Wloch, M. and Pamula, E. 2015. Injectable gellan gum-based nanoparticles-loaded system for the local delivery of vancomycin in osteomyelitis treatment. *Journal of Materials Science: Materials in Medicine*. **27**: 9.

173. Rayar, A., Babaladimath, G., Ambalgi, A. and Chapi, S. 2020. An eco-friendly synthesis, characterisation and antibacterial applications of gellan gum based silver nanocomposite hydrogel. *Materials Today: Proceedings*. **23**: 211–220.

174. Razali, M. H., Ismail, N. A. and Amin, K. A. M. 2019. Fabrication and characterization of antibacterial titanium dioxide nanorods incorporating gellan gum films. *Journal of Pure and Applied Microbiology* **13**: 1909–1916.

175. Pacelli, S., Paolicelli, P., Moretti, G., Petralito, S., Di Giacomo, S., Vitalone, A. and Casadei, M. 2016. Gellan gum methacrylate and laponite as an innovative nanocomposite hydrogel for biomedical application. *European Polymer Journal.* **77**: 114–123.

176. Alallam, B., Altahhan, S., Taher, M., Mohd Nasir, M. H. and Doolaanea, A. A. 2020. Electrosprayed alginate nanoparticles as CRISPR plasmid DNA delivery carrier: Preparation, optimization, and characterization. *Pharmaceuticals.* **13**: 158.

177. Rukmanikrishnan, B., Jo, C., Choi, S., Ramalingam, S. and Lee, J. 2020. Flexible ternary combination of gellan gum, sodium carboxymethyl cellulose, and silicon dioxide nanocomposites fabricated by quaternary ammonium silane: Rheological, thermal, and antimicrobial properties. *ACS Omega.* **5**: 28767–28775.

178. Xiao, G., Zhu, Y., Wang, L., You, Q., Huo, P. and You, Y. 2011. Production and storage of edible film using gellan gum. *Procedia Environmental Sciences.* **8**: 756–763.

179. Criado, M. P., Fraschini, C., Shankar, S., Salmieri, S. and Lacroix, M. 2021. Influence of cellulose nanocrystals gellan gum-based coating on color and respiration rate of Agaricus bisporus mushrooms. *Journal of Food Science.* **86**: 420–425.

180. Yang, Z., Zhai, X., Zou, X., Shi, J., Huang, X., Li, Z., Gong, Y., Holmes, M., Povey, M. and Xiao, J. 2021. Bilayer pH-sensitive colorimetric films with light-blocking ability and electrochemical writing property: Application in monitoring crucian spoilage in smart packaging. *Food Chemistry.* **336**: 127634.

181. De Filpo, G., Palermo, A. M., Munno, R., Molinaro, L., Formoso, P. and Nicoletta, F. P. 2015. Gellan gum/titanium dioxide nanoparticle hybrid hydrogels for the cleaning and disinfection of parchment. *International Biodeterioration & Biodegradation.* **103**: 51–58.

182. De Filpo, G., Palermo, A. M., Tolmino, R., Formoso, P. and Nicoletta, F. P. 2016. Gellan gum hybrid hydrogels for the cleaning of paper artworks contaminated with Aspergillus versicolor. *Cellulose.* **23**: 3265–3279.

183. Abdullah, A. and Mohammed, A. 2019. Scanning electron microscopy (SEM): A review. *Proceedings of the 2018 International Conference on Hydraulics and Pneumatics—HERVEX,* Băile Govora, Romania.

184. Luo, Y. and Wang, Q. 2014. Recent development of chitosan-based polyelectrolyte complexes with natural polysaccharides for drug delivery. *International Journal of Biological Macromolecules.* **64**: 353–367.

185. Edens, R. E. 2005. *Polysaccharides: Structural Diversity and Functional Versatility,* 2nd ed Edited by Severian Dumitriu (University of Sherbrooke, Quebec). Marcel Dekker, New York. 2005. xviii + 1204 pp. $269.95. ISBN 0-8247-5480-8. J. Am. Chem. Soc. 127: 10119–10119.

186. García-Ochoa, F., Santos, V. E., Casas, J. A. and Gómez, E. 2000. Xanthan gum: Production, recovery, and properties. *Biotechnology Advances.* **18**: 549–579.

187. Bergmann, D., Furth, G. and Mayer, C. 2008. Binding of bivalent cations by xanthan in aqueous solution. *International Journal of Biological Macromolecules.* **43**: 245–251.

188. Dário, A. F., Hortêncio, L. M. A., Sierakowski, M. R., Neto, J. C. Q. and Petri, D. F. S. 2011. The effect of calcium salts on the viscosity and adsorption behavior of xanthan. *Carbohydrate Polymers.* **84**: 669–676.

189. Xu, W., Jin, W., Huang, K., Huang, L., Lou, Y., Li, J., Liu, X. and Li, B. 2018. Interfacial and emulsion stabilized behavior of lysozyme/xanthan gum nanoparticles. *International Journal of Biological Macromolecules.* **117**: 280–286.

190. Santoshi Kumari, A., Venkatesham, M., Ayodhya, D. and Veerabhadram, G. 2015. Green synthesis, characterization and catalytic activity of palladium nanoparticles by xanthan gum. *Applied Nanoscience.* **5**: 315–320.

191. Muddineti, O., Kumari, P., Ajjarapu, S., Lakhani, P., Bahl, R., Ghosh, B. and Biswas, S. 2016. Xanthan gum stabilized PEGylated gold nanoparticles for improved delivery of curcumin in cancer. *Nanotechnology.* **27**: 325101.

192. Comba, S. and Sethi, R. 2009. Stabilization of highly concentrated suspensions of iron nanoparticles using shear-thinning gels of xanthan gum. *Water Research.* **43**: 3717–3726.

193. Husain, F. M., Hasan, I., Qais, F. A., Khan, R. A., Alam, P. and Alsalme, A. 2020. Fabrication of zinc oxide-xanthan gum nanocomposite via green route: Attenuation of quorum sensing regulated virulence functions and mitigation of biofilm in gram-negative bacterial pathogens. *Coatings.* **10**: 1190.

194. Xu, W., Jin, W., Lin, L., Zhang, C., Li, Z., Li, Y., Song, R. and Li, B. 2014. Green synthesis of xanthan conformation-based silver nanoparticles: Antibacterial and catalytic application. *Carbohydrate Polymers.* **101**: 961–967.

195. Kennedy, J. R. M., Kent, K. E. and Brown, J. R. 2015. Rheology of dispersions of xanthan gum, locust bean gum and mixed biopolymer gel with silicon dioxide nanoparticles. *Materials Science and Engineering: C.* **48**: 347–353.

196. Kumar, N., Labille, J., Bossa, N., Auffan, M., Doumenq, P., Rose, J. and Bottero, J.-Y. 2017. Enhanced transportability of zero valent iron nanoparticles in aquifer sediments: Surface modifications, reactivity, and particle traveling distances. *Environmental Science and Pollution Research*. **24**: 9269–9277.

197. Pooja, D., Panyaram, S., Kulhari, H., Rachamalla, S. S. and Sistla, R. 2014. Xanthan gum stabilized gold nanoparticles: Characterization, biocompatibility, stability and cytotoxicity. *Carbohydrate Polymers*. **110**: 1–9.

198. Raval, N., Maheshwari, R., Kalyane, D., Youngren-Ortiz, S., Chougule, M. and Tekade, R. 2019. Importance of physicochemical characterization of nanoparticles in pharmaceutical product development, pp. 369–400. In: *Advances in Pharmaceutical Product Development and Research*. doi: 10.1016/B978-0-12-817909-3.00010-8.

199. Ebrahiminezhad, A., Najafipour, S., Kouhpayeh, A., Berenjian, A., Rasoul-Amini, S. and Ghasemi, Y. 2014. Facile fabrication of uniform hollow silica microspheres using a novel biological template. *Colloids and Surfaces B: Biointerfaces*. **118**: 249–253.

200. Basha, S. K., Lakshmi, K. V. and Kumari, V. S. 2016. Ammonia sensor and antibacterial activities of green zinc oxide nanoparticles. *Sensing and Bio-Sensing Research*. **10**: 34–40.

201. Katzbauer, B. 1998. Properties and applications of xanthan gum. *Polymer Degradation and Stability*. **59**: 81–84.

202. Dave, P. N. and Gor, A. 2018. Chapter 3- Natural polysaccharide-based hydrogels and nanomaterials: Recent trends and their applications, pp. 36–66. In: *Handbook of Nanomaterials for Industrial Applications*, (Mustansar Hussain, Chaudhery eds.) Elsevier, Amsterdam.

203. Kothari, P. and Setia, H. 2017. Silver nanoparticle filled HPMC and xanthan films for food packaging and safety. *Indian Journal of Science and Technology*. **10**: 1–6.

204. Fan, Y., Yang, J., Duan, A. and Li, X. 2021. Pectin/sodium alginate/xanthan gum edible composite films as the fresh-cut package. *International Journal of Biological Macromolecules*. **181**: 1003–1009.

205. Rukmanikrishnan, B., Ismail, F. R. M., Manoharan, R. K., Kim, S. S. and Lee, J. 2020. Blends of gellan gum/xanthan gum/zinc oxide based nanocomposites for packaging application: Rheological and antimicrobial properties. *International Journal of Biological Macromolecules*. **148**: 1182–1189.

206. Laffleur, F. and Michalek, M. 2017. Modified xanthan gum for buccal delivery—A promising approach in treating sialorrhea. *International Journal of Biological Macromolecules*. **102**: 1250–1256.

207. Bhaskar, K., Anbu, J., Ravichandiran, V., Venkateswarlu, V. and Rao, Y. M. 2009. Lipid nanoparticles for transdermal delivery of flurbiprofen: Formulation, in vitro, ex vivo and in vivo studies. *Lipids in Health and Disease*. **8**: 6.

208. Samia, O., Hanan, R. and Kamal, E. T. 2012. Carbamazepine Mucoadhesive Nanoemulgel (MNEG) as brain targeting delivery system via the olfactory mucosa. *Null*. **19**: 58–67.

209. Madhusudana Rao, K., Kumar, A. and Han, S. S. 2017. Polysaccharide based bionanocomposite hydrogels reinforced with cellulose nanocrystals: Drug release and biocompatibility analyses. *International Journal of Biological Macromolecules*. **101**: 165–171.

210. Shekarforoush, E., Ajalloueian, F., Zeng, G., Mendes, A. C. and Chronakis, I. S. 2018. Electrospun xanthan gum-chitosan nanofibers as delivery carrier of hydrophobic bioactives. *Materials Letters*. **228**: 322–326.

211. Oh, J.-W., Chun, S. and Murugesan, C. 2019. Preparation and in vitro characterization of chitosan nanoparticles and their broad-spectrum antifungal action compared to antibacterial activities against phytopathogens of tomato. *Agronomy*. **9**: 21.

212. Sundar, S., Kundu, J. and Kundu, S. C. 2010. Biopolymeric nanoparticles. *Science and Technology of Advanced Materials*. **11**: 014104–014104.

213. Khalid, S., Yu, L., Meng, L., Liu, H., Ali, A. and Chen, L. 2017. Poly(lactic acid)/starch composites: Effect of microstructure and morphology of starch granules on performance. *Journal of Applied Polymer Science*. **134**: 45504.

214. Majoinen, J., Kontturi, E., Ikkala, O. and Gray, D. 2012. SEM imaging of chiral nematic films cast from cellulose nanocrystal suspensions. *Cellulose*. **19**: 1599.

215. Maia, J., Evangelista, M., Gil, H. and Ferreira, L. 2014. Dextran-based materials for biomedical applications, pp. 31–53. In: *Carbohydrates Applications in Medicine*, (M. H. Gil ed.) ISBN: 978-81-308-0523-8.

216. Ramasubba Reddy, P., Shimoga, G., Kroneková, Z., Sláviková, M., Saha, N. and Saha, P. 2017. Green synthesis of silver nanoparticles and biopolymer nanocomposites: A comparative study on physico-chemical, antimicrobial and anticancer activity. *Bulletin of Materials Science*. **41**: 55.

217. Dhar, S., Murawala, P., Shiras, A., Pokharkar, V. and Prasad, B. L. V. 2012. Gellan gum capped silver nanoparticle dispersions and hydrogels: Cytotoxicity and in vitro diffusion studies. *Nanoscale*. **4**: 563–567.

218. Helmiyati, H. and Wahyuningrum, K. D. 2018. Synthesis and photocatalytic activity of nanocomposite based on sodium alginate from brown algae with ZnO impregnation. *AIP Conference Proceedings.* **2023**: 020107.
219. Kumar, A., Negi, Y. S., Choudhary, V. and Bhardwaj, N. 2014. Characterization of cellulose nanocrystals produced by acid-hydrolysis from sugarcane bagasse as agro-waste. *Journal of Materials Physics and Chemistry.* **2**: 1–8.
220. Pakravan, M., Heuzey, M.-C. and Ajji, A. 2012. Core-shell structured PEO-chitosan nanofibers by coaxial electrospinning. *Biomacromolecules.* **13**: 412–21.
221. Diem, P. H. N., Thao, D. T. T., Phu, D. V., Duy, N. N., Quy, H. T. D., Hoa, T. T. and Hien, N. Q. 2017. Synthesis of gold nanoparticles stabilized in dextran solution by gamma Co-60 ray irradiation and preparation of gold nanoparticles/dextran powder. *Journal of Chemistry.* **2017**: 6836375.
222. Liu, C., Li, M., Ji, N., Liu, J., Xiong, L. and Sun, Q. 2017. Morphology and characteristics of starch nanoparticles self-assembled via a rapid ultrasonication method for peppermint oil encapsulation. *Journal of Agricultural and Food Chemistry* **65**: 8363–8373.
223. Kalia, S., Dufresne, A., Cherian, B. M., Kaith, B. S., Avérous, L., Njuguna, J. and Nassiopoulos, E. 2011. Cellulose-based bio- and nanocomposites: A review. *International Journal of Polymer Science.* **2011**: 837875.
224. Wang, H., Tingting, D., Zhou, S., Huang, X., Li, S., Sun, K., Zhou, G. and Dou, H. 2017. Self-assembly assisted fabrication of dextran-based nanohydrogels with reduction-cleavable junctions for applications as efficient drug delivery systems. *Scientific Reports.* **7**: 40011.
225. Zhu, Q., Zhou, R., Liu, J., Sun, J. and Wang, Q. 2021. Recent progress on the characterization of cellulose nanomaterials by nanoscale infrared spectroscopy. *Nanomaterials.* **11**: 1353.
226. Nam, S., MubarakAli, D. and Kim, J.-W. 2016. Characterization of alginate/silver nanobiocomposites synthesized by solution plasma process and their antimicrobial properties. *Journal of Nanomaterials.* **2016**. doi: 10.1155/2016/4712813.
227. Luo, K., Kim, N., You, S.-M. and Kim, Y.-R. 2019. Colorimetric determination of the activity of starch-debranching enzyme via modified tollens' reaction. *Nanomaterials.* **9**: 1291.
228. Jaggi, N. and Vij, D. R. 2007. Fourier transform infrared spectroscopy, pp. 411–450. In: *Handbook of Applied Solid State Spectroscopy.* Springer, Boston, MA.
229. Rahdar, A. and Almasi-Kashi, M. 2016. Dynamic light scattering of nano-gels of xanthan gum biopolymer in colloidal dispersion. *Journal of Advanced Research.* **7**: 635–641.
230. Boluk, Y. and Danumah, C. 2014. Analysis of cellulose nanocrystal rod lengths by dynamic light scattering and electron microscopy. *Journal of Nanoparticle Research.* **16**: 2174.
231. Venkateshaiah, A., Padil, V. V. T., Nagalakshmaiah, M., Waclawek, S., Černík, M. and Varma, R. S. 2020. Microscopic techniques for the analysis of micro and nanostructures of biopolymers and their derivatives. *Polymers.* **12**: 512.
232. Sharma, S., Jaiswal, S., Duffy, B. and Jaiswal, A. K. 2019. Nanostructured materials for food applications: Spectroscopy, microscopy and physical properties. *Bioengineering.* **6**: 26.
233. Golding, C. G., Lamboo, L. L., Beniac, D. R. and Booth, T. F. 2016. The scanning electron microscope in microbiology and diagnosis of infectious disease. *Scientific Reports.* **6**: 26516.
234. Malzahn, K., Jamieson, W. D., Dröge, M., Mailänder, V., Jenkins, A. T. A., Weiss, C. K. and Landfester, K. 2014. Advanced dextran based nanogels for fighting Staphylococcus aureus infections by sustained zinc release. *Journal of Materials Chemistry B.* **2**: 2175–2183.
235. Mourdikoudis, S., Pallares, R. M. and Thanh, N. T. K. 2018. Characterization techniques for nanoparticles: Comparison and complementarity upon studying nanoparticle properties. *Nanoscale.* **10**: 12871–12934.
236. Mayeen, A., Shaji, L. K., Nair, A. K. and Kalarikkal, N. 2018. Chapter 12- Morphological characterization of nanomaterials, pp. 335–364. In: *Characterization of Nanomaterials*, (S. Mohan Bhagyaraj, O. S. Oluwafemi, N. Kalarikkal and S. Thomas, eds.) Woodhead Publishing, Sawston.
237. Wang, Z. L. 1999. Transmission electron microscopy and spectroscopy of nanoparticles, pp. 37–80. In: *Characterization of Nanophase Materials.* John Wiley & Sons, Ltd., Hoboken, NJ.
238. Williams, D. B. and Barry Carter, C. 2009. *Transmission Electron Microscopy.* New York: Springer.
239. Gopiraman, M., Deng, D., Saravanamoorthy, S., Chung, I.-M. and Kim, I. S. 2018. Gold, silver and nickel nanoparticle anchored cellulose nanofiber composites as highly active catalysts for the rapid and selective reduction of nitrophenols in water. *RSC Advances.* **8**: 3014–3023.
240. Tomoda, B., Yassue Cordeiro, P., Ernesto, J. V., Lopes, P., Péres, L., Silva, C. and Agostini de Moraes, M. 2020. Characterization of biopolymer membranes and films: Physicochemical, mechanical, barrier, and biological properties, pp. 67–95. In: *Biopolymer Membranes and Films.* doi: 10.1016/B978-0-12-818134-8.00003-1.

241. Fan, T. W.-M. and Lane, A. N. 2016. Applications of NMR spectroscopy to systems biochemistry. *Progress in Nuclear Magnetic Resonance Spectroscopy.* **92–93**: 18–53.

242. Alavi, S., Thomas, S., Sandeep, K., Kalarikkal, N., Varghese, J. and Yaragalla, S. 2014, Analytical techniques for structural characterization of biopolymer-based nanocomposites, Polymers and packaging applications, CRC press, Taylor and Francis.

243. Mohamed, M., Mohd Hir, Z. A., Mokthar, W. and Osman, N. 2020. Features of metal oxide colloidal nanocrystal characterization, pp. 83–122. In: *Colloidal Metal Oxide Nanoparticles.* doi: 10.1016/B978-0-12-813357-6.00008-5.

244. Silva, M. M., Calado, R., Marto, J., Bettencourt, A., Almeida, A. J. and Gonçalves, L. M. D. 2017. Chitosan nanoparticles as a mucoadhesive drug delivery system for ocular administration. *Marine Drugs.* **15**: 370.

245. Emwas, A.-H., Roy, R., McKay, R. T., Tenori, L., Saccenti, E., Gowda, G. A. N., Raftery, D., Alahmari, F., Jaremko, L., Jaremko, M. and Wishart, D. S. 2019. NMR spectroscopy for metabolomics research. *Metabolites.* **9**: 123.

246. Guo, C. 2017. Nuclear Magnetic Resonance (NMR) Spectroscopic Characterization of Nanomaterials and Biopolymers, Arizona state University, PhD thesis.

247. Lin, P.-C., Lin, S., Wang, P. C. and Sridhar, R. 2014. Techniques for physicochemical characterization of nanomaterials. *Biotechnology Advances.* **32**: 711–726.

248. Saxena, A., Tripathi, R. and Singh, R. 2010. Biological synthesis of silver nanoparticles by using onion (Allium cepa) extract and their antibacterial activity. *Digest Journal of Nanomaterials and Biostructures.* **5**: 427–432.

249. Stuart, B. 2005. *Infrared spectroscopy: Fundamentals and Applications.* Wiley, Chichester.

250. Wang, X., Ramström, O. and Yan, M. 2011. Dynamic light scattering as an efficient tool to study glyconanoparticle–lectin interactions. *The Analyst.* **136**: 4174–8.

251. Nimesh, S. 2013. Tools and techniques for physico-chemical characterization of nanoparticles, pp. 43–63. In: *Gene Therapy.* doi: 10.1533/9781908818645.43.

252. Karow, A., Götzl, J. and Garidel, P. 2014. Resolving power of dynamic light scattering for protein and polystyrene nanoparticles. *Pharmaceutical Development and Technology.* **20**: 84.

3 Enhancement of Permeation Rate of Rifabutin Using Encapsulation of Lipid-Based Nanoparticles through Chicken Ileum by Statistical Optimization Approach

Om M. Bagade
Vishwakarma University School of Pharmacy Pune

Priyanka E. Doke-Bagade
D.Y. Patil International University School of Pharmacy Pune

CONTENTS

DOI: 10.1201/9781003319153-4

LIST OF ABBREVIATIONS

SLNs   Solid lipid nanoparticles
RBT   Rifabutin
MAC   Mycobacterium avium complex
HIV   Human immunodeficiency virus
RNA   Ribonucleic acid
HPLC   High-performance liquid chromatography
GMS   Glyceryl monostearate
PVA   Polyvinyl alcohol
CP   Crospovidone
SLS   Sodium lauryl sulphate
ESD   Emulsification solvent diffusion
RSA   Response surface analysis
RSP   Response surface plot
CP   Contour plot
EE   Entrapment efficiency
DL   Drug loading
PS   Particle size
ZP   Zeta potential
UV   Ultraviolet
FTIR   Fourier transforms infrared spectroscopy
TEM   Transmission electron microscopy

PXRD   Powder X-ray diffraction
DSC   Differential scanning calorimetry
AUC   Area under the curve
nm   Nanometre

3.1 INTRODUCTION

Solid lipid nanoparticles (SLNs), which are lipid nanocrystals in water with a solid core into which pharmaceuticals are integrated, are a new concept in nanotechnology. The SLNs utilize an assortment of far more conventional drug carriers, including liposomes and polymeric nanoparticles, while removing some of their drawbacks, such as the concerns of burst release and long-term stability in liposomes and residual solvents and bulk formation in polymeric nanoparticles [1]. These lipid nanocarriers initially invented in around 1991 to merge the rewards of other colloidal carriers while attempting to avoid the upsides and downsides have garnered considerable consideration in modern existence and are now deemed as a viable substitute for time honoured colloidal carrier approach vis-a-vis micro- and nanoparticulate system, liposomes, and emulsions [2–6].

Throughout the mid-1990s, SLNs were initially invented using two separate methodologies: hot homogenization and heated microemulsion. And since then, the number of research practitioners in the industry has continued to rise, to the point where there are now about forty of them, mostly academics; at the same time, the number of lawsuits has elevated as methods of synthesis such as the solvent emulsification strategy and the solvent diffusion method have already been filed [7].

3.1.1 GASTROINTESTINAL PHYSIOLOGY PERTAINING TO PARTICULATE UPTAKE

The GIT is responsible for food digestion and nutritional, water, and electrolyte absorption and serves as a protective layer in between the systemic circulation and environment. Whereas the GIT is aimed at preventing particle matter (possibly hazardous substances and germs) from the environment, it is not a flawless obstacle. The acceptability of macromolecules and particulates is partly attributable to unique pathways. Mucus, which is mostly made up of mucin molecules and surrounds the intestine's porous enterocytes, serves as a deterrent to the uptake of foreign chemicals, including nanostructures, through the mouth [8,9]. Because it is equipped for transcytosis/endocytosis of foreign particulates and pathogens to the codified glandular tissue within that M cell with basal layer and mucosa, FAE tends to offer a big contribution to enabling epitope-to-tissue and cellular-to-tissue interrelations throughout an innate immunity.

In contrast to neighbouring absorptive cells, M cells lack largely accomplished microvillae and shuttle the particles currently occupied to the lymphatic, where they are subsequently discharged into the systemic circulation in an area-wise manner; refer Figure 3.3 [10–12].

3.1.2 PATHWAYS OF UPTAKE

To transport their pharmacological content to the target organs, lymph, or blood, SLNs must pass through the gastrointestinal barrier, whether by facilitated diffusion either through paracellular or transcellular pathway or through the elements of the project constrained by membrane-bound vehicles or membrane-based organelles; refer Figure 3.1 [13,14].

3.1.3 LYMPHATIC ABSORPTION

Sanders and Ashworth revealed the endocytosis of 200 nm microparticles in typical enterocytes transferred to the liver via imaging techniques in 1961. Upon oral administration, even just a

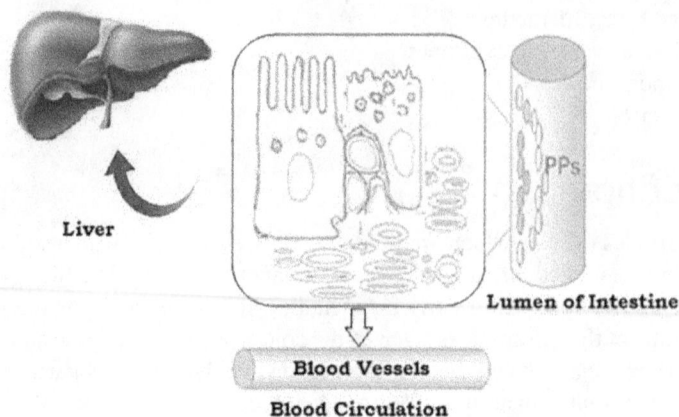

FIGURE 3.1 Uptake principle.

comparatively tiny fraction of particles dose seems to be in the blood, which could be due to particulate encapsulation in M cell pockets stuffed with lymphocytes and macrophages, non-fenestrated capillary endothelium in PP, and lymph node sequestration, all of which obstruct direct particle access to circulation.

3.1.4 Considerations of Toxicity and *in vivo* Fate

Considering SLNs are synthesized from physiological components, metabolism is projected to be well tolerated in biological systems. Obviously, the emulsifiers' cytotoxicity must be acknowledged; however, their principal drawback also pertains towards other targeted delivery. Challenges could emerge because SLNs would interact with pyrogen examinations (limulus test) and, in whatsoever event, induce gelation. The human species could apparently withstand such a quantity of bigger microparticles. Similar assumption really shouldn't be generalized to SLNs even though a solid form of lipid is not malleable like oil, and thereby, with exception of nanoemulsions, capillary obstruction will ensue unless the molecular structure surpasses the blood vessel dimension.

SLNs have become less lethal than polymeric form of nanoparticles, as per the outcomes of cytotoxicity assessments (MTT test). The findings of the toxicity investigations on cells in vitro corroborate the carrier's projected minimal toxicity [15].

3.1.5 Mode of Administration Including *in vivo* Fate

The SLN molecules' *in vivo* fate will be substantially influenced by three factors:

Delivery route
SLN interacts with the physiological environment, which contains:

a. Dispersion pathways (physiological component adsorption on the surface of particles and SLN constituent discharges into physiological environment).
b. Enzyme-mediated approach (e.g. lipases and esterases degrade lipids).
   Lipases, which are prevalent across many organs and tissues, are responsible for SLN breakdown. Lipases break down the ester bond, resulting in incomplete free fatty acids, glycerol, and glycerides. An oil/water interaction is required to initiate most lipases and thereby access the catalytic core (opening of lid).

3.1.5.1 Administration through the Oral Cavity

Administration through the peroral way aqueous colloids or SLN-loaded typical dosage forms such as capsules, tablets, or pellets may be used for peroral delivery of SLNs. Because of its acidification and high ionic potency, the extracellular matrix of the stomach facilitates particle aggregation. Food is likely to have a substantial impact on SLN effectiveness; nevertheless, to our understanding, no experimental values were already reported upon the matter. The influence of gut and pancreatic lipases on lipid nanoparticles degradation *in vivo* is also a challenge that is to be explored; refer Figure 3.2.

3.1.5.2 Administration via Parenteral Way

SLNs were fed intravenously to mammals. Upon i.v. infusion in rats, a pharmacokinetic investigation of doxorubicin encompassed into SLNs revealed greater blood levels than a marketed product. In considerations of anatomical distribution, SLNs were reported to induce increased effective concentration in the brain, spleen, and lungs; meanwhile, the fluid culminated in a somewhat more concentrated dissemination in the kidney and liver. Deposition of a bloodstream protein across the surface of nanoparticles is estimated to be associated with SLN transport by the brain via influencing attachment to the blood–brain barrier's endothelial cells [16].

3.1.5.3 Application on the Skin

SLN microemulsion having negligible content of lipid has the tiniest droplet size (up to 5%). A dermal administration through skin is hampered by the trace amount of disseminated lipid as well as the moderate fluidity. Across most circumstances, the SLN dispersion should always be integrated into gel or an ointment in an attempt to develop a dosage that can be applied to the skin. The rheological properties are related to those of conventional cutaneous preparations. The findings suggest that steeply

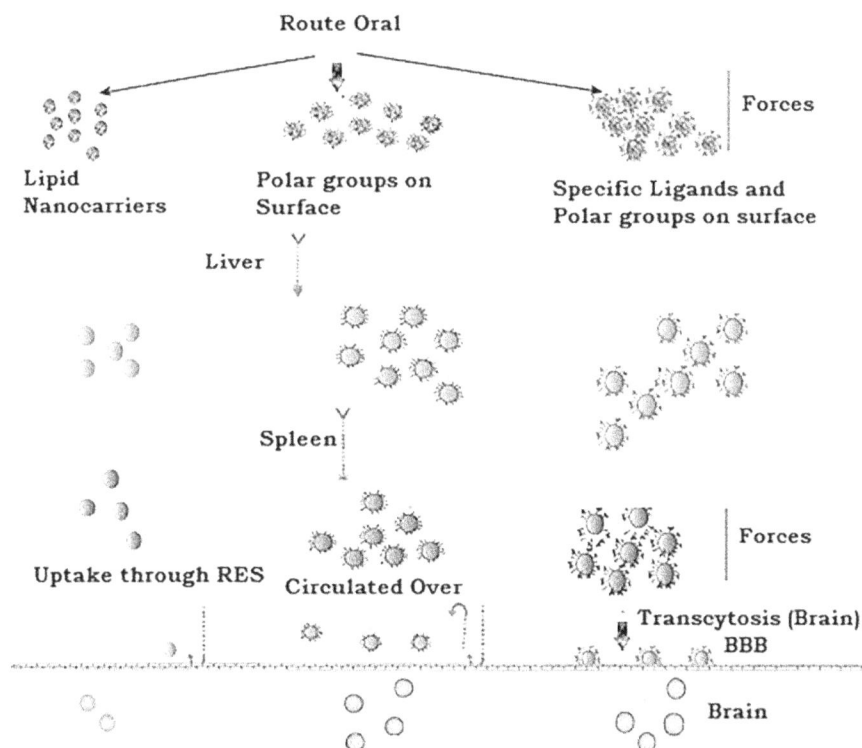

FIGURE 3.2 Fate of SLNs after oral administration.

lipid colloids in the nanometric average diameter could be synthesized in a single step. As a corollary, certain manufacturing stages (such as dilution of SLN in gel or cream) can sometimes be eliminated.

3.1.6 Steps in the Direction of the Pharmaceutical Sector

The very first products of cosmetics based on NCLs have entered the market over the forecast period. Their advent into the lipid-based cosmetics sector might be interpreted as a stepping stone to the pharmaceutical market [17].

The following are the most essential requirements:

a. Legislative approval of the ingredients
b. Massive production feasibility
c. Sufficient therapeutic loading to accomplish clinical efficacy.

Solid lipid nanoparticles (SLNTM; Muller and Lucks, 1996) are biodegradable components that are toxicologically tolerable, such as nutritious glycerides. There is a broad array of lipids and emulsifiers with the Generally Recognized as Safe (GRAS) designation. Including both hydrophobic and hydrophilic medicines, SLNs can be utilized as nanocarriers. Drug-loaded SLN devices can be used intravenously, orally, or topically [18]. Rifabutin is an antituberculous medication with a significant lipophilicity with45-hour half-life. While nearly half of the amount taken orally is absorbed, absolute bioavailability is just approximately 20% due to first-pass metabolism. Lymphatic distribution is an approach to peroral medication delivery that avoids first-pass metabolism. Although intestinal lymph arteries drain straight into the thoracic duct, downstream into the venous blood, and so bypass the portal circulation, augmented lymphatic transport of medicines lessens hepatic first-pass metabolism and promotes bioavailability. The lymphatic system's key purpose seems to be to promote the absorption of long-chain fatty acids through the creation of chylomicrons. The fabrication of a rising number of lipophilic prodrugs and placement of the drug in a lipid transporter are two separate lipid-based strategies required to enhance lymphatic transport. The uptake of nano- and microparticles by M cells of Peyer's patches seems to be another way for lymphatic transport [19–23]. SLNs that have been encased with hydrophobic lipids are more likely to be acquired by lymphatic cells in the body [24].

Rifabutin-equipped SLNs were developed using the emulsification solvent diffusion (ESD) approach and a 3^2 factorial model to acquire the optimal point in formulation of monoglyceride (glyceryl monostearate) and polyvinyl alcohol (PVA) as a surfactant to boost rifabutin bioavailability. Several other methods have also been used to synthesize the SLNs, such as high-pressure homogenization, membrane emulsification method, and solvent injection method. But above all, the solvent diffusion approach was opted due to its high yield, ease of formulation, and practicality. Likewise, the factorial design was applied in the quest for nanoparticles optimization. Experimental runs are proven to provide reliable items in a short space of time while also conserving vital resources. As a result, this study used a 3^2 factorial design [25].

To assess the state of the lipid and the active drug, investigators applied transmission electron microscopy (TEM), powder X-ray diffractometry (PXRD), differential scanning calorimetry (DSC), and Fourier transform infrared (FTIR) spectroscopy; the formulations' *in vitro* stability and bioavailability were tested.

3.2 TECHNIQUES AND MATERIALS

3.2.1 Materials

Rifabutin was acquired as a gift sample from Lupin Ltd. (Aurangabad, India), and glyceryl monostearate (GMS), polyvinyl alcohol (PVA), and D-mannitol were acquired as gift specimens from Loba Chemie, Bombay. Some other chemicals used in the investigation were of analytical grade.

Organic solution:

Polymer + Drug in
partially water soluble
solvent

H₂O

Solvent
Elimination

Aqueous solution:

Stabilizer in water

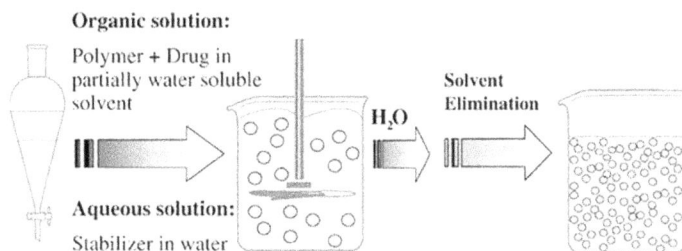

FIGURE 3.3 Emulsion solvent diffusion techniques.

3.2.2 METHODOLOGY

3.2.2.1 Fabrication of SLNs

The SLNs were synthesized using the 'emulsion solvent diffusion' (ESD) approach with distilled water as a basis [1,26]. In a water bath, drug and GMS (in varying proportions) were immersed in a combination of acetone/ethanol (15 mL each) and heated to 60°C–70°C. The entire drug:lipid ratio was kept at 1:3, 1:5, and 1:7 percent W/W, respectively. The recovered sample was prepared into a 25 mL watery PVA solution that was refrigerated at 4°C–8°C with vigorous stirring. The SLNs were synthesized in an instant and collected after centrifugation at 35,000 rpm for 30 minutes at 4°C–8°C. Pellets were created as part of this. Three times with distilled water, the pellets were thoroughly cleaned. Nanoparticles were reconstituted in 5 mL water that contains 5% w/v mannitol following final washing and then freeze-dried. During freeze-drying, it was deep frozen for 12 hours at −40°C and then freeze-dried for 48 hours at −55°C; refer Figure 3.3.

3.2.2.2 Complete Factorial Design

The dosage framework was developed using a 3^2 randomized full factorial design. Throughout this investigation, two factors were examined at three levels each, and experimental trials were conducted on all nine potential combinations. As independent variables, the amount of drug: lipid ratio (X1) and the amount of aqueous phase (X2) were determined. To investigate the responses, the entrapment efficiency, drug loading, and particle size were chosen as dependent factors in a 3^2 randomized complete factorial design. As illustrated in Tables 3.1 and 3.2, the design matrix and coded levels are specified in real values. Table 3.3 shows the nine formulations developed based on the factorial design [27].

3.2.2.2.1 Surface Plots of Response

Response surface plots were constructed by each response to examine the effect of both factors on response.

3.3 DEPICTION OF SOLID FORM OF LIPID NANOPARTICLES

3.3.1 DRUG LOAD AND ENTRAPMENT EFFICIENCY

Efficiency of encapsulation (EE) and drug loading (DL) were estimated by placing an accurate and suitable amount of solid lipid nanoparticles (25 mg) in a 25 ml volumetric flask and adding enough 0.01N HCl to bring the volume up to 25 ml. The suspension was forcefully shaken before being left at room temperature for 24 hours with periodic shaking. Investigation of the filtrate was done for drug content computation by a UV/Vis spectrophotometer set to an appropriate frequency (279 nm). Formulas (3.1) and (3.2) were used to assess entrapment efficiency for each batch in terms of percent entrapment.

TABLE 3.1
Design Matrix of Independent Variables

Formulation	Coded Levels	
	X_1 (mg)	X_2 (mL)
F1	−1	−1
F2	1	0
F3	−1	+1
F4	0	−1
F5	0	0
F6	0	+1
F7	+1	−1
F8	+1	0
F9	+1	+1

X_1, drug: lipid ratio; X_2, organic phase (PVA solution).

TABLE 3.2
Coded Levels in Actual Values

Coded Levels	Actual Values	
	X_1 (mg)	X_2 (mL)
1	1:3	150
0	1:5	300
+1	1:7	450
−1	1:5	150
0	1:5	300
+1	1:5	450
−1	1:7	150
0	1:7	300
+1	1:7	450

$$EE = (Wa - Ws/Wa) \times 100 \qquad (3.1)$$

$$DL = (Wa - Ws) / (Wa - Ws + WI) \times 100 \qquad (3.2)$$

Wa, Ws, and WI were the quantities of the drug delivered to the system, the assessed mass of the drug there in the supernatant, and the mass of added lipid to the system, respectively.

3.3.2 Investigations Using the Fourier Transform Infrared (FTIR) Spectrometer

The KBr disc method (Perkin Elmer, Spectrum BX-2, USA) was being used to acquire Fourier transform infrared spectra of rifabutin, glyceryl monostearate, and a physical mixing of rifabutin

TABLE 3.3

Formulation of SLNs Using Factorial Design

Formulations	Drug: Lipid Ratio	Weight of Drug (mg)	Weight of Lipid (mg)	Organic Phase (acetone: ethanol) mL	Aqueous Phase (1% PVA Solution) mL
F1	1:3 (1)	150	450	30	150
F2	1:3 (2)	150	450	30	300
F3	1:3 (3)	150	450	30	450
F4	1:5 (1)	150	750	30	150
F5	1:5 (2)	150	750	30	300
F6	1:5 (3)	150	750	30	450
F7	1:7 (1)	150	1,050	30	150
F8	1:7 (2)	150	1,050	30	300
F9	1:7 (3)	150	1,050	30	450

(1): Aqueous phase 150 mL, (2): aqueous phase 300 mL, (3): aqueous phase 450 mL

and glyceryl monostearate in the region of 4,000 to $400 \, cm^{-1}$. Drug-lipid interactions were investigated using the spectra.

3.3.3 DIFFERENTIAL SCANNING CALORIMETRY (DSC)

DSC analysis was executed on 1 mg samples using the Shimadzu-Thermal Analyzer DSC 60 (Kyoto, Japan). Under the nitrogen flow of 50 mL/min, specimens were exposed for heating at a pace of 10°C/min in an open aluminium pan throughout the temperature range of 30–350°C.

3.3.4 SURFACE MORPHOLOGY AND PARTICLE SIZE ANALYSIS

The characterization and geometry of SLNs were investigated using the following methods.

3.3.4.1 Malvern Zetasizer

Samples were diluted adequately with the formulation's aqueous phase for measurement. Diluted samples were adjusted to a pH of 6.9–7.2. To disintegrate the droplets into the nanoscale range, an adequate input was required. The size distribution of the nanoparticles population was assessed using the polydispersity index. The average of at least three measurements was used to calculate each value.

3.3.4.2 PHILIPS CM-200 London Transmission Electronic Microscope (TEM)

The operating voltage is maintained at 20–200kv (2.4 Å resolutions). Transmitted electrons (rather than visible light) are used to create TEM pictures, which may yield magnifying facts up to 1,000,000× with a resolution of greater than 10 Å. A fluorescent screen or photographic film can be used to interpret the images. Importantly, the elemental composition of the sample may be assessed with great spatial precision by analysing the X-ray due to the interaction of the accelerated electrons with the sample.

3.3.5 X-RAY DIFFRACTOMETRY (XRD)

Patterns of X-ray powder diffraction were captured using a Bruker AXS, DH Advance, Germany. Over a 10°–80° divergence angle (2θ) range, the scanning rate was 6° min^{-1}.

3.3.6 Study on the Release of Drugs

In vitro medication discharge in simulated GI secretions was investigated on the SLNs [28–32]. In 0.1N HCL (900mL),the drug release of rifabutin from freeze-dried SLN formulations was investigated using a USP Apparatus I (compiled according to the USP) [33] with rotating basket at 50rpm and temperature kept at 37°C±0.5°C. Each dissolving investigation comprised samples of formulations containing 150 mg of rifabutin. Samples of 5 mL were taken and strained over Whatman (No. 41) filter paper at periods of 0, 15 minutes, and 12 hours, with rifabutin concentration quantified by means of a UV/Vis spectrophotometer at a maximum of 279 nm.

3.3.7 Ex vivo Permeability and Simultaneous Dissolution Investigations

The diffusion of rifabutin-loaded solid lipid nanoparticles (SLNs) through the chicken ileum was studied utilizing everted tubes and the USP dissolution equipment I [34]. Figure 3.4 illustrates an in vitro concurrent dissolution–absorption technique that permits dissolution and permeability to be monitored simultaneously.

3.3.7.1 Ileum Preparation for a Permeation Investigation

A freshly slaughtered chicken's GIT was extracted. The tissues of the ileum were purified by flushing with distilled water and phosphate buffer pH 7.2 preparations followed by dissecting an ileum part of the small intestine about 15cm in length. The ileum was tethered on one end to a tube or thin rod and pushed enough that it telescoped on the knotted end. The ileum was pressed through the tube until it was absolutely flipped. The tube's knotted end was unlocked, and the ileum was

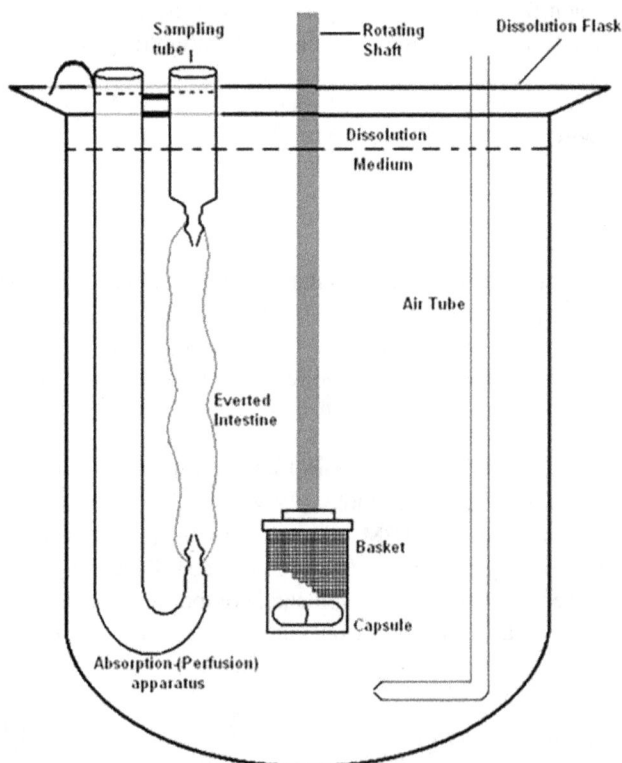

FIGURE 3.4 Simultaneous in vitro continuous dissolution–absorption systems to study permeability.

extracted. The connected to both ends of the everted sac assembly were attached to whichever extremity of the inverted chick ileum. Forty-five millilitres of phosphate buffer pH7.2 solution was placed into the everted sac assembly.

3.3.7.2　Methodology for Permeation Investigation

Permeation studies are time-consuming. The USP dissolving equipment I in a dissolution device was used to test permeation with *invitro* liberation of rifabutin-encapsulated solid form of lipid nanoparticles (SLNs) (Electrolab, TDT-08L). Baskets were utilized in this approach, and 1,000 mL of phosphate buffer pH 6.8 served as the dissolution medium, which was retained at 37°C ±0.5°C and whirled at 50 rpm. Each jar with aeration tubes was filled with the prepared everted sac assembly filled with appropriate quantity of pH7.2 phosphate buffer. The ileum's viability was maintained with appropriate quantity of pH7.2 phosphate buffer and considerable aeration. Specimens are taken from the jar and the everted sac assembly, correspondingly. Exact 10 and 3 mL specimens were pipette from each jar and cannulated sac assembly at 10-minute intervals for the first half hour and afterwards at 30-minute intervals up to 3 hours. Absorption specimens were taken 3 minutes upon their comparable dissolution samples to provide time for medication to flow from the dispersion vessel to the anterior end of intestinal membrane. By refilling the proportion comparable to the quantity evacuated with new medium, the sink condition was perpetuated in both the jars and the everted sac assembly. Aliquots of the jar's components were swirled for 5 minutes at 2,300 rpm. The supernatant was used to make effective dilutions. Consequently, the solutions from the everted sac assembly were implemented. A UV spectrophotometer was used to assess these solutions. Each experiment was carried out in triplicate. The standard calibration curves had been used to compute the inoculative release and drug absorbed.

3.3.8　Stability Testing

The nanoparticles were investigated for stability by storing them at 40°C for three months. The Malvern zetasizer was then employed to estimate particle size and drug content in the sample [35, 36].

3.3.8.1　Post-Stability Study Characterization by HPLC

The optimized formulations were checked for any degradation by analysing with a HPLC system. The chromatograms of formulations were compared with those obtained after forceful degradation of the pure drug by acid and alkali hydrolysis and oxidative degradation.

3.4　RESULTS AND DISCUSSION

The 3^2 factorial designs have been used to synthesize solid lipid nanoparticles. To evaluate the impact of variables from a smaller to a higher concentration of drug, three levels were carefully selected: GMS and 1% PVA solutions. To find out the level necessary for the optimization studies, preliminary trials were conducted employing varying concentrations of drug: GMS and 1 percent PVA solutions.

3.4.1　Response Surface Analysis (RSA)

3.4.1.1　Calculation of Coefficients

The parameters of the nonlinear regression for efficiency of entrapment, size of particle, and drug loading of the rifabutin-loaded SLNs investigated, as well as the values of r^2, are displayed in table. With B0 as the intercept, coefficients such as B1–B7 were computed. Because of their simplicity, the coefficients B1 through B7 consider multiple quadratic and interaction components in equation (3.3).

The common equation in terms of modified process order was

$$Y = \beta_0 + \beta_1 X_1 + \beta_2 X_2 + \beta_3 X_1 X_2 + \beta_4 X_1^2 + \beta_5 X_2^2 + \beta_6 X_1 X_2^2 + \beta_7 X_2 X_1^2 \quad (3.3)$$

where β_0 is the slope and $\beta_1, \ldots, \beta7$ are the coefficients of components, which pointer quadratic and interaction elements but are expressed as being in expression (1) ease of operation, and $X_1 X_2$ are the output parameters.

Applying linear regression model, the resulting polynomial expression for entrapment efficiency was calculated by means of coded components.

$$\%\textbf{Entrapment Efficiency}(\textbf{EE}) = +91.00 + 0.34 X_1 - 1.15 X_2 + 3.12 X_1 X_2 - 4.15 X_1^2 - 0.91 X_2^2 \quad (3.4)$$

Applying multiple linear regressions investigation, the resultant polynomial equation for particulate size in terms of coded components was

$$\textbf{Particle Size}(\textbf{PS}) = +197.33 + 260.50 X_1 - 2.67 X_2 - 2.00 X_1 X_2 + 429.50 X_1^2 - 1.00 X_2^2 \quad (3.5)$$

Considering multiple regression analysis, the final polynomial equation for medication loading in terms of coded inputs was

$$\%\textbf{Drug Loading}(\textbf{DL}) = +15.23 - 5.98 X_1 - 0.84 X_2 - 0.085 X_1 X_2 + 1.30 X_1^2 - 0.33 X_2^2 \quad (3.6)$$

In quadratic models, negatively or positively indications until a coefficient indicate yet if the element has a synergistic or antagonistic influence. The results of the data analysis of variance (ANOVA) applied to build the statistical model are shown in Table 3.4.

3.4.1.2   Plot Lines of Rational Responses on a Response Variable

Figures 3.5, 3.7, and 3.9 depict the three-dimensional surface response plots, and Figures 3.6, 3.8, and 3.10are of EE, PS, and DL. The outcomes of multiple regression models demonstrate that EE increases with increasing proportions of drug: GMS ratios, but there is no substantial shift

TABLE 3.4

Results of Analysis of Variance for Measured Response

Parameters	DF	SS	MS	F	Significance (P)
		For EE			
Regression	5	83.60	16.72	4.41	0.0001
Residual	3	11.38	3.79		
Total	8	94.98			
		For PS			
Regression	5	7.762	1.552	9439.82	0.0001
Residual	3	49.33	16.44		
Total	8	7.762			
		For DL			
Regression	5	222.35	44.47	167.13	0.0001
Residual	3	0.80	0.27		
Total	8	223.15			

DF, degree of freedom; SS, sum of squares; MS, mean sum of squares; P, Fischer's ratio.

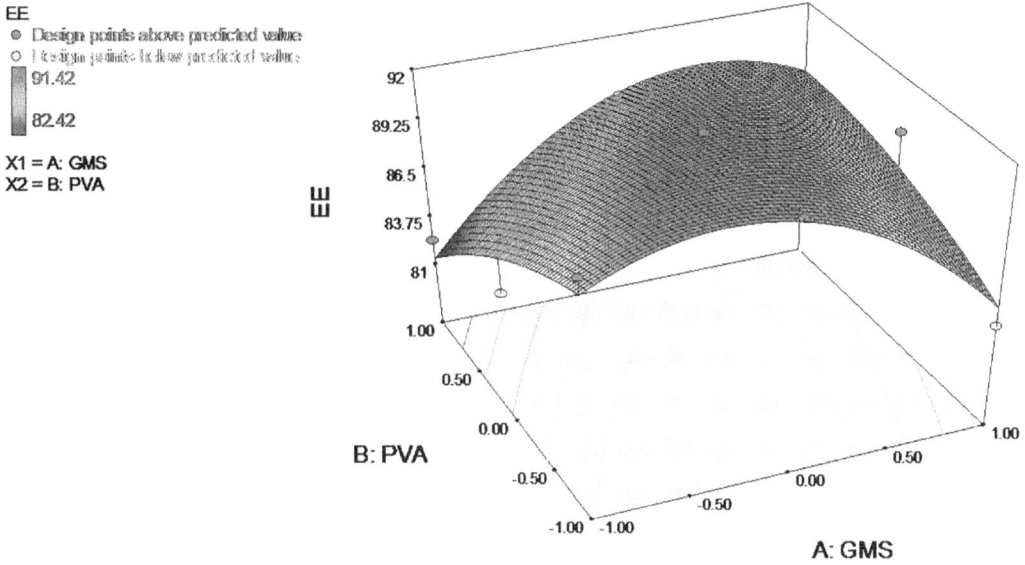

FIGURE 3.5 Response surface plots showing the effect of drug: GMS and OP: PVA ratios on the % EE from formulation.

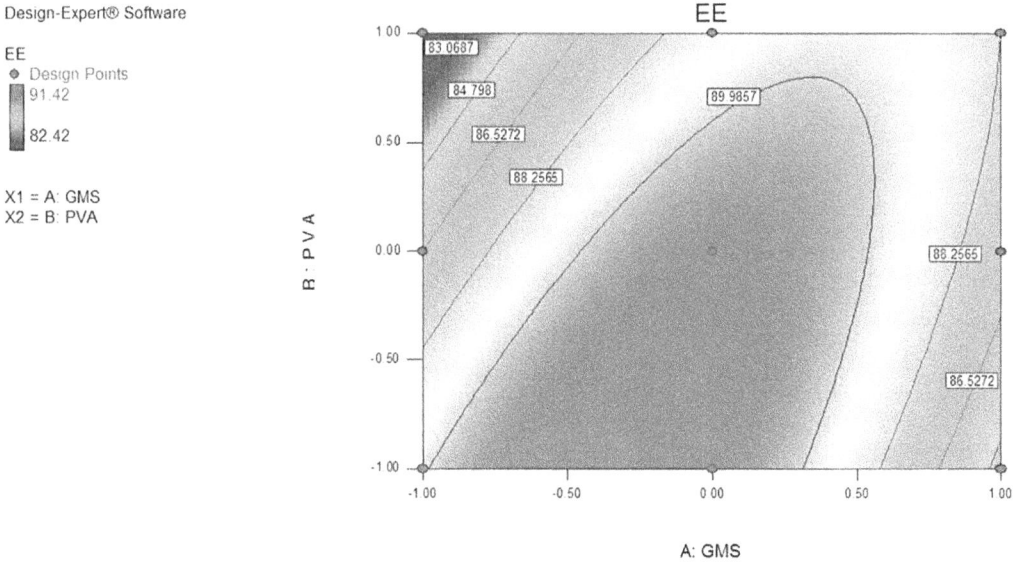

FIGURE 3.6 Contour plots showing the effect of drug: GMS and OP:PVA ratios on the % EE from formulation.

in EE with increasing concentrations of organic phase (OP): polyvinyl alcohol (PVA) ratios; refer Figures 3.5and 3.6. The outcomes of regression analysis for PS, shown in Figures 3.7and 3.8, reveal that PS increases with increasing drug: GMS ratios, even though there was no big variation in PS with expanding OP: PVA ratios in the context of DL; refer Figures 3.9 and 3.10 the outputs of linear regression indicated that DL lowers with rising OP:PVA ratios, along with DL reducing with growing drug:GMS ratios.

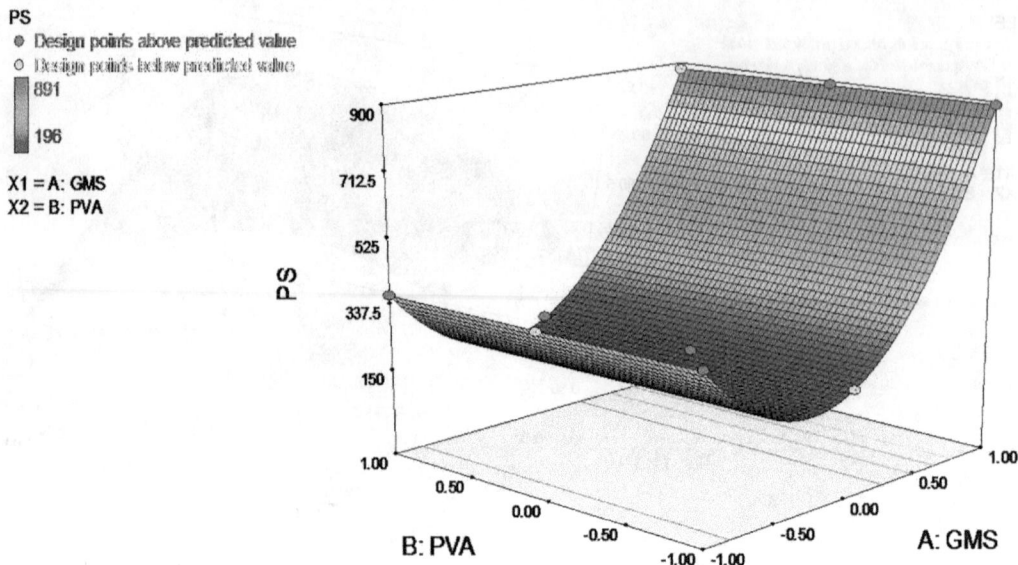

FIGURE 3.7   Response surface plots showing the effect of drug:GMS and OP:PVA ratios on the PS of SLN.

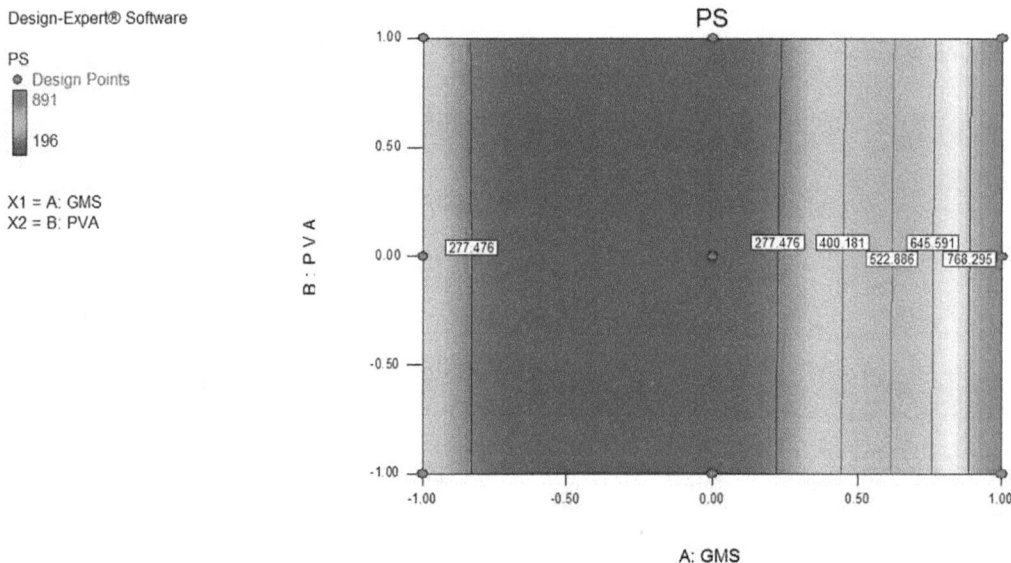

FIGURE 3.8   Contour plots showing the effect of drug: GMS and OP: PVA ratios on the PS of SLN.

3.4.2 Efficiency of Entrapment and Drug Loading

The % encapsulation efficiency of rifabutin nanocarriers varied as 82.51%–91.42%, and drug loading was from 9.92% to 23.31%, which are depicted in Table 3.5.The encapsulation efficiency increased with increasing concentration of lipid, but at higher lipid concentration batches, it was found to be decreased. The highest drug encapsulation was found with batch F5. Drug loading was found to decrease with increasing drug: lipid ratio as well as with increasing OP: PVA ratio. Nonetheless, as F5 accomplished the requisite size of the particles and encapsulation effectiveness, it must have been modified while its drug loading remained lower than F1.

Design-Expert® Software

DL
⊚ Design points above predicted value
○ Design points below predicted value
23.31

8.99

X1 = A: GMS
X2 = B: PVA

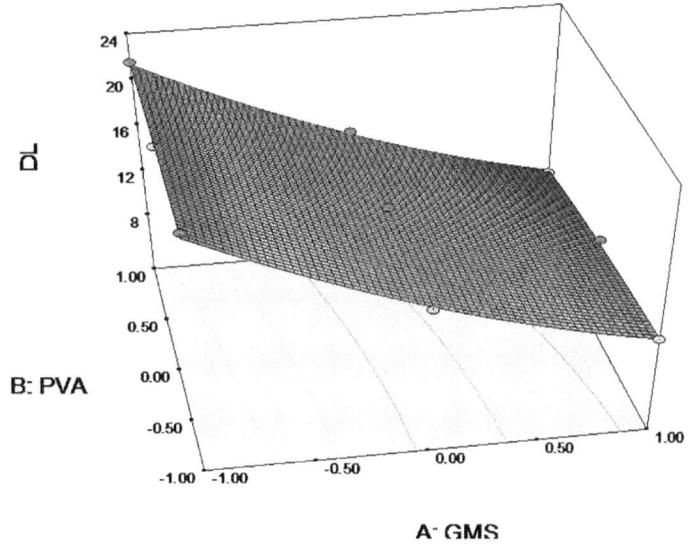

FIGURE 3.9   Response surface plots showing the effect of drug: GMS and OP: PVA ratios on the % DL from formulation.

Design-Expert® Software

DL
● Design Points
23.31

8.99

X1 = A: GMS
X2 = B: PVA

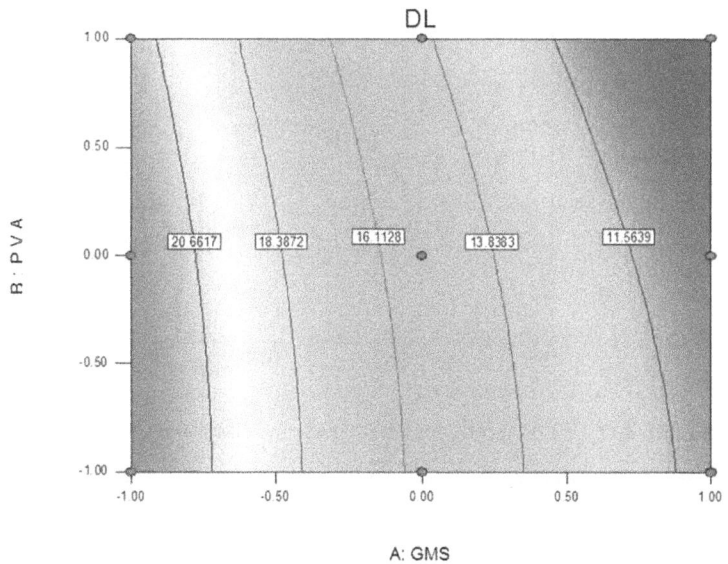

FIGURE 3.10   Contour plots showing the effect of drug: GMS and OP: PVA ratios on the % DL from formulation.

TABLE 3.5

Characterization of Nanoparticles

Run	Batch	Entrapment Efficiency (%)	Particle Size (nm)	Drug Loading (%)
F1	1:3(1)	82.51	365	23.31
F2	1:3(2)	84.55	363	22.96
F3	1:3(3)	90.76	369	21.02
F4	1:5(1)	91.08	199	15.44
F5	1:5(2)	91.42	198	15.28
F6	1:5(3)	91.06	199	14.46
F7	1:7(1)	89.10	879	11.08
F8	1:7(2)	88.37	890	10.76
F9	1:7(3)	86.22	891	9.92

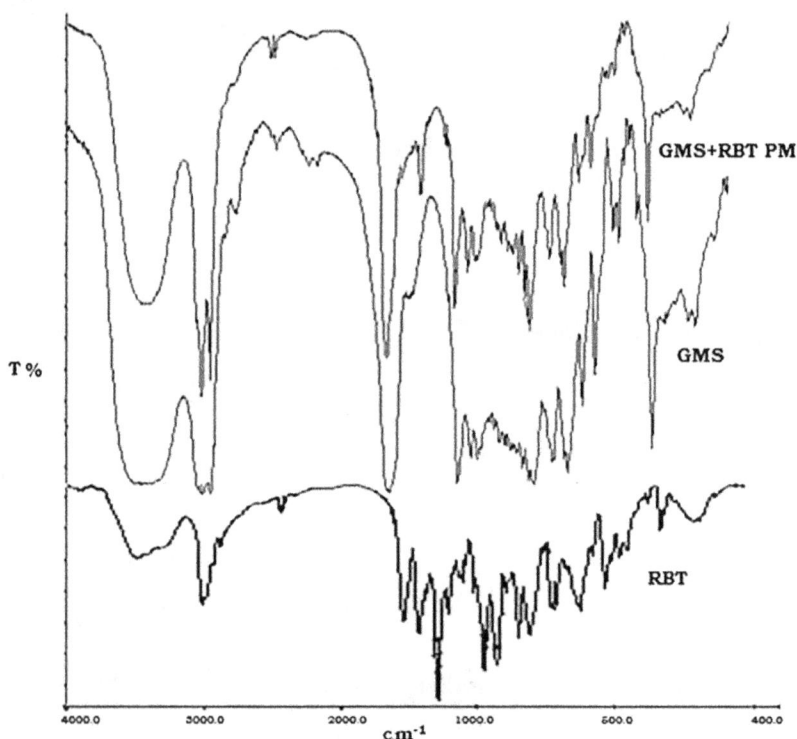

FIGURE 3.11   FTIR spectra for RBT, GMS, and GMS+RBT PM.

3.4.3   INVESTIGATIONS USING THE FOURIER TRANSFORM INFRARED (FTIR) SPECTROMETER

The IR band indicates that the first two peaks of pure drug were shifted, and the third peak disappeared in the IR spectra of F2, F5, and F7 formulation. Mostly in IR spectra of the preparation, a peak intensity of GMS was seen. Remaining peaks of drug either disappeared or shifted in the IR spectrum of formulations; refer Figures 3.11and 3.12. This indicates that there may be interaction between lipid and some of the functional groups of the drug or these groups may be covered by the lipidic coat. Hence, it can be suggested that the drug has been embedded or physically adsorbed onto the lipidic matrices of all formulations.

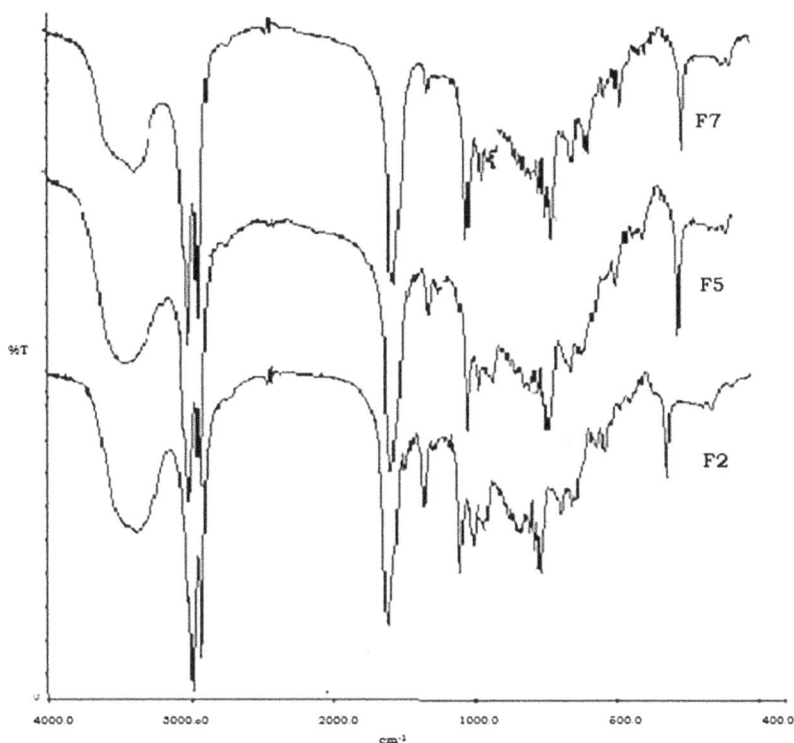

FIGURE 3.12 FTIR spectra for F2, F5, and F7.

3.4.4 DIFFERENTIAL SCANNING CALORIMETRY (DSC)

The DSC thermogram of purified rifabutin exhibits a glass transition at approximately 145°C equivalent to melting, accompanied by an exotherm equivalent to recrystallization of the molten that subsequently decomposes exothermically from about 240°C. The softening signal of pure drug was entirely absent in the thermogram of loaded nanoparticles preparations F2, F5, and F7,shown in Figure 3.13, demonstrating that crystalline drug was not detected in any of the three formulations of solid lipid nanoparticles specimen. Hence, the thermogram of all formulations (F2, F5, and F7) revealed that the drug has successfully been embedded into the lipid matrices.

3.4.5 PARTICLE SIZE ANALYSIS AND SURFACE MORPHOLOGY

3.4.5.1 Malvern Zetasizer

Particle size analysis was carried out using a HELOS particle size analyser (Model SYMPA –TEC). The blank formulation of SLN, i.e. placebo (without drug) was tested, and it was found that 90% of particles were found to be of 545 nm in size and 50% of particles were found to be of 700 nm. The SLNs of F2, F5, and F7 batches, respectively, were tested for particle size distribution. According to the particle size distribution, in F2 90% of particles were found to be of 363 nm in size and 50% of particles were found to be of 902 nm in size, in F5 90% of particles were found to be of 199 nm in size and 50% of particles were found to be of 879 nm in size and in F7 90% of particles were found to be of 890 nm in size and 50% of particles were found to be of 1.01 μm in size. This indicated that F7 had many particles that were in the 'μ' size range and not in the nano-range, while in F5, the maximum of particles were in the nano-range and was also the best formulation from the view of smallest size of the particle. Table 3.6 represents the typical size of particles of all SLN batches;

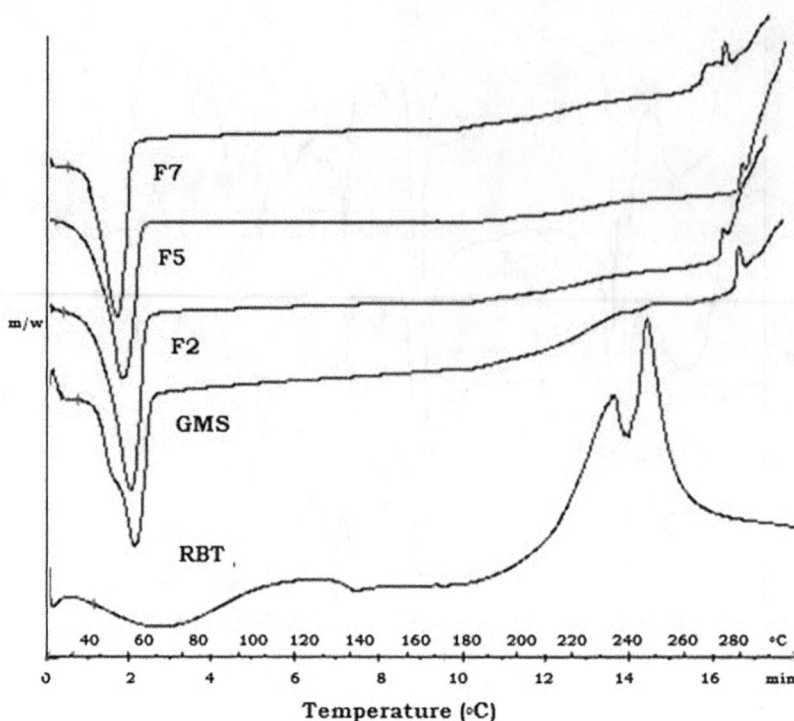

FIGURE 3.13   DSC of RBT, GMS, F2, F5, and F7.

TABLE 3.6
Different r^2 Values for EE, PS, and DL

Response Variables	R-Squared	Adj. R-Squared	Pred. R-Squared	Adeq. Precision
EE	0.9941	0.9870	0.9842	6.234
PS	0.9999	0.9998	0.9992	210.609
DL	0.9964	0.9905	0.9575	32.402

however, Figures 3.14–3.17 depict the comprehensive distribution of particle size of the blank, F2, F5, and F7 batches, correspondingly.

3.4.5.2   Transmission Electron Microscope (TEM)

Transmission electron microscopy (TEM) in Figure 3.18 shows that the particles (RBT: lipid – 1:5) had almost round and uniform shape with an average particle size of 196 nm (based on the scale). In addition to this, the particle size analysis (using Malvern zetasizer) revealed that 90% of the particles had a particle size around 199 nm, which perfectly matched the TEM photomicrographs.

3.4.6   PXRD (Powder X-ray Diffraction)

Figure 3.19 illustrates the PXRD characteristics of pure drug and nanostructure lipid carriers. The structure of pure drug exhibited five peaks, with a measured maximum intensity of 1,580; meanwhile, the trends of all slab system, F2, F5, and F7 had only three peaks, with maximum peak intensities of 7,100, 8,500, and 7,300, accordingly.

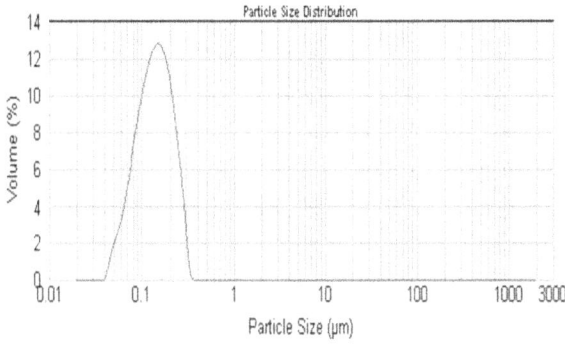

FIGURE 3.14 Particle size analysis of SLN (blank). (Particle size: d 0.9–545 nm; d 0.5–700 nm; polydispersity index – 1.25.)

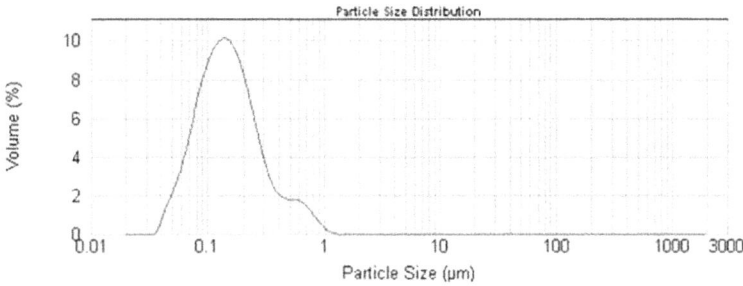

FIGURE 3.15 Particle size analysis of SLN (F2). (Particle size: d 0.9–363 nm; d 0.5–902 nm; polydispersity index – 1.99.)

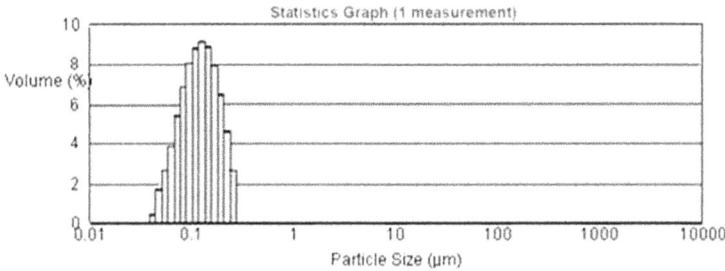

FIGURE 3.16 Particle size analysis of SLN (F5). (Particle size: d 0.9–199 nm; d 0.5–879 nm; polydispersity index – 2.1.)

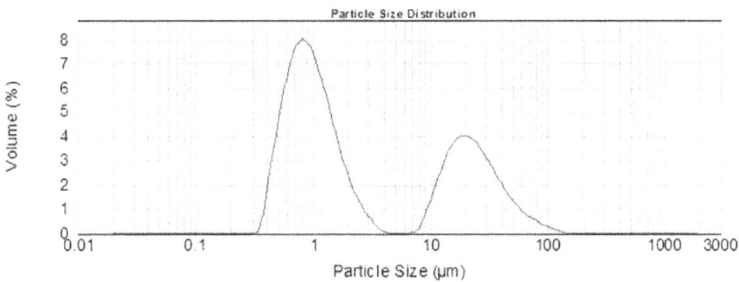

FIGURE 3.17 Particle size analysis of SLN (F7). (Particle size: d 0.9–890 nm; d 0.5–10.32 μm; polydispersity index – 1.32.)

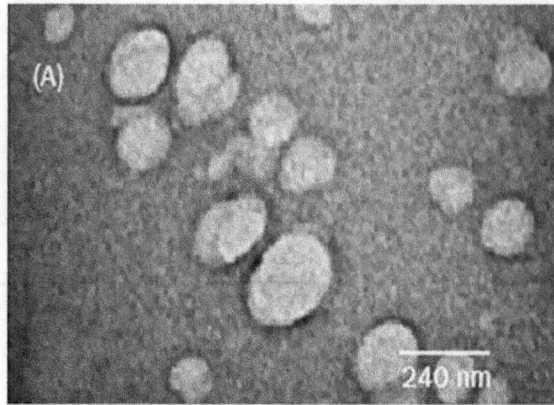

FIGURE 3.18   Micrographs of SLN by transmission electron microscope for F5.

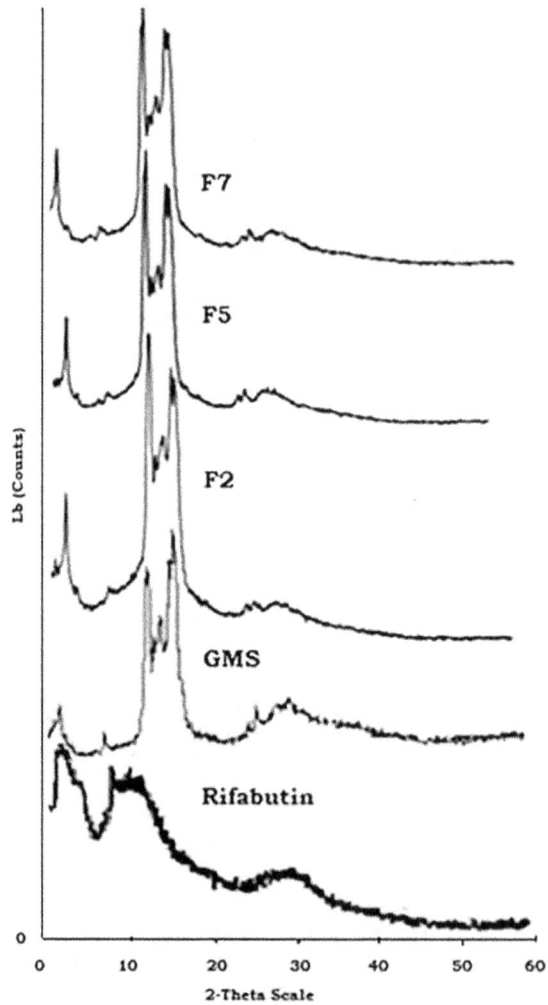

FIGURE 3.19   XRD of RBT, GMS, F2, F5, and F7.

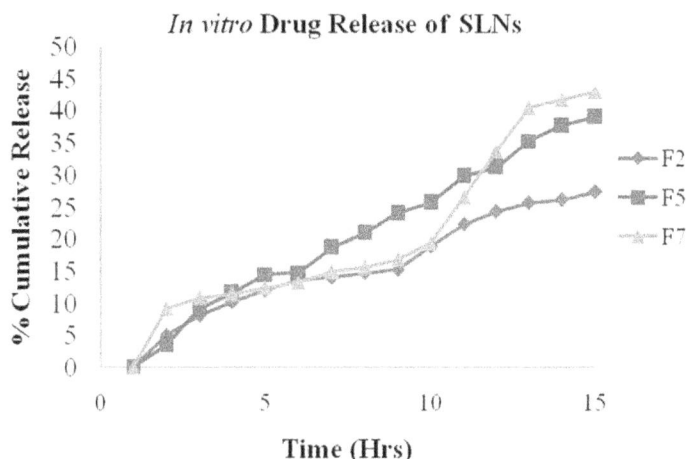

FIGURE 3.20 Release profiles of different formulations of SLNs.

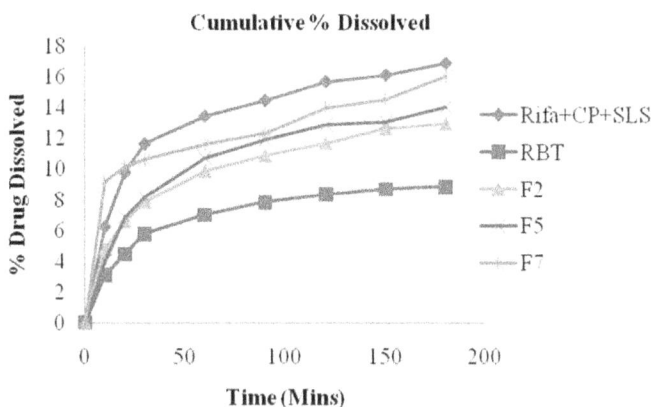

FIGURE 3.21 Cumulative % drug dissolved.

There were no characteristic peaks of rifabutin observed in diffraction patterns of all the three formulations, i.e. F2, F5, and F7. The peaks of RBT are completely masked by the peaks of the lipid. The peaks and the pattern exactly matched that of the lipid (GMS).

3.4.7 Investigation of *in vitro* Uptake and Kinetics

The uptake by *in vitro* profile of rifabutin from nanoparticles is shown in Figure 3.20.The percent drug release was not the same for all the formulations. Initially, the solid lipid nanoparticles showed immediate release of rifabutin, followed by slow and controlled release of drug.

3.4.8 *Ex vivo* Penetration and Simultaneous Dissolution Investigations

Figures 3.21 and 3.22 illustrate the dissolution and ingestion profiles of active pharmaceutical and solid lipid nanostructures for RBT, correspondingly. The cumulative % release profiles of the drug as well as of the SLNs of RBT from continuous dissolution–absorption system were found to be similar to the release profiles obtained in normal dissolution conditions.

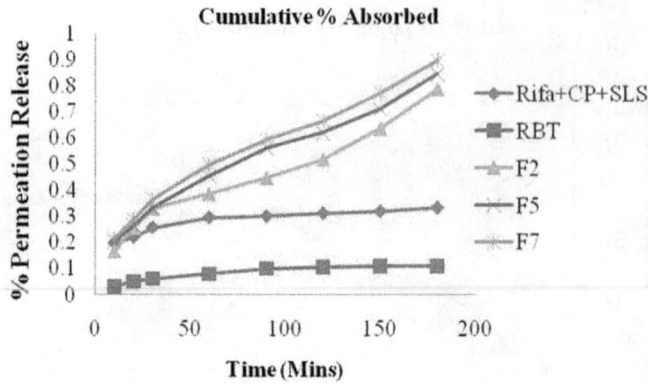

FIGURE 3.22   Release profile % drug absorbed.

To see the effect of formulation of SLNs on the permeation (absorption) of RBT through chick ileum, it was compared with pure RBT as well as against a product containing sodium lauryl sulphate (SLS) and crospovidone (CP) along with RBT, to prove that SLNs have superior release and permeation profile. All three formulations were filled in capsules for the study. It was observed from the permeation studies that the SLN-based product (F5) shows the highest permeation through ileum as compared to other products.

3.4.9   STABILITY STUDIES

During the period of stability studies, the colour, integrity and the drug content of the formulations remained unchanged. To check for any changes in particle size, studies were also conducted, which did not show any major differences in release patterns over the period of 3 months of stability studies. The chromatograms of both the RBT formulations showed no signs of degradation after stability studies of 3 months, indicating that the formulations were quite stable.

3.4.9.1   Drug Content of SLNs Containing Rifabutin

The chromatogram of suspension formulation at zero days was recorded as shown in Figure 3.23. The chromatogram of optimized formulation was recorded after three months as shown in Figure 3.24, and it was found that the chromatogram shows the same retention time and area under curve (AUC) as that of formulation at zero day as shown in Figure 3.25. At the end of 90 days, the drug content of rifabutin-loaded solid lipid nanoparticles was found to be 99.14 ± 1.20, which was not different from the initial drug content (Table 3.7). This indicated the stability of rifabutin in the presence of the excipients at specified temperature and humidity.

3.4.9.2   Particle Size Analysis during Stability Studies

The stability data of an optimized batch of rifabutin-loaded SLNs revealed that 90% (d 0.9) of particles were of size 199 nm and about 50% (d 0.5) were of 879 nm, while after 90 days, it was found that 90% (d 0.9) of particles had a particle size of 217 nm and about 50% (d 0.5) of particles had a particle size of 892 nm. Comparing it at room temperature for 90 days, particle size was analysed after 30, 60, and 90 days, which is depicted in Figures 3.26–3.29. The particle size did not undergo any significant change in the entire three-month period, which is depicted in Figure 3.30.

The particle size did not undergo any significant change in the entire three-month period, which is depicted in Figure 3.30.

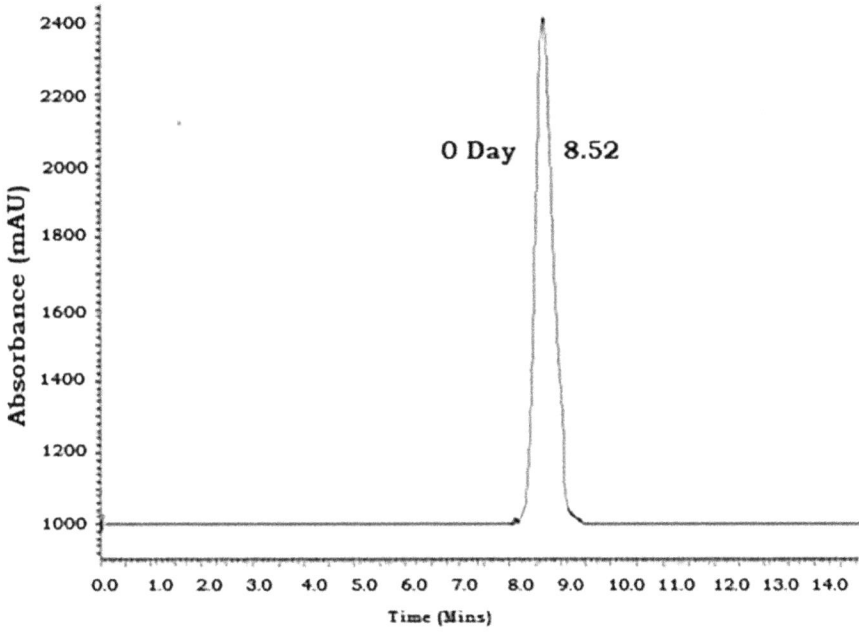

FIGURE 3.23   The chromatogram of the optimized formulation (F5) of rifabutin-loaded SLNs at 0 days.

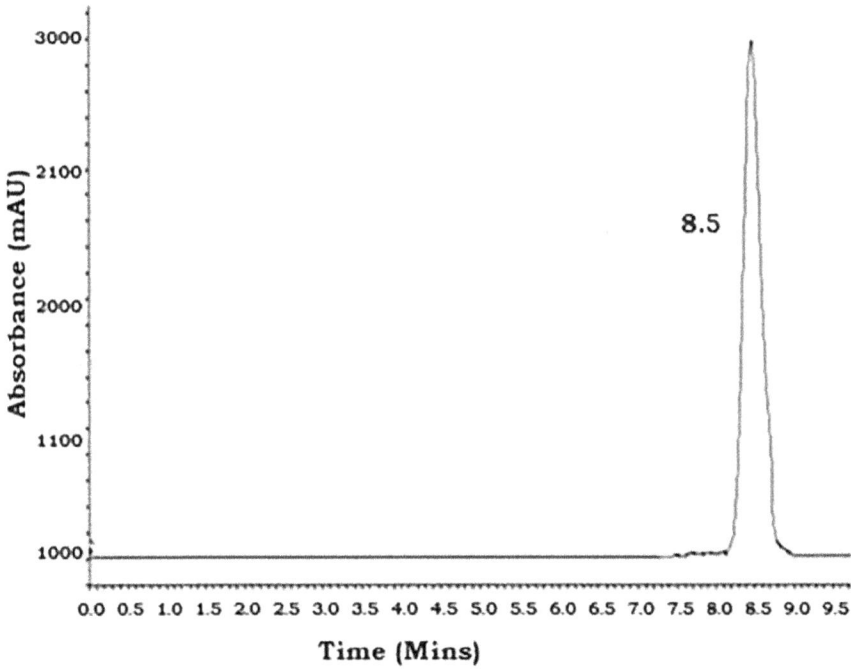

FIGURE 3.24   The chromatogram of the optimized formulation (F5) of rifabutin-loaded SLNs at 90 days.

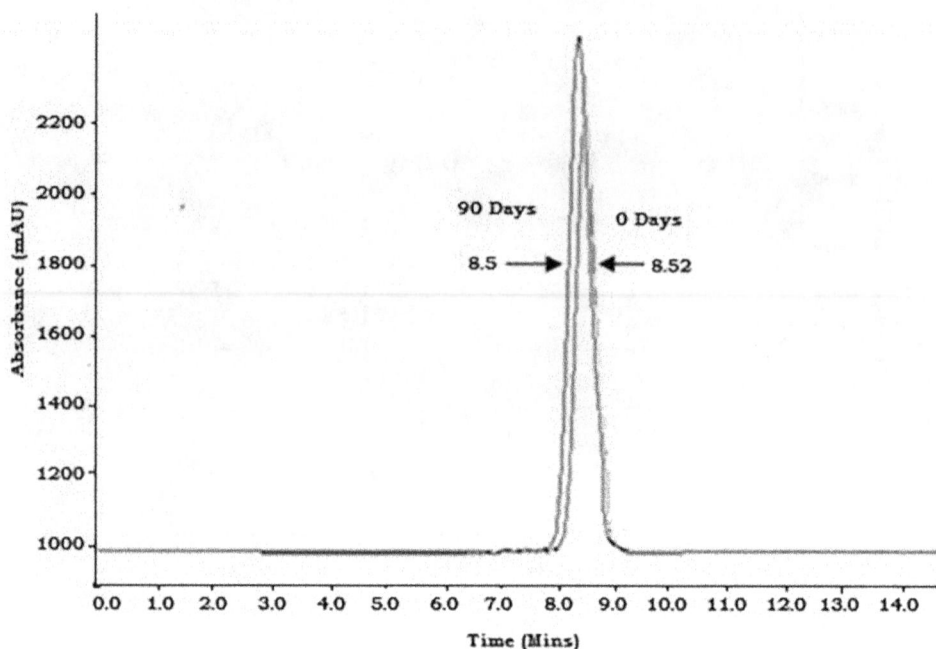

FIGURE 3.25 The chromatogram of the optimized formulation (F5) of rifabutin-loaded SLNs at 0 and 90 days.

TABLE 3.7
Drug Content of Formulation on Stability Studies

Time (days)	Drug Content at 40°C±2°C & 75% ±5% RH (Rifabutin Suspension)	Drug Content at 40°C±2°C & 75% ±5% RH (Formulation Containing RBT SLNs)
0	100±0.44	99.13±0.98
7	99.13±0.75	99.65±3.87
15	97.58±2.97	99.89±1.89
30	95.48±0.91	99.12±1.58
60	91.32±0.89	98.18±1.56
90	88.58±0.72	99.14±1.20

3.5 DISCUSSION

Rifabutin is considered as one of the highly effective anti-tubercular drugs used in the treatment of different stages of tuberculosis individually and as a combination therapy. The major problem associated with rifabutin is its poor bioavailability, which is a major hurdle in anti-tubercular activity. The literature survey revealed that there are very few approaches undertaken to improve its bioavailability. In aspects of biocompatibility and scale-up, SLNs are among the delivery systems that outperform alternative colloidal delivery technologies. In light of this, rifabutin-loaded SLNs were formulated by emulsification solvent diffusion (ESD) method, which is considered as one of the reliable and preferable methods for the formulation of SLNs. For the optimization of these formulations, the ESD method was accompanied with a 3^2 factorial design. The surfactants such

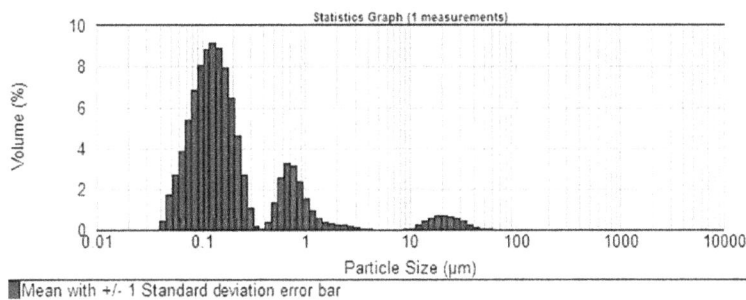

FIGURE 3.26 Particle size analysis of SLN (F5) initially. (Particle size: d 0.9–199 nm; d 0.5–879 nm; polydispersity index– 2.1.)

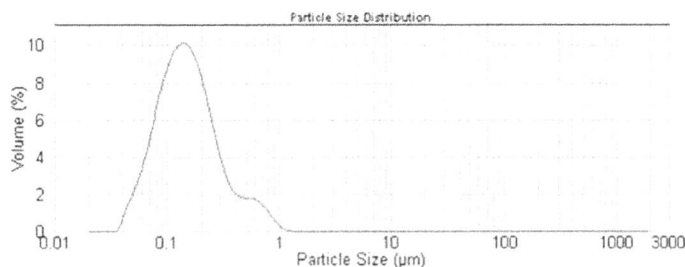

FIGURE 3.27 Particle size analysis of SLN (F5) after 30 days. (Particle size: d 0.9–207 nm; d 0.5–876 nm; polydispersity index – 1.56.)

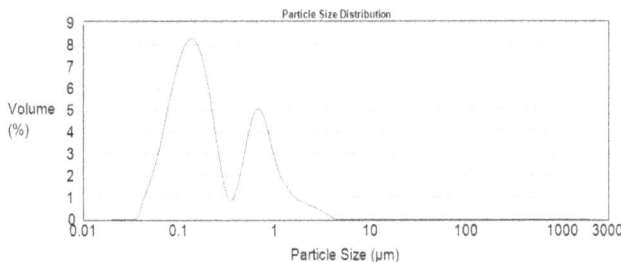

FIGURE 3.28 Particle size analysis of SLN (F5) after 60 days. (Particle size: d 0.9–215 nm; d 0.5–910 nm; polydispersity index – 1.72.)

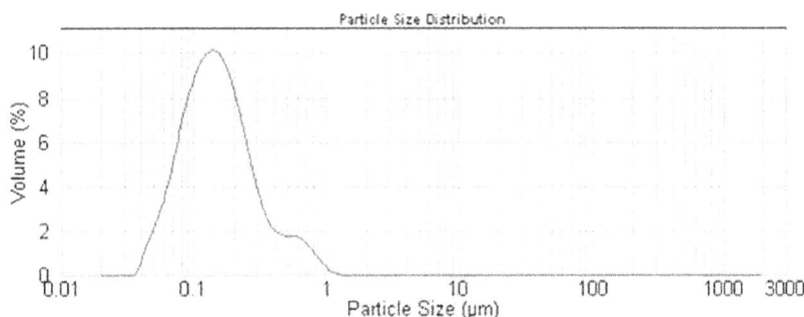

FIGURE 3.29 Particle size analysis of SLN (F5) after 90 days. (Particle size: d 0.9–217 nm; d 0.5–892 nm; polydispersity index – 1.99.)

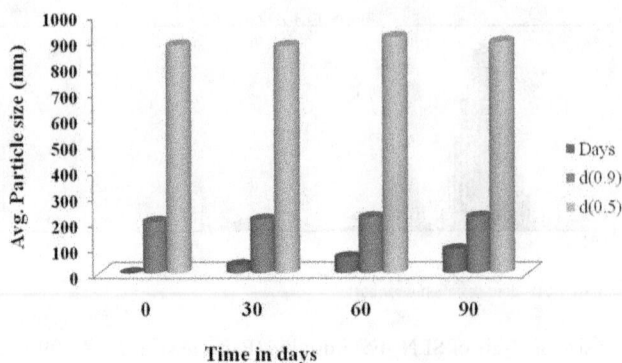

FIGURE 3.30 Particle size (nm) at various time intervals.

as monoglyceride (glyceryl monostearate) and polyvinyl alcohol (PVA) were used in this regard to assess the influence of variables from a lower concentration to a better concentration of drug; three components were opted: GMS and 1% PVA liquids. To find out the level needed for the optimization studies, exploratory tests were performed employing sufficient amounts of drug: GMS and 1 percent PVA solutions.

Response surface analysis was carried out, and coefficients of variables were calculated using regression analysis, which were used to determine the effect of different surfactants on drug load, size of particle, and drug entrapment of the rifabutin-loaded SLNs. The results of response surface analysis revealed that glyceryl monostearate highly affected the entrapment efficiency as well as particle size of SLNs, which were found to be increased with an increase in the concentration of GMS, whereas polyvinyl alcohol did not show such an effect. As far as drug loading was concerned, it was found to be decreased with increased concentrations of both the surfactants. These results indicated that GMS was the only surfactant found to alter different characteristics of SLNs.

Nine batches (F1–F9) of rifabutin-loaded SLNs were formulated using different drug: surfactant ratios, and they were characterized and compared for different parameters such as efficiency of entrapment, size of particles, and loading of drug. The amount of drug in the dispersed phase can be used to assess the loading capacity of the SLNs system. The quantification of the active agent is consequently performed. For the SLN formulations, a lower lipid concentration results in a lower entrapment efficiency and the cooling process after diffusion leads to the distribution of the active agent into the shell of the nanoparticles lipid core. In the present study, the encapsulation/entrapment efficiency was found to increase with increased lipid concentration to some extent, but was further found to be decreased at higher concentration of lipid, which indicated that optimum concentration of lipid was required for the formulation of SLNs. This efficiency was found to be highest with batch F5, which showed the optimum concentration.

An important point in the assortment of a specific drug delivery method is its loading competence and also its intended use and route of administration, but in the present investigation, all the batches of SLNs showed decreased drug loading capacities, which showed that the drug loading capacity decreased with increased lipid and surfactant concentrations.

Infrared (IR) findings indicated no interaction between the drug as well as the lipid, while differential scanning calorimetry (DSC) research revealed that perhaps the softening peak of pure drug was entirely absent in the thermogram of loaded nanoparticle formulations F2, F5, and F7, showing the absence of crystalline drug throughout all three possible preparations of solid lipid nanoparticles sample.

To determine the uniformity and desired surface characteristics, all the batches of SLNs were subjected to particle size analysis and surface morphology. The results indicated that the batch F7 had many particles that were in the 'μ' size range and not in the nano-range, while in F5, the maximum of particles were in the nano-range and was also the best and optimized formulation from the view of smallest particle size.

The SLNs were then subjected to transmission electron microscopy (TEM) to determine exact particle size, shape as well as morphology of SLNs. The results exhibited that the SLNs of batch F5 had almost round and uniform shape with average particle size of 196 nm (based on the scale). In addition to this, the particle size analysis (using Malvern zetasizer) revealed that 90% of the particles had a particle size around 199 nm, which perfectly matched the TEM photomicrographs.

Powder X-ray diffraction examination of active drug and SLNs was carried out to check the entrapment of rifabutin in the SLNs. The PXRD pattern did not show any characteristic peaks of rifabutin in the case of F2, F5, and F7. The peaks of RBT were completely masked by the peaks of the lipid. The peaks and the pattern exactly matched that of the lipid (GMS) which further proved that RBT was present either as molecule dispersion or as a free material or in a solid solution state in the lipid matrix. This further indicated that the rifabutin was successfully entrapped in the solid nucleus of the lipid.

One of the major criteria to be considered during the formulation of SLNs is the *in vitro* release and kinetics study, which must be carried out to determine their pharmacokinetic characteristics. The investigation indicated that the concentration of drug and lipid impacted the release rate of rifabutin from nanoparticles. Initially, the rifabutin was immediately released, indicating the liberation of drug as of the surface of solid lipid nanoparticles, which was followed by slow and controlled release of drug encapsulated in solid lipid nanoparticles. Conversely, as the lipid concentration was increased, the rate and amount of drug discharge were found to be decreased. This may be due to the augmentation in the mass of the lipid medium and also an increase in the diffusion corridor span that the drug fragments have to travel.

To corroborate and correlate with the in vitro permeation investigations, concomitant dissolution and ex vivo permeability analyses were performed. The permeation of rifabutin-loaded solid lipid nanoparticles (SLNs) through the chick ileum was studied utilizing an everted sac and the USP dissolution apparatus I. In vitro uptake studies employing secluded intestinal sacs are usual, but the chick ileum might be a good concept for intestinal absorption predicated on the assumption that drug membrane permeability is not species-dependent because the plasma membrane composition of intestinal epithelial cells is equivalent across species. As a result, permeability throughout the chicken intestine section should be similar. This framework incorporates several cell types such as the mucus layer, and it is generally quick and inexpensive, making it a better selection than the others. This modelling approach was appropriate for calculating kinetic parameters with great accuracy and repeatability.

In this investigation, the % drug discharge profiles of the pure drug over and above of the SLNs of RBT were found to be similar to the release profiles obtained in normal dissolution conditions. To see the effect of formulation of SLNs on the permeation (absorption) of RBT through chick ileum, it was compared with pure RBT as well as against a product containing sodium lauryl sulphate (SLS) and crospovidone (CP) along with RBT, to prove that SLNs have superior release and permeation profile. All three formulations were filled in capsules for the study. It was observed from the permeation studies that the SLN-based product (F5) shows the highest permeation through ileum as compared to other products. This indicated that if the permeability is improved, it may be reflected in increased absorption *in vivo*.

The stability studies of SLNs were finally carried out by the virtue of drug content chromatogram, suspension, particle size, etc., and compared with that of pure drug. The stability studies of SLNs showed that the colour, integrity, particle size, drug release, and drug content of the formulations remained unchanged during the 3-month period of stability studies at specified temperature and humidity, indicating that the formulations were highly stable, which fulfilled one of the criteria for SLNs. This may be attributed to the protection provided by the lipid coating

3.6 CONCLUSIONS

Consequently, the present investigation have used emulsion solvent diffusion (ESD) approach to substantially design and manufacture rifabutin-loaded solid lipid nanoparticles(SLNs),suggesting that SLNs might be used as a drug carrier for rifabutin. The permeability of rifabutin from its formulation was found to be significantly higher than the drug alone. Hence, from this observation it is concluded that the bioavailability of rifabutin could be significantly increased by preparing SLNs.

ACKNOWLEDGEMENTS

We are grateful to M/s. Lupin Ltd., Aurangabad for providing gift samples for this study. The authors are also thankful to AICTE, India, for providing financial assistance.

REFERENCES

1. Pandey R., Sharma S., Khuller G.K., Oral solid lipid nanoparticle-based antituberculer chemotherapy. *Tuberculosis*, 2005; 85:415–420.
2. Muller R.H., Karsten M., Sven G., Solid lipid nanoparticles (SLNs) for controlled drug delivery – a review of the state or the art. *Eur J Pharm Biopharm*, 2000; 50:161–177.
3. Vringer D., Topical preparation containing a suspension of solid lipid particles. U.S. Patent 1999; 5:904–932.
4. Westesen K., Particles with modified physico-chemical properties, their preparation and uses. U.S. Patent 2000; 1:97, 349.
5. Muller R.H, Ruge S.A., Solid lipid nanoparticles (SLNs) for controlled drug delivery. In: S. Benita (ed.) *Submicron Emulsions in Drug Targeting and Delivery*. Amsterdam: Harwood Academic Publishers, 1998, pp. 219–34.
6. Gulati M., Grover M., Singh S., Singh M., Lipophilic drug derivatives in liposomes. *Int J Pharm*, 1998; 165:129–168.
7. Manjunath K., Reddy J.S., Venkateswarlu V., Solid lipid nanoparticles as drug delivery systems. *Methods Find Exp Clin Pharmacol*, 2005; 27:127–144.
8. Leo E., Angela V.M., Cameroni R., Forni F., Doxorubicin-loaded gelatin nanoparticles stabilized by glutaraldehyde: Involvement of the drug in the cross-linking process. *Int J Pharm*, 1997, 155:75–82.
9. Gabor F., Bogner E., Weissenboeck A., Wirth M., The lectin-cell interaction and its implications to intestinal lectin-mediated drug delivery, *Adv Drug Deliv Rev*, 2004; 56:459–480.
10. Chen H., Langer R., Oral particulate delivery: Status and future trends, *Adv Drug Deliv Rev*, 1998; 34:339–350.
11. Yeh P., Ellens H., Smith P.L., Physiological considerations in the design of particulate dosage forms for oral vaccine delivery. *Adv Drug Deliv Rev*, 1998; 34:123–133.
12. LeFevre M.E., Warren J.B., Joel D.D., Particles and macrophages in murine Peyer's patches. *Exp Cell Biol*, 1985; 53:121–129.
13. Lehr C.M., Lectin-mediated drug delivery: The second generation of bioadhesives. *J Control Release*, 2000; 65:19–29.
14. Hussain N., Jaitley V., Florence A.T., Recent advances in the understanding of uptake of microparticulates across the gastro-intestinal lymphatics. *Adv Drug Deliv Rev*, 2001, 50:107–142.
15. Metha R.C., Head L.F., Hazrati A.M., Parr M., Rapp R.P., DeLuca P.P., Fat emulsion particle-size distribution in total nutrient admixtures. *Am J Hosp Pharm*, 1992; 49: 2749–2755.
16. Zara G.P., Cavalli R., Fundaro A., Bargoni A., Caputo O., Gasco M.R., Pharmacokinetics of doxorubicin incorporated in solid lipid nanospheres (SLN). *Pharm Res*, 1999; 44: 281–286.
17. Alyautdin R.N., Petrov V.E., Langer K., Berthold A., Kharkevich D.A., Kreuter J., Delivery of loperamide across thenblood–brain barrier with Polysorbate 80-coated poly-butylcyanoacrylate nanoparticles, *Pharm Res*, 1997; 14:325–328.

18. Freitas C., Muller R.H., Effect of light and temperature on zeta potential and physical stability in solid lipid nanoparticle (SLN™) dispersions. *Int J Pharm*, 1998; 168:221–229.

19. Delie F. Evaluation of nano- and microparticle uptake by the gastrointestinal tract. *Adv Drug Deliv Rev*, 1998; 34:221–233.

20. Bargoni A., Cavalli R., Caputo O., Fundaro A., Gasco M.R., Solid lipid nanoparticles in lymph and plasma after duodenal administration to rats. *Phar Res*, 1998; 15:745–750.

21. Eldridge J.H., Hammond C.J., Meulbrock J.A., Staas J.K., Controlled vaccine release in the gut-associated lymphoid tissue, I: Orally administered biodegradable microspheres target the Peyer's patches. *J Control Release*, 1990; 11:205–214.

22. Pappo J., Ermak T.M., Steger H.J., Monoclonal antibody directed targeting of fluorescent polystyrene microspheres to Peyer's patches M cells. *Immunology*, 1991; 73:277–280.

23. Jepson M.A., Simmons N.L., O'Hagan D.T., Hirst B.H., Comparison of poly (DL-lactide-co-glycolide) and polystyrene microspheres targeting to intestinal m cells. *J Drug Target*, 1993; 1:245–249.

24. Nishiok Y., Yoshino H., Lymphatic targeting with nanoparticulates system. *Adv Drug Deliv Rev*, 2001; 47:55–64.

25. Bolton S., *Factorial Designs, in 'Pharmaceutical Statistics: Practical and Clinical Application*. 3rd ed, Marcel Dekker, New York, 1997; pp. 326–351.

26. Hu F.Q., Yuan H., Zhang H.H., Fang M., Preparation of solid lipid nanoparticles with clobetasol proprionate by a novel solvent diffusion method in aqueous system and physiochemical characterization. *Int J Pharm*, 2002; 239:121–128.

27. Shivkumar H.N., Patel P.B., Desai B.G., Design and statistical optimization of glipizide loaded lipospheres using response surface methodology, *Acta Pharm*, 2007; 57: 269–285.

28. Lovrecich M., Nobile F., Rubessa F., Zingone G., Effect of ageing on the release of indomethacin from solid dispersions with Eudragits. *Int J Pharm*, 1996; 131:247–255.

29. Nazzala S., Khan M.A., Controlled release of a self-emulsifying formulation from a tablet dosage form: Stability assessment and optimization of some processing parameters. *Int J Pharm*, 2006; 315:110–121.

30. Singh B., Agarwal R., Design development and optimization of controlled release microcapsules of dilitiazem hydrochloride. *Indian J Pharm Sci*, 2002; 64:378–385.

31. Yang M.S., Cui F.D., You B.G., Fan Y.L., Wang L., Yue P., Yang H., Preparation of sustained-release nitrendipine microspheres with Eudragit RS and Aerosil using quasi-emulsion solvent diffusion method. *Int J Pharm*, 2003; 259:103–113.

32. Nandy B.C., Mazumder B., Formulation and characterizations of delayed release multi-particulates system of indomethacin: Optimization by response surface methodology. *Current Drug Delivery*, 2014; 10:72–86.

33. United States Pharmacopeia. USP 26/NF21, United States Pharmacopoeial Convention, Rockville, MD, 2003, p. 2528.

34. Kale V., Kasliwal R.H., Avari J.G., Attempt to design continuous dissolution-absorption system using everted intestine segment for in-vitro absorption studies of slow drug release formulations. *Dissolut Technol*, 2007; 14 (2):336.

35. Cavalli R., Gasco M.R., Chetoni P., Buralassi S., Saettone M.F., Solid lipid nanoparticles (SLN) as ocular delivery system for tobramycin. *Int J Pharm*, 2002; 238:241–245.

36. Bagade O., et al., Solid lipid nanoparticles: A critical appraisal. *Int J Pharm Sci Rev Res*, 2014; 29(1): 110–121.

4 Green Synthesis of Visible Light-Driven g-C$_3$N$_4$-Based Composites for Wastewater Treatment

Akash Balakrishnan and Mahendra Chinthala
National Institute of Technology Rourkela

CONTENTS

4.1 INTRODUCTION

The entire world is under considerable water stress due to the shortage in the availability of drinking water. This may arise due to increasing population, natural disasters, and overuse of resources, which cause severe damage to the ecosystem.[1–3] The effluent streams released from the industries and other human-made activities exploit natural resources. Different industries such as textile, fertilizers, pesticides, paper and pulp, batteries, food processing, and pharmaceuticals discharge tons of effluent water into the ecosystem.[4]

The wastewater constitutes a wide variety of pollutants such as dyes, pesticides, pharmaceuticals, phenols, inorganic contaminants such as heavy metals, nutrients, agricultural run-off, microorganisms, and pathogens.[5–10] The elimination of organic pollutants from the water streams is mandatory as they can persist in the natural ecosystem and become a potential threat to entire manhood.[4,11,12] So, suitable technology is highly required for removing these organic pollutants.

Different treatment methods such as adsorption, coagulation, membrane process, and photocatalysis are adopted to eliminate organic pollutants.[13–16] The conventional treatment methods suffer from several problems. For example, coagulation concentrates the pollutants by transferring them from one phase to another and cannot eliminate the pollutants completely. Sedimentation and chemical oxidation are also incapable of removing organic pollutants and require high quantities of chemicals and prolonged treatment time and also produce toxic by-products.[16] On the other hand, biological degradation needs more extended time and highly sophisticated instruments for maintenance, making the system costlier.[17] Compared to other processes, photocatalysis effectively mineralizes organic pollutants. Different catalysts such as TiO$_2$, WO$_3$, ZnO, ZnS, and Fe$_2$O$_3$ showed superior results in degrading organic contaminants.[18–22] The main advantages of photocatalysis are

DOI: 10.1201/9781003319153-5

environmentally friendly methods as they do not produce any secondary pollutants. However, the cost of preparation of hetero-photocatalysts is limiting its commercial applications. In recent times, the green synthesis of GCN has also been studied to reduce the cost of production.

This chapter describes the preparation of different graphitic carbon nitride-based photocatalysts using green methods. The preparation of green nanomaterials using the leaf extracts is also emphasized. The photocatalytic applications of GCN-based materials are explained in detail.

4.2 GRAPHITIC CARBON NITRIDE

The graphitic carbon nitride g-C_3N_4 (GCN) is a metal-free polymer n-type semiconductor.[23] Because of its higher activity and visible light absorption, GCN is evolved as a trending material in photocatalysis. The structure of GCN is composed of only two elements, namely carbon and nitrogen.[24] The higher chemical and thermal stability is ascribed to the existence of the strong covalent bond between carbon and nitrogen in the conjugated structure of the GCN.[25] The bandgap of 2.7 eV makes it favourable for visible light harvesting.[26] The GCN also possesses higher electrical conductivity due to the delocalized conjugated structure composed of stacked carbon nitride layers connected by means of tertiary amines.[25] Generally, the GCN is prepared using urea, melamine, thiourea, cyanamide, dicyandiamide, etc., due to higher carbon and nitrogen contents.[27]

The main advantages of the GCN are non-toxicity, greater stability, easy fabrication, abundance, higher activity in visible light, and stronger reduction ability.[28] These special properties permit the usage of GCN for different applications in the areas of catalysis for both energy and environmental remediation, biomedical applications, hydrogen fuel generation, chemical transformation, etc.[29-34] The higher recombination ratio of electron-hole pairs, incomplete utilization of the visible light absorption, the lower surface area of the GCN, lesser availability of the active sites for the interfacial reactions, greater degree of the condensation of monomers, and slow surface kinetics limit the usage of GCN as a photocatalyst.[24,35,36] Several studies emphasized different methods to overcome the problems associated with the GCN-based photocatalysts. For example, the developments of heterojunctions are beneficial in promoting charge transfer, mobility, and separation efficiency.[31,37] The doping using metals and non-metals improves visible light absorption to a greater extent.[26,38]

4.2.1 PHOTOCATALYTIC DEGRADATION MECHANISM

The heterogeneous photocatalysis occurs in five different steps as follows:[39,40]

    I. Diffusion of organic pollutants from the bulk liquid phase to the surface of GCN.
    II. Adsorption of organic pollutants onto the surface of GCN.
    III. Photochemical reaction due to the interaction between light and catalyst producing highly reactive species such as oxygen radicals.
    IV. Desorption of products from the surface of GCN.
    V. Removal of products from the bulk of the solution.

The slowest step determines the rate of the chemical reaction for any system. So, Steps I and V are faster than reaction Steps 2, 3, and 5. However, Steps 2 and 3 will not influence the overall reaction rate due to the highly powerful OH oxidizing agent involved in the reaction.[41]

The schematic representation of the photocatalytic degradation mechanism is illustrated in Figure 4.1. Upon the illumination of the visible light, the electrons are moved from the valence band to the conduction band with simultaneous generation of holes in the valence band. These photoinduced holes are easily trapped by hydroxyl ions and converted to hydroxyl radicals. The produced hydroxyl radicals are strong oxidants accountable for the mineralization of POPs.[42] So, the trapped oxygen may suppress the recombination of photoinduced holes. The limited oxygen supply may cause faster recombination of electrons and holes.[24,42,43]

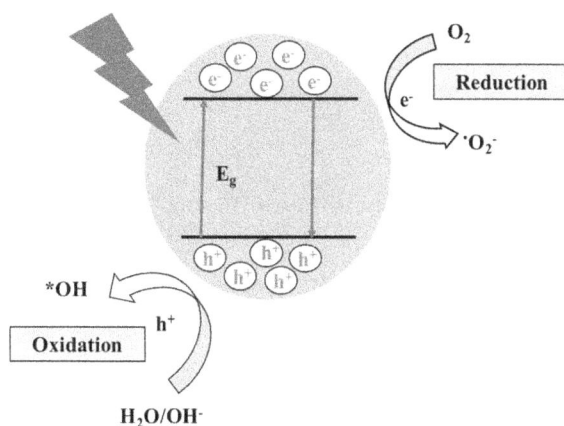

FIGURE 4.1 Schematic representation of the photocatalytic degradation mechanism.

4.3 GREEN SYNTHESIS OF g-C$_3$N$_4$-BASED MATERIALS

The three most essential criteria for preparing any nanoparticle are the selection of an eco-friendly solvent, an efficient reducing agent, and a non-toxic material for stabilization. Generally, physical, chemical, and biological routes are used to synthesize nanoparticles.[44] The chemical methods typically consume hazardous and toxic chemicals that severely threaten the environment and are also expensive.[45] On the other hand, biological routes are considered safe, biocompatible, and eco-friendly routes for the preparation of the nanoparticles using plants or microorganisms.[46] The main advantages of a biological method over chemical methods are the following:[47]

 i. Cost-effectiveness and easy scale-up for larger production
 ii. Mild operating conditions
 iii. Eco-friendly method.

One of the prominent methods to develop a sustainable photocatalyst is through green synthesis. Plant extracts such as leaves, stems, roots, flowers, and microorganisms such as yeast, algae, and fungi are used as the sources for green synthesis (see Figure 4.2).[48] The utilization of different plant extracts has garnered research importance due to their abundance and low cost. Additionally, the presence of various chemicals such as amino acids, alkaloids, terpenoids, and polyphenolic compounds exhibits a critical role in producing nanoparticles. The main merits of using the plant extracts are the following[49]:

• Choice of a non-toxic capping agent
• Choice of an environmentally friendly reducing agent
• Selection of a suitable solvent.

The generalized preparation of green synthesis of photocatalysts is shown in Figure 4.3. Sundaram et al. reported the preparation of Cu$_2$O nanospheres-decorated GCN using the extracts of *Citrus limon* leaves. The presence of leaf extract is attributed to the reducing and the stabilization effect during biosynthesis. The TEM images reported that the Cu$_2$O nanoparticles were dispersed on the GCN structure with aggregated Cu$_2$O/GCN. The average particle size of the Cu$_2$O was in the range of 2–10 nm that dispersed into the framework of GCN.[50] Khan et al. reported the *Eriobotrya japonica*-assisted GCN preparation. Leaves were pulverized and extracted with the help of absolute ethanol. Later, it was filtered and refined using a rotary vacuum evaporator, and finally, the extracts

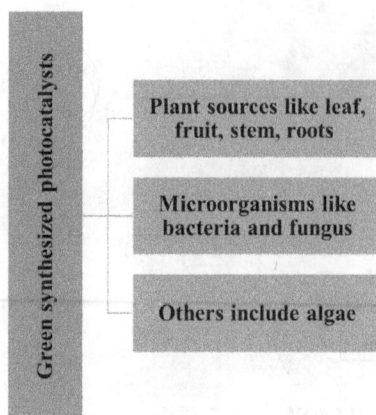

FIGURE.4.2 Different sources used for green synthesis.

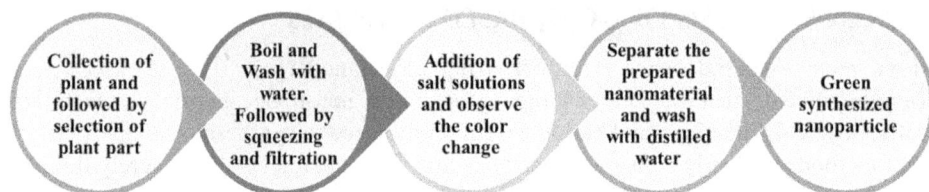

FIGURE 4.3 Generalized preparation of green synthesized photocatalysts.

were stored. The GCN nanosheets were mixed with leaf extracts under ultrasound, followed by thermal treatment, aging, washing, and drying.[51]

Sarangapany and Kaustubha Mohanty reported the preparation of Ag@GCN composites using the bark extract of *Sapindus emarginatus* and anchored over the surface of the CN. The *Sapindus emarginatus* stem bark was dried and grounded into a fine powder using a grinder; followed by the preparation of aqueous bark extract, the powder was added and refluxed at 65°C. Later, the filtered prepared GCN powder and the $Ag(NO_3)$ were mixed, and the bark extract was added to form the nanoparticles. The bark extract acted as a reducing agent. Finally, the resultant solution is dried and grounded.[52]

Liu et al. (2016) synthesized GCN/Ag nanocomposites with grape seed extract via one-pot green synthesis. The GCN supernatant was mixed with the silver nitrate solution to which 2 mL of the grape extract was added and heated at 95°C for 20 minutes for the nanocomposites to form. The grape seed behaved as a reducing agent and the stabilizing agent.[53] Different from these, the sol-gel method was adopted to prepare $DyFeO_3/CuO/GCN$ using the *Capsicum annuum* extract. The *Capsicum annuum* acted as a chelating agent. These chelating agents are composed of biomolecules that can control the nucleus and growth of the formed crystals by developing a space barrier and controlling the accumulation of the nano-products.[54]

The tertiary $CeO_2/GCN/Ag$ photocatalyst was developed using a green method by employing *Piper betle* as the plant extract. Similar to the above papers, after collecting the extract, it was mixed with GCN, silver nitrate, and cerium nitrate and kept for combustion at 500°C. SEM images reported that the GCN is blended with Ce nanoparticles, and Ag is assumed to mix them. The HR-TEM reported a lattice spacing of 0.33 nm to 002 plane of GCN, 0.31 nm to [011] plane of CeO_2, and 0.27 nm to 111 plane of Ag.[55]

The one-pot GCN/Bi_2O_3 was fabricated through an efficient biogenic method by utilizing the surfactant and *Eichhornia crassipes* leaf extract. The scanning electron microscopy affirms the heterojunction formation between the GCN and bismuth oxide, which is ascribed to the scaffolding effect of the surfactant and GCN. The transmission electron microscopy reported a spacing of 0.33 nm between the planes of Bi$_2$O$_3$ and GCN belonging to 120 and 200 planes, respectively. It is also noticed that the heterojunction between the bismuth oxide and GCN is the driving force for the charge transfer mechanism.[56]

The bio-surfactant-mediated preparation of Au/GCN using *Averrhoa carambola* leaf was proposed by Devi et al. (2022). The gold nanoparticles were synthesized using the leaf extract of *Averrhoa carambola*, which was mixed with HAuCl$_4$, followed by a proper colour change to Amaranth. Later, it was mixed with GCN suspension and yielded a light violet colour. Spherical gold nanoparticles with an average size of 25 nm were finally obtained. The average particle size was found to be 15 nm. The decrement in the particle size is due to the growth inhibition during the synthesis in the presence of biomolecules of the leaf extract and the GCN nanosheets.[57] So, most of the studies reported the utilization of green lead as a plant source for the preparation of GCN-based composites.

4.4 APPLICATION OF GREEN SYNTHESIZED g-C$_3$N$_4$-BASED COMPOSITES IN WASTEWATER TREATMENT

The green synthesized photocatalysts are versatile materials because of their higher surface area, altered bandgap, and hindered recombination ratio. These properties make photocatalysts unique in the field of environmental remediation. The green nanomaterials are employed as a photocatalyst or adsorbent to eliminate dyes, pesticides, pharmaceutical compounds, and heavy metals from the wastewater.[58–60] However, very few studies are conducted on the green synthesized GCN for wastewater treatment.

Adithya et al. reported the utilization of green synthesized Gd$_2$O$_3$/GCN to remove organic dyes such as Amaranth and Congo red. The photocatalytic studies reported that 85% of the Amaranth and 95% of the Congo red dyes were removed within 1 hour of visible light irradiation. The scavenger studies reported that the holes and superoxide play a phenomenal role in pollutant removal than the hydroxyl radicals. The reusability studies showed the stability of the photocatalyst up to five cycles. The phenomenal catalytic activity is due to the inhibited electron-hole recombination and better charge carrier separation.[61] The GCN/Bi$_2$O$_3$ heterojunction was employed to remove malachite green under visible light. The experimental studies reported 98.7% removal of malachite green using visible light irradiation. The enhanced removal is ascribed to the effective interaction between the GCN and bismuth oxide. Total organic carbon analysis revealed 78% reduction in malachite green (20 ppm) using visible light irradiation. The catalyst demonstrated the reusability of GCN/Bi$_2$O$_3$ for four cycles with a slight decline in the removal efficiency. The higher removal efficiency of the photocatalyst is due to the easy transfer and higher separation of electron-hole pairs at the interface of the photocatalyst.[56]

The GCN/Ag$_3$PO$_4$ prepared using a green synthesis showed remarkable efficiency against the removal of methylene blue. About 96% of methylene blue was degraded within 10 minutes of visible light irradiation. The catalyst also exhibited reusability of up to five cycles by maintaining the catalyst activity.[62] The green GCN structures exhibited a 100% removal of rhodamine B under the illumination of visible light within 15 minutes with extended reuse of up to five times.[63] Khan et al. reported that the *Eriobotrya japonica*-assisted green synthesized GCN exhibited 2.6 times higher removal efficiency than bare GCN towards the destruction of rhodamine B. The better removal efficiency is due to the prolonged lifetime of charge carriers due to the SnO$_2$ coupling.[51] The green wet ball milling-assisted preparation of GCN nanosheets exhibited 95% rhodamine B elimination within 120 minutes. The greater removal efficiency may be due to the creation of more active sites due to ball milling peeling.[64] Green synthesized Au@GCN exhibited 86% methyl orange removal within 180 minutes using visible light irradiation.[65]

Savunthari et al. described the green synthesized lignin nanorods/GCN nanocomposites to eliminate triclosan from the aqueous water stream. The experimental studies reported 99.5% removal of triclosan within 90 minutes. However, after five cycles, the catalyst removal efficiency was reduced from 99.2% to 89.8%. The higher efficiency is related to the synergistic effect between the lignin and GCN in the prepared photocatalyst. The decrease in bandgap from 2.86 to 2.83 eV is attributed to the redshift, which results from the change in morphology, crystallite size, and microstructures present on the surface.[66]

Dou et al. emphasized that the green synthesized sulphur-doped GCN eliminates oxytetracycline under visible light illumination. Photocatalytic studies reported the complete elimination of the oxytetracycline within 40 minutes of visible irradiation at an initial pH of 7 and catalyst dosage of 1 g/L. The degradation followed a pseudo-first-order reaction.[67] The green synthesized Co-MOF-based/GCN composite-activated peroxymonosulfate significantly removed antidepressant venlafaxine with 100% removal efficiency within 120 minutes with mineralization of 51%. The studies proved the peroxymonosulfate activation was helpful in yielding a higher removal efficiency. The hydroxyl radicals and sulphate radicals were crucial in eliminating the pollutant.[68]

The Au/GCN plasmonic hybrid nanocomposite developed using a bio-surfactant-assisted room-temperature route was utilized to eliminate mono-nitrophenols. The photocatalytic activity showed the degradation of 99% of nitrophenol, 97% of nitrophenol, and 98% of 2-nitrophenol. The adsorption of nitrophenol is reduced in the basic pH due to the electrostatic repulsion of nitrophenol and the photocatalyst's cationic surface, which ends up in the inhibition of reduction rate. The catalyst demonstrated a very high stability of ten cycles. However, a slight decline in rate constant is observed and it is attributed to the congregation of the nitrophenol on the surface of the photocatalyst during recovery. The leaching of the catalyst may also lead to the same. Overall, the plasmonic catalyst's potentially high efficiency is indicated by the higher charge transfer and strong interaction between the gold nanoparticles and GCN due to surface plasmon resonance.[57] The GCN/CuONP/LDH composites reported a very low phenol removal (22%) within 4 hours using a 400 W halogen lamp.[69]

The Bi$_2$S/GCN nanosheets were effective against different organic pollutants present in the wastewater. Experimental studies reported 97% Cr(VI) reduction within 180 minutes and 94% reactive yellow dye removal in 120 minutes. The enhanced removal efficiency is ascribed to the lowered bandgap of the photocatalyst along with the higher charge transfer and enhanced separation of electron-hole pairs in the structure of the nanosheets.[9]

4.5   FOCUS ON WASTE AND BIO-DERIVED MATERIALS

One of the most important approaches for synthesizing green nanomaterials includes the utilization of biomass or another waste during the preparation route. Unlike the previous route, this method is described on the different sources of materials for the synthesis of photocatalysts. It improves sustainability as it exploits abundantly available biomass or waste to prepare the catalyst. The best biomass-based photocatalyst preparation uses different carbonaceous materials such as activated carbon, carbon dots, graphene, and fullerene. Incorporating carbon into the GCN-based composites will enhance both photostability and photoactivity. The formation of a synergistic effect also benefitted in the increment in the surface area, active sites, narrowed bandgap, and better surface charge utilization.

Activated carbon is a well-known adsorbent and catalyst support with a large surface area and porosity.[70–72] Its low cost, ease of availability, and sustainability have attracted researchers over these years in wastewater treatment.[73,74] Chen et al. described the preparation of GCN-assisted activated carbon using melamine and activated carbon as the starting material through pyrolysis. The obtained photocatalyst was able to degrade the phenol completely within 160 minutes. The higher surface area of the composite improved the adsorption of organic pollutants on the surface of activated carbon. Eventually, it enhanced mass transfer and pollutants accumulation on the catalyst surface for further reactions.[75]

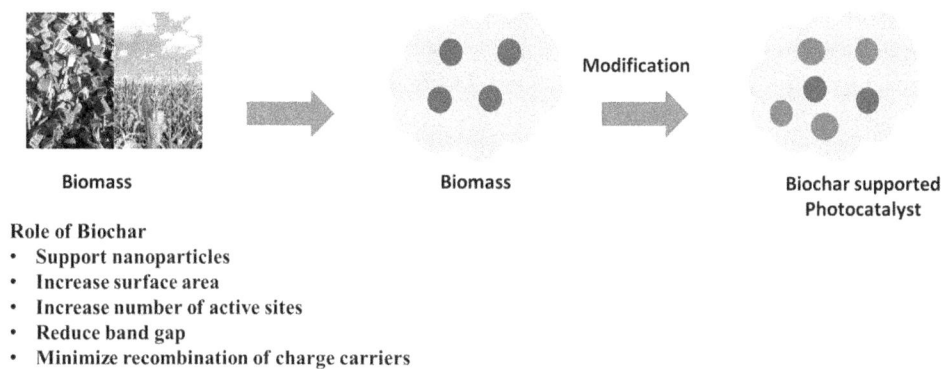

Role of Biochar
- **Support nanoparticles**
- **Increase surface area**
- **Increase number of active sites**
- **Reduce band gap**
- **Minimize recombination of charge carriers**

FIGURE 4.4 Advantages of biochar as a support for photocatalysts.

Several studies emphasized the biochar-supported GCN to redeem organic pollutants present in the wastewater.[76] Figure 4.4 emphasizes the advantages of biochar as a support for photocatalysts. Porous biochar/GCN was utilized for the remediation of formaldehyde from the waste water. The biochar/GCN was prepared using one-step co-thermal treatment using wood-derived bleached kraft pulp and melamine as the precursor. The catalyst displayed a higher visible light absorption, ascribed to the incorporation of biochar into GCN. The studies reported an 84.6% removal of formaldehyde using a 250 W high-pressure mercury lamp.[77] The incorporation of biochar/GCN by utilizing glycine and arginine as the carbon source could remove orange G and Cr(VI).[78]

Pi et al. (2015) developed a GCN-modified biochar to remove MB. The preparation step includes the carbonization of chestnut leaf biomass followed by the thermal polycondensation of melamine at 520°C. Photocatalytic studies reported 91% MB removal due to the higher biochar content.[79] Lin et al. emphasized the facile preparation of modified biochar-based supramolecular self-assembled GCN to remove phenanthrene. The modified biochar was prepared using KOH and sunflower straw powder as the carbon source, followed by supramolecular self-assembly using melamine and cyanuric acid. Photocatalytic studies reported 76.27% removal of phenanthrene under visible light illumination and exhibited reusability of up to four cycles.[80] The improved removal efficiency is due to the higher surface area and faster charge transfer rate. The π-π interacted biochar/GCN exhibited 90% tetracycline removal within 1 hour. The *Enteromorpha* biochar could supply electron-withdrawing groups to adjust the electronic structure of GCN, and the π-π interaction minimizes the recombination of electron-hole pairs.[81]

The core-shell P-laden biochar/ZnO/CN was prepared using the rice straw and melamine as the chief precursor. Studies reported 85.3% of atrazine removal within 260 minutes. The coating of GCN on the surface of P-laden biochar-ZnO improved the visible light activity and increased the separation efficiency. The biochar also acted as a bridge for promoting electron-hole pairs.[82] The biochar/GCN was successfully prepared using the waste *Camellia oleifera* shells and melamine. The studies showcased that the hexavalent chromium ion present in the water is completely removed by the adsorption-combined photocatalysis.[83] Sun et al. (2020) reported the preparation of GCN/ferrite/biochar hybrid photocatalyst using the biomass bamboo fibre as the carbon source. Studies reported a 96.7% removal of MB in 120 minutes, which is ascribed to the formation of low graphitization of biochar.[84]

Zhu et al. (2020) emphasized the preparation of Ce-C@GCN using the plant-growing guide and calcination method. The biochar was based on *Alternanthera philoxeroides Griseb.*, capable of reunion and stacking. The presence of biochar enhanced the number of active sites and the electron-hole separation efficiency. Because of this, Ce-C@GCN exhibited 96% removal of 2-mercaptobenzothiazole using visible light irradiation.[85] Wen et al. (2020) reported the preparation of GCN nanosheets dispersed on the rice-husk-derived carbon using the one-step sealed calcination method. The surface area of the catalyst was increased from 7.7 to $17 \, m^2/g$. The dangling C-C bond defects in the rice husk carbon were highly responsible for inhibiting the recombination of photo-generated electron-hole pairs. Experimental studies reported 98% methylene blue removal in 4.5 hours using a 5 W LED lamp.[86]

Wu et al. reported the utilization of wood as a carbon source to prepare GCN@C. The wood-derived photocatalyst possesses several advantages: wide pore distribution, high ordered degree range, and greater mechanical strength. Degradation studies reported a 96.2% removal of methyl orange, which decreased to 78.2% after the fifth cycle.[87] Panneri et al. reported 95% tetracycline removal within 90 minutes using the spray granulated GCN nanosheets. The in situ carbon doping improved the properties of the pristine GCN and increased the surface area to $150 \, m^2/g$.[88]

Zhao et al. described the preparation of C-doped GCN/WO_3 using cellulose as the carbon source was synthesized using a combination of hydrothermal impregnation, calcination, and self-assembly methods. Due to the porous structure, GCN exhibited a surface area of $57 \, m^2/g$. The narrower bandgap, improved light absorption, and faster interfacial charge transfer were attributed to the phenomenal catalytic activity towards the elimination of tetracycline.[89]

Similarly, the one-pot construction of carbon/GCN by employing chitin as the carbon source was proposed by Bin He. The photocatalytic studies reported a 93.7% removal of rhodamine B within 3 hours of visible light irradiation. The incorporation of carbon into the GCN reduced the bandgap from 2.7 to 2.5 eV, showing the carbon-based photocatalyst's stronger visible light absorption. Studies also indicated that the chitin could alter the microstructure of GCN due to the unbalanced electron density, and the interaction among urea and chitin ends up in a higher surface area.[90]

4.6   FUTURE PERSPECTIVES

The green synthesis of GCN-based composites is not highly explored compared to other photocatalytic materials. The necessary developments in this area are essential to enhance photocatalytic performance.

i. Detailed studies are necessary for understanding the reaction mechanism of green materials with carbon precursors in the synthesis of GCN photocatalysts.
ii. The toxicological effects of green extracts were not reported in most studies.
iii. The stability of the green synthesized GCN photocatalysts in different environments needs to be addressed.
iv. The large-scale implementation of these green materials is still a dream as it is economically unfeasible and inconvenient from the perspective of an industrialist. So, detailed studies are necessary for economically efficient green synthesis of photocatalysts by overcoming all the troubles related to the different physical and chemical properties of biomass.
v. The detailed exploration of the reusability and recovery of the photocatalysts must be studied to know the end-of-life of photocatalysts.

4.7   SUMMARY

This chapter discussed the green synthesis of GCN-based photocatalysts for wastewater remediation. GCN nanostructures are among the most important visible light-driven photocatalytic technology. However, GCN faces problems due to a higher recombination ratio, hindering efficiency. The modification of GCN-based photocatalysts using green synthesis is beneficial in enhancing the photocatalyst's electrical, optical, and morphological properties. Plant or plant extract utilization is essential to develop environmentally safe nanoparticles. During the photocatalyst synthesis, the green extracts are used as stabilizing or capping agents to avoid agglomeration and enhance the nanomaterials' stability. A comprehensive review is conducted on the photocatalytic applications of green synthesized GCN to redeem organic pollutants in wastewater. The green synthesized GCN's reaction mechanism and stability need in-depth attention for the technology's scale-up.

REFERENCES

1. Paul, P.; Al Tenaiji, A. K.; Braimah, N. A review of the water and energy sectors and the use of a nexus approach in Abu Dhabi. *Int. J. Environ. Res. Public Health* 2016, 13 (4). https://doi.org/10.3390/ijerph13040364.

2. Zhang, P.; Zou, Z.; Liu, G.; Feng, C.; Liang, S.; Xu, M. Socioeconomic drivers of water use in China During 2002–2017. *Resour. Conserv. Recycl.* 2020, 154, 104636. https://doi.org/10.1016/j.resconrec.2019.104636.

3. Azubuike, C. C.; Chikere, C. B.; Okpokwasili, G. C. Bioremediation techniques–classification based on site of application: Principles, advantages, limitations and prospects. *World J. Microbiol. Biotechnol.* 2016, 32 (11). https://doi.org/10.1007/s11274-016-2137-x.

4. Gusain, R.; Gupta, K.; Joshi, P.; Khatri, O. P. Adsorptive removal and photocatalytic degradation of organic pollutants using metal oxides and their composites: A comprehensive review. *Adv. Colloid Interface Sci.* 2019, 272, 102009. https://doi.org/10.1016/J.CIS.2019.102009.

5. Nasrollahzadeh, M.; Sajjadi, M.; Iravani, S.; Varma, R. S. Starch, cellulose, pectin, gum, alginate, chitin and chitosan derived (nano)materials for sustainable water treatment: A review. *Carbohydr. Polym.* 2021, 251, 116986. https://doi.org/10.1016/J.CARBPOL.2020.116986.

6. Balakrishnan, A.; Appunni, S.; Gopalram, K. Immobilized TiO$_2$/chitosan beads for photocatalytic degradation of 2,4-dichlorophenoxyacetic acid. *Int. J. Biol. Macromol.* 2020, 161, 282–291. https://doi.org/10.1016/j.ijbiomac.2020.05.204.

7. Khanzada, N. K.; Farid, M. U.; Kharraz, J. A.; Choi, J.; Tang, C. Y.; Nghiem, L. D.; Jang, A.; An, A. K. Removal of organic micropollutants using advanced membrane-based water and wastewater treatment: A review. *J. Membr. Sci.* Elsevier B.V. March 15, 2020, 117672. https://doi.org/10.1016/j.memsci.2019.117672.

8. Philip, J. M.; Aravind, U. K.; Aravindakumar, C. T. Emerging contaminants in Indian environmental matrices – A review. *Chemosphere* 2018, 190, 307–326. https://doi.org/10.1016/j.chemosphere.2017.09.120.

9. Abdel-Moniem, S. M.; El-Liethy, M. A.; Ibrahim, H. S.; Ali, M. E. M. Innovative Green/Non-Toxic Bi$_2$S$_3$@g-C$_3$N$_4$ nanosheets for dark antimicrobial activity and photocatalytic depollution: Turnover assessment. *Ecotoxicol. Environ. Saf.* 2021, 226, 112808. https://doi.org/10.1016/J.ECOENV.2021.112808.

10. Gusain, R.; Kumar, N.; Ray, S. S. Recent advances in carbon nanomaterial-based adsorbents for water purification. *Coord. Chem. Rev.* 2020, 405, 213111. https://doi.org/10.1016/j.ccr.2019.213111.

11. Alharbi, O. M. L.; Basheer, A. A.; Khattab, R. A.; Ali, I. Health and environmental effects of persistent organic pollutants. *J. Mol. Liq.* 2018, 263, 442–453. https://doi.org/10.1016/J.MOLLIQ.2018.05.029.

12. Gaur, N.; Narasimhulu, K.; PydiSetty, Y. Recent advances in the bio-remediation of persistent organic pollutants and its effect on environment. *J. Clean. Prod.* 2018, 198, 1602–1631. https://doi.org/10.1016/J.JCLEPRO.2018.07.076.

13. Ahmad, N.; Sultana, S.; Khan, M. Z.; Sabir, S. Chitosan based nanocomposites as efficient adsorbents for water treatment. In *Modern Age Waste Water Problems*; Springer International Publishing, 2020, pp. 69–83. https://doi.org/10.1007/978-3-030-08283-3_4.

14. Vaya, D.; Surolia, P. K. Semiconductor based photocatalytic degradation of pesticides: An overview. *Environ. Technol. Innov.* 2020, 20, 101128. https://doi.org/10.1016/j.eti.2020.101128.

15. Kim, S.; Chu, K. H.; Al-Hamadani, Y. A. J.; Park, C. M.; Jang, M.; Kim, D. H.; Yu, M.; Heo, J.; Yoon, Y. Removal of contaminants of emerging concern by membranes in water and wastewater: A review. *Chem. Eng. J.* Elsevier B.V. March 1, 2018, 896–914. https://doi.org/10.1016/j.cej.2017.11.044.

16. Dong, H.; Zeng, G.; Tang, L.; Fan, C.; Zhang, C.; He, X.; He, Y. An overview on limitations of TiO$_2$-based particles for photocatalytic degradation of organic pollutants and the corresponding countermeasures. *Water Res.* Elsevier Ltd, August 2015, 128–146. https://doi.org/10.1016/j.watres.2015.04.038.

17. Trivedi, N. S.; Mandavgane, S. A. Fundamentals of 2, 4 dichlorophenoxyacetic acid removal from aqueous solutions. *Sep. Purif. Rev.* 2018, 47 (4), 337–354. https://doi.org/10.1080/15422119.2018.1450765.

18. Chinthala, M.; Balakrishnan, A.; Venkataraman, P.; Manaswini Gowtham, V.; Polagani, R. K. Synthesis and applications of nano-MgO and composites for medicine, energy, and environmental remediation: A review. *Environ. Chem. Lett.* Springer International Publishing, 2021; 19. https://doi.org/10.1007/s10311-021-01299-4.

19. Ouyang, K.; Xu, B.; Yang, C.; Wang, H.; Zhan, P.; Xie, S. Synthesis of a Novel Z-Scheme Ag/WO$_3$/g-C$_3$N$_4$ Nanophotocatalyst for degradation of oxytetracycline hydrochloride under visible light. *Mater. Sci. Semicond. Process.* 2022, 137, 106168. https://doi.org/10.1016/J.MSSP.2021.106168.

20. Guo, F.; Shi, W.; Guan, W.; Huang, H.; Liu, Y. Carbon Dots/g-C$_3$N$_4$/ZnO nanocomposite as efficient visible-light driven photocatalyst for tetracycline total degradation. *Sep. Purif. Technol.* 2017, 173, 295–303. https://doi.org/10.1016/j.seppur.2016.09.040.

21. Cheng, L.; Xiang, Q.; Liao, Y.; Zhang, H. CdS-based photocatalysts. *Energy Environ. Sci.* 2018, 11 (6), 1362–1391. https://doi.org/10.1039/c7ee03640j.

22. Jbeli, A.; Ferraria, A. M.; Botelho do Rego, A. M.; Boufi, S.; Bouattour, S. Hybrid chitosan-TiO$_2$/ZnS prepared under mild conditions with visible-light driven photocatalytic activity. *Int. J. Biol. Macromol.* 2018, 116, 1098–1104. https://doi.org/10.1016/j.ijbiomac.2018.05.141.

23. Vigneshwaran, S.; Preethi, J.; Meenakshi, S. Removal of chlorpyrifos, an insecticide using metal free heterogeneous graphitic carbon nitride (g-C$_3$N$_4$) incorporated chitosan as catalyst: Photocatalytic and adsorption studies. *Int J Biol Macromol.* Elsevier B.V, 2019, 132. https://doi.org/10.1016/j.ijbiomac.2019.03.071.

24. Wen, J.; Xie, J.; Chen, X.; Li, X. A review on g-C$_3$N$_4$-based photocatalysts. *Appl. Surf. Sci.* 2017, 391, 72–123. https://doi.org/10.1016/J.APSUSC.2016.07.030.

25. Ismael, M. A review on graphitic carbon nitride (g-C3N4) based nanocomposites: Synthesis, categories, and their application in photocatalysis. *J. Alloys Compd.* 2020, 846, 156446. https://doi.org/10.1016/j.jallcom.2020.156446.

26. Jiang, L.; Yuan, X.; Pan, Y.; Liang, J.; Zeng, G.; Wu, Z.; Wang, H. Doping of graphitic carbon nitride for photocatalysis: A review. *Appl. Catal. B: Environ.* Elsevier B.V. November 15, 2017, 388–406. https://doi.org/10.1016/j.apcatb.2017.06.003.

27. Majdoub, M.; Anfar, Z.; Amedlous, A. Emerging chemical functionalization of g-C$_3$N$_4$: Covalent/noncovalent modifications and applications. *ACS Nano* 2020, 14 (10), 12390–12469. https://doi.org/10.1021/acsnano.0c06116.

28. Reddy, K. R.; Reddy, C. V.; Nadagouda, M. N.; Shetti, N. P.; Jaesool, S.; Aminabhavi, T. M. Polymeric graphitic carbon nitride (g-C$_3$N$_4$)-based semiconducting nanostructured materials: Synthesis methods, properties and photocatalytic applications. *J. Environ. Manage.* 2019, 238, 25–40. https://doi.org/10.1016/J.JENVMAN.2019.02.075.

29. Liao, G.; He, F.; Li, Q.; Zhong, L.; Zhao, R.; Che, H.; Gao, H.; Fang, B. Emerging graphitic carbon nitride-based materials for biomedical applications. *Prog. Mater. Sci.* Elsevier Ltd, July 1, 2020, 100666. https://doi.org/10.1016/j.pmatsci.2020.100666.

30. Ding, F.; Yang, D.; Tong, Z.; Nan, Y.; Wang, Y.; Zou, X.; Jiang, Z. Graphitic carbon nitride-based nanocomposites as visible-light driven photocatalysts for environmental purification. *Environ. Sci. Nano* 2017, 4 (7), 1455–1469. https://doi.org/10.1039/c7en00255f.

31. Acharya, R.; Parida, K. A review on TiO$_2$/g-C$_3$N$_4$ visible-light-responsive photocatalysts for sustainable energy generation and environmental remediation. *J. Environ. Chem. Eng.* 2020, 8 (4), 103896. https://doi.org/10.1016/j.jece.2020.103896.

32. Yuda, A.; Kumar, A. A review of g-C$_3$N$_4$ based catalysts for direct methanol fuel cells. *Int. J. Hydrog. Energy* 2022, 47 (5), 3371–3395. https://doi.org/10.1016/J.IJHYDENE.2021.01.080.

33. Chen, X.; Shi, R.; Chen, Q.; Zhang, Z.; Jiang, W.; Zhu, Y.; Zhang, T. Three-dimensional porous g-C$_3$N$_4$ for highly efficient photocatalytic overall water splitting. *Nano Energy* 2019, 59, 644–650. https://doi.org/10.1016/J.NANOEN.2019.03.010.

34. Zhang, Y.; Cheng, Y.; Yang, F.; Yuan, Z.; Wei, W.; Lu, H.; Dong, H.; Zhang, X. Near-infrared triggered Ti$_3$C$_2$/g-C$_3$N$_4$ heterostructure for mitochondria-targeting multimode photodynamic therapy combined photothermal therapy. *Nano Today* 2020, 34, 100919. https://doi.org/10.1016/J.NANTOD.2020.100919.

35. Sudhaik, A.; Raizada, P.; Shandilya, P.; Jeong, D. Y.; Lim, J. H.; Singh, P. Review on fabrication of graphitic carbon nitride based efficient nanocomposites for photodegradation of aqueous phase organic pollutants. *J. Ind. Eng. Chem.* Korean Society of Industrial Engineering Chemistry, November 25, 2018, 28–51. https://doi.org/10.1016/j.jiec.2018.07.007.

36. Zhang, S.; Gu, P.; Ma, R.; Luo, C.; Wen, T.; Zhao, G.; Cheng, W.; Wang, X. Recent developments in fabrication and structure regulation of visible-light-driven g-C$_3$N$_4$-based photocatalysts towards water purification: A critical review. *Catal. Today* 2019, 335, 65–77. https://doi.org/10.1016/J.CATTOD.2018.09.013.

37. Fu, J.; Yu, J.; Jiang, C.; Cheng, B. g-C$_3$N$_4$-based heterostructured photocatalysts. *Adv. Energy Mater.* 2018, 8 (3), 1–31. https://doi.org/10.1002/aenm.201701503.

38. Patnaik, S.; Sahoo, D. P.; Parida, K. Recent advances in anion doped g-C$_3$N$_4$ photocatalysts: A review. *Carbon N. Y.* 2021, 172, 682–711. https://doi.org/10.1016/J.CARBON.2020.10.073.

39. Paumo, H. K.; Das, R.; Bhaumik, M.; Maity, A. Visible-light-responsive nanostructured materials for photocatalytic degradation of persistent organic pollutants in water. 2020, 1–29. https://doi.org/10.1007/978-3-030-16427-0_1.

40. Dong, H.; Zeng, G.; Tang, L.; Fan, C.; Zhang, C.; He, X.; He, Y. An overview on limitations of TiO$_2$-based particles for photocatalytic degradation of organic pollutants and the corresponding countermeasures. *Water Res.* 2015, 79, 128–146. https://doi.org/10.1016/j.watres.2015.04.038.

41. Al-Mamun, M. R.; Kader, S.; Islam, M. S.; Khan, M. Z. H. Photocatalytic activity improvement and application of UV-TiO$_2$ photocatalysis in textile wastewater treatment: A review. *J. Environ. Chem. Eng.* 2019, 7 (5). https://doi.org/10.1016/j.jece.2019.103248.

42. Aanchal; Barman, S.; Basu, S. Complete removal of endocrine disrupting compound and toxic dye by visible light active porous g-C$_3$N$_4$/H-ZSM-5 nanocomposite. *Chemosphere* 2020, 241. https://doi.org/10.1016/j.chemosphere.2019.124981.

43. Meenakshi, S.; Farzana, M. Synergistic effect of chitosan and titanium dioxide on the removal of toxic dyes by photodegradation technique. *Ind. Eng. Chem. Res.* 2013, 53, 55–63.

44. Khan, Z.; Al-Thabaiti, S. A. Green synthesis of zero-valent Fe-nanoparticles: Catalytic degradation of rhodamine B, interactions with bovine serum albumin and their enhanced antimicrobial activities. *J. Photochem. Photobiol. B Biol.* 2018, 180, 259–267. https://doi.org/10.1016/j.jphotobiol.2018.02.017.

45. Jadoun, S.; Arif, R.; Jangid, N. K.; Meena, R. K. Green synthesis of nanoparticles using plant extracts: A review. *Environ. Chem. Lett.* 2021, 19 (1), 355–374. https://doi.org/10.1007/s10311-020-01074-x.

46. Parveen, K.; Banse, V.; Ledwani, L. Green synthesis of nanoparticles: Their advantages and disadvantages. *AIP Conf. Proc.* 2016, 1724. https://doi.org/10.1063/1.4945168.

47. Ahmed, S.; Ahmad, M.; Swami, B. L.; Ikram, S. A review on plants extract mediated synthesis of silver nanoparticles for antimicrobial applications: A green expertise. *J. Adv. Res.* 2016, 7 (1), 17–28. https://doi.org/10.1016/J.JARE.2015.02.007.

48. Sasidharan, S.; Raj, S.; Sonawane, S.; Sonawane, S.; Pinjari, D.; Pandit, A. B.; Saudagar, P. Nanomaterial synthesis: Chemical and biological route and applications. In: *Nanomaterials Synthesis.* Elsevier Inc., 2019. https://doi.org/10.1016/B978-0-12-815751-0.00002–X.

49. Nagajyothi, P. C.; Prabhakar Vattikuti, S. V.; Devarayapalli, K. C.; Yoo, K.; Shim, J.; Sreekanth, T. V. M. Green synthesis: Photocatalytic degradation of textile dyes using metal and metal oxide nanoparticles-latest trends and advancements. *Crit. Rev. Environ. Sci. Technol.* 2020, 50 (24), 2617–2723. https://doi.org/10.1080/10643389.2019.1705103.

50. Sundaram, I. M.; Kalimuthu, S.; Ponniah, G. P. Highly active ZnO modified G-C3N4 nanocomposite for dye degradation under UV and visible light with enhanced stability and antimicrobial activity. *Compos. Commun.* 2017, 5, 64–71. https://doi.org/10.1016/j.coco.2017.07.003.

51. Khan, S.; Wan, C.; Chen, J.; Khan, I.; Luo, M.; Wang, C. Eriobotrya japonica assisted green synthesis of g-C$_4$N$_4$ nanocomposites and its exceptional photoactivities for doxycycline and rhodamine B degradation with mechanism insight. *J. Chinese Chem. Soc.* 2021, 68 (11), 2093–2102. https://doi.org/10.1002/jccs.202100276.

52. Sarangapany, S.; Mohanty, K. Facile green synthesis of Ag@g-C$_3$N$_4$ for enhanced photocatalytic and catalytic degradation of organic pollutant. *J. Clust. Sci.* 2020, 32. https://doi.org/10.1007/s10876-020-01816-5.

53. Liu, C.; Wang, L.; Xu, H.; Wang, S.; Gao, S.; Ji, X.; Xu, Q.; Lan, W. "One Pot" Green Synthesis and the antibacterial activity of g-C$_3$N$_4$/Ag nanocomposites. *Mater. Lett.* 2016, 164, 567–570. https://doi.org/10.1016/J.MATLET.2015.11.072.

54. Valian, M.; Masjedi-Arani, M.; Salavati-Niasari, M. Sol-gel synthesis of DyFeO$_3$/CuO nanocomposite using capsicum annuum extract: Fabrication, structural analysis, and assessing the impacts of g-C$_3$N$_4$ on electrochemical hydrogen storage behavior. *Fuel* 2021, 306, 121638. https://doi.org/10.1016/J.FUEL.2021.121638.

55. Pinheiro, D.; Sunaja Devi, K. R.; Jose, A.; Karthik, K.; Sugunan, S.; Krishna Mohan, M. Experimental design for optimization of 4-nitrophenol reduction by green synthesized CeO$_2$/g-C$_3$N$_4$/Ag catalyst using response surface methodology. *J. Rare Earths* 2020, 38 (11), 1171–1177. https://doi.org/10.1016/J.JRE.2019.10.001.

56. Sunaja Devi, K. R.; Mathew, S.; Rajan, R.; Georgekutty, J.; Kasinathan, K.; Pinheiro, D.; Sugunan, S. Biogenic synthesis of G-C$_3$N$_4$/Bi$_2$O$_3$ heterojunction with enhanced photocatalytic activity and statistical optimization of reaction parameters. *Appl. Surf. Sci.* 2019, 494 (July), 465–476. https://doi.org/10.1016/j.apsusc.2019.07.125.

57. Devi, A. P.; Padhi, D. K.; Mishra, P. M.; Behera, A. K. Bio-surfactant mediated synthesis of Au/g-C$_3$N$_4$ plasmonic hybrid nanocomposite for enhanced photocatalytic reduction of mono-nitrophenols. *J. Ind. Eng. Chem.* 2022. https://doi.org/10.1016/J.JIEC.2021.12.030.

58. Arularasu, M. V.; Devakumar, J.; Rajendran, T. V. An innovative approach for green synthesis of iron oxide nanoparticles: Characterization and its photocatalytic activity. *Polyhedron* 2018, 156, 279–290. https://doi.org/10.1016/j.poly.2018.09.036.

59. Nasrollahzadeh, M.; Sajjadi, M.; Iravani, S.; Varma, R. S. Green-synthesized nanocatalysts and nano-materials for water treatment: Current challenges and future perspectives. *J. Hazard. Mater.* 2021, 401, 123401. https://doi.org/10.1016/J.JHAZMAT.2020.123401.

60. Bolade, O. P.; Williams, A. B.; Benson, N. U. Green synthesis of iron-based nanomaterials for environmental remediation: A review. *Environ. Nanotechnol. Monit. Manag.* 2020, 13, 100279. https://doi.org/10.1016/j.enmm.2019.100279.

61. Aditya, M. N.; Chellapandi, T.; Prasad, G. K.; Venkatesh, M. J. P.; Khan, M. M. R.; Madhumitha, G.; Roopan, S. M. Biosynthesis of rod shaped Gd_2O_3 on g-C_3N_4 as nanocomposite for visible light mediated photocatalytic degradation of pollutants and RSM optimization. *Diam. Relat. Mater.* 2022, 121, 108790. https://doi.org/10.1016/J.DIAMOND.2021.108790.

62. Zhang, J.; Lv, J.; Dai, K.; Liu, Q.; Liang, C.; Zhu, G. Facile and green synthesis of novel porous g-C_3N_4/Ag_3PO_4 composite with enhanced visible light photocatalysis. *Ceram. Int.* 2017, 43 (1), 1522–1529. https://doi.org/10.1016/J.CERAMINT.2016.10.125.

63. Fattahimoghaddam, H.; Mahvelati-Shamsabadi, T.; Lee, B. K. Efficient photodegradation of rhodamine B and tetracycline over robust and green g-C_3N_4 nanostructures: Supramolecular design. *J. Hazard. Mater.* 2021, 403, 123703. https://doi.org/10.1016/J.JHAZMAT.2020.123703.

64. Ma, Z.; Zhou, P.; Zhang, L.; Zhong, Y.; Sui, X.; Wang, B.; Ma, Y.; Feng, X.; Xu, H.; Mao, Z. g-C_3N_4 nanosheets exfoliated by green wet ball milling process for photodegradation of organic pollutants. *Chem. Phys. Lett.* 2021, 766, 138335. https://doi.org/10.1016/J.CPLETT.2021.138335.

65. Manjari Mishra, P.; Pattnaik, S.; Prabha Devi, A. Green synthesis of bio-based Au@g-C_3N_4 nanocomposite for photocatalytic degradation of methyl orange. *Mater. Today Proc.* 2021, 47, 1218–1223. https://doi.org/10.1016/J.MATPR.2021.03.712.

66. Venkatesan Savunthari, K.; Arunagiri, D.; Shanmugam, S.; Ganesan, S.; Arasu, M. V.; Al-Dhabi, N. A.; Chi, N. T. L.; Ponnusamy, V. K. Green synthesis of lignin nanorods/g-C_3N_4 Nanocomposite materials for efficient photocatalytic degradation of triclosan in environmental water. *Chemosphere* 2021, 272, 129801. https://doi.org/10.1016/J.CHEMOSPHERE.2021.129801.

67. Dou, Y.; Yan, T.; Zhang, Z.; Sun, Q.; Wang, L.; Li, Y. Heterogeneous activation of peroxydisulfate by sulfur-doped g-C_3N_4 under visible-light irradiation: Implications for the degradation of spiramycin and an assessment of n-nitrosodimethylamine formation potential. *J. Hazard. Mater.* 2021, 406, 124328. https://doi.org/10.1016/j.jhazmat.2020.124328.

68. Luo, J.; Dai, Y.; Xu, X.; Liu, Y.; Yang, S.; He, H.; Sun, C.; Xian, Q. Green and efficient synthesis of Co-MOF-based/g-C_3N_4 composite catalysts to activate peroxymonosulfate for degradation of the antidepressant venlafaxine. *J. Colloid Interface Sci.* 2022, 610, 280–294. https://doi.org/10.1016/J.JCIS.2021.11.162.

69. Mureseanu, M.; Radu, T.; Andrei, R. D.; Darie, M.; Carja, G. Green synthesis of G-C_3N_4/CuONP/LDH composites and derived g-C_3N_4/MMO and their photocatalytic performance for phenol reduction from aqueous solutions. *Appl. Clay Sci.* 2017, 141, 1–12. https://doi.org/10.1016/J.CLAY.2017.02.012.

70. Jin, H.; Lee, Y. S.; Hong, I. Hydrogen adsorption characteristics of activated carbon. *Catal. Today* 2007, 120 (3–4 SPEC. ISS.), 399–406. https://doi.org/10.1016/j.cattod.2006.09.012.

71. Yang, K.; Peng, J.; Srinivasakannan, C.; Zhang, L.; Xia, H.; Duan, X. Preparation of high surface area activated carbon from coconut shells using microwave heating. *Bioresour. Technol.* 2010, 101 (15), 6163–6169. https://doi.org/10.1016/j.biortech.2010.03.001.

72. Budinova, T.; Petrov, N.; Parra, J.; Baloutzov, V. Use of an activated carbon from antibiotic waste for the removal of Hg(II) from aqueous solution. *J. Environ. Manage.* 2008, 88 (1), 165–172. https://doi.org/10.1016/j.jenvman.2007.02.005.

73. Asiltürk, M.; Şener, Ş. TiO_2-activated carbon photocatalysts: Preparation, characterization and photo-catalytic activities. *Chem. Eng. J.* 2012, 180, 354–363. https://doi.org/10.1016/j.cej.2011.11.045.

74. Mohan, D.; Singh, K. P.; Singh, V. K. Trivalent chromium removal from wastewater using low cost activated carbon derived from agricultural waste material and activated carbon fabric cloth. *J. Hazard. Mater.* 2006, 135 (1–3), 280–295. https://doi.org/10.1016/j.jhazmat.2005.11.075.

75. Chen, X.; Kuo, D. H.; Lu, D. Nanonization of g-C_3N_4 with the assistance of activated carbon for improved visible light photocatalysis. *RSC Adv.* 2016, 6 (71), 66814–66821. https://doi.org/10.1039/c6ra10357j.

76. Xiang, W.; Zhang, X.; Chen, J.; Zou, W.; He, F.; Hu, X.; Tsang, D. C. W.; Ok, Y. S.; Gao, B. Biochar technology in wastewater treatment: A critical review. *Chemosphere* 2020, 252, 126539. https://doi.org/10.1016/j.chemosphere.2020.126539.

77. Li, X.; Qian, X.; An, X.; Huang, J. Preparation of a novel composite comprising biochar skeleton and "Chrysanthemum" g-C_3N_4 for enhanced visible light photocatalytic degradation of formaldehyde. *Appl. Surf. Sci.* 2019, 487, 1262–1270. https://doi.org/10.1016/J.APSUSC.2019.05.195.

78. Jeon, P.; Lee, M. E.; Baek, K. Adsorption and photocatalytic activity of biochar with graphitic carbon nitride (g-C$_3$N$_4$). *J. Taiwan Inst. Chem. Eng.* 2017, 77, 244–249. https://doi.org/10.1016/J.JTICE.2017.05.010.

79. Pi, L.; Jiang, R.; Zhou, W.; Zhu, H.; Xiao, W.; Wang, D.; Mao, X. g-C$_3$N$_4$ modified biochar as an adsorptive and photocatalytic material for decontamination of aqueous organic pollutants. *Appl. Surf. Sci.* 2015, 358, 231–239. https://doi.org/10.1016/j.apsusc.2015.08.176.

80. Lin, M.; Li, F.; Cheng, W.; Rong, X.; Wang, W. Facile preparation of a novel modified biochar-based supramolecular self-assembled g-C$_3$N$_4$ for enhanced visible light photocatalytic degradation of phenanthrene. *Chemosphere* 2022, 288, 132620. https://doi.org/10.1016/J.CHEMOSPHERE.2021.132620.

81. Tang, R.; Gong, D.; Deng, Y.; Xiong, S.; Deng, J.; Li, L.; Zhou, Z.; Zheng, J.; Su, L.; Yang, L. Enhanced visible photocatalytic degradation towards atrazine via peroxymonosulfate activation. *Chem. Eng. J.* 2022, 427 (May 2021), 131809. https://doi.org/10.1016/j.cej.2021.131809.

82. An, X.; Wang, H.; Dong, C.; Jiang, P.; Wu, Z.; Yu, B. Core-shell P-Laden Biochar / ZnO / g-C$_3$N$_4$ composite for enhanced photocatalytic degradation of atrazine and improved P slow-release performance. *J. Colloid Interface Sci.* 2022, 608, 2539–2548. https://doi.org/10.1016/j.jcis.2021.10.166.

83. Li, K.; Huang, Z.; Zhu, S.; Luo, S.; Yan, L.; Dai, Y.; Guo, Y.; Yang, Y. Removal of Cr(VI) from water by a biochar-coupled g-C$_3$N$_4$ nanosheets composite and performance of a recycled photocatalyst in single and combined pollution systems. *Appl. Catal. B Environ.* 2019, 243 (August 2018), 386–396. https://doi.org/10.1016/j.apcatb.2018.10.052.

84. Sun, J.; Lin, X.; Xie, J.; Zhang, Y.; Wang, Q.; Ying, Z. Facile synthesis of novel ternary g-C$_3$N$_4$/Ferrite/biochar hybrid photocatalyst for efficient degradation of methylene blue under visible-light irradiation. *Colloids Surf. A Physicochem. Eng. Asp.* 2020, 606, 125556. https://doi.org/10.1016/J.COLSURFA.2020.125556.

85. Zhu, Z.; Ma, C.; Yu, K.; Lu, Z.; Liu, Z.; Huo, P.; Tang, X.; Yan, Y. Synthesis Ce-doped biomass carbon-based g-C$_3$N$_4$ via plant growing guide and temperature-programmed technique for degrading 2-mercaptobenzothiazole. *Appl. Catal. B Environ.* 2020, 268, 118432. https://doi.org/10.1016/J.APCATB.2019.118432.

86. Wen, X.; Wang, W.; Ye, Q.; Zhou, Y.; Yang, J.; Sun, N.; Tan, Y.; Wang, W.; Hou, Y.; Yan, C. One-step synthesis of rice husk carbon with dangling CC bonds loaded g-C$_3$N$_4$ for enhanced photocatalytic degradation. *J. Clean. Prod.* 2020, 272, 122625. https://doi.org/10.1016/J.JCLEPRO.2020.122625.

87. Wu, L.; Chen, Y.; Li, Y.; Meng, Q.; Duan, T. Functionally integrated g-C$_3$N$_4$@wood-derived carbon with an orderly interconnected porous structure. *Appl. Surf. Sci.* 2021, 540, 148440. https://doi.org/10.1016/J.APSUSC.2020.148440.

88. Panneri, S.; Ganguly, P.; Mohan, M.; Nair, B. N.; Mohamed, A. A. P.; Warrier, K. G.; Hareesh, U. S. Photoregenerable, bifunctional granules of carbon-doped g-C$_3$N$_4$ as adsorptive photocatalyst for the efficient removal of tetracycline antibiotic. *ACS Sustain. Chem. Eng.* 2017, 5 (2), 1610–1618. https://doi.org/10.1021/acssuschemeng.6b02383.

89. Zhao, C.; Ran, F.; Dai, L.; Li, C.; Zheng, C.; Si, C. Cellulose-assisted construction of high surface area Z-scheme C-doped g-C$_3$N$_4$/WO$_3$ for improved tetracycline degradation. *Carbohydr. Polym.* 2021, 255, 117343. https://doi.org/10.1016/J.CARBPOL.2020.117343.

90. He, B.; Feng, M.; Chen, X.; Zhao, D.; Sun, J. One-pot construction of chitin-derived carbon/g-C$_3$N$_4$ heterojunction for the improvement of visible-light photocatalysis. *Appl. Surf. Sci.* 2020, 527, 146737. https://doi.org/10.1016/J.APSUSC.2020.146737.

5 Plant-Based Antimicrobial Nanofilms

Kavita Kulkarni, Yogesh Chendake, and Anand Kulkarni
Bharati Vidyapeeth Deemed to be University

CONTENTS

5.1 INTRODUCTION

Nanoparticles have shown very large potential for applications in different sectors due to their peculiar properties and possibility for property optimizations by chemical treatment and surface modifications [1]. Large research has been undertaken towards formation, modification and application of metallic nanoparticles. Although these materials have shown excellent properties and potential towards applications in different areas, there are limitations during industrial-scale production and its economics. Moreover, there are issues of formation of harmful compounds, which makes them ecologically unattractive options [2]. Hence, alternative methods of green synthesis are being investigated and being optimized these days, which will help to reduce the harmful compounds.

Various researchers have reported synthesis of nanoparticles for industrial applications by either top-down or bottom-up approach [3]. The top-down technique is based on crushing to get the preferred size and shape of nanoparticles. It is used for the production of nanoparticles applied in computer parts and optical mirrors. The bottom-up technique is utilized in cosmetics, molecular devices and additives used in fuels [4].

The synthetic methods for nanomaterials involve utilization and evolution of noxious chemicals and solvents. Further, there are issues regarding control on properties and purity. This restricts their applicability due to issues with property, purity and waste generation [5]. It has resulted in investigations towards the use of green sources for the synthesis of nanoparticles using environment-friendly, clean and biologically suitable routes over the traditional ones [6]. These routes include application of plant-based materials, microorganisms and organic compounds at various stages and chemical interactions. Such nanoparticles synthesized by green approach showed some of outstanding properties, with the property reproducibility and tuning.

Numerous biomaterials have been used as bioreduction agents for the metal ions during nanoparticles synthesis. Biomaterials such as bacteria (unicellular/multicellular) [7], yeast [8] and plant extract [9] have been reported for property tuning and synthesis of nanomaterials. This synthesis is dependent on three major factors, viz. reducing agent, solvent and environmental safety [10].

In green approach, synthesis of nanoparticles is majorly carried out by two methods of bioreduction and biosorption. In these, bioreduction provides the nanomaterial with higher stability properties. The process is used for the separation of metal nanoparticles from complex samples.

DOI: 10.1201/9781003319153-6

It involves simultaneous reduction of metal ions along with their reduction using oxidization reaction of enzymes [11]. Bioreduction is a useful tool for the recovery of metal nanoparticles from a complex sample. Biosorption involves bonding of metal ions with the cell wall/peptides, which converts into a stable structure of nanoparticles [12].

Different parameters are involved during the selection of biological route for the synthesis of nanoparticles. This selection is mainly affected by the structure of nanoparticles. Second, organisms develop some resistance to nanoparticles, which impose limitations in the selection of organisms. Considering this and other variables such as pH, pressure, solvent, temperature and precursor, biological methods will be selected for the synthesis of nanoparticles. Sources of green solutions and their applications are one the factors that affect this selection, properties and application of the synthesized nanoparticles.

5.2   SOURCES OF GREEN SOLUTIONS

Varieties of green resources are used in the synthesis of nanoparticles from metal and metal salts. These microbial resources can be enumerated as yeast, bacteria, algae, viruses and fungi [13].

1. Bacteria

    Bacterial species are largely used for various biotechnological applications. They are important aspirants in the preparation of nanoparticles [14]. Bacteria are implemented in the synthesis of nanoparticles because of comparative processing simplicity [15]. Metallic and other types of nanoparticles are synthesized by use of many bacterial species. In the synthesis of metal nanoparticles, bacterial species is believed to be the probable biofactory. Out of these, magnetotactic bacteria and S layer (surface layer) bacteria are prominently used for the synthesis of inorganic materials. There is an issue with metal ion toxicity towards bacteria, which limits its applicability. Bacteria develop the protective mechanism by ion reduction or by forming a water-insoluble compound [16], which is used in the formation of metal nanomaterials of biological origin. Various metal nanoparticles synthesis methods are categorized below.

    a.   Silver nanoparticles:

        Bacillus subtilis culture supernatant (media used for cell growth) and microwave irradiation can be used for the green synthesis of silver nanoparticles. Silver nanoparticles of (5–50 nm) can be synthesized with the help of microwave. Microwave provides uniform heating, which is useful for preventing agglomeration of nanoparticles and straight translation of polydispersed nanoparticles in mono dispersed particles [17]. *Bacillus licheniformis* has been reported for the synthesis of silver nanoparticles of various sizes from 40 nm to nanocrystals by Ag^+ ion bio reduction [18]. The synthesis is facilitated by bioreduction outside of the cells, which made the process easy and safe. The non-pathogenic nature of bacterium and formation of stable nanoparticles makes the process more attractive. The bioreduction supported by capping agent to avoid agglomeration helps to enhance the stability and life. *Bacillus flexus* is another bacterium that can be utilized for biosynthesis and shape modification of nanoparticles in triangular and spherical form. The formed Ag+ based bio-nanomaterials showed excellent antibacterial property and resistance to antibiotics [19].

        Bacillus amyloliquefaciens extract is another material used in the amalgamation of silver nitrate used to obtain nanoparticles by solar irradiation. The synthesis was affected by three parameters: solar intensity, concentration of sodium chloride and extract concentration. The triangular and circular crystalline of 14.6 nm mean diameter particles showed high stability due to interaction with protein layer. It has excellent antimicrobial properties against *Bacillus subtilis* (liquid medium) and *E. Coli* (solid medium) [20]. *Bacillus cereus* is another source for the synthesis Ag nanoparticles. It can be used to obtain spherical nanoparticles with 20–40nm size with small aggregation and excellent antimicrobial activity towards pathogenic bacteria [21]. *Pseudomonas stutzeri* AG259 is also used for the biosynthesis of silver nanoparticles. Single crystals were generated near cell poles with clear compositions [22].

In general, the bacterial systems with mesophilic, psychrophilic, *Pseudomonas* and *Bacillus* type bacteria were used for the synthesis and modification of silver nanoparticles. These materials showed excellent synthesis and modification properties with applicability in material and antimicrobial applications. The particles showed excellent stability properties due to encapsulation by protein layers during synthesis procedures. Their properties and applicability can be controlled using bacterial species, chemicals and concentrations along with other synthetic parameters. Almost all these particles are reported to be monodispersed materials with excellent control on the properties and in turn applicability [23].

b. Gold nanoparticles:

The gold nanoparticles are used in many curative applications. These nanoparticles were synthesized by biological routes and have shown applications in various fields such as photocatalysis, medicine, electronic devices and sensors [24]. The biological route for the synthesis of gold nanoparticles is an eco-friendly approach that has shown good results in various applications. The biosynthesis of nanoparticles may be intracellular or extracellular depending upon synthesis location and conditions. Intracellular mechanism takes place inside the cell wall, while extracellular mechanism occurs outside the cell wall. *Rhodomonas capsulate* showed an analogous mechanism for the synthesis of gold nanoparticles. Toxic metals may be converted into volatile states [25]. *Bacillus subtilis*168 was utilized as a reducing agent for the synthesis of Au^{3+} particles with the gold chloride at atmospheric conditions. The process of reduction by using *Bacillus subtilis*168 was different as compared to the synthesis of sliver nanoparticles. The stabilization of the synthesized nanoparticles may be because of surfactin or due to the release of any new biomolecule [26].

Au^{3+} adsorption was seen by *Bacillus megaterium*D01.The synthesized particles showed monodispersed gold nanoparticles having capping of thiol. The mechanism was extracellular. The stability of the particles was found to be good for a number of weeks [27]. Exposure of Lactobacillus strain to gold ions showed the development of gold nanoparticles in the cell wall. Bacteria present in whey of buttermilk and gold ions showed growth of nanoparticles. Nucleation was found on the cell wall surface by using enzymes and sugars. Transportation of the nuclei in the cell was observed [28]. Gold nanoparticles were synthesized by using *S. algae*. In the presence of H_2 gas and Fe(III) citrate (electron donor and electron acceptor, respectively), *S. algae* cells reduced $AuCl^{4-}$ ions in gold within half an hour. Solution pH was the very important factor. At low pH, deposition of gold nanoparticles was observed extracellularly, but at 7 pH, deposition of particles was in the periplasmic gap of bacterium. In the presence of H_2, only the bacterium could reduce Au (III) into gold particles [29]. Liangwei et al. used *Escherichia coli* DH 5α for the synthesis of gold nanoparticles. The synthesized nanoparticles were not uniform in shape and size. Triangular, quasi-hexagon and spherical shapes were observed. The surface of the bacterium was covered by nanoparticles [30]. *R. capsulata* secreted an enzyme, which was responsible for the reduction of Au (III) ions. The initiation of electron transfer from NADH enzyme to NADH was observed [31]. *Plectonemaboryanum* UTEX 485 (filamentous cyanobacteria) controlled the surface morphology of the synthesized particles [32]. Reaction with aqueous solution of $AuCl_{4-}$ and Au $(S_2O_3)_{23-}$ showed production of cubic gold nanoparticles with octahedron gold platelets inside the bacterial cells. At the start, amorphous gold nanoparticles were observed near the cell wall when the bacterium was interacted with, and lastly, metallic gold octahedron nanoparticles were seen close to cell surface [33].

   c. Magnetite Nanoparticles:

     Desulfovibriomagneticus strain RS-1 was used for the synthesis of magnetite nanoparticles intracellularly. The bacterium acted as an electron acceptor. The size of the crystals was bigger than 30 nm [34]. *Thermoanaero bacterethanolicus* (TOR-39) is used for iron reduction during the synthesis of magnetite crystals. Substitution of metals such as Ni, Co and Cr led to octahedral-shaped particles with weak crystal-line stage. *Magnetospirillum magnetotacticum* was used for the synthesis of magnetic nanoparticles. They were obtained by forbidding the movement of *M. magnetotacticum* using magnetic field. It resulted in the formation of controlled nanoparticles in a structured way [35]. Sulphate-reducing bacterium showed the capacity to produce magnetic iron sulphide nanoparticles. The synthesized nanoparticles were used in the adsorption of desorbed radioactive metals and provided appropriate medium for the secure storage [36].

2. Fungi:

     Fungal culture is hazardous, but it can be utilized for the synthesis of nanoparticles by extracellular and intracellular ways. They have been applied in many industries. *Fusarium oxysporum* type of fungus was used for the synthesis of silver nanoparticles of 5–15 nm size, and the capping by proteins made them stable. The synthesis mechanism was extra-cellular [37]. Silver nanoparticles were also synthesized by using *Aspergillus fumigatus*. The size was from 5 to 25nm, which can be controlled with the catalytic action. Here, the fungus acts as a reducing agent for Ag and can synthesize nanoparticles within a short span of time [38]. Extracellular manufacturing of silver nanoparticles can be carried out using the fungus *Trichoderma reesei*. It gave nanoparticles of size 5–50nm, which required a time of 72 hours. *T. reesei* showed the advantage of making large quantity of enzymes and also increased the synthesis of nanoparticles. It is used to synthesize heterogeneous nanoparticles [39]. White-rot fungus can be utilized to synthesize silver nanoparticles intracellularly by optimizing conditions [40]. *Verticillium sp.* was used for the synthesis of gold nanoparticles, where $AuCl_4$ was biologically reduced [41]. The fungus *Neurospora crassa* with PtNPs is utilized for the production of platinum nanoparticles with intracel-lular mechanism, and the diameter of the formed particles was 4–35nm. Additionally, they can be utilized for the formation of nano-agglomerates [42]. *F. oxysporum* was used to produce platinum particles by extracellular and intracellular mechanisms, in a suboptimal amount [43]. Secretion proteins were helpful in the synthesis of nanoparticles with fungi where maximum fungi are phytopathogenic [44].

3. Yeast:

     Toxic metal absorption and accumulation capacity of yeast is high because of large surface [45]. Yeast removes the metals by extracellular sequestration, biosorption, bio-precipitation and chelation. *Schizosaccharomyces pombe* yeast's growth phase is rec-ommended for the synthesis of CdS quantum dots [46]. *Torulopsis sp.*is utilized for the synthesis of Pb_2 ions. Intracellular production was seen for *Pichiajadinii* for Au particles [47]. These biological origin nanomaterials are preferred for potential applications due to safety properties.

4. Algae:

     Luangpipatre commended dehydrated unicellular algae for the synthesis of nanopar-ticles of different shapes. Nanoparticles were accumulated close to the surface [48]. *Chlorella vulgaris* extract along with $CuFe_2O_4$ was used for the synthesis of nanoparticles. This material showed resistivity towards *Staphylococcus aureus*, a round-shaped bacte-rium seen in upper respiratory tract [49]. Proteins play a very important role in stabilizing and reducing action and also in modification of shape Govindaraju and team used marine algae (*Sargassum wightii*) for the extracellular synthesis of two different nanoparticles [50]. Different algae were used for the synthesis of silver and gold nanoparticles, e.g., *Kappaphycusalvarezii* [51], and *Euglena gracilis* microalga [52].

5. Viruses:

These are other potential and smallest living organisms used in metallic nanomaterial synthesis [53]. Scientists found that outer protein shell plays an important role in the synthesis of nanoparticles with extremely reactive surface, which interacts with metallic ions [54]. Inorganic materials were used; for example, silicon dioxide, zinc sulphide, iron oxide and cadmium sulphide were utilized in the synthesis. CdS and ZnS nanoparticles are used in electronic industry; this underlines the value of particles [55]. The outer layer of *Tobacco mosaic virus* was covered by the protein. Proteins make a notch where the material deposits on the surface and can be used in various applications [56]. The size of the nanoparticles was seen decreased when treated with the extract of *Nicotiana benthamiana*/barely and salts of Ag/Au [57].

6. Plants:

Plant diversity makes it the most suitable resource in the synthesis of nanoparticles through research pathways.

Reduction of metal ions in a single stage can be achieved by the presence of biomolecules present in plants [58]. Traditionally also this resource was preferred for being environment-friendly. Plant-based materials are inexpensive, are easily available and can be easily modified as compared to other resources. The synthesis of nanoparticles can be achieved by using whole plant extract [59]. Plant extract is better to control in comparison with whole plant. The synthesis of nanoparticles from plant extract is a simple procedure of mixing with metal salt solution. Biochemical reduction takes place, and zero valent state is achieved from monooxidation state. A change in colour in culture media indicates the synthesis of nanoparticles. It can be used for the synthesis of a variety of nanoparticles [60].

a. Gold nanoparticles:

Various green methods are utilized for the synthesis of gold nanoparticles. Generally, polysaccharides and phytochemicals were utilized for the synthesis of these gold nanoparticles. Reaction of lemongrass plant extract with chloroaurate ions resulted in gold nanoparticles of triangular shape at room temperature. The mobility of electrons was higher in the nano triangles [61]. Gold and silver nanoparticles were prepared by using *Mentha piperita* plant extract. Chloroauric acid and silver nitrate were used for the synthesis of both nanoparticles. The characterization of particles showed spherical shape with a size of 90 and 150 nm for silver and gold nanoparticles, respectively. These materials showed antimicrobial activity against *E. coli* and *S. aureus*. Leaf broth of *Cinnamomum zeylanicum* was used as a reducing agent in the biosynthesis of gold nanoparticles. The shape of the particle was decided by the concentration of extract; in general, it was spherical and prism type structure. A low concentration gave prism-shaped particles, while at high concentrations, a spherical shape was observed with high-quality crystals. These materials showed photo luminescence with an increase in photoemission with plant extract concentration [62]. Bankar et al. used the extract of banana peel powder with chloroauric acid for the synthesis of gold nanoparticles. The properties of the obtained nanoparticles are dependent upon parameters such as temperature, concentration of chloroauric acid, extract content and pH, which affected the type of nanoparticles. The size of the particles was 300nm, with antimicrobial activity against cultures of bacteria and fungi [63]. *Madhuca longifolia* extract was used for the synthesis of gold nanoparticles with anisotropic property. Reduction of gold ion was carried out by tyrosine compounds from plant extract. These nanoparticles can be used as a coating for the absorption of IR radiation [64]. An herb named Barbated Skullcup was used to synthesize nanoparticles extracellularly. The herb worked as a reducing agent, and the fast reduction of ions confirmed the gold nanoparticles formation in the size range of 5–30nm. The coating of nanoparticles was applied on the carbon electrode for the enhancement of transmission rate from p-nitrophenol and electrode [65]. The green synthesis of gold nanoparticles is tried by honey, a natural sweetener. Reduction as well as capping was observed by honey during synthesis. Varying concentrations of both the reactants showed the formation of nanocrystals maybe of spherical shaped and anisotropic type with 15 nm size particles. The amine group was responsible for the strapping proteins on the surface by Au particles [66]. The stem extract of Indian snowberry (*Breyniarhamnoides*) is recommended for the synthesis of Au and Ag nanoparticles. The process showed swift synthesis of nanoparticles, where reduction was done by sugar and biochemicals such as phenolic glycosides from the extract. Extract concentration was the most important parameter responsible for the size of the nanoparticles [67].

Shells of Indian soapberry were tried for the biological synthesis of gold nanoparticles using $HAuCl_4$ as a precursor. The quantity of shells and $HAuCl_4$ concentration were the two important parameters in the synthesis process, which resulted in highly crystalline nanoparticles of varied size as 9, 17 and 19 nm. The stability of the particles was due to the carboxylic group and flavonoids in the shell extract. Excellent catalytic activity was observed by nanoparticles for p-nitroaniline reduction [68]. Krishnaraj et al. studied *Acalypha indica* extract with $HAuCl_4$ for the synthesis of gold and silver nanoparticles. Spherical nanoparticles of size 20–30 nm was obtained, which showed anticancer properties against breast cancer cells [69].

b. Silver nanoparticles:

Silver nanoparticles are widely used in many applications due to their unique properties [70]. As a precursor, silver nitrate is mixed with the plant extract from various plants, which act as a reducer [71]. Silver nitrate (1%) and blackberry fruit extract were used to synthesize nanoparticles. Polyphenols present in the fruit extract worked as a reducer. The anti-inflammatory effect of the synthesized particles was also tested [72]. Lippiacitriodora leaf extract was used by Cruz et al. for the synthesis of nanoparticles. They found that the main reducing component was glycoside verbascoside, which worked as a capping agent for the stability of particles [73]. The extract of *Arbutus unedo* leaf was used for Ag nanoparticles synthesis. They confirmed the reduction and stabilization done by the compounds from extract. Due to the natural capping property, these nanoparticles have shown applications in drug delivery [74]. *Chelidonium majus* leaf extract was recommended for the synthesis of nanoparticles. Reduction of silver ions took place due to alkaloids and flavonoids present in the extract, and also capping was seen due to these chemicals. These materials showed excellent antimicrobial activity against *E coli*. An increase in inhibition zone was observed in green synthesized nanoparticles due to alkanoids and flavonoids [75]. Coseri et al. reported the use of polysaccharide pullulan for the synthesis of silver nanoparticles. The nanoparticle size and properties were controlled by variation in the concentration of silver nitrate and pullulan [76]. For the synthesis of silver nanoparticles, Capulin cherry extract was utilized. The photocatalytic synthesis showed excellent reduction properties and control on nanomaterial formation reaction [77].

c. Copper-based nanoparticles synthesis:

Cu nanoparticles showed toxicity properties during chemical synthesis, which can be eliminated through green synthesis method. These nanoparticles have shown broad applications in every field. CuO nanoparticles and Cu nanoparticles have shown many applications as a catalytic material/antibacterial material [78]. The biological synthesis of copper nanoparticles was recommended by the use of extract from *Magnolia kobus* leaf. From the aqueous solution of copper sulphate, the synthesis of nanoparticles was achieved, which showed excellent antibacterial activity against *E. coli*. The size of the nanoparticles was from 37 to 110nm.Increased antibacterial activity was observed by biologically synthesized particles than the conventional ones [79]. The extract of *Vitis vinifera* leaf is used to prepare copper nanoparticles. Copper solvates reduction took place, leading to the formation of copper nanoparticles. FTIT studies revealed that proteins played an important role in the synthesis of nanoparticles. The formed nanoparticles showed excellent antibacterial activity against *E. coli*, *K. pneumoniae*, *S. typhi* and *B. subtilis* [80]. *Lawsoniainermis* (Henna) leaves were recommended for the synthesis of copper nanoparticles. The extract of leaves and copper sulphate pentahydrate were heated for 15 minutes at 100°C and at pH 01. The formed nanoparticles were cleaned by water and ethanol treatment and calcinated, showing increased electrical conductivity [81]. Copper nanoparticles were biologically synthesized using Citrus Grandis peel extract. Biomolecules present in the peel extract worked as a reducing and stabilizing agent. Absorption band was found at 590 nm in the UV-Vis spectra. The shape of the particles was spherical with 22–27nm size. It showed excellent activity during the degradation of methyl red dye [82]. Clove extract was used for the synthesis of copper nanoparticles by reduction of copper sulphate. The synthesized particles were of spherical shape and of 5.40nm size [83]. *Capparis zeylanica* leaf extract worked as a reducing agent as well as capping agent for the synthesis of copper nanoparticles using copper sulphate solution as a precursor. The characterization showed size in the range of50–100nm. The formed material showed antimicrobial activity against pathogens (gram negative and gram positive) [84]. Hibiscus rosa-sinensis extract was used for the synthesis of copper nanoparticles. The antioxidant activity of the nanoparticles was studied. The antimicrobial activity study was carried out against important pathogens. The synthesized particles were useful in the treatment of lung cancer [85]. Kulkarni et al. used *Eucalyptus sp.* extract for the synthesis of Cu nanoparticles using copper sulphate as a precursor. FTIR study confirmed the presence of proteins, flavonoids and phenolic acid. These biomolecules performed as a reducing and capping agent. Phenolic agent showed strong capacity for binding of metals. Nanoparticles size was 27.65-48.19nm [86]. The lemongrass tea was used for the synthesis of ultrasmall Cu nanoparticles using one-pot method, which provided2.90–0.64 nm size particles. Functional groups containing oxygen were responsible for the synthesis of nanoparticles. Lemon extract is another material used for the synthesis of copper nanoparticles. It worked as a reducing agent, while the addition of curcumin externally worked as stabilizer. This results in the formation of excellent quality nanoparticles with good antimicrobial activity [87]. The biological synthesis of copper nanoparticles was also obtained from gooseberry at 6–10 pH. For four pathogens, the antimicrobial activity was checked. FTIR indicated the presence of proteins, phenolic groups and biomolecules. The synthesized nanoparticle size was 15–30nm. The elements present in plant extract, such as polyphenols and ascorbic acid, were responsible for reducing the metal ions, and quantitative assay confirmed the same [88]. The leaf extract of *Celastruspani culatus* was utilized as a reducing agent in the synthesis of copper nanoparticles. The formed Cu nanoparticles were UV active and 2.10nm in size. It can be used in waste treatment, where the degradation of methylene blue dye with pseudo-first-order kinetic was observed. Also, its application as an antifungal agent for *Fusarium oxysporum* is reported [89].

5.3 SYNTHESIS OF NANOFILMS BY GREEN CHEMISTRY

In the environment, sustainability is a very important factor that is focused by 12 green chemistry principles. They concentrate on saving energy, less toxic waste generation, no harm to environment, use of natural resources and no threat to global warming. Greener route can be applied for the synthesis of nanomaterials that have many applications in various fields. The use of natural solvents and natural precursors is preferred in green synthesis. The green chemistry processes include several methods such as ball milling [90], microwave irradiation [91], photocatalysis [92], hydrothermal synthesis [93], ultrasound-assisted synthesis [94], magnetic field-assisted synthesis [95] and biological methods [96]. Due to less hazard formation and environmentally friendly nature, biological methods are preferred as compared to conventional methods, so the biological methods are focused.

1. Biological method:

In this method, stabilizing, reducing and capping agents are not required externally. Microorganisms or biomolecules present in the extract do the work of reducing, stabilizing and capping. No toxic chemicals are used in the synthesis to control the size or shape of the nanoparticles [97]. Nanoparticles synthesized by this method are eco-friendly, non-toxic, cost-effective and stable. High antimicrobial activity of these particles was observed as compared to conventional methods. Effective stabilization as well as capping was observed by proteins present in the plant extract or microorganisms cell wall. The synthesis of nanoparticles can be rationally divided into two steps:
1. Nucleation
2. Growth.

These two steps are important to control the nanoparticle size and shape. Also, mono dispersity can be decided by these two steps. Reduction, capping and stabilization are the three important steps in the green synthesis. The compounds called reducing agent, stabilizing agent, and capping agent play a very important role in the green synthesis. Electrons are given to ions by reducing agent for the formation of atoms. The atoms combine together to form a particle. Capping agent does the stabilization of nanoparticles by preventing from agglomeration. Capping agent can change the morphology kinetics of the reaction due to their pattern. The influence of capping agent on the growth process was observed in various processes. By reduction of metal ions, the formation of nano-metal ions was observed. Some of the agents can work as a reducing as well as capping agent. In the synthesis of nanoparticles from plants, plant extract is mixed homogenously with metal salt solution. The conversion of metals to zero valent position from mono- or divalent position by biochemical reduction was observed in very less time at atmospheric conditions. A colour change was observed, which indicates the formation of nanoparticles [60]. Every plant has variety of biomolecules responsible for bioreduction. Biomolecules may be flavonoids, phenolic compounds and terpenoids. Plant extract source is the most important factor that decides nanoparticles morphology because of the concentration of biomolecules [98].

Figure 5.1 elaborates the specific as well as selective nanoparticle synthesis that can be achieved by this method. The form of biomolecules can control the nanoparticle size and shape without changing the stability of particles.

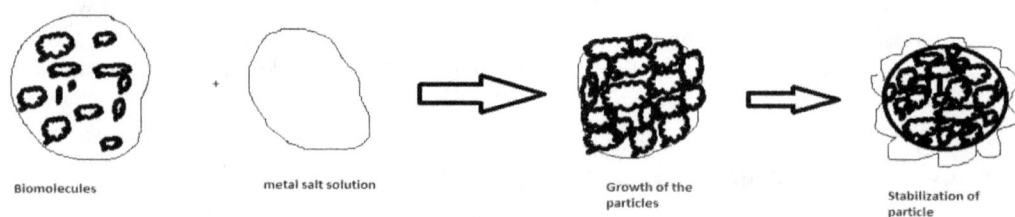

Biomolecules   metal salt solution   Growth of the   Stabilization of
                                                       particles   particle

FIGURE 5.1   Illustration of step-by-step nanoparticles synthesis by green chemistry.

TABLE 5.1

Use of Plant Extract for Nanoparticles Synthesis and Their Applications

S. No.	Name of Plant	Type of Nanoparticle	Application
1	Sapindusmukorossifruit pericarp	Au	p-Nitroaniline reduction
2	Mentha piperita	Au	Antibacterial
3	Cymbopogoncitratus	Au	IR radiation blocker
4	Madhuca longifolia	Au	Infrared reflective coatings on glass
5	Scutellariabarbata	Au	p-Nitroaniline reduction
6	Cinnamomumzeylanicum	Au	Biosensors and labelling for cell
7	Moringa olifera	Au	Photocatalytic activity
8	Sesbaniadrummondii	Au	p-Nitroaniline reduction
9	Acalyphaindica	Au	Treatment of breast cancer cells
10	Breyniarhamnoides	Au	p-Nitroaniline reduction
11	Aloe vera	Au and Ag	Wastewater treatment
12	Psidium guajava	Au	Novel drug delivery system
13	Chelidonium majus L. plant extract	Ag	Antimicrobial activity
14	Casein hydrolytic peptides	Ag	Antimicrobial activity
15	Potato starch	Ag	Antimicrobial activity, biosensors
16	Prunus serotina	Ag	Used as an antioxidant
17	Lippia citriodora	Ag	-
18	Peanut shell	Ag	Antifungal agent
19	Anacardiumothonianum Rizz.	Ag	Antifungal agent and optoelectrical characteristics
20	ά-D-Glucose, sucralose, triethylamine	Ag	Biosensors
21	Illicium verum seed	Ag	Nano-photonic devices
22	Picrasmaquassioides bark	Ag	Catalytic activity
23	Cassia tora leaf	Ag	Antibacterial activity
24	Clerodendruminerme	Ag	Antimicrobial and antioxidant activity
25	Matricaria chamomilla L.	Ag	Catalytic activity
26	Carissa carandas L.	Ag	Antioxidant and antimicrobial activity
27	Silybum marianum	Ag	Biological and clinical activities
28	Corn cob xylan	Ag	Antifungal activity
29	Sisymbrium irio	Ag	Antibacterial activity
30	Magnolia kobus	Cu	Antibacterial activity
31	Lawsoniainermis	Cu	Nano-biocomposites
32	Hibiscusrosa-sinensis	Cu	Antioxidant activity
33	Vitis vinifera	Cu	Antibacterial activity
34	Lemon (Citrus sp.)	Cu	Antibacterial activity
35	Phyllanthus emblica(gooseberry)	Cu	Antibacterial activity
36	Centella asiatica	Cu	Photocatalytic activity
37	Nerium oleander	Cu	Antibacterial activity
38	Gum karaya	Cu	Antibacterial activity
39	G. superba	Cu	Antibacterial activity
40	Celastrus paniculatus.	Cu	Antifungal activity
41	Durantaerecta	Cu	Reduction of toxic dyes
42	Crotalaria candicans	Cu	Antibacterial activity
43	Cuscutareflexa	Cu	Catalytic activity
44	P. granatum seeds	Cu	Catalytic activity
45	Hibiscussabdariffa	Zn	Antibacterial and antidiabetic activity
46	Azadirachta indica	Zn	Antibacterial and photocatalytic activity

(Continued)

TABLE 5.1 (*Continued*)

Use of Plant Extract for Nanoparticles Synthesis and Their Applications

S. No.	Name of Plant	Type of Nanoparticle	Application
47	Lemon (Citrus sp.)	Zn	Photocatalytic activity
48	Caulerpa peltata, Hypneavalencia, S. myriocystum	Zn	Antimicrobial activity
49	Ruta graveolens	Zn	Antibacterial and antioxidant activity
50	Solanum nigrum	Zn	Antibacterial activity
51	Azadirachta indica	Zn	Antimicrobial activity
52	Lycopersicones culentum	Zn	Photovoltaic activity
53	Plectranthusamboinicus	Zn	Photocatalytic activity
54	Moringa oleifera	Zn	Antimicrobial activity
55	Vitex negundo L. extract	Zn	Binding interaction between human serum albumin
56	Ocimumbasilicum L. var. purpurascens Benth.-Lamiaceae	Zn	Wastewater treatment
57	P. trifoliata	Zn	Catalytic activity
58	Trifolium pratense	Zn	Antibacterial activity
59	Pongamia pinnata	Zn	Antibacterial activity
60	Nyctanthes arbor-tristis	Zn	Antifungal activity
61	Acorus calamus	TiO_2	Photocatalytic and antimicrobial activity
62	Pomegranate peel	TiO_2	Antimicrobial activity
63	Orange peel	TiO_2	Antibacterial activity, sensors
64	Dandelion pollen	TiO_2	Photocatalytic activity
65	Jatropha curcas L.	TiO_2	Photocatalytic degradation
66	Cynodondactylon	TiO_2	Antibacterial and anticancer activity
67	Trianthemaportula castrum	TiO_2	Antifungal activity
68	Abelmoschus esculentus	TiO_2	Antifungal activity
69	Glycyrrhiza glabra L.	TiO_2	Antioxidant activity
70	Carica papaya leaves	TiO_2	Photocatalytic activity
71	Syzygiumcumini	TiO_2	Photocatalytic activity
72	Moringa oleifera	CeO_2	Antimicrobial activity
73	Gloriosa superba L.	CeO_2	Antibacterial and optical properties
74	Olea europaea	CeO_2	Antimicrobial activity
75	Calotropis procera	CeO_2	Photocatalytic activity
76	Carica papaya	FeO_2	Photocatalytic and antibacterial activity
77	Phoenix dactylifera	FeO_2	Antioxidant activity
78	Moringa oleifera	FeO_2	Antibacterial activity
79	Peltophorumptero carpum	FeO_2	Catalytic activity
80	Pomegranate seeds	FeO_2	Photocatalytic activity
81	Cynometraramiflora	FeO_2	Catalytic activity
82	Platanus orientalis	FeO_2	Antifungal activity
83	Cynometraramiflora	FeO_2	Antibacterial and catalytic activity
84	Juglans regia green husk	FeO_2	Toxicity
85	Avocado fruit peel	FeO_2	Antimicrobial and catalytic activity
86	Avicennia marina	FeO_2	Catalytic activity for environmental pollution
87	Withania	FeO_2	Antimicrobial and catalytic activity
88	Ruellia tuberosa	FeO_2	Catalytic activity
89	Delonixelata	SnO_2	Photocatalytic activity
90	Psidium guajava	SnO_2	Photocatalytic activity

(Continued)

TABLE 5.1 (*Continued*)
Use of Plant Extract for Nanoparticles Synthesis and Their Applications

S. No.	Name of Plant	Type of Nanoparticle	Application
91	Aloe vera	$Y_2O_3:Dy^{3+}$	Photocatalytic activity
92	Anogeissus latifolia	Pd	Antioxidant and catalytic activity
93	Camellia sinensis	Pd	Antioxidant, antibacterial and antiproliferative activity
94	Hippophae rhamnoides L.	Pd	Catalytic activity
95	Andean blackberry	Pd	Photocatalytic activity
96	Sargassum algae	Pd	Electrocatalytic activity
97	Euphorbia condylocarpa M. Bieb	Pd/Fe_3O_4	Catalytic activity
98	Barberry fruit	Pd	Catalytic activity
99	Cinnamon zeylanicum	Pd	Catalytic activity
100	Garciniapedunculata Roxb.	Pd	Catalytic and antimicrobial activity

5.4 APPLICATIONS OF GREEN NANOFILMS

Nanoparticles have huge potential in sustainable technologies. The green nanotechnology has shown applications in various fields such as food, medicine, agriculture, energy and biomedicine. There is a huge opportunity for the green nanotechnology in future challenges in all aspects.

In biomedical field, the green nanotechnology can provide many opportunities in terms of new drugs, which can be utilized against a number of diseases. The use of nanoparticles in modern medicinal field has opened up new possibilities. In disease finding and delivery of a particular drug, these particles can be used because of nanosize. Currently, the use of biomass for biofuel production is also a demanding topic. To develop new systems for biomass conversion, the nanotechnology is used. For the production of biofuels, many nanocatalysts are used. The specifics of applications in all fields are illustrated in this chapter. The following table illustrates the use of plant extract for nanoparticles synthesis and their applications in various fields (Table 5.1).

Case study

Water hyacinth (*Eichhorniacrassipes*) and Roselle (*Hibiscus sabdariffa*) plant extracts were utilized for the synthesis of nanofilms. Cellulose dissolution was achieved by using polyethylene glycol-600. Zinc acetate with the addition of hexamethylenetetramine (HMTA) was used as a precursor for the synthesis of nanofilms. The nanofilms were prepared in the microwave to get equal heating from all directions. Antimicrobial activity was checked against pathogens for both the films. As ZnO concentration was increased, the antimicrobial activity increased. Further growth of bacteria was prevented [99].

5.5 CONCLUSIONS AND EXPECTATIONS

There is huge potential for the design of green nanomaterials of plant origin for biomedical applications. These materials possess the benefits of excellent biocompatibility along with natural origin and property tuning. Further, these materials also provide the benefit of reduction in pollutions. They can be tuned to extract different materials and properties accordingly. These natural material nanofilms possess very high efficiency in terms of their antimicrobial activity.

Many a time, they can be combined with metal nanoparticles synthesized through microbial or plant material activity. Further, their properties and area of influence can be tuned by methods of synthesis and modifications. This provides large potential in biomedical applications. A huge amount of work is still going on the property tuning and applications of these materials. They possess the potential to provide replacement for many current medicines while avoiding their limitations of side effects. They can be tuned for targeted and controlled medicinal applications with higher efficiency.

REFERENCES

1. Medvedeva, N. V.; Ipatova, O. M.; Ivanov, Y. D.; DRozhzhin, A. I.; Archakov, A. I. Nanobiotechnology and nanomedicine. *Biochemistry. (Moscow) Supplement Series B: Biomedical Chemistry*, 2007, 1, 114–124.

2. Gupta, R.; Xie, Nanoparticles in daily life: Applications, toxicity, and regulations. *Journal of Environmental Pathology, Toxicology, and Oncology*, 2018, 37, 209–230.

3. Charitidis, C. A.; Georgiou, P.; Koklioti, M.A.; Trompeta, A.F.; Markakis, Manufacturing nanomaterials: From research to industry. *Manufacturing Reviews*, 2014, 1, 11.

4. The Royal Society, *Nanoscience and Nanotechnologies*. The Royal Society, London, 2004, pp. 25–33.

5. Hua, S.M.; Metselaar, J. M.; Storm, G. Current trends and challenges in the clinical translation of nanoparticulate nanomedicines: Pathways for translational development and commercialization. *Frontiers in Pharmacology*, 2018, 9,790.

6. Kulkarni, N.; Muddapur, U. Biosynthesis of metal nanoparticles: A review. *Journal of. Nanotechnology*, 2014, 2014, 510246.

7. Joglekar, S.; Kodam, K.; DHaygude, M.; Hudlikar, M. Novel route for rapid biosynthesis of lead nanoparticles using aqueous extract of *Jatropha curcas L.* latex. *Material Letters*, 2010, 65, 3170–3172.

8. Gericke, M.; Pinches, A. Biological synthesis of metal nanoparticles. *Hydrometallurgy*, 2006, 83, 132–140.

9. Shukla, A.K.; Iravani, S. *Green Synthesis, Characterization, and Applications of Nanoparticles*. 2018, Elsevier, Amsterdam.

10. Cheviron, P.; Gouanve, F.; Espuche, E. Green synthesis of colloid silver nanoparticles and resulting biodegradable starch/silver nanocomposites. *Carbohydrate Polymers*, 2014, 108, 291–298.

11. Deplanche, K.; Caldelari, I.; Mikheenko, I.; Sargent, F.; Macaskie, L. Involvement of hydrogenases in the formation of highly catalytic Pd(0) nanoparticles by bioreduction of Pd(II) using *Escherichia coli* strains. *Microbiology*, 2010, 156, 2630–2640.

12. Yong, P.; Rowson, N. A.; Farr, J. P.; Harris, I. R.; Macaskie, L. E. Bioaccumulation of palladium by *Desulfovibrio desulfuricans. Journal of Chemical Technology and Biotechnology*. 2002, 77, 593–601.

13. Gahlawat, G.; Choudhury, A. A review on the biosynthesis of metal and metal salt nanoparticles by microbes. *RSC Advances*. 2019, 9, 12944–12967.

14. Iravani, S. Bacteria in nanoparticle synthesis: Current status and future prospects. *International Scholarly Research Notices*, 2014, 2014, 1–18.

15. Thakkar, K.N.; Mhatre, S.S.; Parikh, R.Y. Biological synthesis of metallic nanoparticles. *Nanomedicine: Nanotechnology, Biology, and Medicine*, 2010, 6, 257–62.

16. He, S.; Guo, Z.; Zhang, Y.; Zhang, S.; Wang, J.; Gu, N. Biosynthesis of gold nanoparticles using the bacteria *Rhodopseudomonas* capsulata. *Materials Letters*, 2007, 61, 3984–3987.

17. Saifuddin, N.; Wong, C. W.; Yasumira, A.A.N. Rapid biosynthesis of silver nanoparticles using culture supernatant of bacteria with microwave irradiation. *E-Journal of Chemistry*, 2009, 6, 61–70.

18. Kalimuthu, K.; Suresh Babu, R.; Venkataraman, D.; Bilal, M.; Gurunathan, S. Biosynthesis of silver nanocrystals by *Bacillus licheniformis*. *Colloids and Surfaces B: Biointerfaces*, 2008, 65, 150–153.

19. Priyadarshini, S.; Gopinath, V.; Meera, P. N.; MubarakAli, D.; Velusamy, P. Synthesis of anisotropic silver nanoparticles using novel strain, Bacillus flexus and its biomedical application. *Colloids and Surfaces B: Biointerfaces*, 2013, 102, 232–237.

20. Wei, X.; Luo, M.; Li, W.; Liangrong, Y.; Xiangfeng, L.; Lin, X.; Peng, K.; Huizhou, L. Synthesis of silver nanoparticles by solar irradiation of cell-free Bacillus amyloliquefaciens extracts andAgNO$_3$. *Bioresource Technology*, 2012, 103, 273–278.

21. Sunkar, S.; Nachiyar, C.V. Biogenesis of antibacterial silver nanoparticles using the endophytic bacterium *Bacillus cereus* isolated from *Garcinia xanthochymus. Asian Pacific Journal of Tropical Biomedicine*, 2012, 2, 953–959.

22. Klaus, T.; Joerger, R.; Olsson, E.; Granqvist, C. Silver-based crystalline nanoparticles, microbially fabricated. *Proceedings of the National Academy of Sciences of the United States of America*, 1999, 96, 13611–13614.

23. Shivaji, S.; Madhu, S; Singh, S. Extracellular synthesis of antibacterial silver nanoparticles using psychrophilic bacteria. *Process Biochemistry*, 2011, 46, 1800–1807.

24. Khan, M.E.; Khan, M.M.; Cho, M.H. Green synthesis, photocatalytic and photo- electrochemical performance of an Au–graphene nanocomposite. *RSC Advances*, 2015, 5, 26897–26904.

25. Khandel, P.; Shahi, S. K. Microbes mediated synthesis of metal nanoparticles: current status and future prospects. *International Journal of Nanomaterials and Biostructures*, 2016, 6, 1–24.

26. Satyanarayana, R.A.; Chen, C.; Jean, J.; Chien-Cheng, C.; Hau-Ren, C.; Min-Jen, T.; Cheng-Wei, F.; Jung-Chen, W. Biological synthesis of gold and silver nanoparticles mediated by the bacteria *Bacillus subtilis*. *Journal of Nanoscience and Nanotechnology*, 2010, 10, 6567–6574.

27. Wen, L.; Lin, Z.; Gu, P.; Jianzhang, Z.; Bingxing, Y.; Chen, G.; Fu, J. Extracellular biosynthesis of monodispersed gold nanoparticles by a SAM capping route. *Journal of Nanoparticle Research*, 2009, 11, 279–288.

28. Ivanova, E.P.; Nedashkovskaya, O.I.; Zhukova, N.V.; Nicolau, D.V.; Christen, R.; Mikhailov, V.V. Shewanella waksmanii sp. nov. isolated from a sipunculla (*Phascolosoma japonicum*). *International Journal of Systematic and Evolutionary Microbiology*, 2003, 53, 1471–1477.

29. Konishi, Y.; Tsukiyama, T.; Tachimi, T.; Saitoh, N.; Nomura, T.; Nagamine, S. Microbial deposition of gold nanoparticles by themetal-reducing bacterium *Shewanella algae*. *Electrochimica Acta*, 2007, 53, 186–192.

30. Du, L.; Jiang, H.; Liu, X.; Wang, E. Biosynthesis of gold nanoparticles assisted by *Escherichia coli* DH5α and its application on direct electrochemistry of hemoglobin. *Electrochemistry Communications*, 2007, 9, 1165–1170.

31. He, S.; Zhang, Y.; Guo, Z.; Gu, N. Biological synthesis of gold nanowires using extract of *Rhodopseudomonas capsulata*. *Biotechnology Progress*, 2008, 24, 476–480.

32. Lengke, M.F.; Fleet, M.E.; Southam, G. Morphology of gold nanoparticles synthesized by filamentous cyanobacteria from gold(I)-Thiosulfate and gold(III)-chloride complexes. *Langmuir*, 2006, 22, 2780–2787.

33. Lengke, M.F.; Ravel, B.; Fleet, M.E.; Wanger, G.; Gordon, R.A.; Southam, G. Mechanisms of gold bioaccumulation by filamentous cyanobacteria from gold(III)-chloride complex. *Environmental Science and Technology*, 2006, 40, 6304–6309.

34. Posfai, M.; Moskowitz, B.M.; Arato, B.; Schuler, D.; Flies, C.; Bazylinski, D. A.; Frankel, R.B. Properties of intracellular magnetite crystals produced by *Desulfovibriomagneticus* strain RS-1. *Earth and Planetary Science Letters*, 2006, 249, 444–455.

35. Lee, H.; Purdon, A.M.; Chu, V.; Westervelt, R.M. Controlled assembly of magnetic nanoparticles from magnetotactic bacteria using microelectromagnets arrays. *Nano Letters*, 2004, 4, 995–998.

36. Watson, J.H.P.; Croudace, I.W.; Warwick, P.E.; James, P.A.B.; Charnock, J.M.; Ellwoods, D.C. Adsorption of radioactive metals by strongly magnetic iron sulfide nanoparticles produced by sulfate-reducing bacteria. *Separation Science and Technology*, 2001, 36, 2571–2607.

37. Ummartyotin, S.; Bunnak, N.; Juntaro, J.; Sain, M.; Manuspiya, H. Synthesis of colloidal silver nanoparticles for printed electronics. *Comptes Rendus Chim*. 2012, 15, 539–544.

38. Bhainsa, K.; D'Souza, S. Extracellular biosynthesis of silver nanoparticles using the fungus *Aspergillus fumigatus*. *Colloids and Surfaces B: Biointerfaces*, 2006, 47, 160–164.

39. Ahmad, A.; Mukherjee, P.; Mandal, D.; Senapati, S.; Khan, M. I.; Kumar, R.; Murali, S. Enzyme mediated extracellular synthesis of CdS nanoparticles by the fungus, *Fusarium oxysporum*. *Journal of American Chemical Society*, 2002, 124, 12108–12109.

40. Sanghi, R.; Verma, P. Biomimetic synthesis and characterization of protein capped silver nanoparticles. *Bioresource Technology*, 2009, 100, 501–504.

41. Mukherjee, P.; Ahmad, A.; Mandal, D.; Senapati, S.; Sainkar, S.; Khan, M.; Renu, P.; Ajaykumar, P.V.; Mansoor, A.; Rajiv, K.; Murali, S. Fungus-mediated synthesis of silver nanoparticles and their immobilization in the mycelial matrix: A novel biological approach to nanoparticle synthesis. *Nano Letters*, 2001, 1, 515–519.

42. Castro, L.; Blazquez, M. L.; Munoz, J.; Gonzalez, F.; Ballester, A. Biological synthesis of metallic nanoparticles using algae. *IET Nanobiotechnology*, 2013, 7, 109–116.

43. Riddin, T. L.; Gericke, M.; Whiteley, C. G. Analysis of the inter- and extracellular formation of platinum nanoparticles by Fusarium oxysporum f. sp. lycopersici using response surface methodology. *Nanotechnology*, 2006, 17, 3482–3489.

44. Spadaro, D.; Gullino, M. Improving the efficacy of biocontrol agents against soil borne pathogens. *Crop Protect.* 2005, 24, 601–613.

45. Mandal, D.; Bolander, M.; Mukhopadhyay, D.; Sarkar, G.; Mukherjee, P. The use of microorganism for the formation of metal nanoparticles and their application. *Applied Microbiology and Biotechnology*, 2006, 69, 485–492.

46. Kowshik, M.; Deshmukh, N.; Vogel, W.; Urban, J.; Kulkarni, S.; Paknikar, K. Microbial synthesis of semiconductor CdS nanoparticles, their characterization, and their use in the fabrication of an ideal diode. *Biotechnology and Bioengineering*, 2002, 78, 583–588.

47. Gericke, M.; Pinches, A. Microbial production of gold nanoparticles. *Gold Bullettin*, 2006a, 39, 22–28.

48. Luangpipat, T.; Beattie, I.; Chisti, Y.; Haverkamp, R. Gold nanoparticles produced in a microalga. *Journal of Nanoparticle Research*, 2011, 13, 6439–6445.

49. Kahzad, N.; Salehzadeh, A. Green synthesis of CuFe$_2$O$_4$@Ag nanocomposite using the *Chlorella vulgaris* and evaluation of its effect on the expression of *norA efflux* pump gene among *Staphylococcus aureus* strains. *Biological Trace Element Research*, 2020, 198, 359–370.

50. Govindaraju, K.; Kiruthiga, V.; Kumar, G.; Singaravelu, G. Extracellular synthesis of silver nanoparticles by a marine alga, Sargassum wightii grevilli and their antibacterial effects. *Journal for Nanoscience and Nanotechnology*, 2009, 9, 5497–5501.

51. Rajasulochana, P.; Dhamotharan, R.; Murugakoothan, P.; Subbiah, M.; Krishnamoorthy, P. Biosynthesis and characterization of gold nanoparticles using the alga *Kappaphycusalvarezii*. *International Journal of Nanoscience*, 2011, 9, 511–516.

52. Dahoumane, S. A.; Yepremian, C.; Djediat, C.; Coute, A.; Fievet, F.;Coradin, T.; Alain, F.; Fernand, C.; Thibaud, B. R. Improvement of kinetics, yield, and colloidal stability of biogenic gold nanoparticles using living cells of *Euglena gracilis* microalga. *Journal of Nanoparticle Research*, 2016, 18, 79.

53. Lee, S.-W.; Mao, C.; Flynn, C. E.; Belcher, A. M. Ordering of quantum dots using genetically engineered viruses. *Science*, 2002, 296, 892–895.

54. Makarov, V.; Love, A. J.; Sinitsyna, O.; Makarova, S.; Yaminsky, I.; Taliansky, M., Kalinina, N.O.; Douglas, T.; Young, M.; Stubbs, G.; Mann, S. Green nanotechnologies: synthesis of metal nanoparticles using plants. *Acta Naturae*, 2014, 6, 35–44.

55. Merzlyak, A.; Lee, S.-W. Phage as templates for hybrid materials and mediators for nanomaterial synthesis. *Current Opinion in Chemical Biology*, 2006, 10, 246–252.

56. Kobayashi, M.; Tomita, S.; Sawada, K.; Shiba, K.; Yanagi, H.; Yamashita, I., Yukiharu, U. Chiral metamolecules consisting of gold nanoparticles and genetically engineered tobacco mosaic virus A new tobacco mosaic virus vector and its use for the systemic production of angiotensin-I-converting enzyme inhibitor intransgenic tobacco. *Optic Express*, 2012, 20, 24856–24863.

57. Love, A.; Makarov, V.; Yaminsky, I.; Kalinina, N.; Taliansky, M. The use of tobacco mosaic virus and cowpea mosaic virus for the production of novel metal nanomaterials. *Virology*, 2014, 449, 133–139.

58. Mukunthan, K.; Balaji, S. Cashew apple juice (*Anacardiumoccidentale* L.) speeds up the synthesis of silver nanoparticles. *International Journal of Green Nanotechnology*, 2012, 4, 71–79.

59. Park, Y.; Hong, Y. N.; Weyers, A.; Kim, Y. S.; Linhardt, R. J. Polysaccharides and phytochemicals: A natural reservoir for the green synthesis of gold and silver nanoparticles. *IET Nanobiotechnology*, 2011, 5, 69–78.

60. Safaepour, M.; Shahverdi, A. R.; Shahverdi, H. R.; Khorramizadeh, M. R.; Reza, G. A. Green synthesis of small silver nanoparticles using geraniol and its cytotoxicity against Fibrosarcoma-Wehi 164. *Avicenna Journal of Medical Biotechnology*, 2009, 1, 111–115.

61. Shankar, S.; Rai, A.; Ankamwar, B.; Singh, A.; Ahmad, A.; Sastry, M. Biological synthesis of triangular gold nanoprisms. *Nature Material*, 2004, 3, 482–488.

62. Smitha, S. L.; Philip, D.; Gopchandran, K. G. Green synthesis of gold nanoparticles using *Cinnamomum zeylanicum* leaf broth. *Spectrochim Acta A*, 2009, 74,735–739.

63. Bankar, A.; Joshi, B.; Kumar, A.; Zinjarde, S. Banana peel extract mediated synthesis of gold nanoparticles. *Colloids and Surfaces B Biointerface*, 2010, 80, 45–50.

64. Fayaz, AM.; Girilal, M.; Venkatesan, R.; Kalaichelvan, P.T. Biosynthesis of anisotropic gold nanoparticles using Maduca longifolia extract and their potential in infrared absorption. *Colloids and Surfaces B Biointerface*, 2011, 88,287–291.

65. Wang, Y.; He, X.; Wang, K.; Zhang, X.; Tan, W. BarbatedSkullcup herb extract-mediated biosynthesis of gold nanoparticles and its primary application in electrochemistry. *Colloids and Surfaces B Biointerface*, 2009, 73, 75–79.

66. Philip, D. Honey mediated green synthesis of gold nanoparticles. *Spectrochim Acta A*, 2009, 73,650–653.

67. Gangula, A.; Podila, R.; Ramakrishna, M.; Karanam, L.; Janardhana, C.; Rao, A.M. Catalytic reduction of 4-nitrophenol using biogenic gold and silver nanoparticles derived from *Breyniarhamnoides*. *Langmuir*, 2011, 27, 15268–15274.

68. Reddy, V.; Ramulu, T.; Sunjong, O.; CheolGi, K. Biosynthesis of gold nanoparticles assisted by *Sapindusmukorossi Gaertn*. Fruit pericarp and their catalytic application for the reduction of *p*-Nitroaniline. *Industrial Engineering Chemistry Research*, 2013, 52, 556–564.

69. Krishnaraj, C.; Muthukumaran, P.; Ramachandran, R.; Balakumaran, M.D.; Kalaichelvan, P.T. *Acalypha indica* Linn: Biogenic synthesis of silver and gold nanoparticles and their cytotoxic effects against MDA-MB-231, human breast cancer cells. *Biotechnology Reports*, 2014, 4, 42–49.

70. Velmurugan, P.; Sivakumar, S.; Young-Chae, S.; Seong-Ho, J.; Pyoung-In, Y.; Jeong-Min, S.; Sung-Chul, H. Synthesis and characterization comparison of peanut shell extract silver nanoparticles with commercial silver nanoparticles and their antifungal activity. *Journal of Industrial Engineering Chemistry*, 2015, 31, 51–54.

71. Luna, C.; Chavez, V.H.G.; Barriga-Castro, E.D.; Nunez, N.O.; Mendoza-Resendez, R. Biosynthesis of silver fine particles and particles decorated with nanoparticles using the extract of *Illicium verum* (star anise) seeds. *Spectrochimica Acta Part A: Molecular and Biomolecular Spectroscopy*, 2015, 141, 43–50.

72. David, L.; Moldovan, B.; Vulcu, A.; Olenic, L.; Perde-Schrepler, M.; Fischer-Fodor, E.; Florea, A.; Crisan, M.; Chiorean, I.; Clichici, S.; Filip, G.A. Green synthesis, characterization and anti-inflammatory activity of silver nanoparticles using European black elderberry fruits extract. *Colloids and Surfaces B Biointerface*, 2015, 122,767–777.

73. Diana, C.; Pedro, L. F.; Ana, M.; Pedro, D. V.; Luisa, M. S.; Ana, R. L. Preparation and physicochemical characterization of Ag nanoparticles biosynthesized by Lippia citriodora (Lemon Verbena). *Colloids and Surfaces B: Biointerfaces*, 2010, 81, 67–73.

74. Kouvaris, P.; Delimitis, A.; Zaspalis, V.; Papadopoulos, D.; Tsipas, S.A.; Michailidis, N. Green synthesis and characterization of silver nanoparticles produced using *Arbutus unedo* leaf extract. *Material Letters*, 2012, 76, 18–20.

75. Barbinta-Patrascu, M. E.; Badea, N.; Ungureanu, C.; Constantin, M.; Pirvu, C.; Rau, I. Silver-based biohybrids "green" synthesized from *Chelidonium majus L. Optic Materials*, 2015, 56, 94–99.

76. Coseri, S.; Spatareanu, A.; Sacarescu, L.; Rimbu, C.; Suteu, D.; Spirk, S.; Harabagiu, V. Green synthesis of the silver nanoparticles mediated by pullulan and 6-carboxypullulan. *Carbohydrate Polymers*, 2015, 116, 9–17.

77. Kumar, B.; Angulo, Y; Smita, K.; Cumbal, L.; Debut, A. Capuli cherry-mediated green synthesis of silver nanoparticles under white solar and blue LED light. *Particuology*, 2016, 24, 123–128.

78. Brumbaugh, A.D.; Cohen, K.A.; Angelo, S.K.S. Ultra small copper nanoparticles synthesized with a plant tea reducing agent. *ACS Sustainable Chemistry and Engineering*, 2014, 2, 1933–1939.

79. Lee, H.; Song, J.; Kim, B.S. Biological synthesis of copper nanoparticles using *Magnolia kobus* leaf extract and their antibacterial activity. *Journal of Chemical Technology and Biotechnology*, 2013, 88, 1971–1977.

80. Mahavinod Angrasan, J.K.V.; Subbaiya, R. Biosynthesis of copper nanoparticles by *Vitis vinifera* leaf aqueous extract and its antibacterial activity. *International Journal of Current Microbiology and Applied Sciences*, 2014, 3, 768–774.

81. Cheirmadurai, K.; Biswas, S.; Murali, R.; Thanikaivelan, P. Green synthesis of copper nanoparticles and conducting nanobiocomposites using plant and animal sources. *RSC Advances*, 2014, 4, 19507–1951.

82. Sinha, T.; Ahmaruzzaman, M. Biogenic synthesis of Cu nanoparticles and its degradation behavior for methyl red. *Material Letters*, 2015, 159,168–171.

83. Subhankari, I.; Nayak, P.L. Synthesis of copper nanoparticles using *Syzygiumaromaticum* (Cloves) aqueous extract by using green chemistry. *World Journal of Nano Science Technology*, 2013, 2, 14–17.

84. Saranyaadevi, K.; Subha, V.; Ravindran, R.S.E.; Renganathan, S. Synthesis and characterization of copper nanoparticles using *Capparis zeylanica* leaf extract. *International Journal of Chemtech Research*, 2014, 6, 4533–4541.

85. Subbaiya, R.; Selvam, M.M. Green synthesis of copper nanoparticles from *Hibicus rosa-sinensis* and their antimicrobial, antioxidant activities. *Research Journal of Pharmaceutical, Biological and Chemical Sciences*, 2015, 6, 1183–1190.

86. Ulkarni, V.; Suryawanshi, S.; Kulkarni, P. Biosynthesis of copper nanoparticles using aqueous extract of *Eucalyptus* sp. plant leaves. *Current Science*, 2015, 109, 255–257.

87. Amer, M.; Awwad, A. Green synthesis of copper nanoparticles by *citrus limon* fruits extract, characterization and antibacterial activity. *Chemistry International*, 2021, 7, 1–8.

88. Caroling, G.; Vinodhini, E.; Ranjitham, A.M.; Shanthi, P. Biosynthesis of copper nanoparticles using aqueous *Phyllanthusembilica* (Gooseberry) extract- characterisation and study of antimicrobial effects. *International Journal of Nanotechnology*, 2015, 1, 53–63.

89. Mali, S.C.; Anita, D.; Githala, C.K.; Trivedi, R. Green synthesis of copper nanoparticles using *Celastruspaniculatus* Willd. leaf extract and their photocatalytic and antifungal properties. *Biotechnology Reports*, 2020, 27, e00518,

90. Ranu, B.; Stolle, A. Ball milling towards Green synthesis: Applications, projects, challenges. *Johnson Matthey Technology Reviews*, 2016, 60, 148–150.

91. Kitchen, HJ.; Simon, R. V.; Jennifer, L. K.; Nuria, T.; Lucia, C.; Andrew, H.; Gavin, A.W.; Timothy, D. D.; Samuel, W. K.; Duncan, H. G. Modern microwave methods in solid-state inorganic materials chemistry: From fundamentals to manufacturing. *Chemical Reviews*, 2014, 114(20), 1170–1206.

92. Ardila, A. A. N.; Arriola, E.; Reyes, C. J.; Berrio, M.E.; Fuentes, Z. G. Mineralizacion de etilenglicol-porfoto-fentonasistido con ferrioxalato. *Revista Internacional de Contaminacion Ambienta*, 2016, 32, 213–226.

93. Taublaender, A.M.J.; Glöcklhofer, F.; Marchetti- Deschmann, M.; Unterlass, M.M. Green and rapid hydrothermal crystallization and synthesis of fully conjugated aromatic compounds. *Angewandte Chemie*, 2018, 57, 12270–12274.

94. Ashokkumar, M. Ultrasonic synthesis of functional materials. In *Ultrasonic Synthesis of Functional Materials* (ed. M. Ashokkumar), Springer, Cham, 2016, 17–40.

95. Xiao, W.; Liu, X.; Hong, X.; Yang, Y.; Lv, Y.; Fang, J.; Ding, J. Magnetic-field-interacted synthesis of magnetite nanoparticles via thermal decomposition and their hyperthermia property. *Cryst Eng Comm*, 2015, 17, 1–3.

96. Siddiqi, K.S.; Husen, A.; Rao, R.A.K. A review on biosynthesis of silver nanoparticles and their biocidal properties. *Journal of Nanobiotechnology*, 2018, 16, 14.

97. Duan, H.; Wang, D.; Li, Y. Green chemistry for nanoparticle synthesis. *Chemical Society Reviews*, 2015, 4, 5778–5792.

98. Mukunthan, K.; Balaji, S. Cashew apple juice (*Anacardium occidentale* L.) speeds up the synthesis of silver nanoparticles. *International Journal of Green Nanotechnology*, 2012, 4, 71–

99. Kulkarni, K.; Wadhvane, Y.; Chendake, Y.; Kulkarni, A. Microwave assisted synthesis of antimicrobial nano-films from Water Hyacinth. *Journal of Biomimetics, Biomaterials and Biomedical Engineering*, 2022, 56, 37–48.

Section II

Green Nanomaterials

Applications in Bio-Medicine, Drug Delivery, Energy, Sensing and Other

6 Natural Origin Biodegradable Polymer Bandages and Structures Loaded with Nanomedicines for Biomedical Applications

Sonali Dhamal, Yogesh Chendake, and Sachin Chavan
Bharati Vidyapeeth Deemed to be University,
Pune, India

CONTENTS

6.1 INTRODUCTION

Today's world need a huge amount of materials for numerous applications. Although numerous synthetic systems are present, the natural materials possess large benefits of being renewable resource, property tuning and biocompatibility properties. Biomedical application is one such area where these properties are highly essential. Here, careful property tuning of materials along with their optimization is highly important. This would affect in multiple fashion in terms of medicinal loading, controlled release, targeted delivery and applicability.

This provides a unique position for natural biodegradable polymers for use in these applications. These materials are obtained from natural (plant or animal) origin and possess excellent biocompatibility properties. Their shaping provides new avenues for the formation of different materials. Although various applications are available and being developed regularly, still this area needs large amounts of investigation into the formation of different components starting from medicinal tablets, scaffolding to bandages.[1] Here, the applications can be controlled by medicinal loading, its

DOI: 10.1201/9781003319153-8

quality and other properties. The loading is dependent upon material compatibility, the sorption properties of base polymer and the technological properties controlling shaping and formation of materials. Here, natural polymers possess the benefits of an interactive material and hence support the medicinal loading. Moreover, these materials being natural in origin can undergo degradation using specific enzymes or body fluid components. This gives an important benefit for targeted and controlled release of drug depending upon disease and medical conditions.[2]

One of the important applications in this case is wound bandages. These are special cases of conditions where a large attention and research is going on to prepare biocompatible and biodegradable bandages. It is of special importance due to variable nature of wounds and their medicinal requirements. The wounds can be superficial, deep cuts, pus forming, oozing and with the infection with other components. Each one of them needs special and different kind of care for early recovery. In the conventional treatment methodology, these wounds are cleaned and debrided to remove the degraded tissues, microbial contamination, foreign material, pus and other components. These are then applied with antimicrobial creams and growth initiators so as to start healing. It is then applied with cotton gauge to hold these materials on wound and avoid its contact with external environment and then stick or hold in place using gauge or sticky tapes.[3]

This process has certain limitations associated with it. These are majorly observed in case of large and oozing wounds. In these cases, there is need for large gauges, which may or may not cover the wounds properly, and some of the parts would get exposed to atmosphere during the application of gauges or movements later. Second, the medicine may not get distributed properly and uneven distribution of medicine would affect the healing process. Third, the wounds needs frequent cleaning by removing gauge in case of large or oozing wounds. This would result in pulling and detaching newly growing tissues from the side of wounds. Further, this cleaning and redressing of wound exposes it to microbial contamination from the atmosphere, thus affecting the recovery process. The oozing wounds have an additional issue of cleaning of pus and dead cells, which can be highly painful depending upon wound condition and size. Moreover, these bandages cause accumulation of sweet under the bandage, which can damage the growing tissues there.

Hence, it is necessary to define a modified bandage system. It can be done by the careful design of wound bandage containing nanomaterial medicines dispersed in the biodegradable polymer layer.[3] Here, the proposed polymeric layer can be applied on the wound surface for bandage. The presence of body enzymes would trigger degradation of polymeric layer and release the medicine imbibed in polymeric layer. This release would be managed depending upon body enzymes, which in turn would be dependent upon wound condition. Further, a careful design and optimization of polymeric layer can be used to introduce microporosity in the wound bandage layer. It would be beneficial for the removal of oozing pus from wound while restricting its exposure to microorganisms, while the pores can avoid sweating and corresponding damage of tissues. This would help in early recovery. Additionally, the use of natural biopolymers for bandage preparation would support during setting up its degradation kinetics along with added biocompatibility. Second, these polymers being natural in origin, their degradation products can be easily absorbed in body tissues without resulting in unwanted reactions.

Hence, the use of natural origin biopolymer with controlled degradation loaded with nanomedicine would be highly beneficial in such applications. A similar case can be considered for nanomedicine-loaded scaffolding. They would provide the benefits of support for controlled and directional bone or tissue growth along with sequential degradation of scaffolding polymer and would provide controlled release of medicine and growth components for growing tissues and bones.[4]

Different biopolymers have been investigated for their applicability in biomedical applications, viz. polylactide, polyglycolide, chitosan, gelatin and alginates. They can be obtained from plant or animal origin, and different applications are being developed using these polymers through medicinal loading. On the other hand, cellulose and curcumin possess a different position altogether. They can be formed into nanofilm layers which can be utilized in biomedical applications. The property modulations and loading capabilities along with compositional variations provide unique positions for these materials in different applications.[5]

6.2 SOURCES OF BIOPOLYMERS

Different materials and biopolymers have been reported for their application in biomedical field as described earlier. Some of them can be grouped as proteins (collagen, gelatin, silk fibroin and albumin), polysaccharides (cellulose, starch, chitosan and alginate) or other materials (curcumin and other biosynthesized nanoparticles-based biopolymers).[6] All these are playing an important role in biomedical applications and drug delivery. Let's discuss some of these examples in detail along with their applicability.

6.2.1 CHITOSAN

Chitosan is one of the important groups of biopolymers with huge applications in the biomedical field. A wide range of natural and synthetic forms of chitosan have been utilized in many applications. In case of natural chitosan, they can be of both plant and animal origin and have the benefit of huge property tuning. Further, they possess benefits of biocompatibility, non-toxicity, non-immunogenicity and biodegradability. Their structure is given below. This structure shows the presence of various active groups such as NH_2, –OH and -O- along with the backbone of enclosed cyclic structure (Figure 6.1).

It supports interactive sorption and loading of the components. This would benefit the loading of nanomaterials and other medicines. Further, the chitosan is known for its controlled degradation in contact with body fluids and enzymes. This can be utilized for the controlled release of nanomaterials and medicinal components. Moreover, the presence of different functional groups provides the unique feasibility of chemical modification of these materials. These chemical modifications can be tuned and controlled to provide desired life and targeted material release.

6.2.2 CELLULOSE

Cellulose is another important material and a type of polysaccharide. Its surface structure can be modified for application in tissue engineering and drug delivery systems. Cellulose has been used in the development of biodegradable polymeric products. Some of the special types of cellulose nanofibers possess higher microbial activity zone compared to synthetic polymers. Their properties and composition may vary from the source of cellulose. It gives the major benefit of controlled property variation, medicinal loading, biocompatibility and degradation. This would help to enhance stability and biomedicinal applicability (Figure 6.2).

6.2.3 PECTINS

Pectins are other important components that can be generally sourced from citrus fruit extracts. They are majorly ethylcellulose-, calcium- and hydroxypropyl methylcellulose-based compounds. Their composition itself is important for medicinal application, and it gives the important benefit of

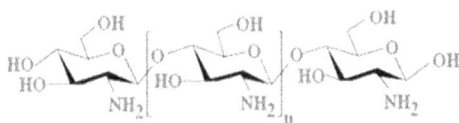

FIGURE 6.1 Structure of chitosan.

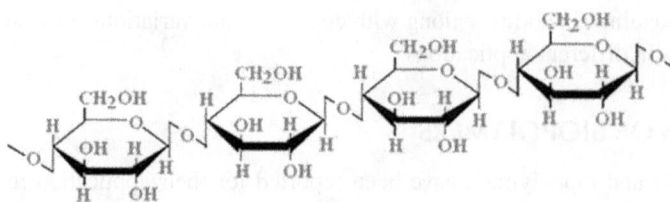

FIGURE 6.2 Structure of cellulose.

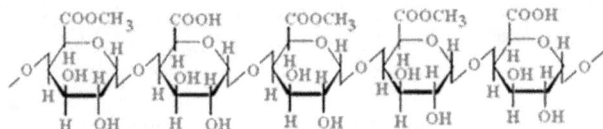

FIGURE 6.3 Structure of pectin.

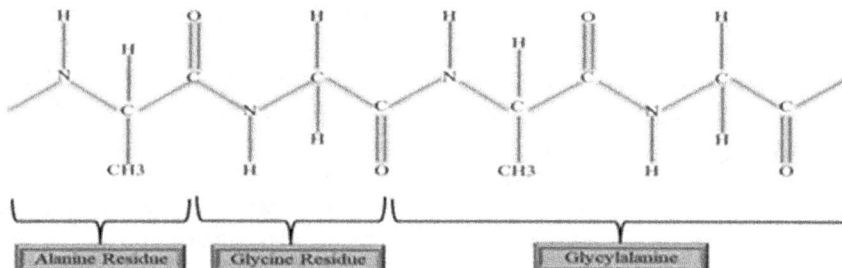

FIGURE 6.4 Structure of silk fibroin.

property tuning for effective loading capacity and sustained release profile for drugs. These biomaterials can be used in the form of pectin for therapeutic area (Figure 6.3).

6.2.4 SILK FIBROIN

It exhibits various properties such as self-assembly, biocompatibility, mechanical hardness and processing malleability. This can be utilized for multiple biomedical applications by loading different medicines. Further, being natural fiber made up of proteins, it can undergo degradation and biosorption in body. Hence, it is responsible for slow and sustained delivery of drug to the target site. The formation of electrospun fibers, hydrogels and films with three-dimensional scaffolds is achievable with silk fibroin. This is classified species-wise as silkworm and spider dragline silk and utilized in the preparation of various nanofibers (Figure 6.4).

6.2.5 COLLAGEN

Collagen is a part of musculoskeletal tissues in mammals. It is one of the proteinaceous materials. It possesses -NH2, -COOH, N, =O and -OH groups along with pyridine group in backbone. These groups assist in the interactive sorption. This would help in the loading of medicine and its capping to avoid any loss of properties or materials. It has been explored for the transmission of antibiotics with low molecular weight in drug delivery systems, which improves the prolonged systematic

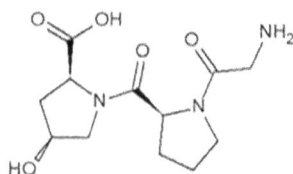

FIGURE 6.5 Structure of collagen.

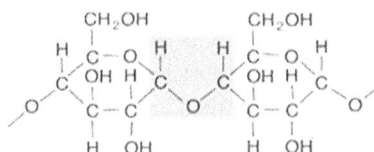

FIGURE 6.6 Structure of simple starch.

exposure of drug delivery. Due to the presence of combination of charges and functional groups, it can be modified and capped to control the release of medicines. Second, the modifications and capping can be controlled in such a way as to control their degradation kinetics. Further, the degradation and medicine release can be designed in such a way to provide targeted and controlled drug delivery (Figure 6.5).

6.2.6 SIMPLE STARCH

Starch is a polysaccharide with the presence of alcohol, acid and ester groups along with carbon backbone. It is the product of metabolism, which gets collected in plants. Hence, it has unique functional and physicochemical characteristics and is easily available from natural sources such as wheat, rice, potato and corn. Chemical modifications are feasible with starch, which enhances the capacity for drug delivery system. Further, they provide the benefits of high blood absorption capacity as being similar to food components. Moreover, their digestion components would result in saccharides, which are similar to those consumed and utilized by body cells for metabolism and energy generation. Thus, these components would be beneficial for body growth without any need for removal as observed in case of synthetic polymers and drug delivery capsule systems (Figure 6.6).

6.2.7 GELATIN

Gelatin is another example of natural protein-based biopolymers. It has benefits of large water absorption and possesses the ability of huge water intake. This would be beneficial in loading of water-soluble drugs and can be carefully designed to encapsulate them. It can be used in the form of flexible scaffolding as well as thin protective layers for multiple applications. Their issues remain in the formation of porous layers, which can act as secondary skin. The significant application of gelatin hydrogels for drug delivery due to their high surface loading capacity improves controlled drug release to the targeted site (Figure 6.7).

6.3 BIOCOMPATIBILITY AND BIODEGRADATION PROPERTIES

The biological activities produce different polymeric structures, which possess numerous applications. They occur as part of skin in human and animals, hairs and their basic components, the web fibers of spider, silk of animal origin; while cellulose, hemicellulose along with jute chitosan and other fibers

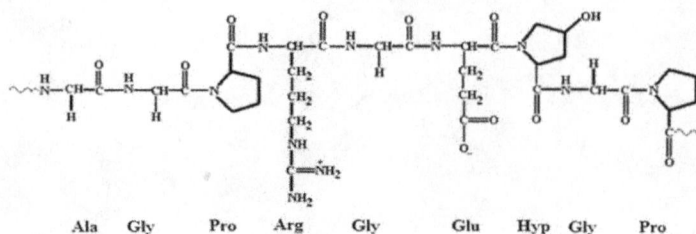

Ala   Gly   Pro   Arg   Gly   Glu   Hyp Gly   Pro

FIGURE 6.7 Structure and composition of gelatin.

of plant origin. These naturally occurring biopolymers are composed of components such as proteins, starch and polysaccharides. These materials are enriched with biocompatibility, biodegradability, antibacterial activity and low immunogenicity. Further, they can be tuned with different property variations, which would help in the loading of drugs and access release locations by interactive digestions. These biopolymers enhance the accessibility of drug particles to targeted tissues and cell walls with significantly low binding capacity and high size distribution in controlled pattern.[7] Further, their degradation can be tuned to occurrence of specific enzymes, so the drug release would be initiated only for target tissues avoiding damage or contamination of other tissue systems.

Moreover, these materials are of natural origin. Their backbone contains proteins, saccharides, starch and other related materials. Their properties can be tuned to provide desired degradation kinetics. It can be degraded by the enzymatic activity and body fluid properties. Thus, the degradation kinetics can be tuned. This degradation would produce the components of proteins, sugars and starch. They are the components of food and can be digested or taken up by human body for growth and energy purpose as a food component. Hence, these materials are highly important for biomedical applications (Figure 6.8).

Research in biodegradable devices for temporary therapeutic applications is becoming the subject of special interest for scientists from allied disciplines. These materials and devices can be implanted or introduced in human body, loaded with medicines. It would provide the targeted and controlled drug delivery. It benefits early recovery and avoids the side effects of medicines. A lager work is targeted towards its development influenced by the vital role of this in human health and human welfare.[8]

Biodegradable devices have overcome long-term biocompatibility issues related to existing permanent implants. They can play the combined role of supporting implants for natural growth and recovery, while providing growth initiators and other medicines in controlled release. Any material with biocompatibility factor is one of the potent prerequisites for the consideration of that material as a biomaterial. Their applicability is dependent upon numerous parameters and myriad factors ranging from biological, chemical and physical properties which support biomaterial to convert into required implant. The biocompatibility of biomaterials has been influenced by inherent properties such as shape, the structure of implant, solubility, lubricity, water absorption, surface energy, hydrophilicity, degradation with erosion mechanism and material chemistry.

A study has recently targeted on the synthesis of biodegradable polymer considering specific properties with respect to applications by (i) unique artificial polymers designed with special compositions to enhance diversity, (ii) biomimetic polymers invented with biosynthetic process and (iii) achieving the design of biomaterial with computational and combinational pathways for the exploration of resorbable polymers.

6.4 LOADING OF NANOMEDICINE

Therapeutic activity has been proceeded with the help of biopolymers, which are important media to carry desired drug material to respective body site in modern drug delivery systems. Loading of

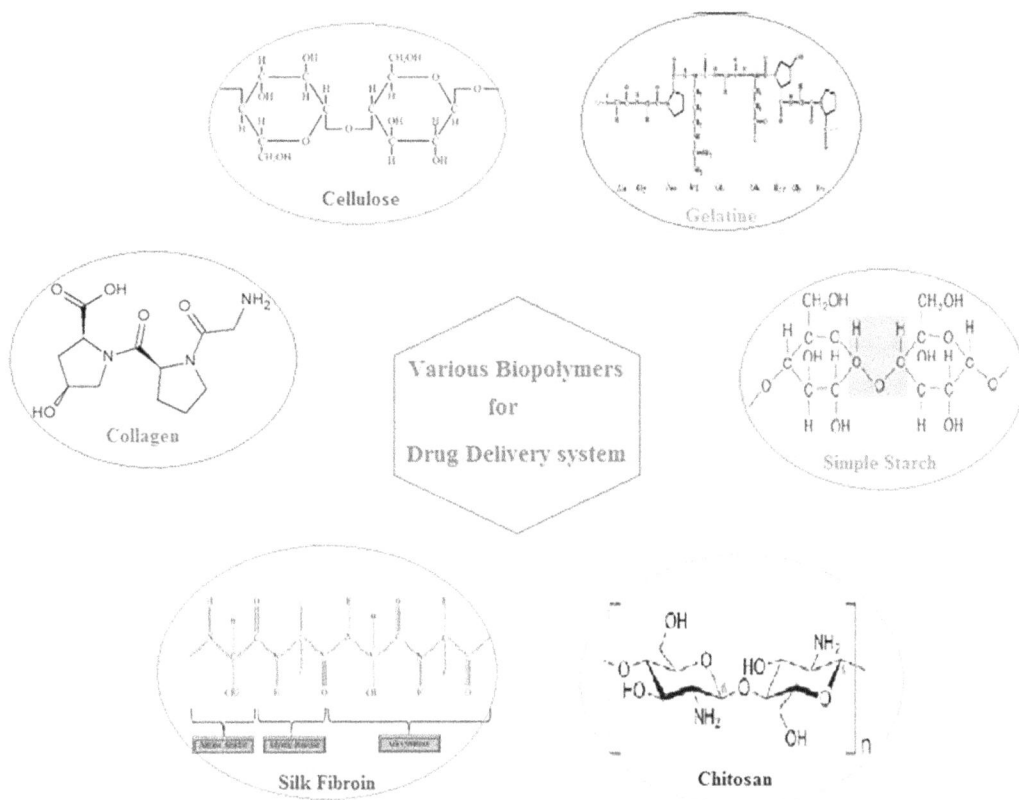

FIGURE 6.8 Depiction of importance and applicability of biopolymers in drug delivery applications.

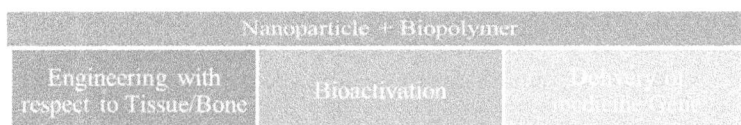

FIGURE 6.9 Depiction of nanoparticle-biopolymer combination for medicinal applications.

drug has been carried out with various carriers for drug, such as beads, gels, hydrogels, biocomposites, microparticles, scaffolds, nanocapsules, nanoparticles, microcapsules, micelles, dendrimers, spheroids, films, implants and patches.[9] Macromolecules are the basic unit of living structures, and biomaterials exhibit a significant change with respect to external factors and thus become an effective strategy to target properties. Diagnostics techniques, antibacterial treatments, drug delivery, wound treatment and cell repair could be programmed with biomaterials in medical applications (Figure 6.9).

6.5 NANOMEDICINE IMPORTANCE, RELEASE SYSTEM AND CONTROL

Biopolymers with relevant significance are applicable to every discipline of medicine. They can be used in the format of extracorporeal devices, temporary implants and permanent implants. The extracorporeal devices would be artificial skin or wound bandages, artificial kidneys, fluid lines and contact lenses. Temporary implants would be, for example, degradable sutures, polymer scaffoldings

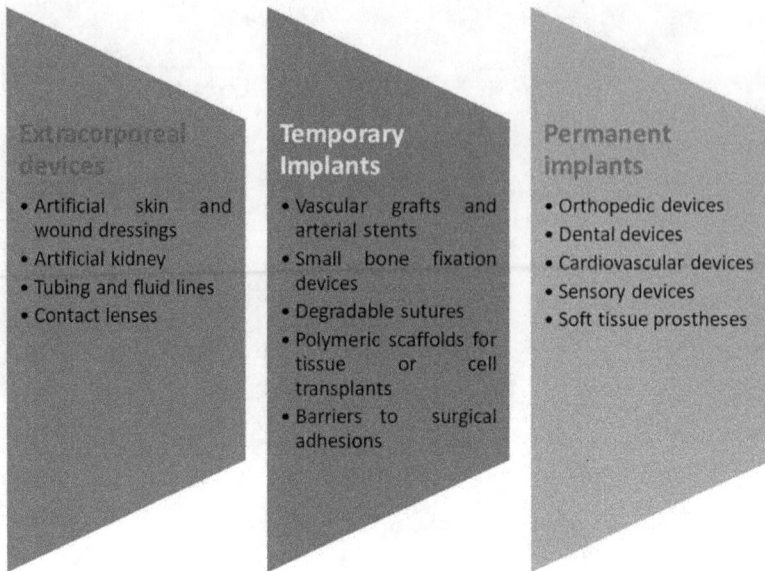

Extracorporeal devices	Temporary Implants	Permanent implants
• Artificial skin and wound dressings • Artificial kidney • Tubing and fluid lines • Contact lenses	• Vascular grafts and arterial stents • Small bone fixation devices • Degradable sutures • Polymeric scaffolds for tissue or cell transplants • Barriers to surgical adhesions	• Orthopedic devices • Dental devices • Cardiovascular devices • Sensory devices • Soft tissue prostheses

FIGURE 6.10   Different biomedical devices and their forms.

or bone grafts, while permanent implants would be, for example, dental, sensory or cardiovascular implants. These components are highly important for biomedical devices or medical applications.[10]

Further, these components would require medicinal components. The loading would require size and property optimization. Nanomaterial-based medicine possesses benefits of smaller size and ease of loading in medicinal devices. Different nanomedicines are being designed such as nano-silver materials and curcumin nanofibers. These are known to possess excellent medicinal applications and higher efficiency and would avoid unnecessary waste formation and side effects (Figure 6.10).[11]

In the development and application of controlled delivery systems, drug dosing profile with respect to rate, time and place influenced with greater convenience, greater effectiveness in the treatment of chronic conditions, immediate-release drugs, decreased side effects and increased levels of patient compliance. The formation of nanomaterial medicine would be able to design and develop the medicine and its delivery system to obtain above properties.[12]

Hence, pharmaceutical development has been investigated on the basis of economic as well as technological perspectives to overcome challenges related to drug delivery systems.

Case study: Bandage and scaffolding applications

Advanced research and developments in the area of bioactive polymers upgraded the importance of various novel and smart drug delivery systems, which improve safety and therapeutic efficacy.

The synthetic bioactive ceramic was reported as a biocompatible material with bone marrow, and it improves osteointegration in vivo tests of rabbits.[1,14] For wound healing and skin engineering, Application of cellulose in formation of Alloderm is the best option that has been studied in the treatment of burn cases, cellulose adhesions and proliferation. Human cadavers are the main source of Alloderm, which is an acellular collagen matrix.[15]

Reconstituted collagen type I matrix was examined in the wound dressing of animals, which has decreased the period of healing with wound contraction.[16] Lee CH et al. reported collagen dressings that have been applied to improve the growth of new skin cells at wound sites.[18] N-acetylglucosamine from chitin applied for wound healing dressing could enhance the rate of tissue repair by overcoming the problem of scars formation with skin contraction.[19,20] The powder form of chitosan and chitin has been incorporated in membranes, films, woven or non-woven dressings and gels.[21]

Oxidized cellulose from rayon with its ethyl, methyl, acetate-phthalate and amino ethyl compounds has been reported acting as membranes in sutures and bandages.[17] Calcium phosphate-based composites with ceramics and PCL are also being reported as suitable scaffolds for bone tissue engineering.[22] The human body has glycosaminoglycans and hyaluronic acid with the presence of highly reactive amino groups along the polymer backbone, which resembles the structural composition of chitosan. Stimulatory properties of chitin and chitosan along with their antibacterial and hemostatic properties improve the potential of chitosan as a natural polymer for wound healing applications.[23–26] For tissue engineering applications, elastic poly(ester urethane) (degrapols) was applied to develop highly porous scaffolds, which have biodegradable property.[27]

6.6 CONCLUSIONS AND PATH FORWARD

Recently, lot of studies have reported on different applications of biopolymers, but these have various challenges related to the flow rate of the recommended biopolymer as per desired therapeutic activity. Apparently, most of the studies investigated with the help of in vitro pathway; hence, it is essential to modify the procedures of fabricated biopolymers as per different demands to overcome deficiencies to enhance accessibility towards the target site. These applications and property designs have resulted in many modifications for property and medicinal applicability modifications. Further property optimizations and drugs and delivery systems need to be optimized as per the medicinal requirements.[13]

REFERENCES

1. Kunduru, K.R., Domb, A.J., Basu, A. Biodegradable polymers: Medical applications. *Encyclopedia Polymer Science and Technology.* doi: 10.1002/0471440264.pst027.pub2.
2. Ulery, B.D., Nair, L.S., Laurencin, C.T. Biomedical applications of biodegradable polymers. *Journal of Polymer Science: Polymer Physics* 2011; 49(12): 832–64. doi:10.1002/polb.22259.
3. Yadav, P., Yadav, H., Shah, V.G., Shah, G., Dhaka, G. Biomedical biopolymers, their origin and evolution in biomedical sciences: A systematic review. *Journal of Clinical and Diagnostic Research* 2015; 9(9): ZE21–5.
4. Mohan, S., Oluwafemi, O.S., Kalarikkal, N., Thomas, S., Songca, S.P. Biopolymers – Application in Nanoscience and Nanotechnology. http://dx.doi.org/10.5772/62225.
5. Hassan, M.E., Bai, J., Dou, D.-Q. Biopolymers; definition, classification and applications. *Egyptian Journal of Chemistry* 2019; 62(9): 1725–37.
6. Altomare, L., Bonetti, L., Campiglio, C.E., De Nardo, L., Draghi, L., Tana, F., Farè, S. Biopolymer-based strategies in the design of smart medical devices and artificial organs. 2018. doi: 10.1177/0391398818765323.
7. Jacob, J., Haponiuk, J.T., Thomas, S., Gopi, S. Biopolymer based nanomaterials in drug delivery systems: A review. *Materials Today Chemistry* 2018; 9: 43e55.
8. Hasnain, Md.S., Ahmed, S.A., Alkahtani, S., Milivojevic, M., Kandar, C.C., Dhara, A.K., Nayak, A.K. *Biopolymers for Drug Delivery.* Springer, Cham, 2020. https://doi.org/10.1007/978-3-030-46923-8_1.
9. Nair, L.S., Laurencin, C.T. Biodegradable polymers as biomaterials. *Progress in Polymer Science* 2007; 32: 762–98.
10. Van de Velde, K., Kiekens, P. Biopolymers: Overview of several properties and consequences on their applications. *Polymer Testing* 2002; 21: 433–42.
11. Vilar, G., Tulla-Puche, J., Albericio, F. Polymers and drug delivery systems. *Current Drug Delivery* 2012; 9: 367–94.
12. Nayak, A.K., Ahmad, S.A., Beg, S., Ara, T.J., Hasnain, M.S. Drug delivery: Present, past, and future of medicine. *Applications of Nanocomposite Materials in Drug Delivery* 2018. doi: https://doi.org/10.1016/B978-0-12-813741-3.00012-1©. Elsevier.

13. Sung, Y.K., Kim, S.W. Recent advances in polymeric drug delivery systems. In: R. Pignatello (ed.) *Biomaterials Applications for Nanomedicine*. ISBN 978-953-307-661-4.

14. George, J., Onodera, J., Miyata, T. Biodegradable honeycomb collagen scaffold for dermal tissue engineering. *Journal of Biomedical Materials Research Part A* 2008, 87A: 1103–11. [PubMed: 18792951]

15. Leipziger, L.S., Glushko, V., DiBernardo, B., Shafaie, F., Noble, J., Nichols, J., Alvarez, O.M. Dermal wound repair: Role of collagen matrix implants and synthetic polymer dressings. *Journal of the American Academy of Dermatology* 1985; 12: 409–10.

16. Devi, K.S., Sinha, T.J.M., Vasudevan, P. Biosoluble surgical material from 2,3-Dialdehyde cellulose. *Biomaterials* 1986; 7: 193–6.

17. Sung, Y.K., Kim, S.W. Recent advances in polymeric drug delivery systems. *Biomaterials Research* 2020; 24:12. https://doi.org/10.1186/s40824-020-00190-7.

18. Lee, C.H., Singlaa, A., Lee, Y. Biomedical applications of collagen. *International Journal of Pharmaceutics* 2001; 221: 1–22. doi: 10.1016/S0378-5173(01)00691-3.

19. Dai, T., Tanaka, M., Huang, Y.Y., et al. Chitosan preparations for wounds and burns: Antimicrobial and wound-healing effects. *Expert Review of Anti-infective Therapy* 2011; 9(7): 857–79.

20. Kofuji, K., Qian, C. J., Nishimura, M., et al. Relationship between physicochemical characteristics and functional properties of chitosan. *European Polymer Journal* 2005; 41(11): 2784–91.

21. Ivanova, E. P., Bazaka, K., Crawford, R. J., 2–Natural polymer biomaterials: Advanced applications. In: *New Functional Biomaterials for Medicine &Healthcare* 2013; 32–70. doi: 10.1533/9781782422662.32.

22. Mondrinos, M.J., Dembzynski, R., Lu, L., Byrapogu, V.K.C., Wootton, D.M., Lelkes, P.I., et al. Porogen-based solid free form fabrication of polycaprolactone—calcium phosphate scaffolds for tissue engineering. *Biomaterials* 2006; 27: 4399–408.

23. Jayakumar, R., New, N., Tokura, S., Tamura, H. Sulfated chitin and chitosan as novel biomaterials. *International Journal of Biological Macromolecules* 2007; 40: 175–81.

24. Suh, J.K.F., Matthew, H.W.T. Application of chitosan-based polysaccharide biomaterials in cartilage tissue engineering: A review. *Biomaterials* 2000; 21: 2589–98.

25. Stone, C.A., Wright, H., Clarke, T., Powell, R., Devaraj, V.S. Healing at skin graft donor sites dressed with chitosan. *British Journal of Plastic Surgery* 2000; 53: 601–6.

26. Burkatovskaya, M., Tegos, G.P., Swietlik, E., Demidova, T.N., Castano, A.P., Hamblin, M.R. Use of chitosan bandage to prevent fatal infections developing from highly contaminated wounds in mice. *Biomaterials* 2006; 27: 4157–64.

27. Saad, B., Hirt, T.D., Welti, M., Uhlscgmid, G.K., Neuenschwander, P., Suter, U.W. Development of degradable polyesterurethanes for medical applications: In vitro and in vivo evaluations. *Journal of Biomedical Materials Research* 1997; 36: 65–74.

7 Callus-Derived Bioactive Nanomaterials for Sustainable Drug Delivery

Neha Saini, Prem Pandey, and Atul Kulkarni
Symbiosis International (Deemed University)

CONTENTS

7.1 INTRODUCTION

Since historical times, human being is using plant products in medicines, food, nutrition and industries. Many plant species got endangered, and some are on the verge of extinction as well, due to limited resources, but expending need of the population [1]. Biotechnology comes as a rescue in this regard providing *in vitro* plant-based tools such as callus, suspension cell and organ culture as well as gene manipulation to get desired plant products [1,2]. Alongside, chemical compounds can be synthesized from secondary plant metabolites by application of external stimulus to plants. The external signals lead to secretion of secondary metabolites due to plant's ability to remould and adapt themselves in extreme conditions; these elicitors (stimuli) could be environmental stimuli, chemicals, pathogens, wounding, nutrient scarcity, etc. [1,3]. The elicitors impact the gene expressions as a defence in response to the stimuli. When wounded, a plant heals it by producing an undifferentiated cell mass called "callus". Callus is undifferentiated cell mass with totipotent/stem cell-like activity. Callus can be produced by any part of the plant and has the potential to re-differentiate itself into full plant body [1,4]. Furthermore, callus culture synthesis when amalgamated with nanotechnological tools has led to unexplored areas of callus-mediated nanomaterials (NMs).

Nanotechnology has a potential impact in the development of sustainable technology. Since nanotechnology seems to be a subsequent industrial revolution of the near future considering its impact on the world economy, industries and other areas of life [5,6], the realm of nanomedicine is bridging the gap between physiology and nanotechnology, thus allowing researchers to explore the field of traditional therapeutics and developing alternatives of contemporary tools and DDS,

DOI: 10.1201/9781003319153-9

overcoming the pitfalls of currents technological drawbacks [7,8]. The method of utilization of plant extracts to synthesize nanomaterials connects nanotechnology with plant biotechnology, via a green chemistry approach [9,10]. These plant components (such as saccharides, terpenoids, polyphenols, alkaloids and proteins) act as a reductant for metal ions and thus form the nanomaterials via a top-down approach [4]. On the contrary, conventional nanotechnology utilizes toxic chemicals that damage the environment [1,11]. Thus, biological approaches are now more in demand due to their eco-friendly approaches [11]. However, more than green synthesis using crude plants extract, the callus extracts and certain phytochemicals from plant cells are considered more effective and rather a sustainable approach [1].

Extensive research has been done in NMs-based drug delivery systems (DDS) as theranostics modalities for chronic diseases such as osteogenic ailments due to their small dimensions and easy passage via bodily barriers to deliver drug to target regions. Nano-DDS are promising as they have long circulation time and higher biodistribution in the system. In the past few years, the scientists have checked the regeneration potential of nano-DDS on tissues [6,7,12]. The synthesis of NMs as DDS via plant extract has major edge over conventional techniques as it is eco-friendly, biocompatible, chemical-free and highly stable [13]. However, the exuberant potential of nanomaterials needs to be explored more as a drug delivery system to combat wide range of ailments [13]. However, callus synthesis is a promising method for sustainable preparation of nanomaterials. Literature shows that fruit peels and other green wastes have also been utilized by the scientists for nanoparticles synthesis [14]. This waste valorization technique using fruit peels also reduces the harmful impact of chemicals used in other conventional nano-synthesis processes. The scientists have prepared gold and silver nanoparticles using fruits wastes [14] for a multitude of applications.

In this book chapter, our prime impetus will be on the biogenic/green nanomaterials synthesized utilizing callus extract and their pros and cons as various DDS to combat chronic diseases including cancer. In biogenic synthesis, the primary focus is on callus-mediated synthesis, which is the most sustainable source as it does not cause any sort of damage to the plants, environment and human beings. The callus-mediated nanomaterial synthesis thoroughly utilizes the power of totipotency of callus cells and its potential to generate huge biomass using a small plant part.

7.1.1 SUSTAINABLE NANOTECHNOLOGY AND NANOMATERIALS

Nanomaterial synthesis and origin has incited the researchers to transform the material science in contemporary times for its huge potential in clinical applications. Although nanotechnology is considered as a contemporary science, it has its deep bounded in history; for example, medicinal properties of silver are well known to our ancestors [15]. Biosynthesis works on the principle of green chemistry, which are eco-friendly, non-toxic and chemical-free [15,16]. These green synthesized nanomaterials have shown great stability, huge dimensions and economy [6,15]. Nanomaterial synthesis can be performed from three methods such as chemical, physical and biological. However, unlike physical and chemical methods, biological methods require less energy, are eco-friendly and are mostly toxin-free [5,15]. Thus, biogenic synthesis using callus extract is a highly reliable procedure and sustainable.

7.1.2 CALLUS-DERIVED BIOACTIVE NANOMATERIALS

The plant cell cultures are an economic method for secondary metabolites production, which possess medicinal properties [1,17]. In plants, the differentiated tissues have a special capacity to de-differentiate and form entire plant; this potential is called totipotency (stem cell-like potential to regenerate any plant part) and the undifferentiated tissue thus formed is called as callus. The addition of precursor molecules to the culture media enhances the biomass production [1]. Green nanotechnology combines the principle of green chemistry with green engineering for the synthesis and fabrication of nanomaterials to overcome threats from various realms of human life, such as agriculture, environment, healthcare and industry. Alongside, amalgamation of site-specific/targeted

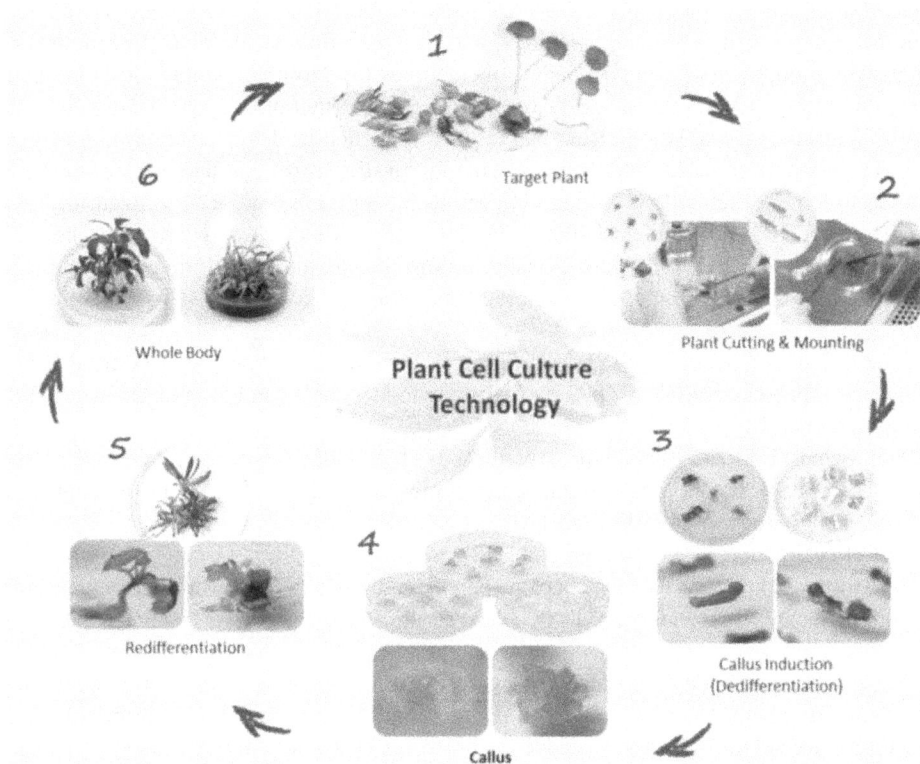

FIGURE 7.1 Plant cell culture technology (callogenesis): Illustration of mechanism of callus extract production for biogenic nanomaterials formation.

TABLE 7.1

Various Abiotic and Biotic Factors Responsible for Callus Induction and Repression [3]

Abiotic Factors		Biotic Factors
Physical	**Chemical**	**Biological**
Photoperiod (the time of exposure to daylight)	Callus-inducing medium (CIM)	Bacterial infection (agrobacterium strain C58)/reacquisition of embryonic or meristematic fate
Ultrasonic waves	Plant hormones (auxin and cytokinin)	Interspecific hybrid crosses
Wound induced	–	Viral infection (natural)

DDS with green nanotechnology has revolutionized the realm of green nanomedicines and made possible the formation of various forms of DDS including solid lipid NPs (SLNs) and dendrimers [13] (Figure 7.1).

Plant tissue culture was discovered by Gottlieb Haberlandt (1854–1945) and got advanced with technological development. After 1960s, secondary metabolites were a major centre of attraction because of their potential as therapeutic agents such as anticancer drug [1]. The phytochemicals can be induced in higher concentrations via various abiotic and biotic stresses to the plants either naturally or by human interferences [3]. Various abiotic and biotic factors, i.e. physical, chemical or biological factors, are mentioned in Table 7.1 with various examples that lead to callus induction and regression.

A plethora of biotechnological tools are available in the contemporary times having applications in the wide domains of nanotechnology, and *in vitro* callus synthesis (callogenesis) is used for large-scale production of plant secondary metabolites [5]. However, callus formation and suspension cell cultures have a multitude of benefits in clinical theranostics [1]. Callus induction is formed by translational and posttranslational regulators, which leads to varied protein expression. Medicinal plants-derived callus have been used in multiple diseases treatment in a sustainable manner. Callus which is genetically modified via gene technology has also been utilized for bioactive metabolites/secondary metabolites synthesis. However, the callus technology needs to be explored to its largest extent to get benefitted by its exclusive sustainable potential [1] (Figure 7.2).

Pros of callus extract (with respect to whole-plant extracts):

1. In callus culture, secondary metabolites can be synthesized ignoring the external stimuli in a controlled environment. Thus, exclusive phytochemicals can be synthesized as per the requirements [1].
2. Microbe/pest infections can be prevented easily [1].
3. Even secondary metabolites of endangered species can be synthesized in large volumes [1].
4. Increased production at lower costs [1].
5. Cost-effective, non-toxic and high yield [18].
6. Nevertheless, conventional nanotechnology poses health and environmental concerns due to the production of toxic secondary products, chemicals utilized and negligible biocompatibility [7].

Recently, AgNPs have been synthesized using callus from various plants such as *Catharanthus roseus, Sesuvium portulacastrum, Centella asiatica* and *Cucurbita maxima,* primarily because callus cultures are more efficient and reduce the problems of whole-plant extracts [11]. Alongside a wide range of applications, active plant extract-derived nanomaterials are highly biocompatible and cause lesser side effects. Various plants have been used till date for their rapid and cost-effective synthesis of gold and silver nanomaterials, such as lemongrass leaf extract (*Cymbopogon flexuosus*),

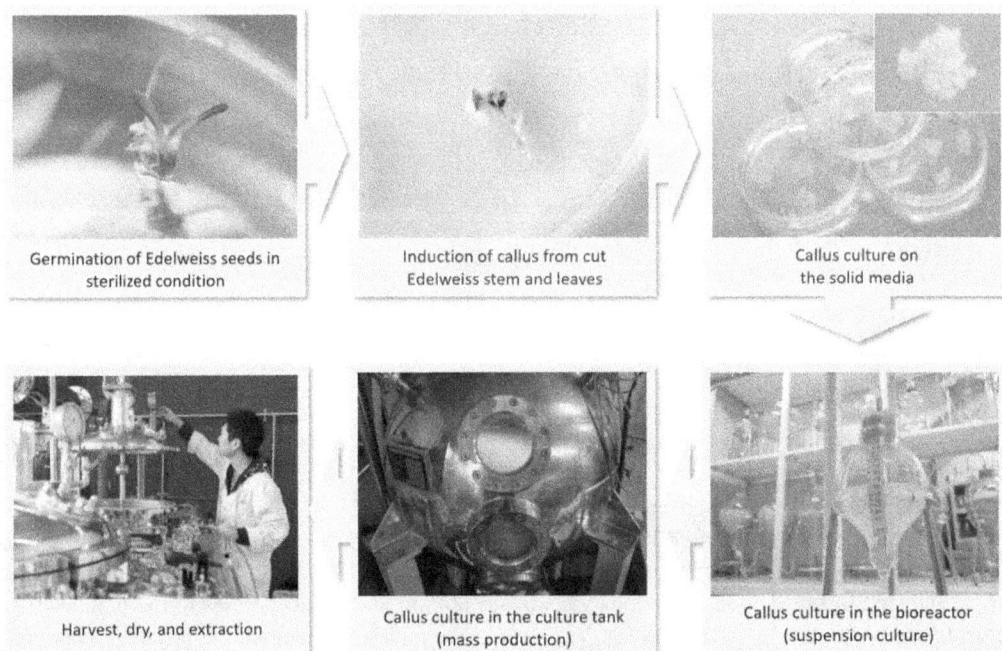

Germination of Edelweiss seeds in sterilized condition

Induction of callus from cut Edelweiss stem and leaves

Callus culture on the solid media

Harvest, dry, and extraction

Callus culture in the culture tank (mass production)

Callus culture in the bioreactor (suspension culture)

FIGURE 7.2 Callogenesis (procedure of callus synthesis) in the balloon bioreactor via suspension culture.

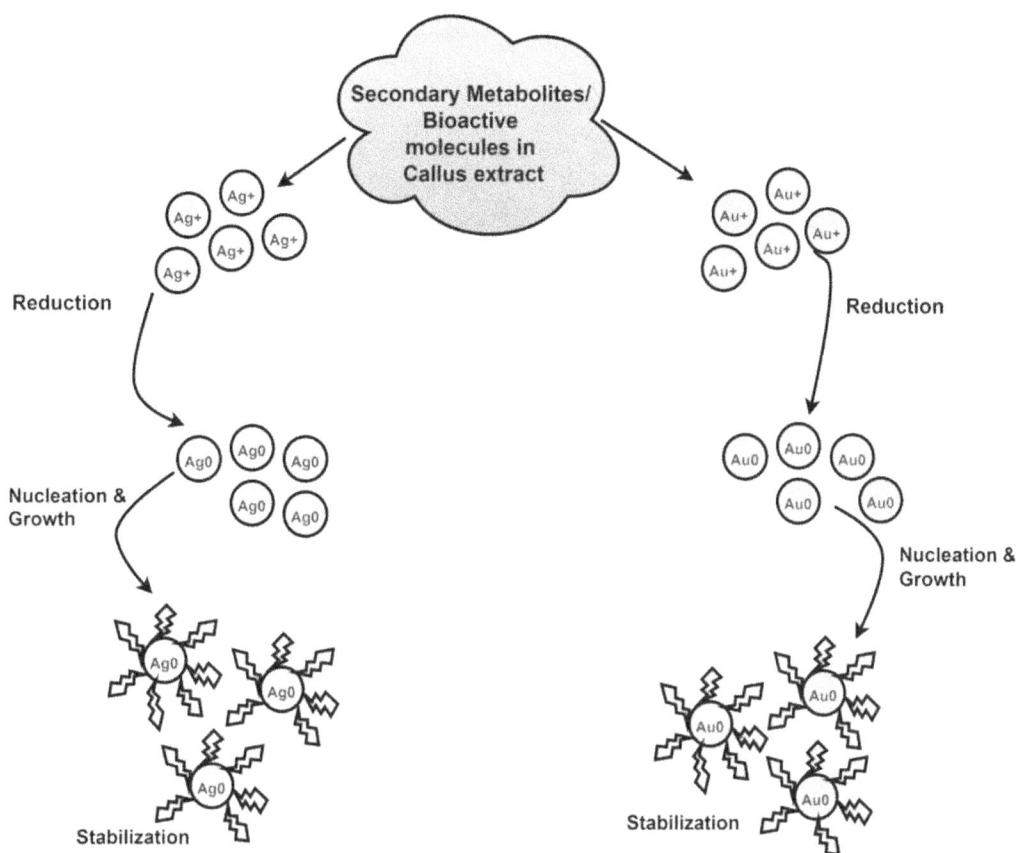

FIGURE 7.3   Mechanism of nanoparticles synthesis via callus extract: Role of callus extract as reductant, nucleation agent and stabilizing agent.

neem (*Azadirachta indica*), aloe vera (*Aloe barbadensis*), tamarind (*Tamarindus indica*) [1], *Alhagi maurorum* [19], callus, *Fagonia indica* [18] and *Carica papaya* callus [20,21]. Alongside, transgenic callus such as *Nicotiana tabacum* callus and *Artemisia annua* callus [22] have been utilized. Various citrus species callus have also been used (Figure 7.3).

The secondary metabolites of citrus, which are responsible for reducing monovalent silver to zero valent atoms, include various flavonoids, catechin and anthraquinones, which provide these AgNPs their antimicrobial, antioxidant and scavenging activities [22]. Out of various types of metallic nanoparticles, silver nanoparticles are the most explored and biogenically synthesized ones, due to their wide range of applications and biocompatibility with human body [10]; other than AgNPs, gold [23] and zinc oxide nanoparticles [24] have also been explored and synthesized using callus extract. The biosynthetic potential of various extracts varies widely depending on the individual biomolecules they contain, which act as the metal ion reductants. However, some phytochemicals (mostly, proteins, saccharides and low weight secondary metabolites) do act as a reductant, capping agent and stabilizer.

7.2   CALLUS-DERIVED BIOMASS AND BIOACTIVE NANOMATERIALS

Plant cell culture technologies include three kinds, of which callus culture is the most prominently used type of plant cell culture. A callus is undifferentiated mass of cells with totipotency and higher plasticity, which can further differentiate and form a whole plant body. These callus masses can be

FIGURE 7.4   Manufacturing of plant cell culture via balloon reactors.

triggered to release/express high concentration of phytochemicals/secondary metabolites [4] via application of physical, chemical or biotic stress as external stimuli. Callus-derived biomass having stem cell-like differentiation potential. The callus biomass thus produced is used to extract the secondary metabolites/phytochemicals using various bioreactors. These extracts are utilized to reduce and stabilize the metal bulk into nanoparticles, thus acting as the reducing and stabilizing agents [2] (Figure 7.4).

The whole process starting from callus induction, screening, culture preparation, harvest and extraction takes around 6–12 months for development and optimization. Using a bioreactor, approximately 2 tons of plant cell culture can be obtained.

Callus extracts of various plants have been noted to reduce the silver metal to form AgNPs. These include *Alternanthera dentata, Acorus calamus, Sesuvium portulacastrum, Tribulus terrestris, Cocos nucifera, A. indicum, Mangifera indica, Ziziphora tenuior, Ficus carica, Cymbopogon citratus, Acalypha indica,* geranium leaves, mushroom extract, *Coleus aromaticus, Chenopodium album, , Pistacia atlantica,* [16], *Leptadenia reticulata, Cassia didymobotrya, Andrographis paniculata, Prunus japonica, Talinum triangulare, Euphorbia antiquorum, Thymbra spicata, Cleome viscosa* [11], *Chlorophytum borivilianum* L. (Safed musli) callus extract [11], *Carica papaya, Citrullus colocynthis, Centella asiatica, Artemisia annua, Allophylus serratus, Chlorophytum borivilianum* L., *Vigna radiata* [22], *Sesuvium portulacastrum* (callus) [25], *Costus speciosus* callus extract [26], *Rosmarinus officinalis* [27] and *Cinnamomum camphora* [5]. Alongside, leaf-derived callus extracts of *Allophylus serratus* have been used to synthesize AgNPs and have shown antimicrobial properties [16]. Medicinal plants have phytochemicals that provide additional properties to the green synthesized AgNPs [16]. Bio-inspired noble metallic NPs are very potent antimicrobial agents. However, not all AgNPs and AuNPs are antimicrobial; thus, the factors that add their antimicrobial properties need to be identified [25]. Other than AgNPs and AuNPs, oxides of other metals are quite economic and are lucrative options [1].

7.2.1   Synthesis of Callus-Derived Bioactive Nanomaterials

Nanomaterials have enhanced properties than their bulk counterparts because of these properties; their applications are expanding in the fields of electronics, electrochemistry, IT, IoT, medicines, diagnostics [5], drug delivery [5,18], drug-gene delivery [5], optoelectronics [5], chemical and biosensors and multi-drug resistance [18], etc. However, the metallic NMs are most potent due to their antimicrobial efficacy and, out of all, AgNPs are most versatile due to their good conductivity, stability, antimicrobial activities [16], large amount of metabolites, non-pathogenic nature, and cost-effectiveness [5] (Figure 7.5).

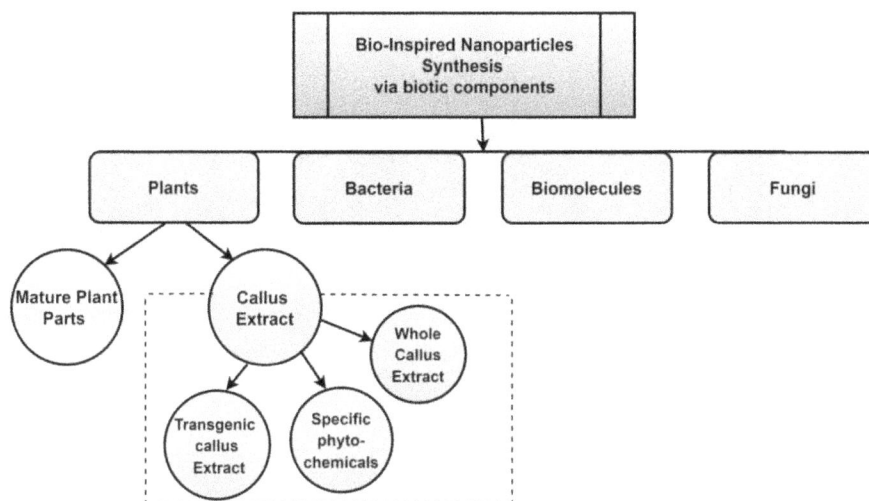

FIGURE 7.5 Callus and other biotic components for the synthesis of bio-inspired nanoparticles.

Lately, silver nanoparticles (AgNPs) have gained huge attention due to their wide range of applications in diagnostics and properties such as anti-inflammation, anti-microbial, wound healing, anti-angiogenesis and cytotoxicity against cancer cells [20,28], thus making AgNPs vital in biomedical and clinical domains. Alongside, a transgenic tobacco callus culture has also been used to synthesize AgNPs, which have a 3.6 times higher reduction potential than the original callus culture [20]. It indicates another realm of transgenic callus culture for producing a higher yield of phytochemicals to produce nanomaterials.

7.3 ADVANTAGES OF CALLUS-DERIVED BIOACTIVE NANOMATERIALS

Amalgamation of green chemistry and green engineering for nanomedicine synthesis has already shown a bright path to combat fatal human ailments. The NMs synthesized utilizing the green route have an upper hand over the ones synthesized via conventional methods of physical and chemical routes. These green NMs could be utilized as site-specific/targeted DDS [9,29]. Callus-derived NMs follow the principles of green chemistry in a sustainable manner; it reduces the harmful impact of chemical substances over human health and environment by eliminating the use of chemical reductants to form nanomaterials. Thus, these NMs are sustainable and with maximized efficiency [2,3] (Figure 7.6).

Advantages of callus culture synthesis in bioreactors:

1. Sustainable and eco-friendly.
2. No environmental variation (as the conditions can be maintained inside) in weather, sunlight, soil and water.
3. Rare/essential active phytochemicals of interest can be synthesized or concentrated.
4. Reproducible biomass and concentrated/pure metabolites.
5. High biological safety.
6. More stable [30].
7. GMO-free.
8. Phytochemicals from endangered plant species can also be produced.
9. Higher efficacy as compared to crude extracts.
10. Green synthesized AgNPs act as potential anti-tumour agents by controlling the progression of tumour development [30] (Figure 7.7).

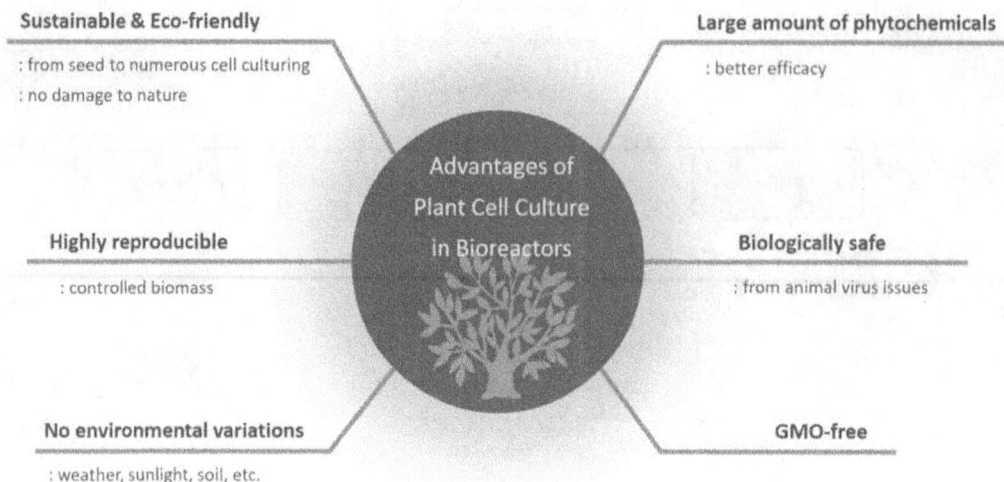

Sustainable & Eco-friendly

: from seed to numerous cell culturing
: no damage to nature

Large amount of phytochemicals

: better efficacy

Advantages of Plant Cell Culture in Bioreactors

Highly reproducible

: controlled biomass

Biologically safe

: from animal virus issues

No environmental variations

: weather, sunlight, soil, etc.

GMO-free

FIGURE 7.6   Advantages of plant/callus culture synthesis over crude plant extract.

Gallic acid

Caffeic acid

Natural Phytochemicals

Shikimic acid

Resveratrol

FIGURE 7.7   Illustration of prominent natural medicinal phytochemicals.

Phytochemicals are the naturally occurring biologically active chemical compounds found in plants. These protect plants from disease, environmental stress and animals. Alongside, these phytochemicals possess certain medical properties for the fatal human ailments. These have higher efficacy than the crude herbal extracts [31] and crude callus extract [4] and are environmentally friendly.

7.4   SUSTAINABLE DRUG DELIVERY SYSTEMS USING CALLUS-DERIVED BIOACTIVE NANOMATERIALS

Drug delivery systems/vehicles (DDS/DDV) are the engineered formulation, devices or the tool used for the targeted, sustained and controlled delivery of therapeutic agents to the affected body part [13,32]. It is usually assisted by controlled release and transport of the pharmaceutically active

TABLE 7.2

Biogenic Metal Nanomaterials (silver, gold, zinc, etc.) Synthesized Using Various Medicinal Callus Extracts, Their Dimensions and Their Clinical Applications in Microbiology and Oncology

Callus Extract	Dimensions of Nanomaterials	Properties	Ref.
Leontopodium alpinum (edelweiss)	–	Anti-inflammatory, anti-ageing	[31]
Hyptis suaveolens	Spherical (12–25 nm), dispersed AgNPs	Anticancer, antimicrobial, antioxidant	[30]
Carica papaya	–	Anticancer, drug delivery, biosensors	[21]
Centella asiatica	–	Antimicrobial	[35]
Citrullus colocynthis	–	Bactericidal, wound healing	[36]
Catharanthus roseus	10–20 nm	Antimicrobial	[37,38]
Nicotiana tabacum	Spherical	Antimicrobial	[20]
Allophylus serratus	40–50 nm	Antimicrobial	[16]
Sesuvium portulacastrum	Spherical, 5–20 nm	Antimicrobial	[39]
Chlorophytum borivilianum	35–138 nm	Antimicrobial and anticancer (colon cancer)	[11,40]
Taxus yunnanensis	6.4–27.2 nm	Antimicrobial and anticancer (hepatoma)	[41]
Cucurbita maxima	AuNPs and AgNPs	–	[24]
Cinnamomum camphora	5.47–9.48 nm	Antimicrobial	[5]
Sweet lime	3–270 nm	Antimicrobial, antioxidant, antimycotic	[22]
Vigna radiate	Spherical, ellipsoid	–	[42]
Fagonia indica	–	Antibacterial (MDR strain)	[18]
Costus speciosus	–	Antibacterial	[26]
Solanum incanum	–	Antimicrobial and anticancer	[43]
Linum usitatissimum	19–24 nm, more scattered	Antimicrobial	[44]
Celastrus paniculatus	–	Antibacterial	[45]
Centella asiatica	32.9 nm	Antioxidant	[46]
Viola canescens	Zinc oxide, 9 nm	Antimicrobial	[23]
Jatropha curcas	2–50 nm	–	[47]

compounds to the desired site with high efficacy and low side effects to the surrounding tissues [32,33]. With the advancement of nanotechnology, various forms of nanomaterials have been used as drug delivery systems to treat fatal ailments. The nanodrug and DDS thus can easily overpass the systemic clearance and metabolism from body machinery [9,12]. Many polymeric delivery systems and complex molecules have been used by the scientists to achieve triggered release such as mesoporous silica [28]. Certain other tools and techniques such as hybrid patches using drug delivery systems have also been developed by the scientists for a controlled drug release [34]. Alongside, amalgamation of site-specific/targeted DDS with green nanotechnology has revolutionized the realm of green nanomedicines and made possible the formation of various forms of DDS including solid lipid NPs (SLNs) and dendrimers [13] (Table 7.2).

Even though nanotechnology is compatible for large-scale production of various nanomaterials, the feasibility poses a major challenge in synthesis of DDS at commercial scale. The Wurster wuester fluid bed technique for the synthesis of polymeric DDS is considered as the best technique so far [32]. The impact of nano-based DDS is deepening with passing time in theranostics realms for combating human ailments, especially via green route. Furthermore, the potential to target stimuli-sensitive, sustained and targeted delivery to target site is incredible [32]. Buckminsterfullerene (C-60) is a

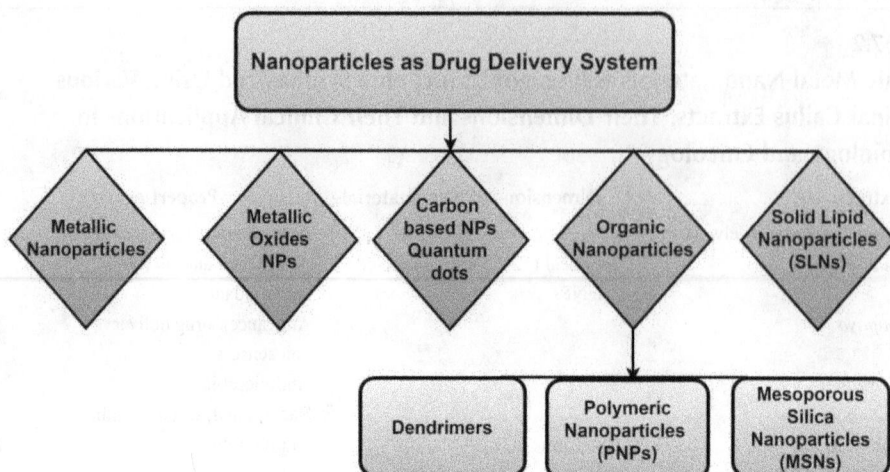

FIGURE 7.8 Flow chart depicting various green synthesized sustainable nanoparticles as drug delivery systems/vehicles.

biocompatible, non-toxic, ball-like 1 nm molecule. It's a potent drug delivery carrier and can deliver oligopeptides and drug molecules to target sites [15]. Dendrimers are another polymeric 3D DDS where the core has dendrites (tree) like branches and can be conjugated with drugs and easily cleared by excretory mechanism of body [15]. Particularly, AgNPs are highly explored [22] and preferred among other MNPs due to their high surface area and catalytic activity and wide range of applications in biomedical domain [5], diagnostics and biomedicine [22]. Alongside, the dimensions (shape and size) of NPs define their properties [18] and their mode of action [20]. The smaller the size, the more is the antimicrobial activity of AgNPs and vice versa [18] (Figure 7.8).

7.4.1 Applications in Oncology

Nanomagnets can be used as a DDS to target tumour cells without disturbing the extracellular tissues [15]. Various biomolecules and phytochemicals such as gelatine and legumin can be associated with DDS and need a lesser drug load. These DDS can also contribute to the enhancement of the bioavailability of drugs [15]. Safed musli callus extract contains numerous phytochemicals whose functional groups associate themselves with the silver ions imparting high cytotoxic efficacy to AgNPs. Alongside, green synthesized AgNPs have nanoregime nature, reduction by phytochemicals, minute size and shape which provide them their anticancer activity [11]. The chemical composition of the phytochemical and secondary metabolites of plant extracts, however, provides AgNPs their main anticancer properties, which are effective, economic and eco-friendly. This exhibition of anticancer property indicates that these NPs can be used for the fabrication of nanodrugs in cancer such as colon cancer [11]. Thus, phytofabricated AgNPs are better than the chemically synthesized ones. These AgNPs have been proven effective against cancer cell line HT-29, and the cell viability of cancer cells decreased with increased time of exposure and concentration of AgNPs, indicating their time- and dose-dependent cytotoxicity [5,11]. These NPs possess lethal toxicity at high doses [11].

Green synthesis via plant extracts, phytochemicals, microbes, algae, etc., has also been used to synthesize gold NPs that were utilized in cosmetics, medicines, etc. Alongside, AuNPs have anticancer properties and, when conjugated with thiol group, show enhanced efficacy. AuNPs are versatile, can be functionalized with variety of organic molecules and are safe with tunable optical properties. Thus, they are good candidate as anti-cancer agents. Alongside, nanomaterials have also been used as DDS for anticancer agents [48].

DDS helps to enhance the efficacy of active pharmaceutical compounds (APCs) such as vaccines, antibodies and drugs. However, conventional DDS have multitude of limitations; thus, advancement is needed in the DDS [6]. Further, research to reduce the dosage amount, enhancing the bioavailability and selectivity, is needed. The first nano-DDS used was based on the biochemical and biophysical attributes of the drug chosen. However, the green synthesized nanomaterials further reduce the toxicity posed by nanomaterials and also the drug used and thus are highly recommended. Traditional metal nanomaterials (NMs) produce a huge amount of toxins; however, the green synthesized metal NMs (silver, gold and platinum) have shown good results in drug delivery, imaging and therapy [23,24].

7.4.2 Applications in Other Chronic Diseases

A drug delivery system is the carrier that helps to transport pharmaceutically active compounds to the target tissue with minimum side effects [7]. Nanotechnology provides drug delivery platforms to treat osteogenic and bone-related disorders. Osteological disorders have been prevented and treated in the past via DDS. In osteology, DDS belong to two categories based on targeted tissue, such as bone site/whole skeletal system and specific bone cellular locations. DDS help the drugs to reach the sites and trigger regeneration [7].

Green nanomedicines can serve a multitude of applications for cancer theranostics, other chronic diseases and neurodegenerative diseases [1,15]. In Future NMs will be of further minute size because of which will easily trick body and remain unidentified, thus will be able to cross even the blood-brain barriers and have a huge potential as drug delivery and drug targeting systems and can treat diseases such as cancer [15]. *Chlorophytum borivilianum L.* (Safed musli) has copious secondary metabolites such as flavonoids, alkaloids, steroids and phenols, which contribute to their huge medicinal properties. These metabolites help to cure diseases such as high blood pressure, diabetes, arthritis and cancer due to their antioxidant and antimicrobial properties. Alongside, biological methods produce more stable and uniform AgNPs with enhanced pharmacological efficacy and thus are recommended over other synthesis methods [11]. Alongside, AgNPs derived from *Cinnamomum camphora* [5] have the potential to treat cardiac ailments. Thus, the secondary metabolites-mediated nanomaterials have opened a new path in clinical and theranostics applications of human ailments treatment, with lesser side effects.

7.5 CONCLUSIONS AND FUTURE PROSPECTS

Sustainable/biogenic nanomaterials or DDS are highly recommended by the scientists and researchers of medical realm. Considering the systemic toxicity and the impact that the conventional nanotechnology poses on the environment, the biogenic synthesis is preferred over physical and chemical synthesis methods. In the biogenic synthesis, the prime focus of this chapter was on callus-mediated synthesis, which is the most sustainable source as it does not cause any sort of damage to the plants themselves. Bone regeneration is possible with the effective drug targeting the damaged site. Even when modern DDS are successful to some extent, they face certain challenges technically. However, non-effective penetration of systemic DDS poses risk of toxicity as high dosage amount is required. Current nano-delivery systems are successful, but have multitude of side effects, other than toxicity. Hence, improvement of DDS is the prime need before their broad spectrum applications in biomedical domains. Nevertheless, the conventional nanotechnology poses environmental, social and health concerns. Thus, delivery systems could be synthesized using green/biogenic nanotechnological approaches (i.e. callus- or suspension cell-derived bioactive compounds), making it more biocompatible and reducing the systemic toxicity and chances of rejection from the biological systems. Even though green nanomaterials have a variety of applications in various realms, the biogenic nano-DDS need to be explored more to meet their highest potential and further research to reduce the dosage amount, enhancing the bioavailability and selectivity needs to be performed.

ACKNOWLEDGEMENTS

This work was supported by Symbiosis International (Deemed University). We thank our colleagues who provided their meaningful insights and expertise that assisted us to do justice to this article.

DECLARATION OF COMPETING INTEREST

The authors declare that they have no known competing financial interests or personal relationships that could have appeared to influence the work reported in this paper.

REFERENCES

1. Efferth, T. (2019). Biotechnology applications of plant callus cultures. *Engineering, 5*(1), 50–59.
2. Tran, M. T., Nguyen, L. P., Nguyen, D. T., Cam-Huong, L., Dang, C. H., Chi, T. T. K., & Nguyen, T. D. (2021). A novel approach using plant embryos for green synthesis of silver nanoparticles as antibacterial and catalytic agent. *Research on Chemical Intermediates, 47*(11), 4613–4633.
3. Ikeuchi, M., Sugimoto, K., & Iwase, A. (2013). Plant callus: Mechanisms of induction and repression. *The Plant Cell, 25*(9), 3159–3173.
4. Benjamin, E. D., Ishaku, G. A., Peingurta, F. A., & Afolabi, A. S. (2019). Callus culture for the production of therapeutic compounds. *American Journal of Plant Biology, 4*(4), 76–84.
5. Aref, M. S., & Salem, S. S. (2020). Bio-callus synthesis of silver nanoparticles, characterization, and antibacterial activities via Cinnamomum camphora callus culture. *Biocatalysis and Agricultural Biotechnology, 27*, 101689.
6. Dhingra, R., Naidu, S., Upreti, G., & Sawhney, R. (2010). Sustainable nanotechnology: Through green methods and life-cycle thinking. *Sustainability, 2*(10), 3323–3338.
7. Medina-Cruz, D., Mostafavi, E., Vernet-Crua, A., Cheng, J., Shah, V., Cholula-Diaz, J. L., ... & Webster, T. J. (2020). Green nanotechnology-based drug delivery systems for osteogenic disorders. *Expert Opinion on Drug Delivery, 17*(3), 341–356.
8. Patra, J. K., Das, G., Fraceto, L. F., Campos, E. V. R., Rodriguez-Torres, M. D. P., Acosta-Torres, L. S., ... & Shin, H. S. (2018). Nano based drug delivery systems: Recent developments and future prospects. *Journal of Nanobiotechnology, 16*(1), 1–33.
9. Karn, B., & Wong, S. S. (2013). Ten years of green nanotechnology. In: N. Shamim, V. K. Sharma (eds.) *Sustainable Nanotechnology and the Environment: Advances and Achievements* (pp. 1–10). American Chemical Society, Washington, D.C.
10. Kuppusamy, P., Yusoff, M. M., Maniam, G. P., & Govindan, N. (2016). Biosynthesis of metallic nanoparticles using plant derivatives and their new avenues in pharmacological applications–An updated report. *Saudi Pharmaceutical Journal, 24*(4), 473–484.
11. Huang, F., Long, Y., Liang, Q., Purushotham, B., Swamy, M. K., & Duan, Y. (2019). Safed Musli (Chlorophytum borivilianum L.) callus-mediated biosynthesis of silver nanoparticles and evaluation of their antimicrobial activity and cytotoxicity against human colon cancer cells. *Journal of Nanomaterials, 2019*, 1–8.
12. Patra, P., & Chattopadhyay, D. (2020). Sustainable release of nanodrugs: A new biosafe approach. In: B. K. Banik (ed.) *Green Approaches in Medicinal Chemistry for Sustainable Drug Design* (pp. 603–615). Elsevier, Amsterdam.
13. Kanwar, R., Rathee, J., Salunke, D. B., & Mehta, S. K. (2019). Green nanotechnology-driven drug delivery assemblies. *ACS Omega, 4*(5), 8804–8815.
14. Ali, S., Chen, X., Shah, M. A., Ali, M., Zareef, M., Arslan, M., ... & Chen, Q. (2021). The avenue of fruit wastes to worth for synthesis of silver and gold nanoparticles and their antimicrobial application against foodborne pathogens: A review. *Food Chemistry, 359*, 129912.
15. Ganachari, S. V., Yaradoddi, J. S., Somappa, V., Mogre, P., Tapaskar, R. P., Salimath, B., ... & Viswanath, V. J. (2019). Green nanotechnology for biomedical, food, and agricultural applications. In: Martínez, L., Kharissova, O., Kharisov, B. (eds.) *Handbook of Ecomaterials*. Springer, Cham.
16. Jemal, K., Sandeep, B. V., & Pola, S. (2017). Synthesis, characterization, and evaluation of the antibacterial activity of Allophylus serratus leaf and leaf derived callus extracts mediated silver nanoparticles. *Journal of Nanomaterials, 2017*, 1–11.
17. Baque, M. A., Moh, S. H., Lee, E. J., Zhong, J. J., & Paek, K. Y. (2012). Production of biomass and useful compounds from adventitious roots of high-value added medicinal plants using bioreactor. *Biotechnology Advances, 30*(6), 1255–1267.

18. Adil, M., Khan, T., Aasim, M., Khan, A. A., & Ashraf, M. (2019). Evaluation of the antibacterial potential of silver nanoparticles synthesized through the interaction of antibiotic and aqueous callus extract of Fagonia indica. *AMB Express*, *9*(1), 1–12.

19. Sehrawat, A. R. (2018). Biosynthesis of silver nanoparticles from embryogenic calli of alhagi maurorum. *Asian Journal of Pharmaceutics (AJP)*, *12*(01), S1–S4.

20. Shkryl, Y. N., Veremeichik, G. N., Kamenev, D. G., Gorpenchenko, T. Y., Yugay, Y. A., Mashtalyar, D. V., … & Zhuravlev, Y. N. (2018). Green synthesis of silver nanoparticles using transgenic Nicotiana tabacum callus culture expressing silicatein gene from marine sponge Latrunculia oparinae. *Artificial Cells, Nanomedicine, and Biotechnology*, *46*(8), 1646–1658.

21. Mude, N., Ingle, A., Gade, A., & Rai, M. (2009). Synthesis of silver nanoparticles using callus extract of Carica papaya—a first report. *Journal of Plant Biochemistry and Biotechnology*, *18*(1), 83–86.

22. Kalia, A., Manchanda, P., Bhardwaj, S., & Singh, G. (2020). Biosynthesized silver nanoparticles from aqueous extracts of sweet lime fruit and callus tissues possess variable antioxidant and antimicrobial potentials. *Inorganic and Nano-Metal Chemistry*, *50*(11), 1053–1062.

23. Khajuria, A. K., Bisht, N. S., Manhas, R. K., & Kumar, G. (2019). Callus mediated biosynthesis and antibacterial activities of zinc oxide nanoparticles from Viola canescens: An important Himalayan medicinal herb. *SN Applied Sciences*, *1*(5), 1–13.

24. Iyer, R. I., & Panda, T. (2018). Biosynthesis of gold and silver nanoparticles using extracts of callus cultures of pumpkin (Cucurbita maxima). *Journal of Nanoscience and Nanotechnology*, *18*(8), 5341–5353.

25. Zinjarde, S. (2012). Bio-inspired nanomaterials and their applications as antimicrobial agents. *Chronicles of Young Scientists*, *3*(1), 74–74.

26. Malabadi, R. B., Meti, N. T., Mulgund, G. S., Nataraja, K., & Kumar, S. V. (2012). Synthesis of silver nanoparticles from in vitro derived plants and callus cultures of Costus speciosus (Koen.); Assessment of antibacterial activity. *Research in Plant Biology*, *2*(4), 32–42.

27. Hussein, E. A., Aref, M. S., & Ramadan, M. M. (2017). Physical elicitation of Rosmarinus officinalis callus culture for production of antioxidants activity. *International Journal of Innovative Science, Engineering & Technology*, *4*, 238–247.

28. Zhang, L., Bei, H. P., Piao, Y., Wang, Y., Yang, M., & Zhao, X. (2018). Polymer-brush-grafted mesoporous silica nanoparticles for triggered drug delivery. *ChemPhysChem*, *19*(16), 1956–1964.

29. Krishnaswamy, K., & Orsat, V. (2017). Sustainable delivery systems through green nanotechnology. In: A. M. Grumezescu (ed.) *Nano-and Microscale Drug Delivery Systems* (pp. 17–32). Elsevier, Amsterdam.

30. Botcha, S., & Prattipati, S. D. (2020). Callus extract mediated green synthesis of silver nanoparticles, their characterization and cytotoxicity evaluation against MDA-MB-231 and PC-3 Cells. *BioNanoScience*, *10*(1), 11–22.

31. Giannouli, E., & Karalis, V. (2022). In the pursuit of longevity: Anti-aging substances, nanotechnological preparations, and emerging approaches. *medRxiv*. doi: 10.1101/2022.03.20.22272670.

32. Domingo, C., & Saurina, J. (2012). An overview of the analytical characterization of nanostructured drug delivery systems: Towards green and sustainable pharmaceuticals: A review. *Analytica Chimica Acta*, *744*, 8–22.

33. Nabipour, H., & Hu, Y. (2020). Sustainable drug delivery systems through green nanotechnology. In: M. Mozafari (ed.) *Nanoengineered Biomaterials for Advanced Drug Delivery* (pp. 61–89). Elsevier, Amsterdam.

34. Evdokimova, O. L., Svensson, F. G., Agafonov, A. V., Håkansson, S., Seisenbaeva, G. A., & Kessler, V. G. (2018). Hybrid drug delivery patches based on spherical cellulose nanocrystals and colloid titania—synthesis and antibacterial properties. *Nanomaterials*, *8*(4), 228.

35 Netala, V. R., Kotakadi, V. S., Nagam, V., Bobbu, P., Ghosh, S. B., & Tartte, V. (2015). First report of biomimetic synthesis of silver nanoparticles using aqueous callus extract of Centella asiatica and their antimicrobial activity. *Applied Nanoscience*, *5*(7), 801–807.

36. Satyavani, K., Ramanathan, T., & Gurudeeban, S. (2011). Green synthesis of silver nanoparticles by using stem derived callus extract of bitter apple (Citrullus colocynthis). *Digest Journal of Nanomaterials and Biostructures*, *6*(3), 1019–1024.

37. Barkat, M. A., Mujeeb, M., Samim, M., & Verma, S. (2014). Biosynthesis of silver nanoparticles using callus extract of Catharanthus roseus var. alba and assessment of its antimicrobial activity. *British Journal of Pharmaceutical Research*, *4*(13), 1591–1603.

38. Osibe, D. A., Chiejina, N. V., Ogawa, K., & Aoyagi, H. (2018). Stable antibacterial silver nanoparticles produced with seed-derived callus extract of Catharanthus roseus. *Artificial Cells, Nanomedicine, and Biotechnology*, *46*(6), 1266–1273.

39. Nabikhan, A., Kandasamy, K., Raj, A., & Alikunhi, N. M. (2010). Synthesis of antimicrobial silver nanoparticles by callus and leaf extracts from saltmarsh plant, Sesuvium portulacastrum L. *Colloids and surfaces B: Biointerfaces, 79*(2), 488–493.

40. Charl, R. K., Sinniah, U. R., & Swamy, M. K. (2017). Antimicrobial activity of safed musli (Chlorophytum borivilianum L.) callus extract and GC-MS based chemical profiling. *Bangladesh Journal of Botany, 46*(1), 305–310.

41. Xia, Q. H., Ma, Y. J., & Wang, J. W. (2016). Biosynthesis of silver nanoparticles using Taxus yunnanensis callus and their antibacterial activity and cytotoxicity in human cancer cells. *Nanomaterials, 6*(9), 160.

42. Iyer, R. I., Selvaraju, C., & Santhiya, S. T. (2016). Biosynthesis of silver nanoparticles by callus cultures of Vigna radiata. *Indian Journal of Science and Technology, 9*(9), 1–5.

43. Lashin, I., Fouda, A., Gobouri, A. A., Azab, E., Mohammedsaleh, Z. M., & Makharita, R. R. (2021). Antimicrobial and in vitro cytotoxic efficacy of biogenic silver nanoparticles (Ag-NPs) fabricated by callus extract of Solanum incanum L. *Biomolecules, 11*(3), 341.

44. Anjum, S., & Abbasi, B. H. (2016). Thidiazuron-enhanced biosynthesis and antimicrobial efficacy of silver nanoparticles via improving phytochemical reducing potential in callus culture of Linum usitatissimum L. *International Journal of Nanomedicine, 11*, 715.

45. Solanki, A., Rathod, D., Patel, I. C., & Panigrahi, J. (2021). Impact of silver nanoparticles as antibacterial agent derived from leaf and callus of Celastrus paniculatus Willd. *Future Journal of Pharmaceutical Sciences, 7*(1), 1–9.

46. Rashmi, V., Prabhushankar, H. B., & Sanjay, K. R. (2021). Centella asiatica L. callus mediated biosynthesis of silver nanoparticles, optimization using central composite design, and study on their antioxidant activity. *Plant Cell, Tissue and Organ Culture (PCTOC), 146*(3), 515–529.

47. Demissie, A. G., & Lele, S. S. (2013). Phytosynthesis and characterization of silver nanoparticles using callus of jatropha curcas: A biotechnological approach. *International Journal of Nanoscience, 12*(02), 1350012.

48. Díaz, M. R., & Vivas-Mejia, P. E. (2013). Nanoparticles as drug delivery systems in cancer medicine: Emphasis on RNAi-containing nanoliposomes. *Pharmaceuticals, 6*(11), 1361–1380.

8 Green and Environment-Friendly Graphene Quantum Dots (GQDs) with State-of-the-Art Performance for Sustainable Energy Conversion and Storage

Shikha Gulati and Sanjay Kumar
University of Delhi

Suvidha Sehrawat
Chandigarh University

CONTENTS

DOI: 10.1201/9781003319153-10

8.1 INTRODUCTION

Quantum dots (QDs) are nanomaterials that are currently having a significant impact on research in a wide range of domains, including physical, chemical, and biological sciences [1,2]. The studies have proven that carbonaceous and carbon-based nanomaterials have attracted the interest of researchers due to their unique properties, which include non-toxicity, good biocompatibility, relative ease of functionalization, and high thermal/mechanical properties [3]. Carbon-based dots are known as fluorescent carbon due to their significant features that, when compared to the conventional fluorescent dyes, demonstrate superior fluorescence, including photobleaching, photostability resistance, and non-blinking [4]. Carbon dots also have attractive properties such as water solubility, good chemical stability, low toxicity, and ease of derivatization. Carbon dots have vital and widespread uses in biotechnology, energy, catalysis, bioimaging, biological labeling, and gene/drug delivery due to all of these unique characteristics [5]. Nanoscience and nanotechnology have opened new frontiers in materials science and engineering, allowing for the creation of new materials for energy generation and storage. Graphitic carbon materials have gained a lot of interest for energy-related applications due to their abundance on the planet, low cost, structural tunability, huge surface area, and unique physicochemical features. However, without functionalization, pure graphene materials are intractable (infusible and insoluble), which has restricted their practical applications. As a result, a significant amount of research has gone into developing functionalized graphene materials with desirable properties for certain purposes, such as energy conversion and storage. Functionalized graphene materials with variable work functions were shown to be effective charge extraction materials, while those with strong electrocatalytic activities could be used to increase solar cell performance. Therefore, in this chapter, we will discuss graphene quantum dots (GQDs), their synthesis, and modifications along with the real-life practical applicability in energy storage devices having a future perspective.

8.2 GRAPHENE QUANTUM DOTS (GQDS): AN EXCEPTIONAL NANOMATERIAL

Graphene is a novel material of the carbon family made up of several carbon bonds having sp^2 hybridization and shows unique properties such as high mechanical stiffness, high electronic conductivity, high thermal stability, strong elasticity, and large surface area [6–8]. Owing to this exceptional behavior, research on several areas of graphene and its applications is extensively increasing. However, graphene is a zero-bandgap material with minimal solubility in aqueous solutions and has low absorptivity but its size, surface functional groups, and the introduction of heteroatoms as defects into the graphene lattice could all be adjusted. In order to overcome this gap, graphene quantum dots (GQDs), a new quasi-zero-dimensional nanomaterial from the graphene family, have attracted much interest in recent years. Therefore, some modifications in the structure were made and GQDs were synthesized by Geim and Ponomarenko in 2008 [9]. GQDs are small graphene fragments with lateral dimensions less than 100 nm that have features derived from both graphene and carbon points [10]. GQDs possess a crystalline structure of one or more graphene layers connected via functional groups on edges with elliptical or circular shapes and are anisotropic in nature. In addition to high fluorescence quantum yield, low toxicity [11], and biocompatibility [12], GQDs have properties that make them promising options for use in a wide range of applications such as they have been used as electrode modifiers due to their large surface area and abundant functional groups, as well as their ease of functionalization with inorganic, organic, or biological substances [13] (Figure 8.1). They are also durable, water-soluble, chemically stable, inert, and photo-stable against photobleaching and blinking [14]. Furthermore, when compared to other carbon allotropes such as fullerene, graphene, and carbon nanotubes (CNTs), the ability to alter the necessary chemical and physical properties of GQDs with ease during the synthesis process makes them more interesting and versatile. Photoluminescence (PL) is one of the many appealing features of GQDs that

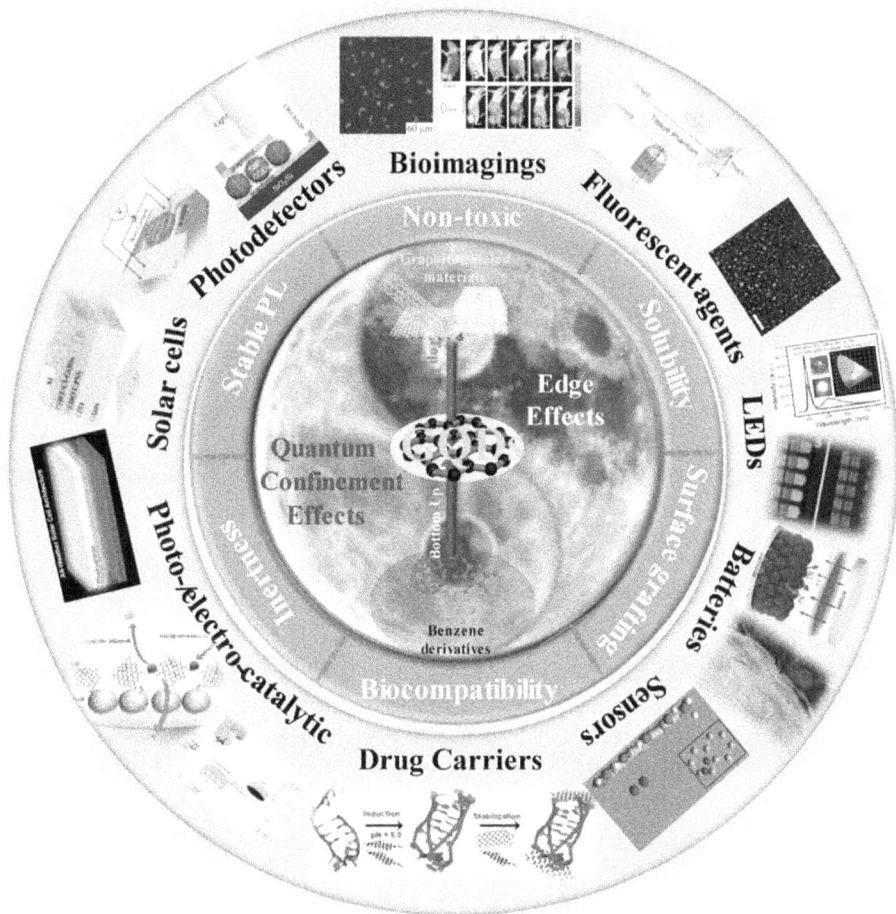

FIGURE 8.1 The wide range of applications of GQDs [63].

has caught the interest of scientists worldwide. Due to its rapid electron movement and high-speed reactivity, GQDs can operate as a good sensing material, making them ideal possibilities for sensing applications. Moreover, due to their aromatic structure, GQDs have peroxidase mimetic activity, which explains the interest in the production of electrocatalytic H_2O_2 detectors [15]. As a result of the above electrochemical properties, there has recently been a lot of interest in using GQDs to design novel electrode materials, not only in the fields of fuel cells [16], supercapacitors [17], and photovoltaic cells [18], but also in the field of electrochemical immuno-sensors for biomedical applications [14] and biosensors [19]. Although GQDs have several desirable properties, their poor product and quantum yield along with narrow spectral coverage limit their practical applicability in various industries. Therefore, heteroatom doping has emerged as a more effective technique for modifying the optical and electrical properties of GQDs to overcome this limitation [20,21]. Chen et al., for example, studied the effect of nitrogen (N) functionalities in N doped GQDs on photo-generated electron transitions and quantum yield [22]. According to the literature, GQDs can be generated by cutting carbon precursors into smaller pieces by hydrothermal method or chemical oxidation under harsh conditions in presence of strong acids [23] that result in higher costs, environmental pollution, human health hazards [24] that lead to the scientists to develop some greener methods for preparation. Recently, GQDs are basically synthesized using two methods classified as top-down and bottom-up that are having a healthy approach toward the environment which are discussed in the next section in detail.

8.3 ENVIRONMENT-FRIENDLY SYNTHESIS OF GQDS: GREENER AND SUSTAINABLE APPROACHES

Green synthetic routes are a growing industry in nanotechnology that offers both economic and environmental benefits as an alternative to traditional approaches. It is well known that the preparation of GQDs frequently requires the use of organic solvents or strong acids, and their green production via sustainable strategies involving non-toxic and safe reagents still faces major challenges. As a result, eco-friendly synthetic approaches with easy separation and no complicated post-processes should be designed and developed. Also, due to the large specific surface area and high electrical conductivity, several carbon nanomaterials have drawn the interest of researchers for the development of energy devices and as innovative electrode materials for electrochemical and environmental sensing applications. Because of its properties such as efficiency and scalability of synthesis, subsequent functionalization, intensive surface chemistry, and high biocompatibility, graphene is one of the most valuable carbon-based products, with tremendous potential for electrochemical and biosensing applications [25–27]. However, the strong bonding and van der Waals interactions result in a drastically reduced surface area due to irreversible graphene sheet clustering that limits its practical applications [28].

8.3.1 DIVERSE ROUTES FOR THE GREENER SYNTHESIS OF GQDs

According to recent studies, the synthesis of GQDs can be categorized into two primary types based on the current fabrication methods reported in the literature, namely top-down and bottom-up preparation methods. The top-down approach uses electron beam lithography and liquid exfoliation techniques to directly cleave bulk carbon materials into nanoscale GQDs. This approach benefits from abundant raw materials and typically produces oxygen-containing functional groups at the edge, enhancing their solubility and functionalization. However, this method has a number of drawbacks, including low yield, high defect density, and non-controllable size and form, among others. While the bottom-up technique focuses on hydrothermal, microwave-assisted hydrothermal, soft-template, and metal-catalyzed technologies to generate appropriate chemical precursors, such as polymers and small molecules into nano-sized GQDs. In contrast to the top-down approach, the bottom-up approach has advantages such as fewer defects and adjustable size and morphology. The bottom-up method, on the other hand, results in low solubility, aggregation issues, small dot size, etc. In the next section, the preparation methods are discussed in a detailed manner. The synthesis of carbonaceous nanomaterials and the choice of precursor materials are both significant since the product will have diverse properties based on the procedure and the feedstock, affecting its future applications. Furthermore, they rely on yields that ensure large-scale manufacturing. As stated in the introduction, there has been a growing interest in exploiting adequate renewable resources in recent years in order to reduce dependence on non-renewable resources and environmental safety and increase energy security [29], leading to the development of several attempts to exploit various natural carbon sources for the production of GQDs using both top-down and bottom-up processes. The ability to use a wide range of precursors and a variety of technological techniques [30] is the actual benefit of bio-based GQDs. Wheat straw, wood charcoal, rice husk, coffee ground, forestry processing residue, poultry and livestock manure, organic waste from food processing, and municipal solid waste are among the biomaterials that have been proposed for a wide range of electrochemical applications to obtain biomass-derived GQDs, ranging from simple and natural molecules to complex compounds. Therefore, various green approaches are used for the synthesis of GQDs which mainly focuses on eco-compatibility and their applications based on perspectives for future development.

8.3.1.1 Top-Down Approach for Preparation of GQDs

In this method of synthesis of GQDs, we use different materials such as carbon black, graphene, carbon fiber, carbon nanotubes, and graphene oxides [31] and synthesize nano-sized graphene sheets via a physical cutting and chemical process. Generally, in this method of GQDs synthesis, large molecules or compounds are broken down. The process comprises electrochemical-based, laser passivation, and discharge-based reactions. For example, Dong et al. synthesized simple and multi-layered GQDs through a chemical process of CX-72 carbon black [31]. Next, Ye et al. demonstrated that it is much easier to synthesize GQDs from coal because it contains weaker nano-sized graphene zones which can be easily broken with a chemical process [32]. Various experiments are conducted by different scientists regarding this method of synthesis of GQDs, from graphite a white fluorescent GQDs is synthesized by sonication method, this method is easy, but also costly because of the isolation of GQDs from the strong oxidizing reagent [33], another synthesis method of GQDs is by using carbon nanotubes, graphite, charcoal, and carbon fiber, this method is eco-friendly and requires only ozone assisted oxidizing agents and doesn't require strong acids [34]. Another technique also developed by a scientist is an acid-free pathway and only requires coal tar and hydrogen peroxide (H_2O_2) and its yield is much better (about 80 wt.%) than other syntheses [35]. A different type of approach is also followed by scientists in which they can control the size and shape of the GQDs. Some of them are using graphene sheets and exfoliating them electrochemically to produce shape and size effective GQDs; another method is by taking multi-layered carbon nanotubes (MWCNTs) and graphite flakes to produce highly fluorescent GQDs with a diameter of approx. 20 nm and high yield by exfoliation and disintegration of reactants [36,37]. Another experiment is performed in which a photo-Fenton reaction is introduced for the production of GQDs by simple control over the reaction parameters, and a GQD with an enriched carboxylic group is formed [38].

Apart from these oxidative methods, another strategy is also used in reducing agents are used for the production of these GQDs. For example, size-controlled GQDs is made by a scientist by reducing and breaking graphene oxide into small molecules and passing them through supercritical water [39]. Another example is synthesizing GQDs by a hydrothermal process of graphene oxide and reactant [40]. Similarly, GQDs with bright color fluorescence amino-functional are also synthesized from oxidized graphene having optical properties under control and adjustment [41]. Also, various research groups have performed experiments regarding the synthesis of GQDs, but many limiting factors such as low yield, high process cost, high material cost, more reaction time, valuable and costly instruments, and harmful and unfavorable conditions limit them to have large-scale applications in the synthesis field of GQDs.

8.3.1.1.1 Hydrothermal Process

As stated, pyrolysis is the most common thermal process for producing nanoscale carbonaceous compounds from biomass. The downside is that the organic material is gradually transformed into dots through a sequence of heating steps, under extreme circumstances, dehydration, decomposition, and carbonization occur, requiring high temperatures and extended reaction times. Hydrothermal treatment, on the other hand, which involves thermochemical decomposition at mild circumstances while using the high moisture content of the biomass, might be a viable top-down green approach for GQDs synthesis from a variety of natural biomasses. In reality, from basic precursors such as glucose, sucrose, and citric acid to more complicated components such as biomass wastes, the hydrothermal approach is now the most widely utilized green technology. Wang et al. [42] presented a novel fascinating hydrothermal technique for turning biomass into GQDs. By processing rice husk at 150°C for 5 hours in a Teflon-lined autoclave, they were able to make GQDs with a size distribution of ca. 3.9 nm and 2–3 graphene layers. The RH-GQDs can be scattered in water gradually and display strong photoluminescence as well as an extremely selective quenching of Fe^{3+} ions, providing them a viable product for Fe^{3+} ions sensing. In a recent study, Tade and Patil described a straightforward method for converting waste bamboo wood biomass. The synthesis of GQDs was achieved by hydrothermally treating cellulose nanocrystals (CNCs) produced from bamboo wood (Bf) for 8 hours at 180°C [43]. Following that, the carbon-containing product was treated to ultrasonic, filtered, and lastly lyophilized. Because of breaking of the 1,4-glycosidic bond of cellulose and the intramolecular polymerization resulted in the formation of the graphene sheets occurring almost simultaneously, the authors demonstrated the high temperatures and pressures needed by hydrothermal treatment help ensure the catalytic conversion of CNCs into one-pot GQDs [44]. Furthermore, the structural and optical characterization of the produced Bf-GQDs suggested that they may be used as curcumin fluorescence sensors.

8.3.1.1.2 Oxidation Method

One of the most adaptable techniques for the manufacture of GQDs from bigger graphitized carbon sources with reasonably high yields is oxidative cleavage. Nirala et al. [45] created an unusual class of carbon quantum dots (E-GQDs) with a virtually uniform size (5 nm) using efficient and environment-friendly carbon electro-oxidation of woody biomass. Wood-derived coal has shown to be an effective precursor for the production of carbon nanomaterials because it allows for effective electrochemical oxidative cleavage in multi-layered graphene sheets. Electrochemical oxidation was achieved in two scenarios by varying current intensity, and electrochemical cleavage was ensured by water-free radicals and ammonium sulfate peroxide. Water-free radicals function as "scissors," breaking carbon-containing biomolecules into multi-layered sheets of graphene, allowing SO_4 radicals to readily insert into the sheets and, eventually, reduce them to GQDs with unusual structural and optical properties. One of the drawbacks of the oxidative method is that it frequently requires the employment of strong acids or hazardous and damaging heavy oxidants, i.e., non-eco-compatible methods.

8.3.1.1.3 Laser Ablation Techniques

Chemical oxidation procedures demand the use of powerful acids, which can harm equipment over time and be considered environmentally unfriendly. As a result, simpler and more environmentally friendly solutions, such as pulsed laser ablation within the liquid phase, have been developed. Narasimhan et al. [46] employed this approach to create fluorescent GQD probes for bioimaging applications. GQDs were created by laser ablation of a graphite plate immersed in an aqueous solution of polyethylene glycol: A microsecond pulsed laser source was optically directed in the direction of the substrate for 30 minutes, causing the ablation of the graphite plate surface, resulting in GQDs in solution, which was then separated by centrifugation and filtration, with larger graphene sheets sinking to the bottom.

The GQDs manufactured using this technology performed well as fluorescent biomarkers. Kang et al. [47] used pulsed laser ablation in the liquid phase for the single-step manufacture of graphite flakes suspended in ethyl alcohol and 3-mercaptopropionic acid (MPA) solution, focusing the pulsed laser source onto the suspension for 30 minutes at room temperature. The pulsed laser-induced graphite flakes decompose into carbonaceous nanoparticles. Sulfur-doped graphene nanosheets are formed when C binds to S, which is produced via the breakdown of MPA. The expensive cost of the devices is one of the primary limitations of laser ablation, and until now, only a few studies of GQDs synthesis by laser ablation utilizing environmentally favorable raw resources have been recorded.

8.3.1.2 Bottom-Up Approach for Preparation of GQDs

In this approach of preparation of GQDs, the observation is done at the molecular level. In this method, small molecules are taken and combined by different techniques of fusion to make GQDs, such as cage opening, hydrothermal processes, thermal pyrolysis, incomplete carbonization, plasma treatment, and microwave techniques. For example, GQDs having photoluminescent properties are synthesized by modifying the carbonization degree of citric acid [48]. Lu et al. synthesized GQDs by ruthenium-based cage opening reaction from C60 having fluorescent properties [49]. Another synthesis was done by Chua et al. and demonstrated that size manipulated GQDs are synthesized by acids and oxidative reagents [50]. GQDs used in ion sensing and bioimaging having blue fluorescent properties are synthesized through pyrolysis of aspartic acid and NH_4HCO_3 mixture [51]. Qu et al. synthesized special N-GQDs and S, N co-doped GQDs via hydrothermal pathway taking citric acid and urea as reactants [52]. An easier and more effective way is also developed by scientists for the preparation of these N-GQDs and S, N co-doped GQDs by a solvothermal pathway under microwave radiation by simply adjusting different solvents [53]. Another synthesis method by one-pot pyrolysis is also demonstrated by a scientist for N and S co-doped GQDs by citric acid and L-cystine as a reactant [54]. For example, blue fluorescent N-GQDs are also synthesized by hydro-thermal pyrolysis of citric acid and urea which is used for medical purposes such as a carrier for an anticancer drug [55]. Also, using salicylic acid as a reactant also helps in the synthesis of GQDs by free radical polymerization followed by irradiation of UV rays [56].

In some of the above methods hydrothermal/solvothermal and thermal pyrolysis techniques are used which can cause some disturbances in the reactor which affect the size and shape of the synthe-sized GQDs. So, to overcome these problems microwave irradiation techniques are used nowadays. In this microwave technique, the heat generated by them is distributed evenly around the reactor and results in better and more precise GQDs synthesis. This process also helped in the synthesis of func-tionalized GQDs by using different functional groups as starting reactants/materials. In the initial phase of this microwave technique, size-controlled GQDs are synthesized by Kumawat et al. in one step, using ethanolic extract of mango leaves as the reactant [57]. Further, development in this tech-nique leads to the synthesis of N-GQDs and N, S co-doped GQDs by glucosamine and thiourea as a reactant, the prepared GQDs have excellent optical properties in near-IR and visible regions [55].

8.3.1.2.1 Microwave Irradiation Method

Microwaves are the type of electromagnetic radiation that exists between radio and infrared fre-quencies. Microwave heating is an easy, rapid, and cost-effective method used for synthesizing a variety of materials, including GQDs [58]. The MW-assisted technique is more efficient than con-ventional heating because radiant energy is uniformly transmitted to the substrate without direct interaction with the source.

Kumawat et al. [59] presented a one-pot microwave irradiation approach with mango leaves, which they utilized to produce GQDs, which they subsequently employed for in vivo imaging. Mango leaves were shredded, soaked in ethanol for 4 hours, and combined with water. The suspen-sion was microwaved for 5 minutes before being centrifuged and filtered again. The GQDs that were created had good biocompatibility and were effective in detecting intracellular temperature.

8.3.1.2.2 Soft-Template Method

Soft-template is a novel approach for synthesizing GQDs that is simple, low cost, and eco-friendly. It provides a nanoscale reaction cavity without the need for time-consuming separation and puri-fication procedures. As a result, this approach is advantageous for the mass production of GQDs. Therefore, it can be concluded that both of these top-down and bottom-up methods have their advantages and disadvantages and are also extensively used methods for the synthesis of GQDs. In the top-down method, various advantages have easy accessibility of raw material and have a simple reaction process. On the other hand, low yield, use of strong acids, and longer time of syn-thesis are some of the factors that limit the industrial application/large-scale production by the top-down method. The bottom-up approach of GQDs synthesis has various advantages such as size-controlled, doped, and functionalized GQDs being synthesized in a shorter time. And the new microwave technique is proved very efficient, rapid, and environment-friendly in the bottom-up approach of GQDs synthesis.

8.3.1.2.3 Pyrolysis

Many organic materials can be pyrolyzed into graphene sheets materials for further exfoliation of GQDs at extreme reaction temperatures and in inert environments. Pyrolysis is one of the most basic carbonization processes, and it involves the use of high temperatures. Variable renewable starting materials are converted into carbonaceous nanoparticles at high temperatures.

Mahesh et al. [60] demonstrated a simple pyrolysis preparation of monodisperse GQDs using emulsifying agent carbonization of carbohydrates from a honey/water emulsion in the vicinity of butanol for the first time. Honey-derived GQDs have been used as transparent privacy ink and a white-light emission component. Veeramani et al. [61] used carbonization from *Bougainvillea spectabilis* flowers to make perforated graphene sheet-like carbon nanoparticles (GPACs). The wildflowers were mashed, dried, and then pyrolyzed at 200°C for 6 hours. The synthesized material was used to make electrodes for catechin detection.

Therefore, it can be concluded that both of these top-down and bottom-up methods have their advantages and disadvantages and are also extensively used methods for the synthesis of GQDs. In the top-down method, various advantages have easy accessibility of raw material and have a simple reaction process. On the other hand, low yield, use of strong acids, and longer time of synthesis are some of the factors that limit the industrial application/large-scale production by the top-down method. The bottom-up approach of GQDs synthesis has various advantages such as size-controlled, doped, and functionalized GQDs being synthesized in a shorter time. And the new microwave technique is proved very efficient, rapid, and environment-friendly in the bottom-up approach of GQDs synthesis.

8.4 COMPARISON BETWEEN DIFFERENT APPROACHES FOR THE SYNTHESIS OF GQDS

As mentioned, the GQDs can be synthesized by two different approaches: (i) "top-down approach" and (ii) "bottom-up approach." A brief comparison is depicted here in Table 8.1.

8.5 DIVERSE APPROACHES FOR MODIFICATION OF GQDS

Pure GQDs have a number of drawbacks that limit their applications. GQDs can be functionalized in a variety of ways to modify their characteristics for specific applications, leading them to be used in a wide range of industries. GQDs can be functionalized to change their chemical, optical, and electrical properties, allowing them to be utilized in a wide range of applications. Functionalization includes doping with heteroatoms, formation of composites with inorganic materials or polymers, and adjusting the size and shape of GQDs, as shown in Figure 8.2 [63]. The basic features of GQDs can be modified using these methods by improving their characteristic features. The functionalization of GQDs is a hot topic for research in the study of carbonaceous nanomaterials.

8.5.1 HETEROATOM DOPING

Doping of semiconductors modulates the basic chemical, physical, and electronic properties of the materials making them more useful for practical applications. In the same way, functionalization

TABLE 8.1
Comparison of Different Methods of GQDs Synthesis [62]

Properties	Top-Down Approach	Bottom-Up Approach
Starting material	Solid state	Either liquid or gaseous
Basic	Fragmentation or exfoliation of graphene sheets is done for getting nanoparticles	Molecular precursors are utilized step by step for the formation of GQDs
Process used	Physical and electrochemical methods	Physical and chemical methods
Pros	Used for large-scale production	Deposition parameters can be managed or controlled
	Chemical purification is not required	Cheaper as compared to the top-down approach
	Deposition is possible over a large surface	Nanoparticles obtained are ultra-fine
Cons	Deposition parameters cannot be controlled	Chemical purification is a must in this approach
	Expensive approach	Only small-scale production can be carried out
	Broad size distribution	

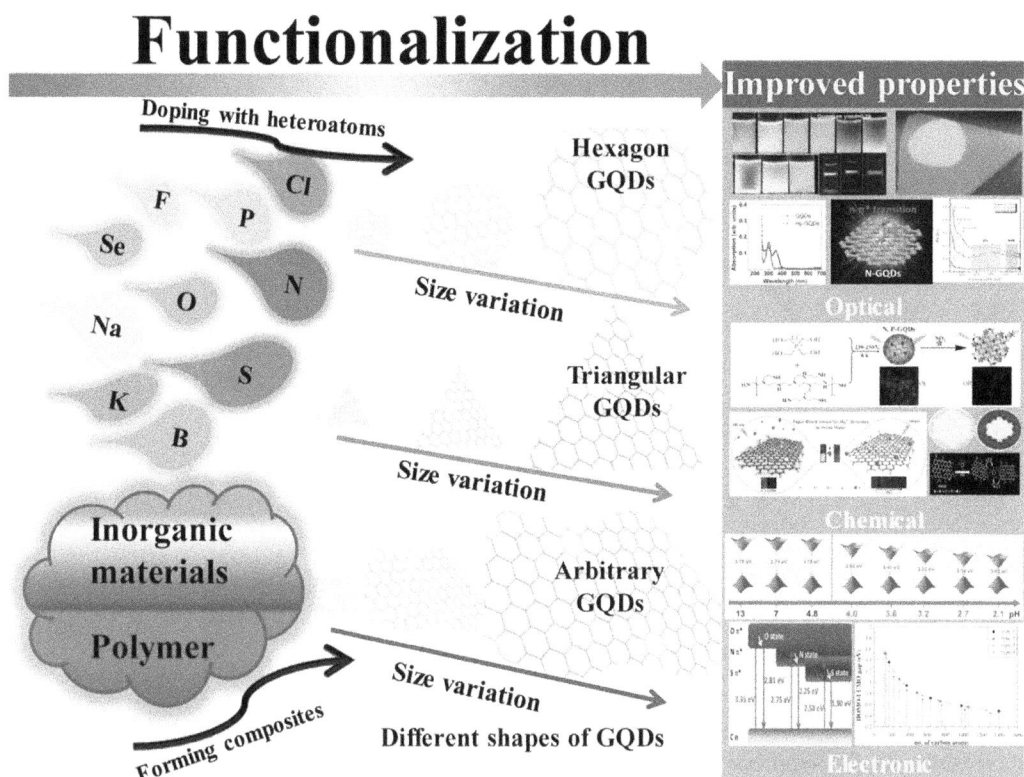

FIGURE 8.2 Representation of functionalization of GQDs with the improvement of various properties [63].

of GQDs by doping can change their chemical, optical, and electrical properties, for novel applications. The doping method for GQDs is categorized into three groups based on the number of doping atoms (Figure 8.3).

The properties of GQDs can all be tuned with effective doping which can be characterized by measuring changes in the color of PL, water solubility, and conductivity of doped GQDs. This section will cover the various heteroatom doping and doping methods. The effect of doping GQDs, which results in a change in their basic properties, will also be discussed. There are various heteroatoms used in doping such as sodium (Na), potassium (K), boron (B), oxygen (O), phosphorus (P), nitrogen (N), selenium (Se), chlorine (Cl), sulfur (S), and fluorine (F). Along with this, the co-doping of these atoms can be done into the GQDs to enhance their different properties, which also changes the basic properties such as electronegativity of the atoms doped.

8.5.1.1 Single Heteroatom Doping

The characteristics of GQDs can be modified by single heteroatom doping in which only one heteroatom is co-doped with the GQDs. This section focuses on the effects of single-atom doping on the properties of GQDs, with less emphasis on doping methods. Wang and his co-workers revealed that doping of boron into GQDs can be done by the MAH method [64] as it is the nearest atom toward carbon so it can be easily doped into the GQDs. The photoluminescence intensity increases after doping of boron which was confirmed by the absorption peaks that occurred at 230, 260, and 320 nm which was differentiating the undoped GQDs. Also, these B-doped GQDs were synthesized by bottom-up approach by Tam et al. in which the electrocatalytic activity has been enhanced [65]. Nitrogen is the most commonly used element in the periodic table in heteroatom doping of GQDs, out of numerous heteroatoms. This is not only because of the stronger electronegativity (3.04) of the N atom compared to the C atom (2.55), but also to the size compatibility (relatively small lattice

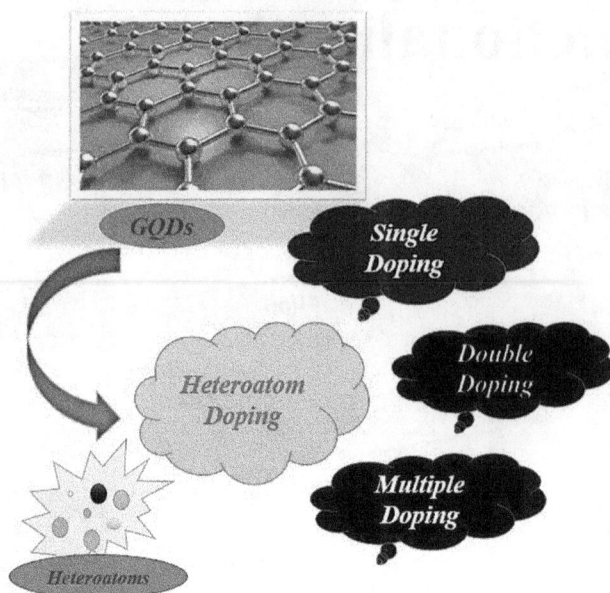

FIGURE 8.3 Schematic representation of different doping methods.

strain) [66,67]. Due to the strong electronegativity, we can assume that the doping concentration decreased with the synthesis temperature, although the band gap was greatly influenced by the doping concentration. In addition, when the N-GQDs were exposed to an acidic environment, the graphitic N were gradually protonated, resulting in a lesser negative induction effect than unprotonated graphitic N. This resulted in a large density of p-electron clouds, which decreased the bandgap. At a doping concentration of 5.5%, this process is decreasing the bandgap of N-GQDs by lowering the pH level in the N-GQDs. Apart from these, doping is done by K and Na atoms that enhance the photoluminescence intensity and tunes the optical absorption [68]. In the case of P doping, the P-GQDs have dispersion and good stability, exhibiting a wide visible light absorption region [69]. All these single-doped GQDs are synthesized by using different approaches such as hydrothermal methods and electrochemical methods and have shown different emissions with stable properties and can be extensively studied in the future.

8.5.1.2 Double Heteroatom Doping

Doping of double atoms was developed to maximize the benefits of various doped elements, and it has been done by researchers all over the world [70–72]. Figure 8.4a shows the UV-Vis absorption spectra of S and N co-doped GQDs (S, N-GQDs) fabricated by Qu and his co-workers [73]. At 338, 467, and 557, three absorption bands were recorded, which differ from the past investigations. The introduction of S and N atoms was implicated for the increasing absorption bands. At wavelengths of 340, 440, and 540 nm, which represent blue, green, and red emission, the PL quantum yields for S, N-GQDs were 61%, 45%, and 8%, respectively. They used the proposed mechanism to explain the PL process of S, N-GQDs through the solvothermal approach of synthesis [73]. The π^* orbital of the C=O (excitation energy was more than 3.10 eV), C=N (excitation energy was more than 2.75 eV), and C=S (excitation energy was more than 1.90 eV) bonds in the S, N-GQDs exhibited O, N, S states. Under the excitation wavelength of 440 nm, the observed blue light could be due to some degree of overlapping of the N and O states. Furthermore, the introduction of S doping resulted in the formation of the S π^* orbital and S state, which resulted in a lower electronegativity. S, N-GQDs can absorb less energy and emit light with longer wavelengths as a result. Using one-step bottom-up molecular fusion in a hydrothermal method, a high quantum yield of

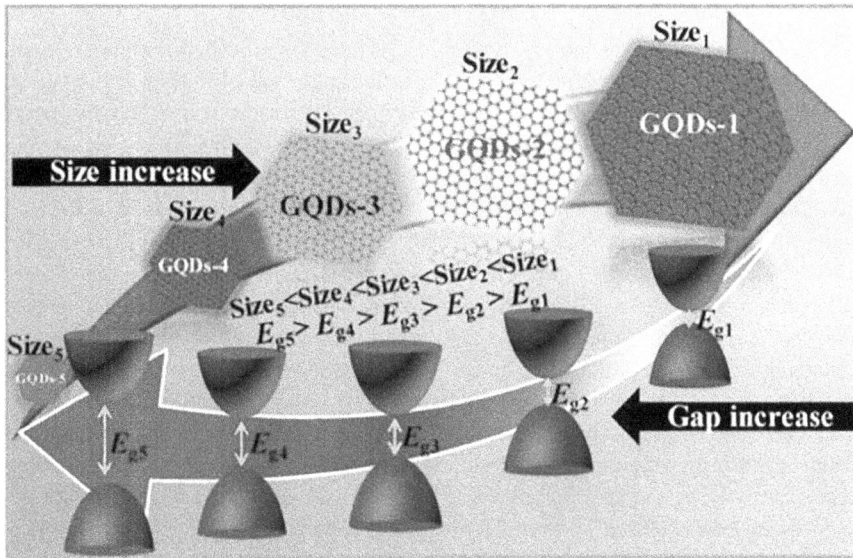

FIGURE 8.4   Analysis of GQDs dependence on shape and size [63].

87.8% for S, N-GQDs was achieved [74]. The prepared S, N co-doped GQDs were observed with bright blue fluorescence having a quantum yield of about 23%. However, Favaro and co-authors synthesized N, B co-doped GQDs to improve electrochemical activity [75], and they observed a significant decrease in the overpotential as a function of dopant concentration in the following order: N>B>B, N. In conclusion, the co-doped approach proved successful in controlling GQDs characteristics. The choice of doping elements, on the other hand, is usually decided by the applications of GQDs.

8.5.1.3   Multiple Heteroatom Doping

In multiple heteroatom doping, more than two heteroatoms are doped into the GQDs, which are important in the field of solar cells, batteries, or other energy storage devices. S, F, N-doped GQDs were fabricated by various scientists for the optimization of dye-sensitized solar cells [76,77]. In a nutshell, multiple-atom-doped GQDs are a fascinating method that should be thoroughly investigated in the future. The absorption peaks of various doped GQDs in the literature were seen against the electronegativity of the dopant and the absorption peaks are largely concentrated in the blue region. Using S, N, or N doping, the absorption peak can be extended from deep UV to red.

8.5.2   Effect of Size and Shape on Tunable Properties of GQDs

Controlling the shape and size of GQDs is another way to tailor their optical, physical, and chemical properties. GQDs size and morphology have been altered using a variety of techniques. The bandgap decreases as the size of the GQDs increases, suggesting a link between energy bandgap and material size [78]. It was proposed in the literature that the embedded tiny sp^2 clusters isolated by sp^3 carbons determined the PL of large GQDs made up of heterogeneously hybridized carbon networks. The degree of adjusting bandgap by modifying the size of GQDs varied slightly between theoretical analysis and experimental work, but the pattern was in good agreement as explored by Ye et al [79]. Therefore, chemical procedures can be used to tune the bandgap of GQDs, which controls their photoluminescence.

8.5.2.1 Modification in Bandgap by Increasing Shape and Size

Many investigations have proved that the size of GQDs can be modified simply by changing the reaction temperature, time, stabilizing agent, and reactant concentration [80]. Kwon et al. [81], for example, used amidative cutting of shredded graphite to create size-controlled GQDs (ranging from 2 to 10 nm) by merely adjusting the amine concentration. Following that, Yeh et al. used ultrafiltration through different pore size membranes to synthesize different size GQDs as part of their investigation into the quantum confinement impact of GQDs [79]. The luminescence of the as-prepared GQDs changes from orange to blue as the size is lowered from 8 to 1 nm, resulting in a shift in the sp^2-domain size and hence the π-π^* gap. The bandgap of GQDs is shown to be dependent on their size and emission. Similarly, Ye et al. came to the same conclusion and developed size-controlled GQDs from anthracite coal by adjusting the reaction temperature and tailoring the level of functionality and bandgap [79]. As a result, it can be said that change in bandgap with an increase in shape and size modifies the GQDs toward better applications by enhancing the electrochemical capacitance and PL intensity. These properties are studied by using different methods such as density of states, Raman scattering, magnetic properties optical characterization, and Faraday optical rotation. The concept of modifying the size and shape of GQDs band gaps has caught the interest of both theoretical [82] and experimental [83] researchers. This is a promising strategy for broadening the scope of GQDs application in the future.

8.6 REAL-WORLD APPLICATIONS OF GQDS FOR THE CONVERSION AND STORAGE OF ENERGY

The exceptional behavior of GQDs such as favorable biocompatibility, ease of functionalization, large surface area, tunable bandgap, the abundance of active sites, and optical properties make GQDs a highly desired and essential material in upcoming advanced technologies. GQDs are extensively used in a variety of applications in our day-to-day life such as drug delivery, biosensing, light-emitting diodes, photodynamic therapy, antibacterial activity, bioimaging, and the storage of energy through solar cells, supercapacitors, rechargeable batteries, photocatalysis, etc. In this section, the real-world and possible future applications for the conversion and storage of energy of GQDs based on the optoelectronics field will be discussed in detail.

8.6.1 ENERGY CONVERSION AND STORAGE-BASED APPLICATIONS

As the population increases and resources become scarce and costly, the search for innovative and sustainable energy sources becomes more essential. Carbon is a great choice for energy-related applications because it is one of the most abundant elements in the world. Solar energy is a clean, sustainable source of electricity that may be used to power our daily lives. Solar energy, on the other hand, is only available during the day and not at night. As a result, energy storage is a hot topic in the energy industry, and it remains one of the biggest difficulties in renewable energy. The usage of GQDs in solar cells and energy storage batteries is discussed in the following sections.

8.6.1.1 Energy Storage Devices

Most renewable energy sources, particularly those obtained from nature, require battery storage, so supercapacitor battery technology with high efficiency, long-term cycling life, low weight, high energy density, and long-term stability is needed to meet the energy requirement. GQDs provide several advantages for energy storage applications due to their superior features, such as ease of functionalization and large surface area. GQDs are used to make an environmentally friendly and high-performance battery and play an important role in improving battery performance. Supercapacitors, also known as electrochemical capacitors, can store energy via redox reactions or the double-layer effect. The energy storage mechanisms enable faster charging and discharging than rechargeable batteries and long-life cycles [84]. In addition, supercapacitors made of nanomaterials have shown relatively high energy density, and, therefore, they are promising candidates as alternative materials for energy-storage-related applications [85]. GQDs have received attention with regard to their application in energy-related fields, but their potential as electrode materials in electrochemical energy storage devices has not yet been studied in depth [86]. Small GQDs have a large number of edge sites in comparison to graphene sheets and, therefore, they provide superior adsorption of ionic charges, which is a crucial property for super capacitance applications [87]. Hassan et al. fine-tuned the characteristics of GQDs using KOH-assisted carbon activation [86,88], an approach inspired by studies on CNTs, and this method significantly enhanced supercapacitor performance [89]. Combining graphene's sp^2 structure and quantum confinement capabilities with activated GQDs' edge effects and edge enrichment enhances the capacity of the materials to store charges through quick and reliable procedures. As previously stated, the introduction of GQDs in batteries has resulted in better battery performance. Importantly, adopting GQDs to improve performance is highly cost-effective. As a result, using GQDs in batteries to improve energy storage is indeed practical and economically feasible.

8.6.1.2 Fuel Cells

The electrochemical process of hydrogen fuel with oxygen in a fuel cell can produce electricity. Fuel cells, unlike batteries, may offer a continuous supply of electricity as long as there is a continuous source of fuel and oxygen (typically from the air) to keep the chemical reaction going. This type of energy technology is both ecologically beneficial and long-lasting. Fuel cells come in a variety of shapes and sizes [90], but they all have an anode and a cathode, which are both necessary components of the fuel cell. Initially, electrode materials such as Pd, Pt, Ag, and Au were primarily made of precious metals, resulting in expensive fuel cells. GQDs have also inspired interest in metal-free electrocatalyst activities, with the goal of replacing conventional Pt-based catalysts in fuel cells for the oxygen reduction reaction (ORR). Nitrogen-doped carbon materials, such as N-graphene and N-CNT, have previously demonstrated promising performance [91]. Zhang et al. were inspired to create N-GQDs coupled to a graphene sheet for the electrocatalytic activity in fuel cell applications. When measured in an oxygen-rich atmosphere, but not under inert conditions, a cathodic peak was observed [92]. Furthermore, the ORR potential and reduction peak are comparable to those of commercial Pt/C catalysts. In contrast to Pt/C electrodes, N-GQD/graphene electrodes have a steady ORR without methanol-specific electroactivity. Because undoped GQD/graphene has negligible electroactivity toward the ORR, doping the GQDs with nitrogen is necessary to obtain these characteristics. The ORR is a first-order reaction that returns a constant current output even after 2 days of continuous cycling in an oxygen-saturated 0.1 m KOH solution, according to electrochemical kinetic tests. This demonstrates GQDs as a powerful metal-free electrocatalyst that could eventually replace Pt-based materials. Functionalized GQDs have been used as fuel cell electrodes in recent investigations [93,94]. The catalytic activity and stability of functionalized GQDs have been improved as a result of inner charge transfer. Since then, there has been a rise in research into using functionalized GQDs as fuel cell cathode materials.

8.6.1.3 Solar Cells

Solar cells, which convert sunlight into electricity, are a clean and renewable source of energy. It is also a vital source of energy for societies with limited resources to create power using traditional methods. The properties of GQDs, such as down-conversion, high fluorescence, intense UV absorption, and ease of functionalization, have led to the development of various solar cells that use GQDs to improve their performance. Traditionally, polymer solar cells, silicon solar cells, and other solar cells were used. Scientists have demonstrated a high open current by incorporation of ZnO nanowires on the solid-state solar cell device with the effect of GQDs [95]. Due to the incorporation of ZnO and GQDs, the power conversion efficiency increases to 1.26%. Besides this, other organic and inorganic materials such as TiO_2 have increased the efficiency by approx. 7.5% [96]. The findings have proved that optical absorptivity was enhanced by the functional groups present on GQDs [97].

In conclusion, GQDs have been used in both inorganic and organic solar cells as hybrid materials. Additionally, the distinctive features of GQDs improved the performance of solar cells. As a result, research into unique GQDs properties will continue to benefit solar cell technology.

8.7 CHALLENGES AND OPPORTUNITIES

Researchers in biomedical applications are interested in the optical and electrochemical detection of GQDs, but there are significant challenges in their production, such as aggregation of carbon-based materials during carbonization, surface properties, size control, and uniformity. Due to their ease of fabrication and integration with other nanomaterials, there is expected to be a sustained increase in interest in the synthesis of these quantum dots. Despite the fact that numerous processes such as microwave irradiation, solvation process, and pyrolysis [23,98,99] and other precursors have been discovered, the nanostructures have still not been made to the marketable products yet. One of the most significant problems is to enhance control over the sizes and morphologies of the generated particles by optimizing production approaches and standardizing fabrication procedures. There are numerous key variations with substantial implications for the functionalities of carbon-based sensors and their biocompatibility. This consistency is critical since the fluorescence properties and quantum yields are correlated to the inclusive composition and the presence of residual chemical groups on the carbon and GQDs surfaces. Scaling up processes require optimized synthetic procedures, and in this case, using simple, greener, environmentally friendly, and cost-effective techniques is vital. The use of synthetic techniques based on renewable raw materials is incredibly fascinating, and it requires further research for commercial and scale-up production. Synthetic procedures at lower temperatures (low energy input) employing earth-abundant starting elements, on the other hand, remain a significant problem for researchers in the field [100–103]. One of the most appealing properties of GQDs is photoluminescence, but this property is highly dependent on the raw materials used in fabrication and the presence of surface functionalities on these particles, which are particularly noticeable in carbon quantum dots derived from organic molecules and biomass. Most importantly, one of the persisting challenges has been to develop and improve the quantum yield of GQDs while getting the benefit of their intrinsic properties and fine-tuning their absorption spectra; the development of GQDs with high fluorescence emission is critical for biomedical imaging without harmful consequences. Future research should focus on improving quantum yield, biocompatibility, and synthesis methods that include green, eco-friendly, low-energy, low-cost, simple, and efficient techniques, appropriate nano-size, and compositional and structural changes. Several approaches for fabricating pure or doped GQDs with distinct shapes and properties have recently been proposed. Although separate sustainable and greener synthetic techniques have been used to synthesize carbon quantum dots, GQDs research is still in its early stages, and more detailed and critical analyses are required. For example, well-defined and atom-precise structures have not been described thoroughly, limiting the in-depth analysis of correlations between structures and features, specific property regulations, and comprehensive assessments of novel synthetic techniques and applications. Additional theoretical evaluations and a better knowledge of the mechanistic features of graphene dots are required; additional theoretical studies and a better understanding of the mechanistic aspects are crucial.

8.8 CONCLUSIONS

Nanoscience advances are of current interest for advanced and developed diagnosis and targeted therapy of several complicated diseases (particularly tumors). Semiconductor quantum dots, in particular, have been created as cutting-edge platforms for high-throughput quantitative studies of numerous biomarkers in cells and clinical tissue samples (ex vivo), in vivo evaluations of cells with illnesses, and potentially targeted and traceable medication delivery. Indeed, quantum dots hold great promise in biomedical, bioimaging, and photoluminescent applications, and they can be used as promising fluorescent probes for imaging with low cytotoxicity, as well as in bioanalysis and related domains. GQDs are also promising fluorophores because of their excellent photoluminescence properties, water solubility, low cost, and toxicity. COOH–GQDs, for example, have shown significant biocompatibility in some situations and could be used in biological and medicinal applications. It appears that significant efforts should be made to develop greener, simpler, more efficient, and environmentally friendly production processes for GQDs, as well as their utilization in bio- and nanomedical applications. More scientific research is needed to scale up and commercialize quantum dots using simple, efficient, cost-effective, and sustainable methodologies that emphasize biocompatibility and minimal cytotoxicity. In the future, an improvement in synthesis processes is expected to improve understanding of both the morphological and dimensional features of the materials created for both applications. GQDs have a bright future ahead of them, and advances in nanotechnology will enable the production of highly selective and stable electrochemical sensors and biosensors using simple and environmentally friendly synthetic techniques.

REFERENCES

1. Miller, K. D., Siegel, R.L., Lin, C.C., Mariotto, A.B., Kramer, J.L., Rowland, J.H., Stein, K.D., Alteri, R., Jemal, A. (2016). Cancer treatment and survivorship statistics, 2016. *CA: Cancer J. Clin.*, 66(4), 271–289.
2. Rana, S., Sidhu, Y.S., Kaur, K., Kaur, K., Sandhu, N.K., Singh, A., Singh, G., Narang, R.K. (2019). Advances in combination of quantum dots and nanotechnology-based carrier systems against cancer – A critical review. *Int. J. Bio-Pharma Res.*, 8(12), 2814–2825. http://dx.doi.org/10.21746/ijbpr.2019.8.12.2.
3. Iravani, S. (2011). Green synthesis of metal nanoparticles using plants. *Green Chem*, 13, 2638–2650. https://doi.org/10.1039/C1GC15386B.
4. Xu, J., Zhou, Y., Liu, S., Dong, M., Huang, C. (2014). Low-cost synthesis of carbon nanodots from natural products used as a fluorescent probe for the detection of ferrum (III) ions in lake water. *Anal. Methods*, 6(7), 2086–2090. https://doi.org/10.1039/C3AY41715H.
5. Baker, S.N., Baker, G.A. (2010). Luminescent carbon nanodots: Emergent nanolights. *Angew. Chem. Int. Ed.*, 49(38), 6726–6744. http://doi.org/10.1002/anie.200906623.
6. Dong, F., Cai, Y., Liu, C., Liu, J., Qiao, J. (2018). Heteroatom (B, N and P) doped porous graphene foams for efficient oxygen reduction reaction electrocatalysis, *Int. J. Hydrogen Energy*, 43, 12661–12670.
7. Kumar, S., Kumar, J., Sharma, S. N. (2020). Investigation of charge transfer properties in MEHPVV and rGO-AA nanocomposites for Green organic photovoltaic application, *Optik (Stuttg).*, 208, 164540.
8. Stoller, M. D., Park, S., Yanwu, Z., Anand, J., Ruoff, R. S. (2008). Graphene-based ultracapacitors, *Nano Lett.*, 8, 3498–3502.
9. Ponomarenko, L.A., Schedin, F., Katsnelson, M.I., Yang, R., Hill, E.W., Novoselov, K.S., Geim, A.K. (2008). Chaotic dirac billiard in graphene quantum dots, *Science*, 320, 356–358.
10. Sun, H., Wu, L., Wei, W., Qu, X. (2013). Recent advances in graphene quantum dots for sensing. *Mater. Today*, 16, 433–442. https://doi.org/10.1016/j.mattod.2013.10.020.
11. Ghosh, D., Kapri, S., Bhattacharyya, S. (2016). Phenomenal ultraviolet photoresponsivity and detectivity of graphene dots immobilized on zinc oxide nanorods. *ACS Appl. Mater. Interfaces*, 8, 35496–35504. https://doi.org/10.1021/acsami.6b13037.
12. Li, N., Than, A., Wang, X.W., Xu, S.H., Sun, L., Duan, H.W., Xu, C.J., Chen, P. (2016). Ultrasensitive profiling of metabolites using tyramine-functionalized graphene quantum dots. *ACS Nano*, 10, 3622–3629. https://doi.org/10.1021/acsnano.5b08103.
13. Mehta, J., Bhardwaj, N., Bhardwaj, S., Tuteja, S., Vinayak, P., Paul, A., Kim, K., Deep, A. (2017). Graphene quantum dot modified screen printed immunosensor for the determination of parathion. *Anal. Biochem.*, 523, 1–9. https://doi.org/10.1016/j.ab.2017.01.026
14. Mansuriya, B.D., Altintas, Z. (2020). Graphene quantum dot-based electrochemical immunosensors for biomedical applications. *Materials*, 13, 96. https://doi.org/10.3390/ma13010096.
15. Xu, Q., Yuan, H., Dong, X., Zhang, Y., Asif, M., Dong, Z., He, W., Ren, J., Sun, Y., Xiao, F. (2018). Dual nanoenzyme modified microelectrode based on carbon fiber coated with AuPd alloy nanoparticles decorated graphene quantum dots assembly for electrochemical detection in clinic cancer samples. *Biosens. Bioelectron.*, 107, 153–162. https://doi.org/10.1016/j.bios.2018.02.026.

16. Islam, M.S., Deng, Y., Tong, L., Roy, A.K., Faisal, S.N., Hassan, M., Minett, A.I., Gomes, V.G. (2017). In-situ direct grafting of graphene quantum dots onto carbon fibre by low temperature chemical synthesis for high performance flexible fabric supercapacitor. *Mater. Today Commun.*, 10, 112–119. https://doi.org/10.1016/j.mtcomm.2016.11.002.

17. Zhang, Z., Zhang, J., Chen, N., Qu, L. (2012). Graphene quantum dots: An emerging material for energy-related applications and beyond. *Energy Environ. Sci.*, 5, 8869–8890. https://doi.org/10.1039/C2EE22982J.

18. Zhao, J., Tang, L., Xiang, J., Ji, R., Hu, Y., Yuan, J., Zhao, J., Tai, Y., Cai, Y. (2015). Fabrication and properties of a high-performance chlorine doped graphene quantum dot-based photo-voltaic detector. *RSC Adv.*, 5, 29222–29229. https://doi.org/10.1039/C5RA02358K.

19. Campuzano, S., Yáñez-Sedeño, P., Pingarrón, J.M. (2019). Carbon Dots and Graphene Quantum Dots in Electrochemical Biosensing. *Nanomaterials*, 9, 634. https://doi.org/10.3390/nano9040634.

20. Park, Y., Yoo, J., Lim, B., Kwon, W., Rhee, S. W. (2016). Improving the functionality of carbon nanodots: Doping and surface functionalization, *J. Mater. Chem. A*, 4, 11582–11603.

21. Lei, H., Diao, H., Liu, W., Xie, J., Wang, Z., Feng, L. (2016). A facile Al(iii)-specific fluorescence probe and its application in biological systems, *RSC Adv.*, 6, 77291–77296.

22. Chen, L. C., Teng, C. Y., Lin, C. Y., Chang, H. Y., Chen, S. J., Teng, H. (2016). Architecting nitrogen functionalities on graphene oxide photocatalysts for boosting hydrogen production in water decomposition process, *Adv. Energy Mater.*, 6, 1600719.

23. Mei, Q., Chen, J., Zhao, J., Yang, L., Liu, B., Liu, R.Y., Zhang, Z.P. (2016). Atomic oxygen tailored graphene oxide nanosheets emissions for multicolor cellular imaging. *ACS Appl. Mater. Interfaces*, 8, 7390–7395. https://doi.org/10.1021/acsami.6b00791.

24. Chen, W., Li, D., Tian, L., Xiang, W., Wang, T., Hu, W., Hu, Y., Chen, S., Chen, J., Dai, Z. (2018). Synthesis of graphene quantum dots from natural polymer starch for cell imaging. *Green Chem.*, 20: 4438–4442. https://doi.org/10.1039/C8GC02106F.

25. Wang, S., Cole, I.S., Li, Q. (2016). The toxicity of graphene quantum dots. *RSC Adv.*, 6, 89867–89878. https://doi.org/10.1039/C6RA16516H.

26. Zheng, X.T., Ananthanarayanan, A., Luo, K.Q., Chen, P. (2015). Glowing graphene quantum dots and carbon dots: Properties, syntheses, and biological applications. *Small*, 11, 1620–1636. https://doi.org/10.1002/smll.201402648.

27. Haque, E., Kim, J., Malgras, V., Reddy, K.R., Ward, A.C., You, J., Bando, Y., Hossain, S.A., Yamauchi, Y. (2018). Recent advances in graphene quantum dots: Synthesis, properties, and applications. *Small Methods*, 2, 1800050. https://doi.org/10.1002/smtd.201800050.

28. Kadian, S., Sethi, S.K., Manik, G. (2021). Recent advancements in synthesis and property control of graphene quantum dots for biomedical and optoelectronic applications. *Mater. Chem. Front.*, 5, 627–658. https://doi.org/10.1039/D0QM00550A.

29. Yaxuan, J., Guo, Y., Qineng, X., Xiaohui, L., Wang, Y. (2019). Catalytic production of value-added chemicals and liquid fuels from lignocellulosic biomass. *Chem*, 5, 2520–2546.

30. Kang, C., Huang, Y., Yang, H., Yan, X.F., Chen, Z.P. (2020). A review of carbon dots produced from biomass wastes. *Nanomaterials*, 10, 2316. https://doi.org/10.3390/nano10112316.

31. Dong, Y., Chen, C., Zheng, X., Gao, L., Cui, Z., Yang, H., Guo, C., Chi, Y., Li, C. M. (2012). Onestep and high yield simultaneous preparation of single- and multi-layer graphene quantum dots from CX-72 carbon black, *J. Mater. Chem.*, 22, 8764–8766. https://doi.org/10.1039/C2JM30658A.

32. Ye, R., Xiang, C., Lin, J., Peng, Z., Huang, K., Yan, Z., Cook, N. P., Samuel, E. L. G., Hwang, C. C., Ruan, G., Ceriotti, G., Raji, A. R. O., Martí, A. A., Tour, J. M. (2013). Coal as an abundant source of graphene quantum dots, *Nat. Commun.*, 4, 1–7. https://doi.org/10.1038/ncomms3943.

33. Luo, Z., Qi, G., Chen, K., Zou, M., Yuwen, L., Zhang, X., Huang, W., Wang, L. (2016). Microwave-assisted preparation of white fluorescent graphene quantum dots as a novel phosphor for enhanced white-light-emitting diodes, *Adv. Funct. Mater.*, 26, 2739–2744. https://doi.org/10.1002/adfm.201505044.

34. Shin, Y., Park, J., Hyun, D., Yang, J., Lee, J. H., Kim, J. H., Lee, H. (2015). Acid-free and oxone oxidant-assisted solvothermal synthesis of graphene quantum dots using various natural carbon materials as resources, *Nanoscale*, 7, 5633–5637. https://doi.org/10.1039/C5NR00814J.

35. Liu, Q., Zhang, J., He, H., Huang, G., Xing, B., Jia, J., Zhang, C. (2018). Green preparation of high yield fluorescent graphene quantum dots from coal-tar-pitch by mild oxidation, *Nanomaterials*, 8, 844. https://doi.org/10.3390/nano8100844.

36. Ananthanarayanan, A., Wang, X., Routh, P., Sana, B., Lim, S., Kim, D. H., Lim, K. H., Li, J., Chen, P. (2014). Facile synthesis of graphene quantum dots from 3D graphene and their application for Fe^{3+} sensing, *Adv. Funct. Mater.*, 24, 3021–3026. https://doi.org/10.1002/adfm.201303441.

37. Lin, L., Zhang, S. (2012). Creating high yield water soluble luminescent graphene quantum dots via exfoliating and disintegrating carbon nanotubes and graphite flakes. *Chem. Commun.*, **48**, 10177–10179. https://doi.org/10.1039/C2CC35559K.

38. Zhou, X., Zhang, Y., Wang, C., Wu, X., Yang, Y., Zheng, B., Wu, H., Guo, S., Zhang, J. (2012). Photo-Fenton reaction of graphene oxide: A new strategy to prepare graphene quantum dots for DNA cleavage, *ACS Nano*, 6, 6592–6599. https://doi.org/10.1021/nn301629v.

39. Tayyebi, A., Akhavan, O., Lee, B. K., Outokesh, M. (2018). Supercritical water in top-down formation of tunable-sized graphene quantum dots applicable in effective photothermal treatments of tissues, *Carbon*, 130, 267–272. https://doi.org/10.1016/j.carbon.2017.12.057.

40. Rajender, G., Giri, P. K. (2016). Formation mechanism of graphene quantum dots and their edge state conversion probed by photoluminescence and Raman spectroscopy, *J. Mater. Chem. C*, 4, 10852–10865. https://doi.org/10.1039/C6TC03469A.

41. Tetsuka, H., Asahi, R., Nagoya, A., Okamoto, K., Tajima, I., Ohta, R., Okamoto, A. (2012). Optically tunable amino-functionalized graphene quantum dots, *Adv. Mater.*, 24, 5333–5338. https://doi.org/10.1002/adma.201201930.

42. Wang, W., Wang, Z., Liu, J., Peng, Y., Yu, X., Wang, W. (2018). One-pot facile synthesis of graphene quantum dots from rice husks for Fe^{3+} sensing. *Ind. Eng. Chem. Res.*, 57, 9144–9150.

43. Tade, R.S., Patil, P.O. (2020). Green synthesis of fluorescent graphene quantum dots and its application in selective curcumin detection. *Curr. Appl. Phys.*, 20, 1226–1236.

44. Eom, Y., Min Son, S., Kim, Y.E., Lee, J.E., Hwang, S., Cha, H.G. (2019). Structure evolution mechanism of highly ordered graphite during carbonization of cellulose nanocrystals. *Carbon*, 150, 142–152.

45. Nirala, N.R., Khandelwal, G., Kumar, B., Vinita, Prakash, R., Kumar, V. (2017). One step electro-oxidative preparation of graphene quantum dots from wood charcoal as a peroxidase mimetic. *Talanta*, 173, 36–43.

46. Narasimhan, A.K., Lakshmi, S.B., Santra, T.S., Ramachandra Rao, M.S., Krishnamurthi, G. (2017). Oxygenated graphene quantum dots (GQDs) synthesized using laser ablation for long-term real-time tracking and imaging. *RSC Adv.*, 7, 53822–53829.

47. Kang, S., Jeong, Y.K., Jung, K.H., Son, Y., Kim, W.R., Ryu, J.H., Kim, K.M. (2020). One-step synthesis of sulfur-incorporated graphene quantum dots using pulsed laser ablation for enhancing optical properties. *Opt. Express*, 28, 21659–21667.

48. Dong, Y., Shao, J., Chen, C., Li, H., Wang, R., Chi, Y., Lin, X., Chen, G. (2012). Blue luminescent graphene quantum dots and graphene oxide prepared by tuning the carbonization degree of citric acid, *Carbon*, 50, 4738–4743. https://doi.org/10.1016/j.carbon.2012.06.002.

49. Lu, J., Yeo, P. S. E., Gan, C. K., Wu, P., Loh, K. P. (2011). Transforming C 60 molecules into graphene quantum dots, *Nat. Nanotechnol.*, 6, 247–252. https://doi.org/10.1038/nnano.2011.30.

50. Chua, C. K., Sofer, Z., Šimek, P., Jankovský, O., Klímová, K., Bakardjieva, S., Hrdličková Kučková, Š., Pumera, M. (2015). Synthesis of strongly fluorescent graphene quantum dots by cage-opening buckminsterfullerene, *ACS Nano*, 9, 2548–2555. https://doi.org/10.1021/nn505639q.

51. Zhang, C., Cui, Y., Song, L., Liu, X., Hu, Z. (2016). Microwave assisted one-pot synthesis of graphene quantum dots as highly sensitive fluorescent probes for detection of iron ions and pH value, *Talanta*, 150, 54–60. https://doi.org/10.1016/j.talanta.2015.12.015.

52. Qu, D., Zheng, M., Du, P., Zhou, Y., Zhang, L., Li, D., Tan, H., Zhao, Z., Xie, Z., Sun, Z. (2013). Highly luminescent S, N co-doped graphene quantum dots with broad visible absorption bands for visible light photocatalysts, *Nanoscale*, 5, 12272–12277. https://doi.org/10.1039/C3NR04402E.

53. Jeon, S. J., Kang, T. W., Ju, J. M., Kim, M. J., Park, J. H., Raza, F., Han, J., Lee, H. R., Kim, J. H. (2016). Modulating the photocatalytic activity of graphene quantum dots via atomic tailoring for highly enhanced photocatalysis under visible light, *Adv. Funct. Mater.*, 26, 8211–8219. https://doi.org/10.1002/adfm.201603803.

54. Zhang, R., Adsetts, J. R., Nie, Y., Sun, X., Ding, Z. (2018). Electrochemiluminescence of nitrogen- and sulfur-doped graphene quantum dots, *Carbon*, 129, 45–53. https://doi.org/10.1016/j.carbon.2017.11.091.

55. Hasan, M. T., Gonzalez-Rodriguez, R., Ryan, C., Faerber, N., Coffer, J. L., Naumov, A. V. (2018). Photo- and electroluminescence from nitrogen-doped and nitrogen–sulfur codoped graphene quantum dots, *Adv. Funct. Mater.*, 28, 1804337. https://doi.org/10.1002/adfm.201804337.

56. Zhu, J., Tang, Y., Wang, G., Mao, J., Liu, Z., Sun, T., Wang, M., Chen, D., Yang, Y., Li, J., Deng, Y., Yang, S. (2017). Green, rapid, and universal preparation approach of graphene quantum dots under ultraviolet irradiation, *ACS Appl. Mater. Interfaces*, 9, 14470–14477. https://doi.org/10.1021/acsami.6b11525.

57. Kumawat, M. K., Srivastava, R., Thakur, M., Gurung, R. B. (2017). Graphene quantum dots from mangifera indica: Application in near-infrared bioimaging and intracellular nanothermometry, *ACS Sustain. Chem. Eng.*, 5, 1382–1391. https://doi.org/10.1021/acssuschemeng.6b01893.

58. Singh, R.K., Kumar, R., Singh, D.P., Savu, R., Moshkalev, S.A. (2019). Progress in microwave-assisted synthesis of quantum dots (graphene/carbon/semiconducting) for bioapplications: A review. *Mater. Today Chem.*, 12, 282–314. https://doi.org/10.1016/j.mtchem.2019.03.001.

59. Kumawat, M.K., Thakur, M., Gurung, R.B., Srivastava, R. (2017). Graphene quantum dots from mangifera indica: Application in nearinfrared bioimaging and intracellular nanothermometry. *ACS Sustain. Chem. Eng.*, 5, 1382–1391.

60. Mahesh, S., Lekshmi, C.L., Renuka, K.D., Joseph, K. (2016). Simple and cost-effective synthesis fluorescent graphene quantum dots from honey: Application as stable security ink and white-light emission. *Particle*, 33, 70–74.

61. Veeramani, V., Sivakumar, M., Chen, S.M., Madhu, R., Alamri, H.R., Alothman, Z.A., Hossain, S.A., Chen, C.K., Yamauehi, Y., Miyamoto, N., et al. (2017). Lignocellulosic biomass-derived, graphene sheet-like porous activated carbon for electrochemical supercapacitor and catechin sensing. *RSC Adv.*, 7, 45668–45675.

62. Al Jahdaly, B. A., Elsadek, M. F., Ahmed, B. M., Farahat, M. F., Taher, M. M., Khalil, A. M. (2021). Outstanding graphene quantum dots from carbon source for biomedical and corrosion inhibition applications: A review. *Sustainability*, 13, 2127.

63. Tian, P., Tang, L., Teng, K.S., Lau, S.P. (2018). Graphene quantum dots from chemistry to applications, *Mater. Today Chem.*, 10, 221–258. ISSN 2468-5194, https://doi.org/10.1016/j.mtchem.2018.09.007.

64. Hai, X., Mao, Q.-X., Wang, W.-J., Wang, X.-F., Chen, X.-W., Wang, J.-H. (2015). An acid free microwave approach to prepare highly luminescent boron-doped graphene quantum dots for cell imaging, *J. Mater. Chem.*, B3, 9109–9114.

65. Tam, V., Kang, S.G., Babu, K.F., Oh, E.-S., Lee, S.G., Choi, W.M. (2017). Synthesis of B-doped graphene quantum dots as a metal-free electrocatalyst for the oxygen reduction reaction, *J. Mater. Chem. A*, 5, 10537–10543.

66. Wang, Y., Shao, Y., Matson, D. W., Li, J., Lin, Y. (2010). Nitrogen-doped graphene and its application in electrochemical biosensing, *ACS Nano*, 4, 1790–1798.

67. Wu, P., Du, P., Zhang, H., Cai, C. (2013). Microscopic effects of the bonding configuration of nitrogen-doped graphene on its reactivity toward hydrogen peroxide reduction reaction, *Phys. Chem. Chem. Phys.*, 15, 6920–6928.

68. Ain, N.-U., Eriksson, M.O., Schmidt, S., Asghar, M., Lin, P.-C., Holtz, P.O., Syvajarvi, M., Yazdi, G.R. (2016). Tuning the emission energy of chemically doped graphene quantum dots, *Nanomaterials (Basel)*, 6, 198.

69. Qian, J., Shen, C., Yan, J., Xi, F., Dong, X., Liu, J. (2018). Tailoring the electronic properties of graphene quantum dots by P doping and their enhanced performance in metal-free composite photocatalyst, *J. Phys. Chem. C*, 122, 349–358.

70. Xie, H., Hou, C., Wang, H., Zhang, Q., Li, Y. (2017). S, N co-doped graphene quantum dot/TiO$_2$ composites for efficient photocatalytic hydrogen generation, *Nanoscale Res. Lett.*, 12, 400.

71. Anh, N.T.N., Chowdhury, A.D., Doong, R.-A. (2017). Highly sensitive and selective detection of mercury ions using N, S-co-doped graphene quantum dots and its paper strip-based sensing application in wastewater, *Sensor. Actuator. B*, 252, 1169–1178.

72. Mondal, T.K., Dinda, D., Saha, S.K. (2018). Nitrogen, sulphur co-doped graphene quantum dot: An excellent sensor for nitroexplosives, *Sensor. Actuator. B*, 207, 586–593.

73. Qu, D., Sun, Z., Zheng, M., Li, J., Zhang, Y., Zhang, G., Zhao, H., Liu, X., Xie, Z. (2015). Three colors emission from S, N co-doped graphene quantum dots for visible light H$_2$ production and bioimaging, *Adv. Opt. Mater.*, 3, 360–367.

74. Shen, C., Ge, S., Pang, Y., Xi, F., Liu, J., Dong, X., Chen, P. (2017). Facile and scalable preparation of highly luminescent N, S co-doped graphene quantum dots and their application for parallel detection of multiple metal ions, *J. Mater. Chem. B*, 5, 6593–6600.

75. Favaro, M., Ferrighi, L., Fazio, G., Colazzo, L., Vaentin, C.D., Durante, C., Sedona, F., Gennaro, A., Agnoli, S., Granozzi, G. (2015). Single and multiple doping in graphene quantum dots: Unraveling the origin of selectivity in the oxygen reduction reaction, *ACS Catal.*, 5, 129–144.

76. Kundu, S., Yadav, R.M., Narayanan, T.N., Shelke, M.V., Vajtai, R., Ajayan, P.M., Pillai, V.K. (2015). Synthesis of N, F and S co-doped graphene quantum dots, *Nanoscale*, 7, 11515–11519.

77. Kundu, S., Sarojinijeeva, P., Karthick, R., Anantharaj, G., Saritha, G., Bera, R., Anandan, S., Patra, A., Ragupathy, P., Selvaraj, M., Jeyakumar, D., Pillai, K.V. (2017). Enhancing the efficiency of DSSCs by the modification of TiO$_2$ photoanodes using N, F and S, co-doped graphene quantum dots, *Electrochim. Acta*, 242, 337–343.

78. Sk, M.A., Ananthanarayanan, A., Huang, L., Lim, K.H., Chen, P. (2014). Revealing the tunable photoluminescence properties of graphene quantum dots, *J. Mater. Chem. C*, 2, 6954–6960.

79. Ye, R., Peng, Z., Metzger, A., Lin, J., Mann, J.K., Huang, K., Xiang, C., Fan, X., Samuel, E.L.G., Alemany, L.B., Marti, A.A., Tour, J.M. (2015). Bandgap engineering of coal-derived graphene quantum dots, *ACS Appl. Mater. Interfaces*, 7, 7041–7048.

80. Yeh, T. F., Huang, W. L., Chung, C. J., Chiang, I. T., Chen, L. C., Chang, H. Y., Su, W. C., Cheng, C., Chen, S. J., Teng, H. (2016). Elucidating quantum confinement in graphene oxide dots based on excitation-wavelength-independent photoluminescence, *J. Phys. Chem. Lett.*, 7, 2087–2092.

81. Kwon, W., Kim, Y. H., Lee, C. L., Lee, M., Choi, H. C., Lee, T. W., Rhee, S. W. (2014). Electroluminescence from graphene quantum dots prepared by amidative cutting of tattered graphite, *Nano Lett.*, 14, 1306–1311.

82. Zhang, P., Hu, Q., Yang, X., Hou, X., Mi, J., Liu, L., Dong, M. (2018). Size effect of oxygen reduction reaction on nitrogen-doped graphene quantum dots, *RSC Adv.*, 8, 531–536.

83. Yang, H., Ku, K.H., Shin, J.M., Lee, J., Park, C.H., Cho, H.-H., Jang, S.G., Kim, B.J. (2016). Engineering the shape of block copolymer particles by surface-modulated graphene quantum dots, *Chem. Mater.*, 28, 830–837.

84. Shen, C. W., Wang, X. H., Zhang, W. F., Kang, F. Y., Power, J. (2011). A high-performance three-dimensional micro supercapacitor based on self-supporting composite materials, *Sources*, 196, 104.

85. Simon, P., Gogotsi, Y. (2008). Materials for electrochemical capacitors, *Nat. Mater.*, 7, 845.

86. Hassan, M., Haque, E., Reddy, K. R., Minett, A. I., Chen, J., Gomes, V. G. (2014). Edge-enriched graphene quantum dots for enhanced photo-luminescence and supercapacitance, *Nanoscale*, 6, 11988.

87. Kalita, H., Harikrishnan, V., Shinde, D. B., Pillai, K. V., Aslam, M. (2013). Hysteresis and charge trapping in graphene quantum dots, *Appl. Phys. Lett.*, 102(14), 3104.

88. Marsh, H., Rodriguez-Reinoso, F. (2006). *Activated Carbon*, Elsevier, London.

89. Raymundo-Pinero, E., Azais, P., Cacciaguerra, T., CazorlaAmoros, D., Linares-Solano, A., Beguin, F. (2005). KOH and NaOH activation mechanisms of multiwalled carbon nanotubes with different structural organization, *Carbon*, 43, 786.

90. Singhal, S.C. (2000). Advances in solid oxide fuel cell technology, *Solid State Ionics*, 135, 305–313.

91. Qu, L. T., Liu, Y., Baek, J. B., Dai, L. M. (2010). Nitrogen-doped graphene as efficient metal-free electrocatalyst for oxygen reduction in fuel cells, *ACS Nano*, 4, 1321.

92. Zhang, M., Bai, L. L., Shang, W. H., Xie, W. J., Ma, H., Fu, Y. Y., Fang, D. C., Sun, H., Fan, L. Z., Han, M., Liu, C. M., Yang, S. H. (2012). Facile synthesis of water-soluble, highly fluorescent graphene quantum dots as a robust biological label for stem cells, *J. Mater. Chem.*, 22, 7461.

93. Fei, H., Ye, R., Ye, G., Gong, Y., Peng, Z., Fan, X., Samuel, E.L.G., Ajayan, P.M., Tour, J.M. (2014). Boron- and nitrogen-doped graphene quantum dots/graphene hybrid nanoplatelets as efficient electrocatalysts for oxygen reduction, *ACS Nano*, 8, 10837–10843.

94. Dhand, A., Suresh, S., Jain, A., Varadan, O.N., Kerawalla, M.A.K., Goswami, P. (2017). Advances in materials for fuel cell technologies-a review, *Int. J. Res. Appl. Sci. Eng. Technol.*, 5, 1672–1682.

95. Tamandani, S., Darvish, G. (2017). Charge transfer modeling in monolayer circular graphene quantum dots-ZnO nanowires system for application in photovoltaic devices, *Int. J. Mod. Phys. B* 31, 1650253.

96. Zhang, Q., Zhang, G., Sun, X., Yin, K., Li, H. (2017). Improving the power conversion efficiency of carbon quantum dot-sensitized solar cells by growing the dots on a TiO_2 photoanode in situ, *Nanomaterials*, 130, 7060130.

97. Kim, J.K., Park, M.J., Kim, S.J., Wang, D.H., Cho, S.P., Bae, S., Park, J.H., Hong, B.H. (2013). Balancing light absorptivity and Carrier conductivity of graphene quantum dots for high-efficiency bulk heterojunction solar cells, *ACS Nano*, 7, 7207–7212.

98. Murugan, N., Prakash, M., Jayakumar, M., Sundaramurthy, A., Sundramoorthy, A.K. (2019). Green synthesis of fluorescent carbon quantum dots from Eleusine coracana and their application as a fuorescence 'turn-of' sensor probe for selective detection of Cu^{2+}. *Appl. Surf. Sci.*, 476, 468–480. https://doi.org/10.1016/j. apsusc.2019.01.090.

99. Wang, L., Li, W., Wu, B., Li, Z., Wang, S., Liu, Y., Pan, D., Wu, M. (2016b). Facile synthesis of fluorescent graphene quantum dots from coffee grounds for bioimaging and sensing. *Chem. Eng. J.*, 300, 75–82. https://doi.org/10.1016/j.cej.2016.04.123.

100. Shen, J., Zhu, Y., Yang, X., Li, C. (2012). Graphene quantum dots: Emergent nanolights for bioimaging, sensors, catalysis and photovoltaic devices. *Chem. Commun.*, 48(31):3686–3699. https://doi.org/10.1039/C2CC00110A.

101. Nekoueian, K., Amiri, M., Sillanpaa, M., Marken, F., Boukherroub, R., Szunerits, S. (2019). Carbon-based quantum particles: An electroanalytical and biomedical perspective. *Chem. Soc. Rev.*, 48, 4281–4316. https://doi.org/10. 1039/C8CS00445E.

102. de Menezes, F.D., Dos Reis, S.R.R., Pinto, S.R., Portilho, F.L., do Vale Chaves, E., Mello, F., Helal-Neto, E., da Silva de Barros, A.O., Alencar, L.M.R., de Menezes, A.S., Dos Santos, C.C., Saraiva-Souza, A., Perini, J.A., Machado, D.E., Felzenswalb, I., Araujo-Lima, C.F., Sukhanova, A., Nabiev, I., Santos-Oliveira, R. (2019). Graphene quantum dots unraveling: Green synthesis, characterization, radiolabeling with 99mTc, in vivo behavior and mutagenicity. *Mater. Sci. Eng. C Mater. Biol. Appl.*, 102, 405–414. https://doi.org/10.1016/j. msec.2019.04.058.

103. Barras, A., Pagneux, Q., Sane, F., Wang, Q., Boukherroub, R., Hober, D., Szunerits, S. (2016). High efciency of functional carbon nanodots as entry inhibitors of herpes simplex virus type 1. *ACS Appl. Mater. Interfaces*, 8, 9004–9013. https://doi.org/10.1021/acsam i.6b01681.

9 Green Nanotechnology
Applications in Medicine

Anam Rais, Itika Varshney, Sunil Kumar,
Vimal Kumar, and Tulika Prasad
Jawaharlal Nehru University

CONTENTS

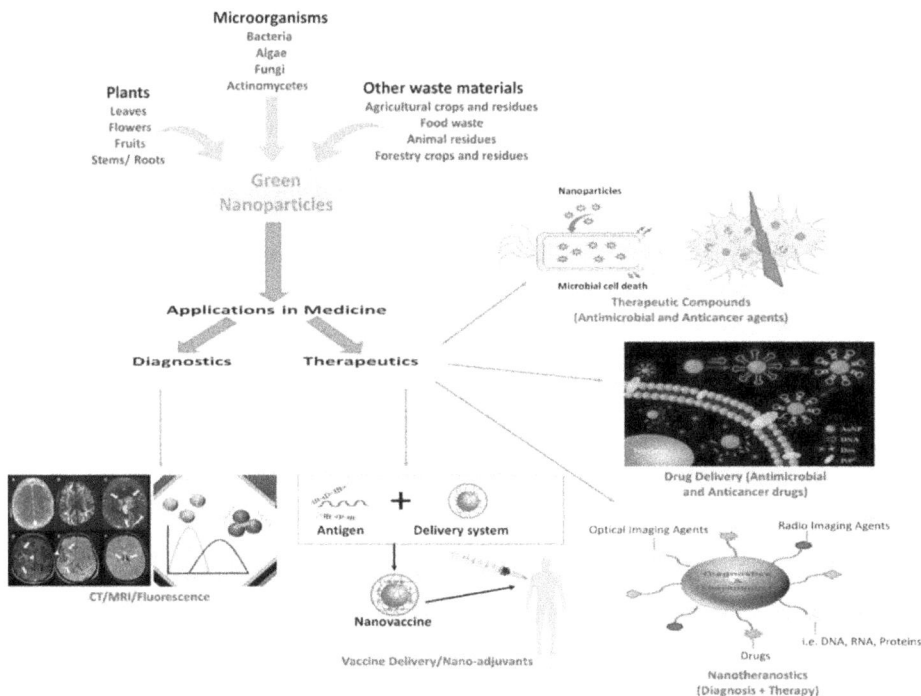

DOI: 10.1201/9781003319153-11

9.1 INTRODUCTION

Nanotechnology is a multidisciplinary field that encompasses engineering, physics, biology and chemistry [1]. The properties of materials are tuned at atomic and molecular levels to impart unique characteristics for high performance in diverse areas from electronics, physics to biological science [2]. Nanomaterials bridge the gap between bulk materials and atomic or molecular structures. Various types of nanoparticles (NPs) such as metallic NPs, organic/inorganic NPs, polymeric NPs, dendrimers, liposomes and carbon nanotubes are used for biomedical applications. Since the last few decades, smart multifunctional nanomaterials have been developed for use in disease diagnosis, therapy, theranostics and as antimicrobial agents in food, cosmetics, agriculture and textile industries. Compared to conventional therapy, functionalized nanomaterials exhibit high performance in accurate diagnosis, detailed bioimaging, targeted drug delivery, sustained and controlled drug release, enabling faster and efficient treatment of various diseases such as cancer [3]. These nanomaterials significantly enhance safety and effectiveness of most commonly used anticancer drugs. Specific targeting, sustained release, prolonged half-life and reduced toxicity are significant benefits of nano-based delivery vehicles [4]. In 1995, United States Food and Drug Administration (FDA) approved the first nano-formulated anticancer drug, Doxil/Caelyx (liposomal doxorubicin) for the treatment of ovarian and HIV-associated Kaposi's sarcoma [5]. Over the years, several nano-based anticancer drugs such as Myocet (liposomal cyclophosphamide) [6], DaunoXome (liposomal daunorubicin) [7], ONPATTRO Patisiran ALN-TTR02 (double-stranded siRNA encapsulated in lipid NPs for treatment of transthyretin (TTR)-mediated amyloidosis) [8] and VYXEOS CPX-351 (liposomal encapsulation of dual drugs, cytarabine and daunorubicin) were approved by European Medicines Agency (EMA) and FDA [9]. Nanotechnology offers other promising applications in medicine such as nanoadjuvants with immunomodulatory properties for delivery of vaccine antigens [10,11], nano-knife as novel, *non*-thermal form of cancer tissue ablation using high-voltage electricity [12] and nanotubes for tissue repair and future nerve regeneration [13]. Such applications require incorporation of green chemistry principles for the synthesis of eco-friendly sustainable nanomaterials.

Chemical synthesis of nanomaterials can generate toxic pollutants (such as heavy metal ions) which can exert adverse effects similar to chemical industries on society, environment and living creatures [14]. The toxicity of nanomaterial is dependent on the physical and chemical properties that are not routinely accounted for in toxicity studies [15]. NPs engineered for medical sciences are highly reactive and easily cross biological barriers, resulting in long persistence in human body and the environment [16]. All major sectors now involve nanotechnology and incorporate nanomaterial into their products [17]. The widespread use of nanomaterials in various consumer products such as food packaging, sunscreens, creams or any cosmetics results in the release of nanomaterials into biosphere, although their performance and fate mostly remain unidentified [18]. Still limited research is available on the adverse effects of nanotechnology on environment. There are abundant opportunities for nanomaterials to pose as potential global hazard between the stages from initial synthesis to application and final disposal.

Green nanotechnology refers to the use of green chemistry principles for cheap, eco-friendly synthesis of nanomaterials for diverse applications and solutions for environment and human health [17,19]. Green nanotechnology plays a fundamental role in bringing forward the potential of nanomaterials across the value chain of products through benign synthesis and tuning unique properties such as size and large surface area-to-volume ratio of final product [17,20]. Thus, eco-friendly approach of green nanotechnology reduces the risks of producing hazardous nanomaterials through chemical reactions and eliminates toxic chemicals, solvents, toxic intermediates and end products. However, potential applications for green nanomaterials are still in laboratory phase or start-up phase and studies are essential to further examine their efficiency and sustainability. The use of non-toxic materials free of heavy metals (e.g., citric acid and amino acids) and materials of biological origin (plants, microbes, agriculture wastes, food wastes or biomass) are preferred alternatives for the synthesis of nanomaterials in an eco-friendly manner [21]. Moreover, surface functionalization of green nanomaterials with biomolecules enhances their biocompatibility as compared to bare NPs prepared by conventional chemical methods. The biocompatibility of green NPs offers very interesting applications in biomedicine and related fields.

The use of nanomaterial life cycle-based approaches combined with risk assessment is absolutely essential to overcome the potential problems and adopt green nano-synthesis methods to reduce the burden on environment and human health for sustainability of applications in medicine. This chapter first gives a brief introduction to green nanotechnology and its advantages. The next section provides an updated list of applications of green nanotechnology in medicine, including diagnostics, therapeutics (as antimicrobial/anticancer agents, drug delivery agents and theranostics) and vaccines (vaccine delivery and as adjuvants). The last section discusses the challenges and limitations in the development of nanomaterials for medicine. Applications of nanoscience in medical area have several projected advantages and remain valuable for entire human population.

9.2 GREEN NANOTECHNOLOGY

NPs have bright prospects for developing sustainable technologies, especially NPs of biological origin, which comprise of green chemistry approach combining nanotechnology and biotechnology. Harmful chemicals used in traditional NPs production are eliminated in green nanotechnology, and green synthesis is an attractive alternative for conventional approaches [22–25]. Conventional methods for the preparation of NPs involve various physical and chemical processes. Although chemical and physical processes manufacture clean, well-defined NPs, they are costly and possibly harmful to environment. Therefore, synthetic treatments employing harmful substances should be replaced by eco-friendly techniques. The most crucial aspect of green nanotechnology is the synthesis of eco-friendly NPs using biological organisms or their biomass [26]. Furthermore, surface functionalization of NPs by biomolecules enhances their biocompatibility and opens a lot of possibilities in biomedicine and other sectors [17]. Top-down and bottom-up synthesis are two major methods for the preparation of NPs (Figure 9.1).

In last few years, nanobiotechnology emerged through the convergence of biology and nanotechnology. This new field involved the biological synthesis of NPs by reduction of metal salts or precursors of biological origin such as plants, viruses, actinomycetes, algae, yeasts, fungi and bacteria

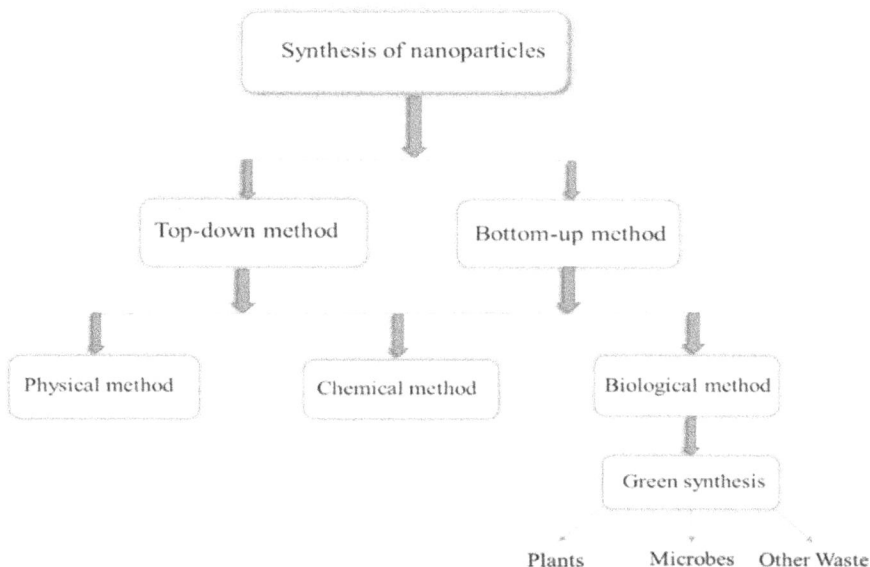

FIGURE 9.1 Methods for synthesis of NPs

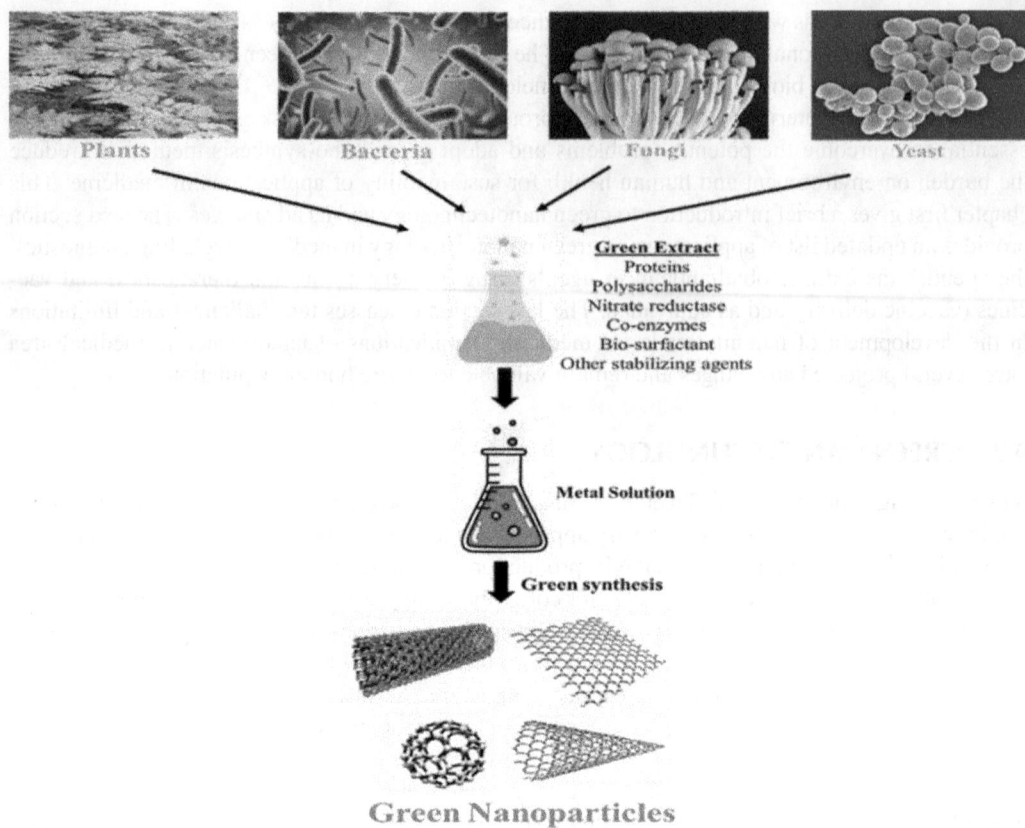

FIGURE 9.2   Synthesis of green NPs using biological entities

(Figure 9.2). Alkaloids, sugars, polyphenols, proteins and phenolic resin acids are all important plant metabolites used for reduction of metal salts into stable NPs [27,28]. Many green NPs, including metals gold (Au), copper (Cu), platinum (Pt), silver (Ag), palladium (Pd), zinc oxide (ZnO) and iron oxide (Fe_2O_3), are successfully synthesized using plants [29]. Plants such as Geranium (*Pelargonium graveolens*), neem (*Azadirachta indica*), lemongrass (*Cymbopogon flexuosus*), tamarind (*Tamarindus indica*), *Emblica officinalis, Cinnamomum camphora* and aloe vera are used for the synthesis of Au and Ag NPs in a cost- and time-efficient manner. Oat (*Avena sativa*), wheat (*Triticum aestivum*), *Chilopsis linearis*, alfalfa (*Medicago sativa*, both natural form and chemically modified hop biomass), legume seedlings and waste water collected from soaked Bengal gram beans (*Cicer arietinum*) are used for the preparation of Au NPs [30–35]. Microbes (bacteria, fungus and algae) are employed for the synthesis of Au and Ag NPs, such as the production of Ag NPs by *Pseudomonas stutzeri* AG259, *Aspergillus niger* and *Aspergillus oryzae*; Au NPs by *Marinobacter pelagius*; and titanium NPs by *Lactobacillus* strains [36–39].

9.2.1 ADVANTAGES OF GREEN NANOTECHNOLOGY

Two fundamental features of NPs make them ideal for various applications in medicine, especially drug delivery. First, owing to their nanosize, NPs can pass through cells and accumulate in target areas. Second, NPs synthesis using perishable ingredients enables to achieve sustainable, controlled and long-term drug release at target sites, ranging between days to even weeks [3]. Nano-based encapsulation prevents drug degradation within body and prolongs drug clearance from body, which enhances therapeutic efficacy and bioavailability. Compared to non-nanoparticulate dosage, nano-based drug delivery increases drug stability and does not impart bio-toxicity [4].

There are two basic goals of green nanotechnology:

1. Synthesis of nanomaterials without endangering human health or environment.
2. Production of nano-based products that address environmental issues.

Green nanotechnology is based on green engineering and green chemistry ideas. The design and process of products where natural resources are conserved with minimum impact on environment is defined as green engineering, whereas the study of chemical engineering and sustainable chemistry that fosters creation of products and process to reduce usage and production of hazardous substances is green chemistry. Furthermore, the use of nanotechnology to create environment-friendly nanomaterials and nano-products is termed as green nanotechnology. Four major factors are taken into account while creating nanomaterials and nano-based products: (i) no use of hazardous components, (ii) lower consumption of energy and use of very low temperatures, (iii) use of renewable inputs and (iv) application of life cycle assessment to all stages of design and engineering. Therefore, high energy efficiency, reduced greenhouse gas emission, prevention of dangerous wastes production, lower consumption of non-renewable raw materials, and elimination of the use and production of hazardous materials and by-products are the main advantages of green nanotechnology.

9.3   APPLICATIONS OF GREEN NANOTECHNOLOGY IN MEDICINE

The cheap, environment-friendly, sustainable green nanotechnology exploits unique physicochemical properties of nanomaterials for innovative applications and exerts an exciting impact on various sectors. Therefore, these green solutions offer opportunities to reduce pressure on chemical raw materials and develop reliable, efficient and safe medical applications such as sensing, bioimaging, drug delivery, therapy and vaccines [40]. Green synthesized NPs are applied for diagnostics, therapeutics (antimicrobial and anticancer agents), vaccine development and theranostics [41], and in the next section, we discuss such applications of green synthesized NPs in medicine.

9.3.1   Diagnostics

Several practical applications of nanotechnology are identified in medicine, such as biomarker-based cancer diagnosis, detection of tumors and new therapeutic approaches, which are hard to achieve via conventional technologies [42,43]. Applications in diagnostics constitute an area where NPs are already promising for future development [9]. Due to unique properties of NPs, the overall performance of existing techniques improves and lowers detection limits by several orders of magnitude. For example, oral cancer is detected by carbon nanotubes and Au NPs-based nanosensors [44] and cancer cells in bloodstream is detected by nanoflares [45]. Ag nanorods are used for separating bacteria, viruses, and other microscopic elements of blood samples [46]. Hence, nanomaterials hold potential for development of applications ranging from cost-effective, more specific, sensitive and fast diagnostic practices, to imaging and visualization.

Various NPs are used for diagnosis of different types of cancer. Such NPs assist in visualization of tumors with reduced side effects and no toxicity. Nanodevices developed for highly specific detection of biomolecules involve Au NPs, quantum dots (QDs), intraoperative magnetic resonance imaging (MRI) Fe_2O_3 NPs, Pt nanoclusters, perfluorocarbon emulsion NPs (PFC) and novel NP-based methods [47,48]. The application of nanotechnology to medical diagnostics promises more efficient diagnostic applications with enhanced sensitivity, higher specificity and intracellular imaging capabilities [49,50] (Figure 9.3). Bioaffinity nanoprobes-based integrated nanodevices have been developed for early cancer screening and detection using molecular and cellular imaging [51–53]. These developments pave the way for personalized medicines, where molecular profiles of

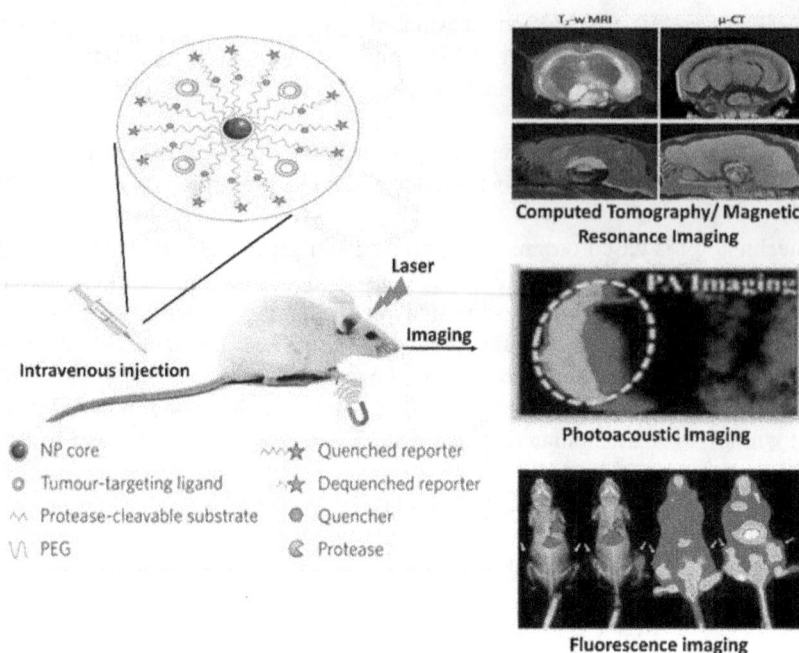

FIGURE 9.3　Schematic illustration for applications of green nanotechnology in cancer diagnostics

individuals (protein and genetic biomarkers) may be used for diagnosis and treatment. However, cost-effectiveness, biocompatibility and economic sustainability impose hurdles for conventional nanotechnology. Therefore, green nanotechnology emerged as latest innovations to develop solutions for medicine that can simultaneously reduce environmental and human health risks of nanomaterials [54,55].

Au NPs attract lot of interest in cancer detection and diagnosis because of their intrinsic properties such as high stability, surface plasmon resonance (SPR), ease of functionalization, lower systemic cytotoxicity and *in vivo* non-immunogenicity [56]. *Olax scandens* (plant leaf extract) was used for green preparation of Ag NPs, which showed bright red fluorescence within cells and could be used to diagnose cancer cells [57]. Anisotropic Au NPs prepared by Fazal et al. using aqueous cocoa extract without any capping agents, served as both reducing and stabilizing agent. The obtained Au NPs exhibited remarkable X-ray contrast when tested via computed tomography (CT) and thus were effective as CT diagnostic imaging and therapeutic (photothermal) agents for cancer [58]. Au NPs conjugated with bovine serum albumin and folic acid (FA) were synthesized for fluorescence imaging in MGC-803 cells. These functionalized Au NPs enhanced CT imaging and improved therapeutic efficacy by targeting FA-overexpressing gastric cancers [59]. Song et al. integrated diagnostics and therapeutics onto a single platform using hollow CoPt alloy (HCPA) NPs. Size-controlled synthesis of these NPs were achieved using plant polyphenols, which can also be used as versatile size-controlled synthesis of various hollow metal alloy NPs, especially due to the diverse metal-chelating ability of polyphenols. Therefore, it was successfully implemented in MRI and photothermal therapy guided by photoacoustic dual-modal imaging [60]. Dextran coated Fe_2O_3 NPs synthesized using green microwave approach, showed excellent superparamagnetic properties, which are desirable for molecular imaging [61]. Fe_2O_3-Au nanohybrid system synthesized using extract of grape seed by Narayanan et al. was used for CT and MRI contrast imaging [62]. This hybrid system showed high-performance stem cell tracking and imaging. Thus, Fe_2O_3-Au nanohybrid material combined both SPR and magnetic properties in a single moiety for effective diagnosis through imaging. Du et al. reported a multimodal bioimaging method for rapid and accurate cancer diagnosis, where fluorescent ZnO NPs coupled with magnetic nanoclusters were used for CT and

MRI imaging [63]. Shevtsov et al. synthesized hybrid chitosan-dextran magnetic NPs, which exhibited high MR contrast for diagnosis of brain tumor [64]. Dragon fruit-like ferritin nanocages were reported by Yang et al., and encapsulation efficiency for iron enabled multimodal imaging through photoacoustic imaging (PAI), MRI and positron emission tomography (PET). They also exhibited targeted therapy against liver and colon cancer in preclinical models [65].

Apart from bioimaging, nano-biosensors are also used for diagnostics. Reduced graphene oxide (rGO)-Au nanocomposites synthesized from rose water was used for electrochemical detection of glucose levels [66]. Zang et al. prepared self-assembled nanohybrid membranes of rGO-Au NPs using one-pot green synthesis for the detection of cancer biomarker, hydrogen peroxide [67]. Wang et al. constructed magnetic virus by binding peptide-viral nanofibers with magnetic Fe_2O_3 NPs for the detection of *C. albicans* in cancer patients. They exploited the genetic ability for display of two functional peptides on a single filamentous human-safe phage virus, one for recognition of infection biomarker (anti-secreted aspartyl proteinase 2 (SAP2) IgG antibody) in cancer patients and the other for binding to MNPs. This method opened avenue for virus-based disease diagnosis and displayed higher specificity and sensitivity than either clinical gold standard method (which takes about 1 week) or the use of antigens or viruses alone (which takes only about 6 hours) for detection [68]. In another study, Au NPs synthesized using *C. albicans* were conjugated to liver cancer-specific antibodies for use as fluorescence-based biosensor to differentiate between cancer and normal cell populations [69]. Biocompatible silk fibroin NPs (SFNPs) synthesized by Xu et al. were encapsulated with indocyanine green (ICG) dye and then cross-linked with proanthocyanidins to obtain ICG-CSFNPs-based nano-platform for photothermal therapy of glioblastoma. As compared to free ICG dye, this nano-platform exerted more stable photothermal effect upon near-infrared (NIR) irradiations [70]. *In vivo* imaging revealed that NPs, after intravenous administration efficiently accumulated in tumor site of mice with C6 glioma and more efficient killing of residual tumor cells was achieved after surgery using NIR radiation. Biocompatible NPs were synthesized by Zhang et al. using albumin-templated method for use as new type of MR contrast agents to diagnose brain glioma [51].

Novel coronavirus disease 2019 (COVID-19) is an ongoing pandemic caused by severe acute respiratory syndrome coronavirus 2 (SARS-CoV-2), which has led to more than 506 million infected cases and 6.22 million deaths globally as of April 22, 2022 [71]. The symptoms of variants of coronavirus strains are often confused with those of influenza and seasonal respiratory viral infections. Therefore, clarity in the development of diagnostic tests is crucial for accurate treatment [72].Sustainable SARS-CoV-2 detection was reported by Li et al. using environment-friendly, bio-compatible green nanomaterials as NIR fluorescence nanosensors [73].

Nanodiagnostics is an upcoming field of molecular diagnostics that holds significance in transforming traditional clinical laboratory procedures, providing improved methods for assessment of patient samples and biomarker-based early detection of diseases with high sensitivity and specificity. Nano-based platforms have been developed and optimized to detect cancer biomarkers and various pathogens, simultaneously making the diagnostic procedures less cumbersome. In addition, the sensitivity of nanoplatforms has been increased by integration onto a simple device the ability for on-spot diagnosis.

9.3.2 THERAPEUTICS

The most common cancer therapies include radiotherapy, surgery and chemotherapy. Unfortunately, various limitations of current therapeutics include inability to cross biological barriers, ineffectiveness against metastatic disease, non-specific drug delivery, poor bio-distribution of drugs, cancer drug resistance and lack of modality for effective monitoring of treatment [74]. Interestingly, a gradual shift has been observed in disease management from traditional approach toward personalized medicine through the use of nanotechnology, where the correct treatment is administered in right amount to right person at right time [75]. Green nanotechnology has immense potential to advance cancer diagnostic methods and therapeutic technologies. Some green NPs platforms such as silica NPs, magnetic NPs, Au NPs and polymeric NPs were widely studied for use as antimicrobial and anticancer agents, drug/vaccine delivery systems and adjuvants.

9.3.2.1   Therapeutic Compounds (Antimicrobial and Anticancer Agents)

In the era of personalized medicine, targeted nanomedicines play a significant role in managing several diseases, including infectious diseases and cancer [76]. The creation of personalized medicine in therapeutics can serve as a solution to overcome the underlying drawbacks of conventional chemotherapy, i.e., release of inadequate drugs in cancer cells and toxicity toward normal cells. Over the past few years, microbes, plants and waste materials were used to synthesize metal NPs for pharmacological applications (e.g., anticancer and antimicrobial agents) (Table 9.1). Due to the increasing incidence of microbial resistance toward common antibiotics and antiseptics, various studies were carried out to develop better antimicrobial agents [77]. *In vitro* antimicrobial studies demonstrated that metal NPs successfully inhibit various microbial species. Due to simultaneous multi-targeted mode of action, metallic NPs can prevent or overcome multidrug resistance and inhibit biofilm formation [78,79]. For example, Ag NPs are the most admired and widely used NPs as effective anti-inflammatory antibacterial, antiviral and antifungal agents [80,81]. Also, Au NPs are highly useful as effective antimicrobial and anticancer agents because of their unique properties, such as biocompatible nature, photothermal activity, polyvalent effects and ability to be functionalized [44].

Researchers showed higher antimicrobial activity of green synthesized NPs as compared to chemically synthesized NPs [79,81]. Sudhasree et al. demonstrated that green nickel NPs obtained from *Desmodium gangeticum* (root extract) exhibited high antioxidant property and antibacterial activity as compared to chemically prepared nickel NPs. Green NPs were non-toxic to epithelial cell lines (LLC PK1) [82]. Mukherjee et al. designed a simple green approach to synthesize Ag NPs using leaf extract (*Olax scandens*), and these NPs exhibited multifunctional activities such as biocompatibility (CHO, HUVEC and H9C2 cell lines), antibacterial and anticancer activity (A549, B16 and MCF7 cancer cell lines). Versatile applications of green NPs have been demonstrated for medicine [83].

In addition to antimicrobial and anticancer activities, green NPs can facilitate targeted delivery of drugs to specific cells, thereby reducing the required drug dosage and minimizing the adverse effects of high drug doses. Furthermore, green metal NPs can also substitute chemically synthesized metal NPs for imaging and photothermal therapy.

9.3.2.2   Drug Delivery (Anticancer and Antimicrobial Drugs)

In recent years, nanomaterials have started receiving attention, especially for the delivery of drugs to diseased cells and improvement of human health. Such engineered NPs hold great significance in nanomedicine as potent drug delivery systems because of improved bioavailability, high drug-loading capacity, sustained and controlled drug release, greater efficacy and reduced side effects. They can achieve targeted delivery of drugs to specific sites or organs, cross cell barriers and enhance intracellular penetration and retention in cells. NPs-based nanomedicines possess the ability for protection against enzymes, physiological pH and moisture and are administrated through several ways such as oral, parenteral, intraocular and nasal routes [4].

To date, for efficient and targeted delivery of drugs, various types of nanocarriers were designed, such as non-metal NPs, metal NPs, magnetic NPs (MNPs), QDs, solid lipid NPs, dendrimers, liposomes, silica NPs, polymeric NPs, nanostructured lipid carriers and carbon nanomaterials [4]. Table 9.2 provides details of different NPs designed to transport drug molecules.

Another type of nanomaterials are viral NPs (VNPs), their genome-free variants and virus-like particles (VLPs), which have rapidly growing applications in green nanomedicine. VLPs are nanoscale structures made from assembled viral proteins that are non-infectious due to the lack of viral genetic material [141]. VLPs are biocompatible, biodegradable and

TABLE 9.1
Green Sources for Synthesis of Metal NPs with Potential Antimicrobial and Anticancer Activity

S. No.	Species	Biological Component	Applications	References
			Ag NPs	
1.	*Tamarindus indica*	Fruit	Antibacterial action against *P. aeruginosa, E. coli, B. cereus, K. pneumonia, S. aureus, M. luteus, Enterococcus* spp., *B. subtilis* and *S. typhi*	[84]
2.	Sea buckthorn	Plant	Antibacterial action on *P. aeruginosa, E. coli* and *K. pneumoniae*	[85]
3.	*Sisymbrium irio*	Plant	Antibacterial action on *E. coli* and multidrug-resistant *Acinetobacter baumannii* and *P. aeruginosa*	[86]
4.	*Lantana camara*	Plant	Antibacterial action on *S. aureus, E. coli* and *P. aeruginosa*	[87]
5.	*Abutilon indicum*	Plant	Anti-proliferative activity on COLO 205 (human colon cancer) cells in a dose-dependent manner	[88]
6.	*Acalypha wilkesiana*	Plant	Antibacterial activity against *E. coli* and *S. aureus*	[89]
7.	*Moringa oleifera, Cucurbita maxima* and *Acorus calamus*	Plant	Anticancer action on A431 skin cancer cell line	[90]
8.	*Fusarium oxysporum*	Fungus	Antibacterial effect on *S. aureus* and *E. coli*	[91]
9.	*Coriandrum sativum*	Plant	*In vitro* anticancer efficacy on MCF-7 (human breast adenocarcinoma) cell line	[92,93]
10.	*Andrographis paniculata*	Plant	Antibacterial activity against *S. aureus, S. typhi, V. cholerae, E. coli, E. faecalis, H. alvei* and *A. baumannii*	[94]
11.	*Citrullus colocynthis*	Plant	Anticancer effect on human cancer cell lines such as HCT-116 (colon), Hep-G2 (liver), Caco-2 (intestine) and MCF-7 (breast)	[95]
12.	*Listeria monocytogenes, Bacillus subtilis* and *Streptomyces anulatus*	Bacteria	Antimicrobial effect on *C. keratinophilum* and mosquitocidal activity on *Anopheles stephensi* and *Culex quinquefasciatus*	[96]
13.	*Euphorbia prostrata*	Plant	Antimicrobial activity (*Leishmania donovani*)	[97]
14.	*Acidophilic actinobacteria SH11*	Bacteria	Antibacterial effect on *B. subtilis, E. coli* and *S. aureus*	[98]
15.	*Ulva lactuca*	Plant	Anticancer effect on HepG2, MCF-7 and HT29 cell lines	[99]
16.	*Weissella oryzae* DC6	Bacteria	Antimicrobial effect on clinical pathogens (*S. aureus, C. albicans, E. coli, Bacillus anthracis, V. parahaemolyticus* and *Bacillus cereus*) and antibiofilm activity against *S. aureus* and *P. aeruginosa*	[100]
17.	*Bacillus funiculus*	Bacteria	Dose-dependent anticancer activity against MDA-MB-231 cancer cell lines	[101]
18.	Fenugreek seed	Plant	Antibacterial action on *S. aureus* and *E. coli*	[102]
19.	*Penicillium oxalicum*	Fungus	Antibacterial activity toward *S. aureus, S. dysenteriae* and *S. typhi*	[103]
20.	*Olax scandens*	Plant	Anticancer activity toward B16, A549 and MCF-7 cells	[83]

(Continued)

TABLE 9.1 (*Continued*)

Green Sources for Synthesis of Metal NPs with Potential Antimicrobial and Anticancer Activity

S. No.	Species	Biological Component	Applications	References
21.	*Ficus religiosa*	Plant	Antibacterial activity toward *B. subtilis, P. fluorescens, S. typhi* and *E. coli*	[104]
22.	*Bacillus pumilus, Bacillus persicus* and *Bacillus licheniformis*	Bacteria	Antiviral activity (bean yellow mosaic virus) and antibacterial activity against pathogen *P. aeruginosa, E. coli, Shigella sonnei, Klebsiella pneumoniae, C. albicans, Streptococcus bovis, S. epidermidis, S. aureus* (methicillin-resistant strain) and *Aspergillus flavus*	[105]
23.	*Brevibacterium frigoritolerans* DC2	Bacteria	Antimicrobial activity against clinical pathogens *S. aureus, E. coli, Bacillus anthracis, C. albicans, Bacillus cereus* and *V. parahaemolyticus*	[100]
24.	*Origanum vulgare*	Plant	Dose-dependent anticancer activity against A549 (lung cancer) cell line and antibacterial effect on *A. hydrophila, E. coli, Salmonella* spp., *S. dysenteriae, Shigella sonnei* and *S. paratyphi*	[106]
25.	*Bacillus methylotrophicus* DC3	Bacteria	Antimicrobial activity against clinical pathogens *S. enterica, E. coli, C. albicans* and *V. parahaemolyticus*	[107]
26.	*Beta vulgaris* L.	Plant	Anticancer effects on human hepatic cancer (HUH-7) cells	[108]
27.	*Nocardiopsis dassonvillei* KY772427	Actinobacteria	Antimicrobial (*P. aeruginosa*) and insecticidal (*Macrosiphum rosae*) activity	[109]
		Au NPs		
28.	*Syzygium aromaticum*	Plant	Anticancer effect on HeLa and MCF-7 cell lines	[110]
29.	*Trichoderma hamatum*	Fungus	Antibacterial action on *B. subtilis, P. aeruginosa, S. aureus* and *Serratia* spp.	[111]
30.	Green tea	Plant	Anticancer activity against MCF-7 cells and Ehrlich ascites carcinoma	[57]
31.	*Deinococcus radiodurans*	Bacteria	Antimicrobial effect against *S. aureus* and *E. coli*	[112]
32.	*Musa paradisiaca*	Fruit	Anticancer effect toward human A549 lung cancer cells	[113]
33.	*Alternanthera bettzickiana*	Plant	Antibacterial action on *B. subtilis, S. aureus, S. typhi, E. aerogenes, M. luteus* and *P. aeruginosa*	[114]
34.	*Pseudomonas veronii* AS41G	Bacteria	Antibacterial activity against human and environmental pathogens (MRSA)	[115]
		Carbon QDs		
35.	Harmful cyanobacteria	Algae	Anticancer therapy (HepG2 and MCF-7 cancer cells)	[116]
		Cu NPs/CuO NPs		
36.	*Shewanella loihica* PV-4	Bacteria	Antibacterial activity against *E. coli*	[117]
37.	*Cystoseira trinodis*	Algae	Antibacterial effect on *E. coli, S. subtilis, S. aureus, S. faecalis, S. typhi* and *E. faecalis*	[118]
38.	*Malva sylvestris*	Plant	Antibacterial activity against Gram-negative and Gram-positive bacteria	[119]
39.	*Acalypha indica*	Plant	Anticancer effect on MCF-7 breast cancer cell line	[120]
40.	*Gloriosa superba*	Plant	Antibacterial activity against *S. aureus* and *Klebsiella aerogenes*	[121]

(Continued)

TABLE 9.1 (*Continued*)

Green Sources for Synthesis of Metal NPs with Potential Antimicrobial and Anticancer Activity

S. No.	Species	Biological Component	Applications	References
41.	*Citrus medica*	Fruit	Antimicrobial against activity *P. aeruginosa, E. coli, Propionibacterium acnes. S. typhi, K. pneumoniae, F. culmorum, F. oxysporum* and *F. graminearum*	[122]
			Se NPs	
42.	*Bacillus licheniformis*	Bacteria	Anticancer activity toward human prostate adenocarcinoma cells	[123]
43.	*Enterococcus faecalis*	Bacteria	Antibacterial effect against *S. aureus*	[124]
44.	Fenugreek	Plant	Dose-dependent anticancer activity against breast cancer MCF-7 cell line	[125]
45.	*Emblica officinalis*	Plant	Antimicrobial activity on foodborne pathogens	[126]
46.	*Allamanda cathartica*	Plant	Antimicrobial activity against *P. marginalis* and *P. aeruginosa*	[127]
47.	*Ceropegia bulbosa*	Plant	Antibacterial activity (*B. subtilis* and *E. coli*), anti-mosquito activity (*Aedes albopictus*) *and* anticancer activity (MDA-MB-231 cells)	[128]
			ZnO NPs	
48.	*Laurus nobilis*	Plant	Anticancer efficacy against human A549 lung cancer cells	[129]
49.	*Glycosmis pentaphylla*	Plant	Antibacterial effect against *S. paratyphi, B. cereus, S. aureus* and *S. dysenteriae*	[130]
50.	*Ziziphus nummularia*	Plant	Anticancer effect against HeLa cancer cell lines	[131]
51.	*Suaeda aegyptiaca*	Plant	Antibacterial action on *P. aeruginosa, B subtilis, S. aureus* and *E. coli*	[132]
52.	*Pichia kudriavzevii*	Fungus	Antibacterial effect against *B. subtilis, S. aureus, E. coli, S. marcescens* and *S. epidermis*	[133]
53.	*Jacaranda mimosifolia*	Plant	Antibacterial activity against *E. faecium* and *E. coli*	[134]
			Graphene NPs	
54.	*Bacillus marisflavi*	Bacteria	Anticancer effect against human breast cancer cells	[135]
			Pt NPs	
55.	Tea polyphenols	Plant	Anticancer effect against human cervical cancer cells	[136]
56.	Dates	Plant	Anticancer effect on colon carcinoma cells (HCT-116), hepatocellular carcinoma (HePG-2) and breast cells (MCF-7)	[137]
			Fe NPs	
57.	*Lawsonia inermis/ Gardenia jasminoides*	Plant	Antibacterial activity against *E. coli, S. enterica, Proteus mirabilis* and *S. aureus*	[138]
58.	*Phoenix dactylifera*	Plant	Antimicrobial action on *B. subtilis, E. coli, Micrococcus luteus* and *Klebsiella pneumoniae*	[139]
59.	*Garcinia mangostana*	Plant	Anticancer activity against HCT116 colon cancer cells	[140]

TABLE 9.2

Applications of Viral NPs (VNPs) in Anticancer and Antimicrobial Drug Delivery

Virus	Drugs	Advantage	References
Bacteriophage MS2 (*Emesvirus zinderi*)	Doxorubicin (DOX), cisplatin, ricin toxin A chain and 5-fluorouracil (5-FU)	Selective cytotoxicity *in vitro* even at very low concentrations Great avidity and specificity toward human hepatocellular carcinoma cells (HCC) and negligible uptake in healthy cells	[150]
Red clover necrotic mosaic virus (RCNMV)	DOX	Ion- and pH-responsive controlled release of encapsulated drug Rapid release of surface-DOX followed by a slow release of encapsulated DOX	[151]
Bacteriophage Qβ	DOX	Increased stability and solubility of polyethylene glycol-functionalized VLPs with photocaged doxorubicin Negligible cytotoxicity with controllable photorelease and cancer cell killing	[152]
Tobacco mosaic virus (TMV), MS2 spheres and nanophage filamentous rods	DOX	Enhanced site-specific drug delivery in vitro to glioblastoma cells Increased healthy cell survival	[92]
TMV conjugate with spherical NPs (SNPs)	DOX	Non-virulent and ease of synthesis Efficient drug release, target-specific cell killing of breast cancer cells (MCF-7 and MDA-MB-231)	[153]
Potato virus X (PVX)	DOX	Higher PVX-DOX efficacy and drug activity against cancer cell lines (breast cancer, cervical cancer and ovarian cancer) Reduced tumor growth in mice bearing human breast (MDA-MB-231) cancer xenografts	[154]
Foot-and-mouth disease virus (FMDV)	DOX	Higher therapeutic efficiency *in vitro* and *in vivo* of HeLa-based tumors Cytotoxicity of integrin-expressing tumor cells via integrin-targeted delivery of DOX	[155]
Hepatitis B core protein (HBc)	DOX	Specific targeting cancer cells through interaction with overexpressed integrin $\alpha v \beta 3$ 90.7% tumor growth inhibition and lower cardiotoxicity in B16F10 tumor-bearing mice	[156]
Macrobrachium rosenbergii nodavirus (MrNVLP)	DOX	Enhanced delivery and higher cytotoxicity of folic acid-functionalized MrNVLP-DOX in cancer cells (HT29) with overexpressed folate receptor, as compared to normal cells (CCD841CoN) and cancer cells (HepG2)	[157]
Adenovirus (AdV)	Bleomycin (BLM)	Improved drug bioavailability in HeLa cells	[158]
TMV	Cisplatin and phenanthriplatin (phenPt)	Improved double-strand DNA breakage and cytotoxicity in platinum-sensitive and platinum-resistant cancer cells than free cisplatin Enhanced loading stability due to heteroaromatic rings on phenPt and efficient encapsulation in TMV	[159–161]

(Continued)

TABLE 9.2 (*Continued*)
Applications of Viral NPs (VNPs) in Anticancer and Antimicrobial Drug Delivery

Virus	Drugs	Advantage	References
TMV and cowpea mosaic virus (CPMV)	Mitoxantrone (MTO)	Effective drug (MTO) loading into TMV (~1000 MTO per TMV carrier) and maintained drug efficacy upon injection in MDA-MB-231 triple-negative breast cancer (TNBC) nude mouse model Increased *in vitro* cytotoxicity upon combining MTO-CPMV with tumor necrosis factor-related apoptosis-inducing ligand (TRAIL)	[162,163]
TMV	Valine-citrulline monomethyl auristatin E (vcMMAE)	Effective *in vitro* cell death of non-Hodgkin lymphoma by TMV-vcMMAE at IC50 250 nM Induces apoptosis via successful cleavage of dipeptide valine-citrulline (vc) to release MMAE	[164]
Bacteriophage Qβ	Azithromycin/ clarithromycin	Substantial accumulation of azithromycin within 2 hours of systemic administration in RAW 264.7 macrophage cells Potential microbial treatment such as *Legionella pneumophilia* and *Mycobacterium tuberculosis* residing in pulmonary macrophages	[165]
Filamentous bacteriophages (fd or M13)	Chloramphenicol	High-capacity drug carrier containing 3,000 copies/p8 major coat protein Lower side effects and retarded growth of bacterial pathogen *Staphylococcus aureus*	[166]

synthesized by molecular farming in the laboratory using a variety of expression systems, including prokaryotic cells [142], insect cell lines [143,144], mammalian cell lines [145,146], yeast [147] and plants [148]. By virtue of hollow interior, VLPs can encapsulate drugs and function as carriers in delivery of bio- and nanomaterials such as vaccines, medicines, QDs and imaging chemicals [53,149].

Although nanomedicine resulted in a medical breakthrough for the treatment of several diseases, there are drawbacks of NPs-based drug delivery vehicles because nano-engineered drugs behave differently as compared to their conventional forms and that cannot be ignored [167]. In addition, some nanomaterials are synthesized using toxic chemicals. In spite of extensive development of nanotechnology in the last 10–20 years, their potential toxicological effects on animals, humans and environment still need to be addressed. There is clearly a need of new approach for design and synthesis of NPs in order to make the application more specific and more effective to promote health, reduce cost and enhance quality of NPs-based drug delivery vehicles. Green synthesis using 12 principles of green chemistry is the best way to accomplish health, social, environmental and economic benefits by designing green novel nanomaterials [21]. Green synthesis uses simple, environment-friendly, inexpensive, non-toxic and readily available raw materials without harmful chemicals. The main goal of green nanotechnology is to develop biocompatible, biodegradable nano-vectors with efficient cellular uptake, prolonged shelf life and minimal toxicity as compared to traditional drugs. Biomedical applications of green synthesized NPs in drug delivery and their advantages are summarized in Table 9.3.

TABLE 9.3

Applications of Green Synthesized NPs in Drug Delivery

Nanovehicles	Drugs	Biomedical Application	Advantage/Potency	References
Lecithin-based mixed polymeric micelle (LMPM)	Quercetin	Anticancer drug	Longer half-life, increased solubility and bioavailability, effective passive tumor-targeting ability through enhanced permeability and retention (EPR) effect	[168]
Poly(D,L-lactide-co-glycolide) (PLGA)–lovastatin–chitosan–tetracycline NPs	Tetracycline	Antibacterial	Decreased implant infection, good biocompatibility	[169]
PLGA-loaded micro- and NPs	Prodigiosin (*Serratia marcescens* and *Serratia marcescens* spp.)	Anticancer drug (breast cancer cells)	Efficient drug delivery, high stability, good compatibility and excellent drug susceptibility	[170]
Poly(L-γ-glutamyl-glutamine)–paclitaxel (PGG–PTX)	Paclitaxel (PTX)	Human lung cancer	Highly water-soluble and maximum sustainability	[171]
Ferritin NPs	Carbachol and atropine	Pancreatic cancer	EPR in tumors and active targeting through binding of transferrin receptor 1 (TfR1) to tumor cells	[172]
Quinapyramine sulfate loaded-sodium alginate NPs	Quinapyramine sulfate	Parasite *Trypanosoma evansi*	Non-toxic and dose-dependent activity	[173]
ROS-cleavable diblock copolymer (MeO–PEG–TK–PLGA)	DOX	Anticancer	Efficiently escape from endosomes and enter into cytoplasm, induce apoptosis	[174]
Polyethylene glycol (PEG) polyester dendrons with anti-angiogenic drug	Combreta-statin-A4	Anticancer	EPR effect	[175]
Folic acid-targeted disulfide-based cross-linking micelle	Curcumin	Anticancer	Tumor cell inhibition, high stability and durable drug accumulation	[176]
Reduced graphene oxide–Ag (RGO–Ag) nanocomposite	-	Antibacterial	Utilized in real-time sensing and enzyme-less detection of H_2O_2	[177]
CdTe QDs (QDs)-attached nanogels (QD-NGs)	Methotrexate (MTX)	Hypoxia-inducible factor-1 (HIF-1) RNA into hypoxic cancer cells	Increased antitumor activity of the drug	[178]

(Continued)

TABLE 9.3 (*Continued*)
Applications of Green Synthesized NPs in Drug Delivery

Nanovehicles	Drugs	Biomedical Application	Advantage/Potency	References
Carbon dots (CDs–Pt [IV]-PEG-[PAH/ DMMA])	Cisplatin (IV)	Human ovarian carcinoma A2780 cells and HeLa cells (*in vivo* and *in vitro* study)	Enhanced internalization, low side effects and high tumor inhibition ability	[179]
Chitosan-capped Au NPs	Rifampicin	Antimicrobial activity (anti-tuberculosis drug)	Prolonged storage and sustained release	[180]
Self-assembly of ginsenoside Rb1 NPs	Betulinic acid, dihydroartemisinin and hydroxycamptothecin	Anticancer therapy	Anticancer drug delivery vehicle	[181]
Zinc oxide (ZnO) NPs using *Borassus flabellifer* fruit extract	DOX	Anticancer therapy	Cytotoxic against HT-29 and MCF-7 and high therapeutic efficacy in a dose-dependent manner	[182]
Graphene nanosheet– carbon nanotube– iron oxide (GN–CNT–Fe$_3$O$_4$) NPs hybrid	5-FU	Anticancer drug delivery systems	High drug-loading capacity, non-toxic for liver cells	[183]
Single-walled carbon nanotubes (SWCNTs) conjugated with PEG-folic acid	DOX	Anticancer treatment (destruction of breast cancer cells)	Excellent stability, selective and effective treatment modality with minimum side effects	[184]
Nanostructured lipid carrier (NLCs)	Tamoxifen	Breast cancer therapy	Long-term stability of drug, optimum drug loading and low cytotoxicity	[185]
Microemulsion type of nanosized colloidal carriers	Naftifine	Antifungal drug delivery	Localized transport, low skin cytotoxicity and effective topical delivery	[186]
Lipid nanotubes (LNTs) conjugated with anthraquinone group	DOX	Anticancer therapy	Non-toxic, biodegradable, stable, biocompatible, controlled and prolonged drug release	[187]
Pluronics-conjugated poly(amidoamine) dendrimer (PAMAM)	5-FU	Anticancer therapy	Highest drug-loading efficiency, non-toxic and reduced side effects of drug	[188]
CLH-PLA and CLH-PLGA nano-formulation	Clindamycin hydrochloride	Antimicrobial activity	Enhanced therapeutic efficacy, high drug-loading efficiency and controlled drug release	[189]
Reconstituted lipoprotein NPs (RLNs)	PTX	Anticancer drug delivery	Minimized adverse effects, efficient intracellular trafficking, targeted PTX accumulation, enhanced anticancer activity and mitigation of drug side effects	[190]

(*Continued*)

TABLE 9.3 (*Continued*)

Applications of Green Synthesized NPs in Drug Delivery

Nanovehicles	Drugs	Biomedical Application	Advantage/Potency	References
Drug conjugating to xyloglucan	DOX	Clinical cancer therapy	More drug release, longer circulation time, high intracellular uptake, higher cytotoxicity and inhibition of growth or volume of tumor	[191]
Emulsion-based nano-formulations	Aspirin (acetylsalicylic acid)	Analgesic and anti-inflammatory drug delivery	Enhanced inhibitory effect against carrageenan-induced inflammation and effective even at lower concentrations	[192]
MTX–Au NPs conjugate	MTX	Anticancer drug carrier	Concentrated effect, efficient transport of drug into tumor cells for higher cytotoxicity	[193]
Anticancer drug with MPA-capped Au NPs	Daunorubicin	Anticancer drug delivery	Inhibition of multidrug resistance in tumor or cancer cells Sensitive for biomarker-based early cancer diagnosis	[194]
Folic acid and PEG-functionalized Au NPs	Docetaxel	Anticancer drug delivery	Concentration-dependent inhibition, reduced side effects and excellent cytotoxicity	[195]
Au NPs-poly(bis(4-carboxyphenoxy)-phosphazene) (PCPP)	Camptothecin	Anticancer drug delivery (breast cancer therapy)	Potentially minimized drug concentration in normal tissues Triggered intracellular drug release in cancer cells	[196]
Green carbon nanotags (G-tags)	DOX	Anticancer drug delivery	Lowered normal cell toxicity, triggered cell death in cancer cells and enhanced anticancer efficacy	[197]
Mitochondria as carriers with CQDs	DOX	Anticancer drug delivery	Systemic biodistribution, prolonged retention time, biosafe and efficient drug carrier system	[70]
Magneto-fluorescent GdNS@CQDs	DOX	Anticancer drug delivery	Excellent stability, biocompatibility, capable of bimodal contrast imaging, targeted drug delivery, pH-responsive drug release and low toxicity in normal cells	[49]
Lipid carrier QDs (CdTe/CdS/ZnS)	PTX	Anticancer drug delivery	Enhanced efficacy of drug, ability for tumor detection and imaging	[198]
Nitrogen-doped graphene QDs (N-GQDs)	MTX	Anticancer drug delivery	Targeted delivery and prolonged cytotoxicity	[199]

(*Continued*)

TABLE 9.3 (*Continued*)
Applications of Green Synthesized NPs in Drug Delivery

Nanovehicles	Drugs	Biomedical Application	Advantage/Potency	References
Fe_3O_4-Ag_2O QDs/cellulose fiber nanocomposites	Etoposide and MTX	Anticancer drug delivery	Low normal cell toxicity, biosafe nanostructures with high potential applications	[200]
GQDs	Berberine hydrochloride	Anticancer drug delivery	G1-phase arrest and apoptosis as confirmed cell death mechanism, suitable as drug-carrying vehicle and dual-fluorescent nanoprobes	[201]
CQDs from aloe vera leaf gel	Vancomycin	Antibacterial	Increased drug-loading efficiency and absorption in gastrointestinal tract	[202]
Polymeric NPs PEG-b-poly(L-lactide) (PEG-PLA-NH-NH$_2$)	Cisplatin	Anticancer	Well-controlled loading of cisplatin, acid-stimulated drug release and high *in vitro* cytotoxicity	[203]
Polymeric NPs PLGA and PLA	Nystatin	Antifungal drug	Increased efficacy of drug, 70% or above encapsulation efficiency, prolonged release of drug and enhanced adhesion to oral mucosa	[204]
Nanospheres with DL lactide/glycolide copolymer (PLGA)	5-FU and indomethacin	Anticancer therapy	Good candidate for encapsulation of drug	[205]
Triblock copolymer of PEG-block-poly(l-lysine)-block-poly aspartyl (N-(N′,N′-diisopropylamino-ethyl)) (PEG-PLL-PAsp (DIP))	Co-delivery of DOX and anti-BCL-2 siRNA	anticancer therapy	Suppressed expression of BCL-2 (anti-apoptotic) in drug-resistant hepatic carcinoma therapy	[206]
Amino-modified mesoporous silica SBA-15	Cefazolin	Antibacterial therapy	Delayed mineralization during formation of bone-mineral-mimicking layer, prolonged drug release and lower systemic side effects	[207]
Polypropylenimine (PPI) dendrimers	MTX sodium, sodium deoxycholate and DOX	Anticancer therapy	Sustained drug release, low cytotoxicity and high drug-loading efficiency	[208]
PEG-poly(-amidoamine) (PEG-PAMAM)	Penicillin V	Antibacterial therapy	Bonding stability, increased circulation time for drug in body and controlled drug release	[209]

(Continued)

TABLE 9.3 (*Continued*)
Applications of Green Synthesized NPs in Drug Delivery

Nanovehicles	Drugs	Biomedical Application	Advantage/Potency	References
Hyaluronic acid, poly(lactide), poly(amidoamine) dendrimer (HA-PALA)	Docetaxel	Anticancer therapy	Outstanding plasma stability, efficient intracellular delivery, biocompatible, high drug-loading capacity and efficient growth inhibition of cancer cells	[210]
Solid lipid NPs containing hydroxypropyl-β-cyclodextrin (smPSH)	PTX	Anticancer therapy	Increased solubility and dissolution of PTX in the form of smPSH and high encapsulation efficiency	[211]
Wheat germ agglutinin (WGA)-conjugated solid lipid NPs (SLNs)	PTX	Anticancer therapy	High conjugation efficiency, higher cancer cell toxicity, PTX-induced apoptosis, improved oral bioavailability and lung targetability of PTX	[212]
Chitosan and folate-grafted copolymer of PEG with drug-loaded SLN	PTX	Anticancer therapy	Limited systemic distribution, improved delivery of PTX at different levels when administered through inhalation, increased anti-tumor activity and lower pulmonary adverse effects	[213]
Lactonic sophorolipid (glycolipid) derived SLNs	Rifampicin and dapsone	Antimicrobial (antileprosy drug)	High biocompatibility, ease of synthesis, sustained release, high entrapment efficiency and increased bioavailability	[214]
Dextran–protamine (Dex–Prot)-coated NLCs	Saquinavir	Antiviral therapy	Increased permeability and mucus penetration of NPs	[215]
NLCs	Curcumin	Antifungal therapy	Enhanced curcumin permeation across excised cornea and no side effects	[216]
NLCs	Voriconazole	Antifungal therapy	High drug-loading capacity, higher retention capacity of voriconazole in deeper regions of affected body parts	[217]
NLCs	Fluconazole	Antifungal therapy	Avoiding drug recognition by efflux pump proteins, preventing drug extrusion by transporters and reduced side effects	[218]

The delivery of drug molecules to the specific target site (diseased area) of body while avoiding possible side effects on non-diseased areas of organs is a major challenge in drug delivery systems [219]. Green nanomedicine approach for drug delivery is beneficial possibly due to efficient loading of drug molecules in nanocarriers and safe, timely delivery of drug to target sites in the body (diseased organ). Such drug formulations can be administered to specific areas in the body by topical, oral, inhalation and parenteral methods.

i. Oral administration

The most common oral route drug delivery is the first choice, but the main challenge is poor bioavailability of drug due to physical (gastrointestinal epithelium) and biochemical barrier (enzymatic degradation) [16]. The presence of villi and microvilli increases the surface area of small intestine because of which maximum absorption of drugs occurs there [220]. Green nanomedicine is a viable approach to overcome the disadvantages of oral administration [221] by facilitating large surface area-to-volume ratio and better adsorption of drug [222]. Green synthesized nanocarriers protect the encapsulated drug and improve the efficiency of drug delivery to the diseased organ.

ii. Inhalation route

Another administrative route for drug delivery is inhalation. It has several advantages such as good epithelial permeability, large surface area of the lungs, highly dispersed nature of an aerosol and lower dose requirement. This route allows rapid drug absorption into systemic circulation by deposition of small drug molecules in the lung [223,224]. Inhaled drug delivery avoids issues associated with oral administration because drug-metabolizing enzymes are present in much lower quantities in the lungs than in GI tract and liver, which prevents drug degradation [225]. The drugs can penetrate deeper after inhalation, enhancing absorption of medication inside lungs and subsequently into circulation. Thus, it facilitates uptake of drug by the cells and prolongs retention of drugs in the lungs. Some therapeutic compounds used for localized delivery through inhalation routes include chemotherapeutics (doxorubicin, fluorouracil and cisplatin) [226,227], antibiotics (cyclosporine, tobramycin, amikacin and fluoroquinolones) [228] and vaccines (measles, influenza, tuberculosis and hepatitis) [229,230].

iii. Topical and transdermal delivery

The skin barrier protects body from external substances such as allergens, bacteria, dust, viruses, fungi and large molecules [231]. Topical drug delivery overcomes the disadvantages of oral (enzymatic degradation) and injectable (site infection) drug administration, making it a more appealing method of drug administration. Green nanodrugs enter the circulation system by hair follicle accumulation and diffusion method [42]. Green nanocarriers (liposomes, transferosomes, dendrimers, ethosomes, polymer NPs, niosomes and nanoemulsions), commonly used for dermal or transdermal drug delivery allow drug molecules to penetrate surface layers and release into deeper layers of skin [232], without causing any skin infection [233]. Biodegradable polymeric nanocarriers deliver drug molecules to skin where it remains for a longer period of time and improves drug absorption [234]. Various advantages of drug delivery via transdermal route include reduction in injection frequency, improved patient compliance and low toxicity profile [235].

9.3.2.3 Theranostics (Imaging and Therapy)

In the last decade, a new field of medicine based on nanotechnology emerged combining diagnostic agents and specific targeted therapy on a single platform. This intersection between diagnosis and therapy is coined as theranostics. Continuous focus on accuracy and personalized medicine based on nano-formulations led to the development of theranostic NPs, which function as both diagnostic and therapeutic agents. Polymeric NPs, lipid-based NPs, dendrimers, nanoemulsions and light-responsive nanomaterials such as QDs and metal NPs are optimal candidates because of excellent optical, magnetic and acoustic properties, superior photoluminescence, small size, ease of modification, biocompatibility and high photostability [54,236] (Figure 9.4). They have vast impact on optimizing drug dosage based on molecular imaging and monitoring treatment.

Moreover, the surface of theranostic NPs can be easily functionalized with therapeutic moieties (e.g., antibodies, small-molecule drugs, peptides, siRNA and vaccines) for use as nanoprobes. These functionalized NPs are capable of detection and specific targeting to cells with high selectivity and sensitivity [54,237]. These nanomaterials are promising next-generation tools that allow simultaneous tracking and therapy. Gao and co-workers synthesized gambogic acid, GA (heat shock protein (HSP90) inhibitor) containing human serum albumin (HSA) NPs for further use in bioimaging and photothermal (dc-IR825) therapy. The nano-formulation HSA/dc-IR825/GA was used for tumor diagnosis and inhibition of tumor growth using photothermal therapy [238]. Another nanoplatform for cancer monitoring and enhancing efficacy of DOX delivery was developed by Xiao et al. PEGylated (PEG: polyethylene glycol) Au nanorods were covalently conjugated to DOX via hydrazone bond for pH-responsive controlled drug release. Tumor-targeting peptides and Cu-chelators were conjugated to PEG arms for active tumor-targeting and PET imaging, respectively [50]. In addition to Au NPs, QDs are also promising candidates for optical imaging and theranostics. Paul et al. developed photoresponsive nanocarrier made of silicon QDs, which phototriggered controlled release of anticancer drug chlorambucil by using o-nitrobenzyl. This o-nitrobenzyl caged chlorambucil and quenched fluorescence of silicon QDs to "off" state through photoinduced electron transfer process. Upon irradiation, drug was released from silicon QDs and fluorescence was switched "on". Using these QDs, *in vitro* drug release using HeLa cells was monitored in real time for 30 minutes [236]. Xu et al. developed Cu/manganese silicate nanospheres (CMSN) coated with lanthanide-doped NPs. $Yb^{3+}/Er^{3+}/Tm^{3+}$ are used for doping, which enable dual-mode up-conversion and down-conversion emissions using NIR laser excitation. Such nanocomposites offer deep tissue penetration, reduce autofluorescence and enhance potential for photodynamic therapy through conversion of O_2 to singlet $^1O_2$ by NIR light emitting from lanthanide element [55,239].

Besides cancer, theranostic nanomaterials are used against variety of infectious agents such as bacteria and virus for diagnosis and eradication of drug-resistant bacterial infections. Zhao et al. designed theranostic nanoprobes based on fluorescent silica NPs coated with vancomycin-modified

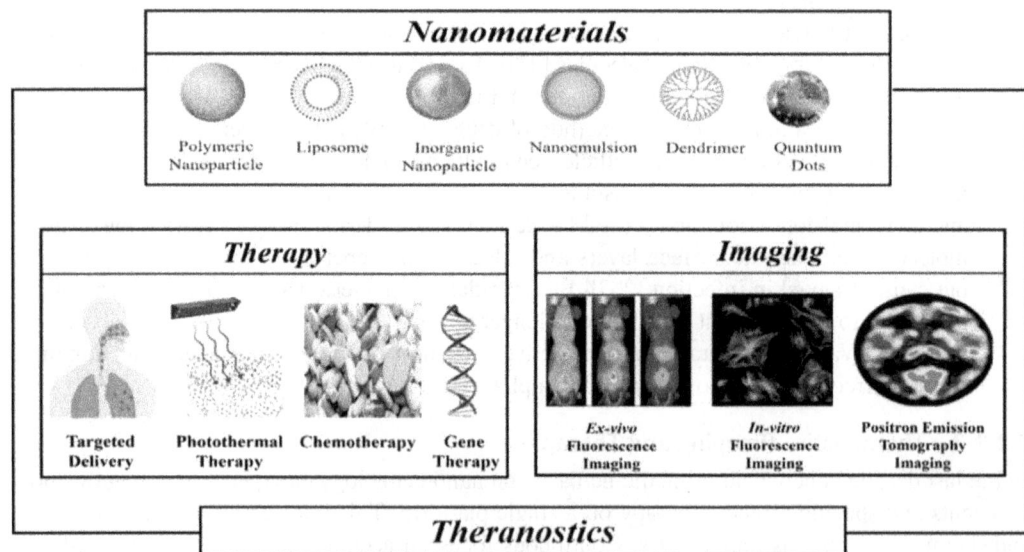

FIGURE 9.4 Role of Nanotechnology in Theranostics

polyelectrolyte-cypate complex (SiO_2-Cy-Van) against methicillin-resistant *Staphylococcus aureus* (MRSA). This nanoprobe leads to MRSA-activated NIR fluorescence *in vitro* and photothermal disruption of bacterial cell envelope. Experiments revealed that the nanoprobe enabled rapid NIR fluorescence imaging 4 hours post-injection for the detection of even 10^5 colony-forming units (CFU) with high sensitivity and further efficient elimination of MRSA by photothermal therapy. This nanoprobe facilitated long-term imaging and tracking of MRSA infection up to 16 days for estimation of bacterial load in real time in infected tissues and monitoring of antimicrobial efficacy [240]. Vancomycin-modified fluorescent silica (Si NPs-Van) NPs were further used by Lu-Zhang and group for designing theranostic agents for rapid, non-invasive diagnosis and treatment of keratitis caused by Gram-positive bacteria. Si NPs-Van aid in rapid imaging and detection of bacterial keratitis within 5 minutes (*in vitro*) and 10 minutes (*in vivo*). Even at the concentration of 0.5 μg/mL, Si NPs-Van exhibit 92.5% antimicrobial activity, which is higher as compared to free vancomycin against *Staphylococcus aureus* [241]. Theranostic NPs are known to significantly reduce the high mortality rates in rare infectious disease caused by a brain-eating amoeba [239]. Dai et al. designed a magnetic core plasmonic shell NPs-based multifunctional platform, where NPs were conjugated to methylene blue aptamer and antibody specific to *Salmonella* for *in vivo* imaging and photo-destruction of drug-resistant *Salmonella* DT104 [242]. The use of theranostic nanomaterials is reported for tracking SARS-CoV-2 viral progression in COVID-19 patients. Moitra et al. explored NIR-II biological window to acquire deep tissue information and used lead sulfide QDs to selectively target N-gene in SARS-CoV-2. Lead sulfide QDs were conjugated to highly specific antisense oligonucleotides using thiol–ene click chemistry triggered by UV for green detection of RNA in SARS-CoV-2. *Ex vivo* imaging confirmed that conjugates of lead sulfide QDs-antisense oligonucleotides exhibited NIR-II emission induced by aggregation, only when RNA of SARS-CoV-2 was present, which might be due to hybridization between antisense oligonucleotides and strands of target RNA [243]. Theranostic nanomaterials enable simultaneous diagnosis and therapy of cancer and other infectious diseases. However, imbibing multiple functions onto single nanomaterial and addressing safety issues are complex challenges for clinical translation and scale-up of industrial production. Optimizing ligand conjugation with cost-effective biomaterials and drug encapsulation efficiency are other challenges for future clinical applications of theranostic nanomaterials. Moreover, applications of theranostic nanomaterials require high-resolution imaging, mitigation of background fluorescence, reduction of photobleaching, elimination of toxic by-products of photobleaching and other limitations.

9.3.2.4 Vaccines Delivery (Antigen Delivery and Antigen Protection)

Vaccines are unprecedented medical products to prevent death and morbidity associated with infectious diseases. Over the years, vaccines conferred long-term immunity against various infectious agents [244]. Diseases such as smallpox, chickenpox, mumps, measles, diphtheria, pertussis, tetanus, hepatitis B, polio or pneumococcal infection are now almost eradicated or brought under control with the help of vaccination [245]. The ongoing pandemic of COVID-19 is under control by vaccination [246]. Beyond infectious diseases, vaccine also displays significant potential in protection against cancer [10]. Vaccines stimulate immune system and are characterized by cellular (lymphocytes) or antibody response. Thus, immunomodulation is at the forefront for the development of vaccines against infectious disease or other kinds of pathologies such as cancer [247,248].

Vaccine development started from traditional vaccines such as first-generation live-attenuated or inactivated vaccines to recent second-generation subunit vaccines and third-generation nucleic acid vaccines [249]. However, first-generation vaccines are associated with virulence reversal (live attenuated vaccines) and weak immune response (inactivated vaccines). Second-generation and third-generation vaccines have disadvantages of low bioavailability, fast degradation of viral antigens and failure in reaching targets. The use of nanomaterials in vaccine advancement to overcome these limitations has attracted interest of researchers. Nanomaterials offer opportunity in design of vaccines to elicit desired immune response by antigen encapsulation and simultaneously protect antigen from proteolytic degradation or rapid clearance. In addition, nanomaterials exhibit inherent adjuvant properties to enhance strong immune response [10,11] (Figure 9.5).

Nanotechnology has caught attention as a promising approach for development of vaccines. Nanomaterials aid smooth delivery of antigen into host cells. Nanomaterials-based antigen delivery has several advantages over conventional vaccine delivery system, which include antigen protection against premature proteolytic degradation by nucleases and proteases, evasion from host immune responses, controlled and targeted delivery [250]. Delivery of antigens using nanomaterials are achieved by encapsulation of antigen in nanomaterial, conjugation or adsorption of antigens on nanomaterial surface or administration of antigen in combination with adjuvant [251]. Nano-based vaccines are easy to manufacture in larger quantity compared to conventional vaccines such as inactivated and live attenuated [252].

Nanocarriers have potential for enhancing safety and efficacy of nanocarrier-based vaccine delivery, better uptake of antigens by antigen-presenting cells and improved immunomodulatory effect [251]. Moreover, nanomaterials have intrinsic adjuvant capacity for loaded antigens and can co-deliver both adjuvant and antigen [253]. Crossing mucosal barrier is also considered highly important in vaccine development, beyond the precondition of persistence of antigen in the internal environment. Studies reveal that nanomaterials of 200 nm size can achieve mucosal delivery in a promising manner [254]. The success of nano-based vaccines depends on rational designing of nanocarriers for antigen loading and effective delivery [255]. Nanomaterials such as polymeric NPs, liposomes and lipid NPs, inorganic NPs (Au, Ag, carbon dots, etc.), virus-like particles (VLPs) and self-assembled protein NPs have been continuously explored as antigen nanocarriers [256]. Polymeric NPs are colloidal systems of 10–500 nm size with high immunogenicity, excellent biocompatibility, large surface area, stability for efficient antigen encapsulation and display. Antigens can either be loaded within the core of nanovehicle or can be conjugated to the surface of vehicle [257]. Polymeric NPs often show significant adjuvant effects, which improve the efficiency of antigen uptake by antigen-presenting cells [258]. Polymeric NPs are utilized in various vaccines. Polylactic-co-glycolic acid/polyethyleneimine (PLGA/PEI) NPs were used for intradermal delivery of DNA vaccine encoding H1N1 influenza virus antigen [259]. Similarly, PLGA of 200–300 nm size was used for the delivery of inactivated H1N2 viral antigen [260]. Biopolymers such as chitosan were explored in addition to synthetic polymers for the targeted and controlled delivery of antigens. Feng et al. used chitosan-based NPs to deliver recombinant HBV antigen against hepatitis B and T cell epitopes of FMS-like tyrosine kinase 3 ligands (FL) and ESAT-6 against *Mycobacterium tuberculosis* [261].

Several vaccines employ lipid nano-formulations for encapsulating genomic material or protein/peptide antigens. mRNA developed by BioNTech/Pfizer and Moderna uses biodegradable lipid NPs of size approximately 100 nm as nanocarriers for encapsulating mRNA strands (encoding SARS-CoV-2 spike glycoprotein) [262,263]. Beside mRNA vaccines, DNA vaccines were developed to combat COVID-19. Entos Pharmaceuticals, the USA, developed "Fusogenix DNA vaccine" by using proprietary proteo-lipid vehicle (PLV) to deliver antigen-encoding plasmid to cytosol of target

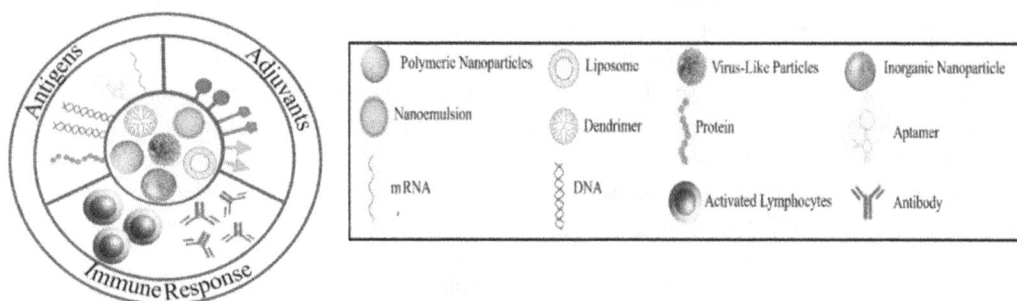

FIGURE 9.5   Role of Nanotechnology in Vaccine Development

cells, which in turn enhanced the immune response [252]. Liposomes were employed in mRNA cancer vaccines, Lipo-MERIT. Lipo-MERIT was developed against melanoma by BioNTech RNA Pharmaceuticals GmbH. Lipo-MERIT is liposomal encapsulation of mRNA encoding 4 tumor antigens (TPTE, tyrosinase, MAGE-A3 and NY-ESO-1) [264].

Besides lipid and polymeric NPs, virus-like particles (VLPs) are highly efficient, safe delivery platforms routinely explored for delivery of antigens. VLPs are virus-derived non-infectious structures with the ability of self-assembly, but without genetic material. They are highly immunogenic and can elicit both cell-mediated and antibody immune responses. Several VLPs-based vaccines, namely Cervarix, Gardasil and Gardasil-9, are approved for the treatment of cervical cancer, and these involve assembly of L1 protein as VLPs. Sci-B-Vac and Mosquirix are other VLP vaccines approved against hepatitis B virus and *Plasmodium falciparum* for the treatment of hepatitis B and malaria, respectively [265].

Biocompatible Au NPs are inorganic NPs, most commonly used for immune activation by vaccines and can be easily internalized by dendritic cells and macrophages. Au NPs are synthesized with strict control over particle size and provide ease of functionalization [266]. The delivery of viral antigen using Au NPs induces robust host immune response against immunodeficiency virus, influenza and foot and mouth disease [256]. Thus, nanomaterials are appealing for vaccine development, and Table 9.4 lists different nanocarriers used in vaccine development against bacterial, viral and protozoan diseases and cancer.

9.3.2.5 Adjuvants (Safe Non-toxic Adjuvant)

The success of vaccines depends not only on protected antigen, but also on adjuvants. Adjuvants are the components of vaccines that are capable of amplifying antigen-specific immune responses [271]. In addition to enhancing antigen immunogenicity, these adjuvants reduce the quantity of antigens as well as the number of immunizations required for protective immunity against the diseases. These adjuvants also act as antigen delivery systems for uptake of antigens by mucosa [272]. Adjuvants currently employed in the approved or under clinical trial human vaccines include aluminum salts (aluminum hydroxide, aluminum phosphate and calcium phosphate) and oil-in-water emulsions (MF59, AF03 and AS03) [273,274]. Among them, MF59 is a nanoemulsion-based adjuvant used in H1N1 influenza vaccines Focetria® and Celture® [275]. Adjuvants currently in use or under development are mostly in nano-range such as polymeric NPs, lipid NPs, liposomes, plant-based nano-formulations, VLPs and inorganic nanomaterials, especially because of their intrinsic immunomodulatory property [276]. Various adjuvants such as polymeric NPs, lipid NPs, virosomes, virus-like particles, colloidally stable Au NPs and calcium phosphate NPs used for vaccine development are summarized below.

Polymeric NPs are widely used as delivery vehicles for various antigens owing to their superior biocompatibility and biodegradability. Polymeric NPs protect antigens from proteolytic degradation and adverse conditions, while maintaining their activity and integrity, and thus boosts the immune response of antigens. PLGA is the most popular material used for fabrication of polymeric NPs. Desai et al. demonstrated comparable adjuvant activity of PLGA and alum [277]. PLGA NPs can play dual role, i.e., as delivery vehicle for antigen and as adjuvant for enhancing immune response. DNA vaccine INO-4800 has recently been developed and customized by Inovio Pharmaceuticals and International Vaccine Institute using PLGA NPs. PLGA NPs were used in the delivery of DNA plasmid pGX9501 encoding full length of S protein of SARS-CoV-2 and also as an adjuvant in vaccine to enhance humoral and cellular immune responses [278]. PLGA NPs are used as adjuvant delivery system in pathogen-mimicking polymeric vaccine NPs against *Plasmodium falciparum*. In this vaccine, PLGA NPs enveloped by lipid membrane display VMP001 malaria antigen (VMP001-NPs) to antigen-presenting cells and then generate a balanced Th1/Th2 humoral response [279].

Chitosan NPs are used both as antigen delivery vehicle and as adjuvant in swine flu DNA vaccine and hepatitis B vaccine. In swine flu DNA vaccine, spherical chitosan NPs protect DNA antigen and

TABLE 9.4

List of Nanocarriers Used in Development of Vaccines against Different Infectious Diseases

Nanocarrier Used	Antigen	Disease/Targeted Pathogen	References
Poly(D,L-lactic-co-glycolic acid) nanospheres	Antigenic protein	Anthrax	[256]
Chitosan NPs	DNA encoding T cell epitopes of FL and ESAT-6	Tuberculosis	[256]
Chitosan NPs	Mycobacterial lipids	Tuberculosis	[256]
Liposomes	Mycobacterium fusion protein	Tuberculosis	[256]
Liposomes	Polysaccharides	Pneumonia	[256]
Liposomes	Parasitic protein and toxic bacterial protein	Malaria and cholera	[256]
Liposomes	Fusion protein	*Helicobacter pylori* infection	[256]
Nanoemulsion	Antigenic protein	Cystic fibrosis	[256]
Nanoemulsion	Antigenic protein	Anthrax	[256]
Fe_2O_3 NPs	Merozoite surface protein	Malaria	[256]
Chitosan NPs	Antigenic protein	Hepatitis B	[256]
Alginate-coated chitosan NPs	Hepatitis B surface antigen	Hepatitis B	[256]
Poly(D,L-lactic-co-glycolic acid) nanospheres	Hepatitis B surface antigen	Hepatitis B	[256]
Au NPs	Viral protein	Foot-and-mouth disease	[256]
Au NPs	Membrane protein	Influenza	[256]
Au NPs	Viral plasmid DNA	HIV	[256]
Lipid NPs	mRNA encoding spike glycoprotein of SARS-CoV-2	COVID-19	[263,267]
M-HBsAgS-N9, M-HBsAgS-N4 VLPs	NANP repeats from small HBV envelope protein (HBaAgS) and circumsporozoite protein (CSP)	Malaria	[268]
Chimeric VLPs (Pfs25 and Pfs230), which are genetically fused to dS of duck HBV	Pfs25 and Pfs230	Malaria	[268]
STh-A14T and STh VLPs	Human heat-stable toxins	Enterotoxigenic *Escherichia coli* (ETEC) infection	[268]
RVFV VLPs	Gc, N and Gn proteins	Rift Valley fever (RVF)	[268]
JEV genotype III (GIII) VLPs	Precursor membrane protein and envelope protein	Japanese encephalitis	[268]
SAG1 VLPs	Surface antigen 1 (SAG1)	*Toxoplasma gondii* infection	[268]
Hepatitis B core (HBc) VLPs and recombinant immune complexes	Minor CP (L2 or L2 with immunoglobulin)	Human papillomavirus (HPV) infection	[268]
HPV16 L1 VLPs	HPV-16 and HPV-18 L1 antigen	Cervical cancer	[269,270]
mRNA lipoplex NPs	NY-ESO-1, tyrosinase, MAGE-A3, and TPTE	Melanoma	[10,264]

generate effective immune response [280], whereas in hepatitis B vaccine, chitosan NPs deliver the antigenic protein at target site to enhance anti-HBsAg IgG levels [281].

The Matrix-M adjuvant attracts global attention and consists of 40 nm honeycomb-like NPs derived from phospholipids, cholesterol and saponin of *Quillaja saponaria* Molina [282]. SARS-CoV-2 subunit vaccine (NVX-CoV2373) is formulated using proprietary Matrix-M adjuvant, where this adjuvant enhances immune response by promoting antigen absorption. Studies demonstrate that Matrix-M adjuvant induces T and B cell response, SARS-CoV-2-neutralizing antibodies and human angiotensin-converting enzyme 2 receptor-blocking antibodies. These studies also support the safety, efficacy and immunogenicity of vaccines [283].

Virosome (liposome) possesses adjuvant properties and is a novel adjuvant delivery system used in hepatitis A vaccine (Epaxal) and influenza vaccine (Inflexal). The uptake of virosome by antigen-presenting cells (APCs) followed by degradation in an endocytic pathway ensures antigen presentation for further activation of T cells and B cells [275].

VLPs, well known for antigen delivery, can exert effects on APCs and antigen-specific lymphocytes. These VLPs can bind to and penetrate the host cells to stimulate both cellular and humoral immunity. Presently, three cervical cancer vaccines are available in the market, viz. Cervarix, Gardasil-9 and Gardasil, to enhance specific anti-L1 VLP immune response. VLP parent virus-specific antibodies produced by VLP vaccines can neutralize the parent virus for protection against infection [275].

Besides organic nanomaterials, inorganic NPs also act as adjuvants. Gold NPs (Au NPs) are routinely exploited in drug delivery because of their exceptional properties, such as stability, ease of functionalization, low toxicity, high biocompatibility and non-immunogenicity. Au NPs can stimulate various immune cells, and their immune efficiency depends on size and shape. Studies revealed that spherical Au NPs of size 40 nm and cubic Au NPs of size (40×40×10 nm) induced inflammatory cytokines such as IL-6, IL-12 and TNFα. However, spherical Au NPs of size 20 nm and Au nanorods of size (40×10 nm) did not enhance immune response in West Nile virus vaccine [284].

Biocompatible calcium phosphate NPs act as adjuvants to enhance Th1 and Th2 humoral responses, which can be used as potential dose-sparing strategy. The adjuvant property and post-vaccination antibody response due to calcium phosphate NPs-based vaccine are dependent on both particle size and route of vaccine administration. Studies demonstrated that nano-calcium phosphate adjuvant (size 73 nm) adsorbed in Human Enterovirus 71 (HEV71) vaccine stimulated higher antibody levels as compared to micro-size or unadsorbed vaccine alone. Furthermore, antibody levels in intradermal route was ten times more than intramuscular route [285].

Vaccine is the most successful biological intervention that reduces disease burden and mortality. With advancements in vaccines, adjuvants have gained enormous interest to enhance stronger and efficacious immune response.

9.4 CHALLENGES OF NANOTECHNOLOGY AND RECOMMENDATIONS FOR MANAGEMENT

The translation of nanotechnology from bench to bedside imposes few challenges. General issues for consideration during nanomedicine product development include precursors, synthesis process, biocompatibility, physicochemical characterization, process control, nanotoxicology evaluation, scale-up reproducibility, pharmacokinetics and pharmacodynamics assessment. A lack of understanding prevails on what makes the patients differ from one another and why the drugs do not have ubiquitous efficacy. This further extends to why nanotechnologies are many times unable to improve the therapeutic outcome of all drugs in each patient. It is therefore, essential to first understand the behavior of nanomedicines after their encounter with physiological characteristics in patients with different disease states.

Nanotechnology also poses risk as health hazard. NPs due to their small size can easily get into human body through inhalation and damage lungs or other organs [286,287]. The synthesis of nanomaterials at times can be expensive due to manufacturing limitations [288]. Although nano-technology has raised the standard of living, at the same time, it has increased the pollution of water, air and soil, more because the long-term effects of this nano-based pollution are still unexplored. The toxicity, risk and life cycle assessment of nanomaterials are not completely known. For example, the extensive use of polyvinyl alcohol as detergent creates toxicity issues. NPs used in drug delivery sometimes result in cytotoxicity and alveolar inflammation. Autonomic dysfunction due to NPs have direct effect on heart and vascular functions. In certain cases, NPs show particle growth, unexpected dynamics of polymeric transmit ions, unpredictable gelation tendency and sometimes burst release, which can be harmful for human body [289].

Emerging governance strategies and mechanisms should aim to foster responsible development of engineered nanomaterials and ensure effective implementation of oversight regulatory methods [290]. Stakeholders should undertake the following steps to develop green nanomaterials [19]:

- Create opportunity to discuss, develop and systematically implement green nano-principles.
- Develop an appropriate life cycle assessment of green nanoproducts.
- Establish specific standards for green nanomaterials and brand as "green nano".
- Extend patent term for green nanoproducts.
- Launch Design for the Environment (DfE) award in "green nano" category.
- Business and tax incentives for innovators to make up for shortage of investment capital and reduce the cost of commercialization of green nanoproducts.
- Encourage public-private partnership by funding and research opportunities for the development of green nanoproducts.

9.5   CONCLUSIONS

The use of biodegradable precursors in the bottom-up synthesis of nanomaterials based on the principles of green chemistry allows us to reduce the associated toxicity risks and improve the properties of nanoproducts for further use in biomedical applications. Seventeen sustainable development goals (SDGs) identified in the year 2015 by United Nations have to be realized by 2030 within a time frame of 15 years to balance social, economic and environmental well-being of mankind globally. Thus, green nanotechnology was developed for the synthesis of non-toxic nanomaterials to eliminate the use and generation of hazardous substances. Such materials can be produced at nanoscale level, and they remain safe for introduction into the body for applications in medicine. This chapter gives an overview of green nanotechnology and its advantages. It also describes various biomedical applications of green nanotechnology, such as diagnostics, therapeutics (as antimicrobial and anticancer agents, drug delivery vehicles and theranostics) and vaccines (vaccine delivery and adjuvants in vaccine). It also presents the challenges and recommendations for the management and development of nanomaterials for use in medicine. This compiled information will be interesting for readers to understand the basics of green nanotechnology; prospects that green nanomaterials can offer in medicine; issues and challenges in the translation of nanomaterials for nanomedicine; and recommendations for the management of challenges to improve healthcare of mankind.

ACKNOWLEDGEMENTS

T.P. gratefully acknowledges ICMR Grants (F. No. 5/8/5/1/Adhoc/2020/ECD-I and F. No. 35/11/2019-Nano/BMS dated 7/10/19), JNU-UPOE II scheme (ID-161) and JNU-DST-PURSE (Phase-II), India. A.R. gratefully acknowledges ICMR, India, for Senior Research Fellowship, S.K. acknowledges CSIR, India, for Senior Research Fellowship, and I.V. and V.K. acknowledge JNU for fellowship.

REFERENCES

1. de Morais, Michele Greque, Vilásia Guimarães Martins, Daniela Steffens, Patricia Pranke, and Jorge Alberto Vieira da Costa. "Biological applications of nanobiotechnology." *Journal of Nanoscience and Nanotechnology* 14, no. 1 (2014): 1007–17.

2. Rogers, John A., and Joseph M. DeSimone. "Novel materials." *Proceedings of the National Academy of Sciences* 113, no. 42 (2016): 11667–69.

3. Bharathala, Subhashini, and Pankaj Sharma. "Biomedical applications of nanoparticles." In: Pawan Kumar Maurya, Sanjay Singh (eds.) *Nanotechnology in Modern Animal Biotechnology*, Elsevier, 2019: 113–32.

4. Khosa, Archana, Satish Reddi, and Ranendra N. Saha. "Nanostructured lipid carriers for site-specific drug delivery." *Biomedicine & Pharmacotherapy* 103 (2018): 598–613.

5. Barenholz, Yechezkel Chezy. "Doxil®—the first FDA-approved nano-drug: Lessons learned." *Journal of Controlled Release* 160, no. 2 (2012): 117–34.

6. Batist, Gerald, Jeremy Barton, Philip Chaikin, Christine Swenson, and Lauri Welles. "Myocet (Liposome-Encapsulated Doxorubicin Citrate): A new approach in breast cancer therapy." *Expert Opinion on Pharmacotherapy* 3, no. 12 (2002): 1739–51.

7. Forssen, Eric A. "The design and development of daunoxome® for solid tumor targeting *In Vivo*." *Advanced Drug Delivery Reviews* 24, no. 2–3 (1997): 133–50.

8. Urits, Ivan, Daniel Swanson, Michael C. Swett, Anjana Patel, Kevin Berardino, Ariunzaya Amgalan, Amnon A. Berger, et al. "A review of Patisiran (Onpattro®) for the treatment of polyneuropathy in people with hereditary transthyretin amyloidosis." *Neurology and Therapy* 9, no. 2 (2020): 301–15.

9. Lancet, Jeffrey E., Geoffrey L. Uy, Jorge E. Cortes, Laura F. Newell, Tara L. Lin, Ellen K. Ritchie, Robert K. Stuart, et al. "CPX-351 (Cytarabine and Daunorubicin) liposome for injection versus conventional cytarabine plus daunorubicin in older patients with newly diagnosed secondary acute myeloid leukemia." *Journal of Clinical Oncology* 36, no. 26 (2018): 2684.

10. Feng, Chan, Yongjiang Li, Bijan Emiliano Ferdows, Dylan Neal Patel, Jiang Ouyang, Zhongmin Tang, Na Kong, Enguo Chen, and Wei Tao. "Emerging vaccine nanotechnology: From defense against infection to sniping cancer." *Acta Pharmaceutica Sinica B* 12 (2022): 2206.

11. Bhardwaj, Prateek, Eshant Bhatia, Shivam Sharma, Nadim Ahamad, and Rinti Banerjee. "Advancements in prophylactic and therapeutic nanovaccines." *Acta Biomaterialia* 108 (2020): 1–21.

12. Jourabchi, Natanel, Kourosh Beroukhim, Bashir A. Tafti, Stephen T. Kee, and Edward W. Lee. "Irreversible electroporation (Nanoknife) in cancer treatment." *Gastrointestinal Intervention* 3, no. 1 (2014): 8–18.

13. Fabbro, Alessandra, Maurizio Prato, and Laura Ballerini. "Carbon nanotubes in neuroregeneration and repair." *Advanced Drug Delivery Reviews* 65, no. 15 (2013): 2034–44.

14. Ray, Paresh Chandra, Hongtao Yu, and Peter P. Fu. "Toxicity and environmental risks of nanomaterials: Challenges and future needs." *Journal of Environmental Science and Health Part C* 27, no. 1 (2009): 1–35.

15. Bystrzejewska-Piotrowska, Grazyna, Jerzy Golimowski, and Pawel L. Urban. "Nanoparticles: Their potential toxicity, waste and environmental management." *Waste Management* 29, no. 9 (2009): 2587–95.

16. Ravichandran, R. "Nanoparticles in drug delivery: Potential green nanobiomedicine applications." *International Journal of Green Nanotechnology: Biomedicine* 1, no. 2 (2009): B108–B30.

17. Subhan, Md Abdus, Kristi Priya Choudhury, and Newton Neogi. "Advances with molecular nanomaterials in industrial manufacturing applications." *Nanomanufacturing* 1, no. 2 (2021): 75–97.

18. Contado, Catia. "Nanomaterials in consumer products: A challenging analytical problem." *Frontiers in Chemistry* 3 (2015): 48.

19. Karn, Barbara P., and Lynn L. Bergeson. "Green nanotechnology: Straddling promise and uncertainty." *Natural Resources & Environment* 24, no. 2 (2009): 9–23.

20. Nasrollahzadeh, Mahmoud, Mohaddeseh Sajjadi, S. Mohammad Sajadi, and Zahra Issaabadi. "Chapter 5- Green nanotechnology." In: Mahmoud Nasrollahzadeh, S. Mohammad Sajadi, Mohaddeseh Sajjadi, Zahra Issaabadi and Monireh Atarod (eds.) *Interface Science and Technology*, 2019, Elsevier, Amsterdam, pp. 145–98.

21. Jahangirian, Hossein, Ensieh Ghasemian Lemraski, Thomas J. Webster, Roshanak Rafiee-Moghaddam, and Yadollah Abdollahi. "A review of drug delivery systems based on nanotechnology and green chemistry: Green nanomedicine." *International Journal of Nanomedicine* 12 (2017): 2957.

22. Ahmad, Naheed, and Seema Sharma. "Green synthesis of silver nanoparticles using extracts of *Ananas Comosus*." *Green and Sustainable Chemistry* 2, no. 4 (2012): 141–7.

23. Albrecht, Matthew A., Cameron W. Evans, and Colin L. Raston. "Green chemistry and the health implications of nanoparticles." *Green Chemistry* 8, no. 5 (2006): 417–32.
24. Allen, Theresa M. and Pieter R. Cullis. "Drug delivery systems: Entering the mainstream." *Science* 303, no. 5665 (2004): 1818–22.
25. Arnold, Michael S., Alexander A. Green, James F. Hulvat, Samuel I. Stupp, and Mark C. Hersam. "Sorting carbon nanotubes by electronic structure using density differentiation." *Nature Nanotechnology* 1, no. 1 (2006): 60–5.
26. Abdelbasir, Sabah M., Kelli M. McCourt, Cindy M. Lee, and Diana C. Vanegas. "Waste-derived nanoparticles: Synthesis approaches, environmental applications, and sustainability considerations." *Frontiers in Chemistry* 8 (2020): 782.
27. Jayapalan, Amal R., Bo Yeon Lee, and Kimberly E. Kurtis. "Can nanotechnology be 'Green'? Comparing efficacy of nano and microparticles in cementitious materials." *Cement and Concrete Composites* 36 (2013): 16–24.
28. Hussain, Imtiyaz, N. B. Singh, Ajey Singh, Himani Singh, and S. C. Singh. "Green synthesis of nanoparticles and its potential application." *Biotechnology Letters* 38, no. 4 (2016): 545–60.
29. Vijayaraghavan, Kuppusamy, and Thirunavukkarasu Ashokkumar. "Plant-mediated biosynthesis of metallic nanoparticles: A review of literature, factors affecting synthesis, characterization techniques and applications." *Journal of Environmental Chemical Engineering* 5, no. 5 (2017): 4866–83.
30. Chau, Chi-Fai, Shiuan-Huei Wu, and Gow-Chin Yen. "The development of regulations for food nanotechnology." *Trends in Food Science & Technology* 18, no. 5 (2007): 269–80.
31. Chinnamuthu, C. and P. Murugesa Boopathi. "Nanotechnology and agroecosystem." *Madras Agricultural Journal* 96, no. 1/6 (2009): 17–31.
32. Christophorou, Loucas G., James Kenneth Olthoff, and David S. Green. "Gases for electrical insulation and arc interruption: Possible present and future alternatives to pure SF6." *National Institute of Standards and Technology (NIST)*. NIST Technical Note 1425 (1997).
33. Damien, Christopher J., and J. Russell Parsons. "Bone graft and bone graft substitutes: A review of current technology and applications." *Journal of Applied Biomaterials* 2, no. 3 (1991): 187–208.
34. Dhingra, Rajive, Sasikumar Naidu, Girish Upreti, and Rapinder Sawhney. "Sustainable nanotechnology: Through green methods and life-cycle thinking." *Sustainability* 2, no. 10 (2010): 3323–38.
35. Ditta, Allah. "How helpful is nanotechnology in agriculture?". *Advances in Natural Sciences: Nanoscience and Nanotechnology* 3, no. 3, (2012): 033002.
36. Kumar, Vineet, and Sudesh Kumar Yadav. "Plant-mediated synthesis of silver and gold nanoparticles and their applications." *Journal of Chemical Technology & Biotechnology: International Research in Process, Environmental & Clean Technology* 84, no. 2 (2009): 151–57.
37. Bhattacharya, Debaditya, and Rajinder K. Gupta. "Nanotechnology and potential of microorganisms." *Critical Reviews in Biotechnology* 25, no. 4 (2005): 199–204.
38. Mohanpuria, Prashant, Nisha K. Rana, and Sudesh Kumar Yadav. "Biosynthesis of nanoparticles: Technological concepts and future applications." *Journal of Nanoparticle Research* 10, no. 3 (2008): 507–17.
39. Guilger-Casagrande, Mariana, and Renata de Lima. "Synthesis of silver nanoparticles mediated by Fungi: A review." *Frontiers in Bioengineering and Biotechnology* 7 (2019): 287.
40. Hussain, Imtiyaz, N. B. Singh, Ajey Singh, Himani Singh, and S. C. Singh. "Green synthesis of nanoparticles and its potential application." *Biotechnology Letters* 38, no. 4 (2016): 545–60.
41. Zhang, Dan, Xin-lei Ma, Yan Gu, He Huang, and Guang-wei Zhang. "Green synthesis of metallic nanoparticles and their potential applications to treat cancer." *Frontiers in Chemistry* (2020): 799.
42. Friedman, Adam. *Nanodermatology: The Giant Role of Nanotechnology in Diagnosis and Treatment of Skin Disease.* CRC Press: Boca Raton, FL, 2013.
43. Alharbi, Khalid Khalaf, and Yazeed A. Al-Sheikh. "Role and implications of nanodiagnostics in the changing trends of clinical diagnosis." *Saudi Journal of Biological Sciences* 21, no. 2 (2014): 109–17.
44. Lee, Kar Xin, Kamyar Shameli, Yen Pin Yew, Sin-Yeang Teow, Hossein Jahangirian, Roshanak Rafiee-Moghaddam, and Thomas J. Webster. "Recent developments in the facile bio-synthesis of gold nanoparticles (AuNPs) and their biomedical applications." *International Journal of Nanomedicine* 15 (2020): 275.
45. Halo, Tiffany L., Kaylin M. McMahon, Nicholas L. Angeloni, Yilin Xu, Wei Wang, Alyssa B. Chinen, Dmitry Malin, et al. "Nanoflares for the detection, isolation, and culture of live tumor cells from human blood." *Proceedings of the National Academy of Sciences* 111, no. 48 (2014): 17104–09.
46. Ratan, Zubair Ahmed, Fazla Rabbi Mashrur, Anisha Parsub Chhoan, Sadi Md Shahriar, Mohammad Faisal Haidere, Nusrat Jahan Runa, Sunggyu Kim, et al. "Silver nanoparticles as potential antiviral agents." *Pharmaceutics* 13, no. 12 (2021): 2034.

47. Hu, Ye, Daniel H. Fine, Ennio Tasciotti, Ali Bouamrani, and Mauro Ferrari. "Nanodevices in diagnostics." *Wiley Interdisciplinary Reviews: Nanomedicine and Nanobiotechnology* 3, no. 1 (2011): 11–32.

48. Puiu, Mihaela, and Camelia Bala. "SPR and SPR imaging: Recent trends in developing nanodevices for detection and real-time monitoring of biomolecular events." *Sensors* 16, no. 6 (2016): 870.

49. Chiu, Sheng-Hui, Gangaraju Gedda, Wubshet Mekonnen Girma, Jem-Kun Chen, Yong-Chien Ling, Anil V. Ghule, Keng-Liang Ou, and Jia-Yaw Chang. "Rapid fabrication of carbon quantum dots as multifunctional nanovehicles for dual-modal targeted imaging and chemotherapy." *Acta Biomaterialia* 46 (2016): 151–64.

50. Howell, Mark, Chunyan Wang, A. Mahmoud, G. Hellermann, Shyam S. Mohapatra, and Subhra Mohapatra. "Dual-function theranostic nanoparticles for drug delivery and medical imaging contrast: Perspectives and challenges for use in lung diseases." *Drug Delivery and Translational Research* 3, no. 4 (2013): 352–63.

51. Zhang, Hong, Tingjian Wang, Yuanyuan Zheng, Changxiang Yan, Wei Gu, and Ling Ye. "Comparative toxicity and contrast enhancing assessments of Gd_2O_3@ BSA and MnO_2@ BSA nanoparticles for MR imaging of brain glioma." *Biochemical and Biophysical Research Communications* 499, no. 3 (2018): 488–92.

52. Li, Wen-Qing, Zhigang Wang, Sijie Hao, Liping Sun, Merisa Nisic, Gong Cheng, Chuandong Zhu, et al. "Mitochondria-based aircraft carrier enhances *In Vivo* imaging of carbon quantum dots and delivery of anticancer drug." *Nanoscale* 10, no. 8 (2018): 3744–52.

53. Steinmetz, Nicole F. "Viral nanoparticles as platforms for next-generation therapeutics and imaging devices." *Nanomedicine: Nanotechnology, Biology and Medicine* 6, no. 5 (2010): 634–41.

54. Liang, Gaofeng, Haojie Wang, Hao Shi, Haitao Wang, Mengxi Zhu, Aihua Jing, Jinghua Li, and Guangda Li. "Recent progress in the development of upconversion nanomaterials in bioimaging and disease treatment." *Journal of Nanobiotechnology* 18, no. 1 (2020): 1–22.

55. Xu, Jiating, Ruipeng Shi, Guanying Chen, Shuming Dong, Piaoping Yang, Zhiyong Zhang, Na Niu, et al. "All-in-one theranostic nanomedicine with ultrabright second near-infrared emission for tumor-modulated bioimaging and chemodynamic/photodynamic therapy." *ACS Nano* 14, no. 8 (2020): 9613–25.

56. Singh, Priyanka, Santosh Pandit, V. R. S. S. Mokkapati, Abhroop Garg, Vaishnavi Ravikumar, and Ivan Mijakovic. "Gold nanoparticles in diagnostics and therapeutics for human cancer." *International Journal of Molecular Sciences* 19, no. 7 (2018): 1979.

57. Mukherjee, Sudeshna, Sayan Ghosh, Dipesh Kr Das, Priyanka Chakraborty, Sreetama Choudhury, Payal Gupta, Arghya Adhikary, Sanjit Dey, and Sreya Chattopadhyay. "Gold-conjugated green tea nanoparticles for enhanced anti-tumor activities and hepatoprotection—synthesis, characterization and *In Vitro* evaluation." *The Journal of Nutritional Biochemistry* 26, no. 11 (2015): 1283–97.

58. Fazal, Sajid, Aswathy Jayasree, Sisini Sasidharan, Manzoor Koyakutty, Shantikumar V. Nair, and Deepthy Menon. "Green synthesis of anisotropic gold nanoparticles for photothermal therapy of cancer." *ACS Applied Materials & Interfaces* 6, no. 11 (2014): 8080–89.

59. Huang, He, Da-Peng Yang, Minghuan Liu, Xiangsheng Wang, Zhiyong Zhang, Guangdong Zhou, Wei Liu, et al. "pH-Sensitive Au–BSA–DOX–FA nanocomposites for combined CT imaging and targeted drug delivery." *International Journal of Nanomedicine* 12 (2017): 2829.

60. Song, Xiao-Rong, Shu-Xian Yu, Gui-Xiao Jin, Xiaoyong Wang, Jianzhong Chen, Juan Li, Gang Liu, and Huang-Hao Yang. "Plant polyphenol-assisted green synthesis of hollow CoPt alloy nanoparticles for dual-modality imaging guided photothermal therapy." *Small* 12, no. 11 (2016): 1506–13.

61. Fernández-Barahona, Irene, Maria Muñoz-Hernando, and Fernando Herranz. "Microwave-driven synthesis of iron-oxide nanoparticles for molecular imaging." *Molecules* 24, no. 7 (2019): 1224.

62. Narayanan, Sreeja, Binulal N. Sathy, Ullas Mony, Manzoor Koyakutty, Shantikumar V. Nair, and Deepthy Menon. "Biocompatible magnetite/gold nanohybrid contrast agents via green chemistry for MRI and CT bioimaging." *ACS Applied Materials & Interfaces* 4, no. 1 (2012): 251–60.

63. Du, Tianyu, Chunqiu Zhao, Lanmei Lai, Xiaoqi Li, Yi Sun, Shouhua Luo, Hui Jiang, Matthias Selke, and Xuemei Wang. "Rapid and multimodal *In Vivo* bioimaging of cancer cells through in situ biosynthesis of Zn & Fe nanoclusters." *Nano Research* 10, no. 8 (2017): 2626–32.

64. Shevtsov, Maxim, Boris Nikolaev, Yaroslav Marchenko, Ludmila Yakovleva, Nikita Skvortsov, Anton Mazur, Peter Tolstoy, Vyacheslav Ryzhov, and Gabriele Multhoff. "Targeting experimental orthotopic glioblastoma with chitosan-based superparamagnetic iron oxide nanoparticles (CS-DX-SPIONS)." *International Journal of Nanomedicine* 13 (2018): 1471.

65. Yang, Min, Quli Fan, Ruiping Zhang, Kai Cheng, Junjie Yan, Donghui Pan, Xiaowei Ma, Alex Lu, and Zhen Cheng. "Dragon fruit-like biocage as an iron trapping nanoplatform for high efficiency targeted cancer multimodality imaging." *Biomaterials* 69 (2015): 30–7.

66. Haghighi, Behzad, and Mahmoud Amouzadeh Tabrizi. "Green-synthesis of reduced graphene oxide nanosheets using rose water and a survey on their characteristics and applications." *RSC Advances* 3, no. 32 (2013): 13365–71.

67. Zhang, Panpan, Xiaoyuan Zhang, Siyu Zhang, Xin Lu, Qing Li, Zhiqiang Su, and Gang Wei. "One-pot green synthesis, characterizations, and biosensor application of self-assembled reduced graphene oxide–gold nanoparticle hybrid membranes." *Journal of Materials Chemistry B* 1, no. 47 (2013): 6525–31.

68. Wang, Yicun, Zhigang Ju, Binrui Cao, Xiang Gao, Ye Zhu, Penghe Qiu, Hong Xu, et al. "Ultrasensitive rapid detection of human serum antibody biomarkers by biomarker-capturing viral nanofibers." *ACS Nano* 9, no. 4 (2015): 4475–83.

69. Chauhan, Arun, Swaleha Zubair, Saba Tufail, Asif Sherwani, Mohammad Sajid, Suri C. Raman, Amir Azam, and Mohammad Owais. "Fungus-mediated biological synthesis of gold nanoparticles: Potential in detection of liver cancer." *International Journal of Nanomedicine* 6 (2011): 2305.

70. Xu, He-Lin, De-Li ZhuGe, Pian-Pian Chen, Meng-Qi Tong, Meng-Ting Lin, Xue Jiang, Ya-Wen Zheng, et al. "Silk fibroin nanoparticles dyeing indocyanine green for imaging-guided photo-thermal therapy of glioblastoma." *Drug Delivery* 25, no. 1 (2018): 364–75.

71. Organization, World Health. "COVID-19 Weekly Epidemiological Update, Edition 84." (2022).

72. Kevadiya, Bhavesh D., Jatin Machhi, Jonathan Herskovitz, Maxim D. Oleynikov, Wilson R. Blomberg, Neha Bajwa, Dhruvkumar Soni, et al. "Diagnostics for SARS-COV-2 infections." *Nature Materials* 20, no. 5 (2021): 593–605.

73. Li, Dan, Zipeng Zhou, Jiachen Sun, and Xifan Mei. "Prospects of NIR fluorescent nanosensors for green detection of SARS-COV-2." *Sensors and Actuators B: Chemical* 362 (2022): 131764.

74. Pucci, Carlotta, Chiara Martinelli, and Gianni Ciofani. "Innovative approaches for cancer treatment: Current perspectives and new challenges." *Ecancermedicalscience* 13 (2019): 961.

75. Sakamoto, Jason, Biana Godin, Ye Hu, Elvin Blanco, Anne L. van de Ven, Adaikkalam Vellaichamy, Matthew B. Murphy, et al. "Nanotechnology toward advancing personalized medicine." In: Ioannis S. Vizirianakis (ed.) *Handbook of Personalized Medicine*. CRC Press, Boca Raton, FL (2013): 1–714.

76. Hejmady, Siddhanth, Rajesh Pradhan, Amit Alexander, Mukta Agrawal, Gautam Singhvi, Bapi Gorain, Sanjay Tiwari, Prashant Kesharwani, and Sunil Kumar Dubey. "Recent advances in targeted nanomedicine as promising antitumor therapeutics." *Drug Discovery Today* 25, no. 12 (2020): 2227–44.

77. Frieri, Marianne, Krishan Kumar, and Anthony Boutin. "Antibiotic resistance." *Journal of Infection and Public Health* 10, no. 4 (2017): 369–78.

78. Sharmin, Shabnam, Md Mizanur Rahaman, Chandan Sarkar, Olubunmi Atolani, Mohammad Torequl Islam, and Oluyomi Stephen Adeyemi. "Nanoparticles as antimicrobial and antiviral agents: A literature-based perspective study." *Heliyon* 7, no. 3 (2021): e06456.

79. Ruddaraju, Lakshmi Kalyani, Sri Venkata Narayana Pammi, Girija Sankar Guntuku, Veerabhadra Swamy Padavala, and Venkata Ramana Murthy Kolapalli. "A review on anti-bacterials to combat resistance: From ancient era of plants and metals to present and future perspectives of green nano technological combinations." *Asian Journal of Pharmaceutical Sciences* 15, no. 1 (2020): 42–59.

80. Yin, Iris Xiaoxue, Jing Zhang, Irene Shuping Zhao, May Lei Mei, Quanli Li, and Chun Hung Chu. "The antibacterial mechanism of silver nanoparticles and its application in dentistry." *International Journal of Nanomedicine* 15 (2020): 2555.

81. Roy, Anupam, Onur Bulut, Sudip Some, Amit Kumar Mandal, and M. Deniz Yilmaz. "Green synthesis of silver nanoparticles: Biomolecule-nanoparticle organizations targeting antimicrobial activity." *RSC Advances* 9, no. 5 (2019): 2673–702.

82. Sudhasree, S., A. Shakila Banu, Pemaiah Brindha, and Gino A. Kurian. "Synthesis of nickel nanoparticles by chemical and green route and their comparison in respect to biological effect and toxicity." *Toxicological & Environmental Chemistry* 96, no. 5 (2014): 743–54.

83. Mukherjee, Sudip, Debabrata Chowdhury, Rajesh Kotcherlakota, and Sujata Patra. "Potential theranostics application of bio-synthesized silver nanoparticles (4-in-1 system)." *Theranostics* 4, no. 3 (2014): 316.

84. Jayaprakash, N., J. Judith Vijaya, K. Kaviyarasu, K. Kombaiah, L. John Kennedy, R. Jothi Ramalingam, Murugan A. Munusamy, and Hamad A. Al-Lohedan. "Green synthesis of Ag nanoparticles using tamarind fruit extract for the antibacterial studies." *Journal of Photochemistry and Photobiology B: Biology* 169 (2017): 178–85.

85. Gondil, Vijay Singh, Thiyagarajan Kalaiyarasan, Vijay K. Bharti, and Sanjay Chhibber. "Antibiofilm potential of Seabuckthorn silver nanoparticles (Sbt@ AgNPs) against *Pseudomonas Aeruginosa*." *3 Biotech* 9, no. 11 (2019): 1–13.

86. Mickymaray, Suresh. "One-step synthesis of silver nanoparticles using saudi arabian desert seasonal plant *Sisymbrium Irio* and antibacterial activity against multidrug-resistant bacterial strains." *Biomolecules* 9, no. 11 (2019): 662.

87. Patil Shriniwas, P. "Antioxidant, antibacterial and cytotoxic potential of silver nanoparticles synthesized using terpenes rich extract of *Lantana Camara* L. leaves." *Biochemistry and Biophysics Reports* 10 (2017): 76.

88. Mata, Rani, Jayachandra Reddy Nakkala, and Sudha Rani Sadras. "Biogenic silver nanoparticles from *Abutilon Indicum*: Their antioxidant, antibacterial and cytotoxic effects *In Vitro*." *Colloids and Surfaces B: Biointerfaces* 128 (2015): 276–86.

89. Dada, Adewumi Oluwasogo, Folahan Amoo Adekola, Fehintoluwa Elizabeth Dada, Adunola Tabitha Adelani-Akande, Micheal Oluwasesan Bello, Chidiogo Rita Okonkwo, Adejumoke Abosede Inyinbor, et al. "Silver nanoparticle synthesis by *Acalypha Wilkesiana* extract: Phytochemical screening, characterization, influence of operational parameters, and preliminary antibacterial testing." *Heliyon* 5, no. 10 (2019): e02517.

90. Nayak, Debasis, Sonali Pradhan, Sarbani Ashe, Pradipta Ranjan Rauta, and Bismita Nayak. "Biologically synthesised silver nanoparticles from three diverse family of plant extracts and their anticancer activity against epidermoid A431 carcinoma." *Journal of Colloid and Interface Science* 457 (2015): 329–38.

91. Figueiredo, Erica Pelegrin, Jhonatan Macedo Ribeiro, Erick Kenji Nishio, Sara Scandorieiro, Amanda Ferreira Costa, Viviane Ferreira Cardozo, Admilton Goncalves de Oliveira, et al. "New approach for simvastatin as an antibacterial: Synergistic effect with bio-synthesized silver nanoparticles against multidrug-resistant bacteria." *International Journal of Nanomedicine* 14 (2019): 7975.

92. Finbloom, Joel A., Ioana L. Aanei, Jenna M. Bernard, Sarah H. Klass, Susanna K. Elledge, Kenneth Han, Tomoko Ozawa, et al. "Evaluation of three morphologically distinct virus-like particles as nanocarriers for convection-enhanced drug delivery to glioblastoma." *Nanomaterials* 8, no. 12 (2018): 1007.

93. Sathishkumar, Palanivel, Johnson Preethi, Raji Vijayan, Abdull Rahim Mohd Yusoff, Fuad Ameen, Sadhasivam Suresh, Ramasamy Balagurunathan, and Thayumanavan Palvannan. "Anti-acne, anti-dandruff and anti-breast cancer efficacy of green synthesised silver nanoparticles using *Coriandrum Sativum* leaf extract." *Journal of Photochemistry and Photobiology B: Biology* 163 (2016): 69–76.

94. Hossain, Md, Shakil Ahmed Polash, Masato Takikawa, Razib Datta Shubhra, Tanushree Saha, Zinia Islam, Sharif Hossain, et al. "Investigation of the antibacterial activity and *In Vivo* cytotoxicity of biogenic silver nanoparticles as potent therapeutics." *Frontiers in Bioengineering and Biotechnology* 7 (2019): 239.

95. Shawkey, Alaa M., Mohamed A. Rabeh, Abeer K. Abdulall, and Ashraf O. Abdellatif. "Green nanotechnology: Anticancer activity of silver nanoparticles using *Citrullus Colocynthis* aqueous extracts." *Advances in Life Science and Technology* 13 (2013): 60–70.

96. Soni, Namita, and Soam Prakash. "Antimicrobial and mosquitocidal activity of microbial synthesized silver nanoparticles." *Parasitology Research* 114, no. 3 (2015): 1023–30.

97. Zahir, Abdul Abduz, Indira Singh Chauhan, Asokan Bagavan, Chinnaperumal Kamaraj, Gandhi Elango, Jai Shankar, Nidhi Arjaria, et al. "Green synthesis of silver and titanium dioxide nanoparticles using *Euphorbia Prostrata* extract shows shift from apoptosis to G_0/G_1 arrest followed by necrotic cell death in *Leishmania Donovani*." *Antimicrobial Agents and Chemotherapy* 59, no. 8 (2015): 4782–99.

98. Wypij, Magdalena, Patrycja Golinska, Hanna Dahm, and Mahendra Rai. "Actinobacterial-mediated synthesis of silver nanoparticles and their activity against pathogenic bacteria." *IET Nanobiotechnology* 11, no. 3 (2017): 336–42.

99. Devi, J. Saraniya, B. Valentin Bhimba, and K. Ratnam. "Anticancer activity of silver nanoparticles synthesized by the seaweed *Ulva Lactuca In Vitro*." *Scientific Report* 1, no. 4 (2012): 242.

100. Singh, Priyanka, Yeon Ju Kim, Hina Singh, Chao Wang, Kyu Hyon Hwang, Mohamed El-Agamy Farh, and Deok Chun Yang. "Biosynthesis, characterization, and antimicrobial applications of silver nanoparticles." *International Journal of Nanomedicine* 10 (2015): 2567.

101. Gurunathan, Sangiliyandi, Jae Woong Han, Vasuki Eppakayala, Muniyandi Jeyaraj, and Jin-Hoi Kim. "Cytotoxicity of biologically synthesized silver nanoparticles in MDA-MB-231 human breast cancer cells." *BioMed Research International* 2013 (2013): 535796.

102. Deshmukh, Aarti R., Arvind Gupta, and Beom Soo Kim. "Ultrasound assisted green synthesis of silver and iron oxide nanoparticles using fenugreek seed extract and their enhanced antibacterial and antioxidant activities." *BioMed Research International* 2019 (2019): 1–14.

103. Feroze, Nosheen, Bushra Arshad, Muhammad Younas, Muhammad Irfan Afridi, Saddam Saqib, and Asma Ayaz. "Fungal mediated synthesis of silver nanoparticles and evaluation of antibacterial activity." *Microscopy Research and Technique* 83, no. 1 (2020): 72–80.

104. Nakkala, Jayachandra Reddy, Rani Mata, and Sudha Rani Sadras. "Green synthesized nano silver: Synthesis, physicochemical profiling, antibacterial, anticancer activities and biological *In Vivo* toxicity." *Journal of Colloid and Interface Science* 499 (2017): 33–45.

105. Elbeshehy, Essam K. F., Ahmed M. Elazzazy, and George Aggelis. "Silver nanoparticles synthesis mediated by new isolates of *Bacillus* spp., nanoparticle characterization and their activity against bean yellow mosaic virus and human pathogens." *Frontiers in Microbiology* 6 (2015): 453.

106. Sankar, Renu, Arunachalam Karthik, Annamalai Prabu, Selvaraju Karthik, Kanchi Subramanian Shivashangari, and Vilwanathan Ravikumar. "*Origanum Vulgare* mediated biosynthesis of silver nanoparticles for its antibacterial and anticancer activity." *Colloids and Surfaces B: Biointerfaces* 108 (2013): 80–4.

107. Wang, Chao, Yeon Ju Kim, Priyanka Singh, Ramya Mathiyalagan, Yan Jin, and Deok Chun Yang. "Green synthesis of silver nanoparticles by *Bacillus Methylotrophicus*, and their antimicrobial activity." *Artificial Cells, Nanomedicine, and Biotechnology* 44, no. 4 (2016): 1127–32.

108. Bin-Jumah, May, AL-Abdan Monera, Gadah Albasher, and Saud Alarifi. "Effects of green silver nanoparticles on apoptosis and oxidative stress in normal and cancerous human hepatic cells *In Vitro*." *International Journal of Nanomedicine* 15 (2020): 1537.

109. Khalil, Maha A., Abd El-Raheem R. El-Shanshoury, Maha A. Alghamdi, Fatin A. Alsalmi, Samia F. Mohamed, Jianzhong Sun, and Sameh S. Ali. "Biosynthesis of silver nanoparticles by marine actinobacterium *Nocardiopsis Dassonvillei* and exploring their therapeutic potentials." *Frontiers in Microbiology* 12 (2021): 705673.

110. Yarramala, Deepthi S., Sejal Doshi, and Chebrolu P. Rao. "Green synthesis, characterization and anticancer activity of luminescent gold nanoparticles capped with APO-A-lactalbumin." *RSC Advances* 5, no. 41 (2015): 32761–67.

111. Abdel-Kareem, M. Marwa, and Abdel-Naser Ahmed Zohri. "Extracellular mycosynthesis of gold nanoparticles using *Trichoderma Hamatum*: Optimization, characterization and antimicrobial activity." *Letters in Applied Microbiology* 67, no. 5 (2018): 465–75.

112. Li, Jiulong, Qinghao Li, Xiaoqiong Ma, Bing Tian, Tao Li, Jiangliu Yu, Shang Dai, Yulan Weng, and Yuejin Hua. "Biosynthesis of gold nanoparticles by the extreme bacterium *Deinococcus Radiodurans* and an evaluation of their antibacterial properties." *International Journal of Nanomedicine* 11 (2016): 5931.

113. Vijayakumar, Sekar, Baskaralingam Vaseeharan, Balasubramanian Malaikozhundan, Narayanan Gopi, Perumal Ekambaram, R. Pachaiappan, Palaniyandi Velusamy, et al. "Therapeutic effects of gold nanoparticles synthesized using *Musa Paradisiaca* peel extract against multiple antibiotic resistant *Enterococcus Faecalis* biofilms and human lung cancer cells (A549)." *Microbial Pathogenesis* 102 (2017): 173–83.

114. Nagalingam, M., V. Nagarajan Kalpana, and A. Panneerselvam. "Biosynthesis, characterization, and evaluation of bioactivities of leaf extract-mediated biocompatible gold nanoparticles from *Alternanthera Bettzickiana*." *Biotechnology Reports* 19 (2018): e00268.

115. Baker, Syed, Kesarla Mohan Kumar, P. Santosh, D. Rakshith, and Saraswathi Satish. "Extracellular synthesis of silver nanoparticles by novel *Pseudomonas Veronii* AS41G inhabiting *Annona Squamosa* L. and their bactericidal activity." *Spectrochimica Acta Part A: Molecular and Biomolecular Spectroscopy* 136 (2015): 1434–40.

116. Lee, Hyun Uk, So Young Park, Eun Sik Park, Byoungchul Son, Soon Chang Lee, Jae Won Lee, Young-Chul Lee, et al. "Photoluminescent carbon nanotags from harmful cyanobacteria for drug delivery and imaging in cancer cells." *Scientific Reports* 4, no. 1 (2014): 1–7.

117. Lv, Qing, Baogang Zhang, Xuan Xing, Yingxin Zhao, Ruquan Cai, Wei Wang, and Qian Gu. "Biosynthesis of copper nanoparticles using *Shewanella Loihica* PV-4 with antibacterial activity: Novel approach and mechanisms investigation." *Journal of Hazardous Materials* 347 (2018): 141–49.

118. Gu, Haidong, Xiao Chen, Feng Chen, Xing Zhou, and Zohreh Parsaee. "Ultrasound-assisted biosynthesis of CuO-NPs using brown alga *Cystoseira Trinodis*: Characterization, photocatalytic AOP, DPPH scavenging and antibacterial investigations." *Ultrasonics Sonochemistry* 41 (2018): 109–19.

119. Awwad, Akl M., Borhan A. Albiss, and Nida M. Salem. "Antibacterial activity of synthesized copper oxide nanoparticles using *Malva Sylvestris* leaf extract." *SMU Medical Journal* 2, no. 1 (2015): 91–101.

120. Sivaraj, Rajeshwari, Pattanathu K. S. M. Rahman, Raju Rajiv, Sadasivam Narendhran, and Rajendran Venckatesh. "Biosynthesis and characterization of *Acalypha Indica* mediated copper oxide nanoparticles and evaluation of its antimicrobial and anticancer activity." *Spectrochimica Acta Part A: Molecular and Biomolecular Spectroscopy* 129 (2014): 255–58.

121. Naika, H. Raja, K. Lingaraju, K. Manjunath, Danith Kumar, G. Nagaraju, D. Suresh, and H. Nagabhushana. "Green synthesis of CuO nanoparticles using *Gloriosa Superba* L. extract and their antibacterial activity." *Journal of Taibah University for Science* 9, no. 1 (2015): 7–12.
122. Shende, Sudhir, Avinash P. Ingle, Aniket Gade, and Mahendra Rai. "Green synthesis of copper nanoparticles by citrus *Medica Linn.*(Idilimbu) juice and its antimicrobial activity." *World Journal of Microbiology and Biotechnology* 31, no. 6 (2015): 865–73.
123. Sonkusre, Praveen. "Specificity of biogenic selenium nanoparticles for prostate cancer therapy with reduced risk of toxicity: An *In Vitro* and *In Vivo* study." *Frontiers in Oncology* 9 (2020): 1541.
124. Shoeibi, Sara, and Mohammad Mashreghi. "Biosynthesis of selenium nanoparticles using *Enterococcus Faecalis* and evaluation of their antibacterial activities." *Journal of Trace Elements in Medicine and Biology* 39 (2017): 135–39.
125. Ramamurthy, C. H., K. S. Sampath, Pitchaimani Arunkumar, M. Suresh Kumar, Venugopal Sujatha, Kumpati Premkumar, and Chinnasamy Thirunavukkarasu. "Green synthesis and characterization of selenium nanoparticles and its augmented cytotoxicity with doxorubicin on cancer cells." *Bioprocess and Biosystems Engineering* 36, no. 8 (2013): 1131–39.
126. Gunti, Lokanadhan, Regina Sharmila Dass, and Naveen Kumar Kalagatur. "Phytofabrication of selenium nanoparticles from *Emblica Officinalis* fruit extract and exploring its biopotential applications: Antioxidant, antimicrobial, and biocompatibility." *Frontiers in Microbiology* 10 (2019): 931.
127. Sarkar, Rajesh Dev, and Mohan Chandra Kalita. "Se nanoparticles stabilized with *Allamanda Cathartica* L. flower extract inhibited phytopathogens and promoted mustard growth under salt stress." *Heliyon* 8 (2022): e09076.
128. Cittrarasu, Vetrivel, Durairaj Kaliannan, Kalaimurugan Dharman, Viji Maluventhen, Murugesh Easwaran, Wen Chao Liu, Balamuralikrishnan Balasubramanian, and Maruthupandian Arumugam. "Green synthesis of selenium nanoparticles mediated from ceropegia *Bulbosa* roxb extract and its cytotoxicity, antimicrobial, mosquitocidal and photocatalytic activities." *Scientific Reports* 11, no. 1 (2021): 1–15.
129. Vijayakumar, Sekar, Baskaralingam Vaseeharan, Balasubramanian Malaikozhundan, and Malaikkarasu Shobiya. "*Laurus Nobilis* leaf extract mediated green synthesis of ZnO nanoparticles: Characterization and biomedical applications." *Biomedicine & Pharmacotherapy* 84 (2016): 1213–22.
130. Vijayakumar, Sekar, C. Krishnakumar, P. Arulmozhi, S. Mahadevan, and N. Parameswari. "Biosynthesis, characterization and antimicrobial activities of zinc oxide nanoparticles from leaf extract of *Glycosmis Pentaphylla* (Retz.) Dc." *Microbial Pathogenesis* 116 (2018): 44–8.
131. Padalia, Hemali, and Sumitra Chanda. "Characterization, antifungal and cytotoxic evaluation of green synthesized zinc oxide nanoparticles using *Ziziphus Nummularia* leaf extract." *Artificial Cells, Nanomedicine, and Biotechnology* 45, no. 8 (2017): 1751–61.
132. Rajabi, Hamid Reza, Reza Naghiha, Mansoureh Kheirizadeh, Hamed Sadatfaraji, Ali Mirzaei, and Zinab Moradi Alvand. "Microwave assisted extraction as an efficient approach for biosynthesis of zinc oxide nanoparticles: Synthesis, characterization, and biological properties." *Materials Science and Engineering: C* 78 (2017): 1109–18.
133. Moghaddam, Amin Boroumand, Mona Moniri, Susan Azizi, Raha Abdul Rahim, Arbakariya Bin Ariff, Wan Zuhainis Saad, Farideh Namvar, Mohammad Navaderi, and Rosfarizan Mohamad. "Biosynthesis of ZnO nanoparticles by a new *Pichia Kudriavzevii* yeast strain and evaluation of their antimicrobial and antioxidant activities." *Molecules* 22, no. 6 (2017): 872.
134. Sharma, Deepali, Myalowenkosi I. Sabela, Suvardhan Kanchi, Phumlane S. Mdluli, Gulshan Singh, Thor A. Stenström, and Krishna Bisetty. "Biosynthesis of ZnO nanoparticles using *Jacaranda Mimosifolia* flowers extract: Synergistic antibacterial activity and molecular simulated facet specific adsorption studies." *Journal of Photochemistry and Photobiology B: Biology* 162 (2016): 199–207.
135. Gurunathan, Sangiliyandi, Jae Woong Han, Vasuki Eppakayala, and Jin-Hoi Kim. "Green synthesis of graphene and its cytotoxic effects in human breast cancer cells." *International Journal of Nanomedicine* 8 (2013): 1015.
136. Alshatwi, Ali A., Jegan Athinarayanan and Periasamy Vaiyapuri Subbarayan. "Green synthesis of platinum nanoparticles that induce cell death and G2/M-phase cell cycle arrest in human cervical cancer cells." *Journal of Materials Science: Materials in Medicine* 26, no. 1 (2015): 1–9.
137. Al-Radadi, Najlaa S. "Green synthesis of platinum nanoparticles using Saudi's dates extract and their usage on the cancer cell treatment." *Arabian Journal of Chemistry* 12, no. 3 (2019): 330–49.
138. Naseem, Tayyaba, and Muhammad Akhyar Farrukh. "Antibacterial activity of green synthesis of iron nanoparticles using *Lawsonia Inermis* and gardenia *Jasminoides* leaves extract." *Journal of Chemistry* 2015 (2015): 1–7.

139. Batool, Faryal, Muhammad Shahid Iqbal, Salah-Ud-Din Khan, Javed Khan, Bilal Ahmed, and Muhammad Imran Qadir. "Biologically synthesized iron nanoparticles (FeNPs) from *Phoenix Dactylifera* have anti-bacterial activities." *Scientific Reports* 11, no. 1 (2021): 1–9.

140. Yusefi, Mostafa, Kamyar Shameli, Ong Su Yee, Sin-Yeang Teow, Ziba Hedayatnasab, Hossein Jahangirian, Thomas J. Webster, and Kamil Kuča. "Green synthesis of Fe_3O_4 nanoparticles stabilized by a *Garcinia Mangostana* fruit peel extract for hyperthermia and anticancer activities." *International Journal of Nanomedicine* 16 (2021): 2515.

141. Bai, Bingke, Qinxue Hu, Hui Hu, Peng Zhou, Zhengli Shi, Jin Meng, Baojing Lu, et al. "Virus-like particles of SARS-like coronavirus formed by membrane proteins from different origins demonstrate stimulating activity in human dendritic cells." *PloS One* 3, no. 7 (2008): e2685.

142. Latham, Theresa, and Jose M. Galarza. "Formation of wild-type and chimeric influenza virus-like particles following simultaneous expression of only four structural proteins." *Journal of Virology* 75, no. 13 (2001): 6154–65.

143. Le, Dinh To, Marco T. Radukic, and Kristian M. Müller. "Adeno-associated virus capsid protein expression in *Escherichia Coli* and chemically defined capsid assembly." *Scientific Reports* 9, no. 1 (2019): 1–10.

144. Joe, Carina C. D., Sayantani Chatterjee, George Lovrecz, Timothy E. Adams, Morten Thaysen-Andersen, Renae Walsh, Stephen A. Locarnini, Peter Smooker, and Hans J. Netter. "Glycoengineered Hepatitis B virus-like particles with enhanced immunogenicity." *Vaccine* 38, no. 22 (2020): 3892–901.

145. Shiri, Farhad, Kevin E. Petersen, Valentin Romanov, Qin Zou, and Bruce K. Gale. "Characterization and differential retention of Q beta bacteriophage virus-like particles using cyclical electrical field–flow fractionation and asymmetrical flow field–flow fractionation." *Analytical and Bioanalytical Chemistry* 412, no. 7 (2020): 1563–72.

146. Mohsen, Mona O., Ariane C. Gomes, Monique Vogel, and Martin F. Bachmann. "Interaction of viral capsid-derived virus-like particles (VLPs) with the innate immune system." *Vaccines* 6, no. 3 (2018): 37.

147. Sailaja, Gangadhara, Ioanna Skountzou, Fu-Shi Quan, Richard W. Compans, and Sang-Moo Kang. "Human immunodeficiency virus-like particles activate multiple types of immune cells." *Virology* 362, no. 2 (2007): 331–41.

148. Zhai, Lukai, Rashi Yadav, Nitesh K. Kunda, Dana Anderson, Elizabeth Bruckner, Elliott K. Miller, Rupsa Basu, Pavan Muttil, and Ebenezer Tumban. "Oral immunization with bacteriophage MS2-L2 VLPs protects against oral and genital infection with multiple Hpv types associated with head & neck cancers and cervical cancer." *Antiviral Research* 166 (2019): 56–65.

149. Chung, Young Hun, Hui Cai, and Nicole F. Steinmetz. "Viral nanoparticles for drug delivery, imaging, immunotherapy, and theranostic applications." *Advanced Drug Delivery Reviews* 156 (2020): 214–35.

150. Ashley, Carlee E., Eric C. Carnes, Genevieve K. Phillips, Paul N. Durfee, Mekensey D. Buley, Christopher A. Lino, David P. Padilla, et al. "Cell-specific delivery of diverse cargos by bacteriophage MS2 virus-like particles." *ACS Nano* 5, no. 7 (2011): 5729–45.

151. Cao, Jing, Richard H. Guenther, Tim L. Sit, Charles H. Opperman, Steven A. Lommel, and Julie A. Willoughby. "Loading and release mechanism of red clover necrotic mosaic virus derived plant viral nanoparticles for drug delivery of doxorubicin." *Small* 10, no. 24 (2014): 5126–36.

152. Chen, Zhuo, Na Li, Luxi Chen, Jiyong Lee, and Jeremiah J. Gassensmith. "Dual functionalized bacteriophage Qβ as a photocaged drug carrier." *Small* 12, no. 33 (2016): 4563–71.

153. Bruckman, Michael A., Anna E. Czapar, Allen VanMeter, Lauren N. Randolph, and Nicole F. Steinmetz. "Tobacco mosaic virus-based protein nanoparticles and nanorods for chemotherapy delivery targeting breast cancer." *Journal of Controlled Release* 231 (2016): 103–13.

154. Le, Duc H. T., Karin L. Lee, Sourabh Shukla, Ulrich Commandeur, and Nicole F. Steinmetz. "Potato virus X, a filamentous plant viral nanoparticle for doxorubicin delivery in cancer therapy." *Nanoscale* 9, no. 6 (2017): 2348–57.

155. Yan, Dan, Zhidong Teng, Shiqi Sun, Shan Jiang, Hu Dong, Yuan Gao, Yanquan Wei, et al. "Foot-and-mouth disease virus-like particles as integrin-based drug delivery system achieve targeting anti-tumor efficacy." *Nanomedicine: Nanotechnology, Biology and Medicine* 13, no. 3 (2017): 1061–70.

156. Shan, Wenjun, Deliang Zhang, Yunlong Wu, Xiaolin Lv, Bin Hu, Xi Zhou, Shefang Ye, et al. "Modularized peptides modified HBC virus-like particles for encapsulation and tumor-targeted delivery of doxorubicin." *Nanomedicine: Nanotechnology, Biology and Medicine* 14, no. 3 (2018): 725–34.

157. Jariyapong, Pitchanee, Charoonroj Chotwiwatthanakun, Monsicha Somrit, Sarawut Jitrapakdee, Li Xing, Holland R. Cheng, and Wattana Weerachatyanukul. "Encapsulation and delivery of plasmid DNA by virus-like nanoparticles engineered from *Macrobrachium Rosenbergii* nodavirus." *Virus Research* 179 (2014): 140–46.

158. Zochowska, Monika, Agnieszka Paca, Guy Schoehn, Jean-Pierre Andrieu, Jadwiga Chroboczek, Bernard Dublet, and Ewa Szolajska. "Adenovirus dodecahedron, as a drug delivery vector." *PLoS One* 4, no. 5 (2009): e5569.

159. Franke, Christina E., Anna E. Czapar, Ravi B. Patel, and Nicole F. Steinmetz. "Tobacco mosaic virus-delivered cisplatin restores efficacy in platinum-resistant ovarian cancer cells." *Molecular Pharmaceutics* 15, no. 8 (2017): 2922–31.

160. Vernekar, Amit A., Gilles Berger, Anna E. Czapar, Frank A. Veliz, David I. Wang, Nicole F. Steinmetz, and Stephen J. Lippard. "Speciation of phenanthriplatin and its analogs in the core of tobacco mosaic virus." *Journal of the American Chemical Society* 140, no. 12 (2018): 4279–87.

161. Czapar, Anna E., Yao-Rong Zheng, Imogen A. Riddell, Sourabh Shukla, Samuel G. Awuah, Stephen J. Lippard, and Nicole F. Steinmetz. "Tobacco mosaic virus delivery of phenanthriplatin for cancer therapy." *ACS Nano* 10, no. 4 (2016): 4119–26.

162. Lin, Richard D. and Nicole F. Steinmetz. "Tobacco mosaic virus delivery of mitoxantrone for cancer therapy." *Nanoscale* 10, no. 34 (2018): 16307–13.

163. Lam, Patricia, Richard D. Lin, and Nicole F. Steinmetz. "Delivery of mitoxantrone using a plant virus-based nanoparticle for the treatment of glioblastomas." *Journal of Materials Chemistry B* 6, no. 37 (2018): 5888–95.

164. Kernan, Daniel L., Amy M. Wen, Andrzej S. Pitek, and Nicole F. Steinmetz. "Featured article: Delivery of chemotherapeutic vcMMAE using tobacco mosaic virus nanoparticles." *Experimental Biology and Medicine* 242, no. 14 (2017): 1405–11.

165. Crooke, Stephen N., Jiri Schimer, Idris Raji, Bocheng Wu, Adegboyega K. Oyelere, and M. G. Finn. "Lung tissue delivery of virus-like particles mediated by macrolide antibiotics." *Molecular Pharmaceutics* 16, no. 7 (2019): 2947–55.

166. Yacoby, Iftach, Marina Shamis, Hagit Bar, Doron Shabat, and Itai Benhar. "Targeting antibacterial agents by using drug-carrying filamentous bacteriophages." *Antimicrobial Agents and Chemotherapy* 50, no. 6 (2006): 2087–97.

167. White, Meghann A., Jeremiah A. Johnson, Jeffrey T. Koberstein, and Nicholas J. Turro. "Toward the syntheses of universal ligands for metal oxide surfaces: Controlling surface functionality through click chemistry." *Journal of the American Chemical Society* 128, no. 35 (2006): 11356–57.

168. Chen, Ling-Chun, Ying-Chen Chen, Chia-Yu Su, Chung-Shu Hong, Hsiu-O Ho, and Ming-Thau Sheu. "Development and characterization of self-assembling lecithin-based mixed polymeric micelles containing quercetin in cancer treatment and *In Vivo* pharmacokinetic study." *International Journal of Nanomedicine* 11 (2016): 1557.

169. Lee, Bor-Shiunn, Chien-Chen Lee, Yi-Ping Wang, Hsiao-Jan Chen, Chern-Hsiung Lai, Wan-Ling Hsieh, and Yi-Wen Chen. "Controlled-release of tetracycline and lovastatin by poly (D, L-Lactide-Co-Glycolide Acid)-chitosan nanoparticles enhances periodontal regeneration in dogs." *International Journal of Nanomedicine* 11 (2016): 285.

170. Obayemi, John D., Karen A. Malatesta, Olushola S. Odusanya, Danyuo Yiporo, Wei Yu, Kathryn E. Uhrich, and Winston O. Soboyejo. "Injectable, biodegradable micro-and nano-particles loaded with prodigiosin-based drug for localized anticancer drug delivery." *Cancer Epidemiology Biomarkers & Prevention* 25, no. 3 (2016): C60.

171. Van, Sang, Sanjib K. Das, Xinghe Wang, Zhongling Feng, Yi Jin, Zheng Hou, Fu Chen, et al. "Synthesis, characterization, and biological evaluation of poly (L-Γ-Glutamyl-Glutamine)-paclitaxel nanoconjugate." *International Journal of Nanomedicine* 5 (2010): 825.

172. Lei, Yifeng, Yoh Hamada, Jun Li, Liman Cong, Nuoxin Wang, Ying Li, Wenfu Zheng, and Xingyu Jiang. "Targeted tumor delivery and controlled release of neuronal drugs with ferritin nanoparticles to regulate pancreatic cancer progression." *Journal of Controlled Release* 232 (2016): 131–42.

173. Manuja, Anju, Balvinder Kumar, Meenu Chopra, Anshu Bajaj, Rajender Kumar, Neeraj Dilbaghi, Sandeep Kumar, et al. "Cytotoxicity and genotoxicity of a trypanocidal drug quinapyramine sulfate loaded-sodium alginate nanoparticles in mammalian cells." *International Journal of Biological Macromolecules* 88 (2016): 146–55.

174. Li, Qing, Yong Wen, Jie Wen, Yun-Peng Zhang, Xiao-Ding Xu, Amanda Victorious, Ryan Zavitz, and Xin Xu. "A new biosafe reactive oxygen species (ROS)-responsive nanoplatform for drug delivery." *RSC Advances* 6, no. 45 (2016): 38984–89.

175. Sumer Bolu, Burcu, Ece Manavoglu Gecici, and Rana Sanyal. "Combretastatin A-4 conjugated antiangiogenic micellar drug delivery systems using dendron–polymer conjugates." *Molecular Pharmaceutics* 13, no. 5 (2016): 1482–90.

176. Zhang, Yumin, Junhui Zhou, Cuihong Yang, Weiwei Wang, Liping Chu, Fan Huang, Qiang Liu, et al. "Folic acid-targeted disulfide-based cross-linking micelle for enhanced drug encapsulation stability and site-specific drug delivery against tumors [Corrigendum]." *International Journal of Nanomedicine* 16 (2021): 7683–84.

177. Bai, Renu Geetha, Kasturi Muthoosamy, Fiona Natalia Shipton, Alagarsamy Pandikumar, Perumal Rameshkumar, Nay Ming Huang, and Sivakumar Manickam. "The biogenic synthesis of a reduced graphene oxide–silver (RGO–Ag) nanocomposite and its dual applications as an antibacterial agent and cancer biomarker sensor." *RSC Advances* 6, no. 43 (2016): 36576–87.

178. Li, Zhenshun, Wei Xu, Yuntao Wang, Bakht Ramin Shah, Chunlan Zhang, Yijie Chen, Yan Li, and Bin Li. "Quantum dots loaded nanogels for low cytotoxicity, pH-sensitive fluorescence, cell imaging and drug delivery." *Carbohydrate Polymers* 121 (2015): 477–85.

179. Feng, Tao, Xiangzhao Ai, Guanghui An, Piaoping Yang, and Yanli Zhao. "Correction to charge-convertible carbon dots for imaging-guided drug delivery with enhanced *In Vivo* cancer therapeutic efficiency." *ACS Nano* 10, no. 5 (2016): 5587–87.

180. Malathi, S., M. D. Balakumaran, P. T. Kalaichelvan, and S. Balasubramanian. "Green synthesis of gold nanoparticles for controlled delivery." *Advanced Material Letter* 4, no. 12 (2013): 933–40.

181. Dai, Lin, Kefeng Liu, Chuanling Si, Luying Wang, Jing Liu, Jing He, and Jiandu Lei. "Ginsenoside nanoparticle: A new green drug delivery system." *Journal of Materials Chemistry B* 4, no. 3 (2016): 529–38.

182. Vimala, Karuppaiya, Shenbagamoorthy Sundarraj, Manickam Paulpandi, Srinivasan Vengatesan, and Soundarapandian Kannan. "Green synthesized doxorubicin loaded zinc oxide nanoparticles regulates the bax and Bcl-2 expression in breast and colon carcinoma." *Process Biochemistry* 49, no. 1 (2014): 160–72.

183. Fan, Xiujuan, Guozheng Jiao, Lei Gao, Pengfei Jin, and Xin Li. "The preparation and drug delivery of a graphene–carbon nanotube–Fe_3O_4 nanoparticle hybrid." *Journal of Materials Chemistry B* 1, no. 20 (2013): 2658–64.

184. Jeyamohan, Prashanti, Takashi Hasumura, Yutaka Nagaoka, Yasuhiko Yoshida, Toru Maekawa, and D. Sakthi Kumar. "Accelerated killing of cancer cells using a multifunctional single-walled carbon nanotube-based system for targeted drug delivery in combination with photothermal therapy." *International Journal of Nanomedicine* 8 (2013): 2653.

185. How, Chee Wun, Abdullah Rasedee, Sivakumar Manickam, and Rozita Rosli. "Tamoxifen-loaded nanostructured lipid carrier as a drug delivery system: Characterization, stability assessment and cytotoxicity." *Colloids and Surfaces B: Biointerfaces* 112 (2013): 393–99.

186. Erdal, M. Sedef, Gül Özhan, M. Cem Mat, Yıldız Özsoy, and Sevgi Güngör. "Colloidal nanocarriers for the enhanced cutaneous delivery of naftifine: Characterization studies and *In Vitro* and *In Vivo* evaluations." *International Journal of Nanomedicine* 11 (2016): 1027.

187. Ilbasmis-Tamer, Sibel, Hande Unsal, Fatmanur Tugcu-Demiroz, Gokce Dicle Kalaycioglu, Ismail Tuncer Degim, and Nihal Aydogan. "Stimuli-responsive lipid nanotubes in gel formulations for the delivery of doxorubicin." *Colloids and Surfaces B: Biointerfaces* 143 (2016): 406–14.

188. Nguyen, Thi Tram Chau, Cuu Khoa Nguyen, Thi Hiep Nguyen, and Ngoc Quyen Tran. "Highly lipophilic pluronics-conjugated polyamidoamine dendrimer nanocarriers as potential delivery system for hydrophobic drugs." *Materials Science and Engineering: C* 70 (2017): 992–99.

189. Rauta, Pradipta Ranjan, Niladri Mohan Das, Debasis Nayak, Sarbani Ashe, and Bismita Nayak. "Enhanced efficacy of clindamycin hydrochloride encapsulated in PLA/PLGA based nanoparticle system for oral delivery." *IET Nanobiotechnology* 10, no. 4 (2016): 254–61.

190. Wang, Ruoning, Xiaochen Gu, Jianping Zhou, Lingjia Shen, Lifang Yin, Peiying Hua, and Yang Ding. "Green design "Bioinspired Disassembly-Reassembly Strategy" applied for improved tumor-targeted anticancer drug delivery." *Journal of Controlled Release* 235 (2016): 134–46.

191. Xie, Xuan, Shiying Luo, Jean Felix Mukerabigwi, Jian Mei, Yuannian Zhang, Shufang Wang, Wang Xiao, Xueying Huang, and Yu Cao. "Targeted nanoparticles from xyloglucan–doxorubicin conjugate loaded with doxorubicin against drug resistance." *RSC Advances* 6, no. 31 (2016): 26137–46.

192. Tang, Siah Ying, Manickam Sivakumar, Angela Min-Hwei Ng, and Parthasarathy Shridharan. "Anti-inflammatory and analgesic activity of novel oral aspirin-loaded nanoemulsion and nano multiple emulsion formulations generated using ultrasound cavitation." *International Journal of Pharmaceutics* 430, no. 1–2 (2012): 299–306.

193. Chen, Yu-Hung, Chiau-Yuang Tsai, Pon-Yu Huang, Meng-Ya Chang, Pai-Chiao Cheng, Chen-Hsi Chou, Dong-Hwang Chen, et al. "Methotrexate conjugated to gold nanoparticles inhibits tumor growth in a syngeneic lung tumor model." *Molecular Pharmaceutics* 4, no. 5 (2007): 713–22.

194. Li, Jingyuan, Xuemei Wang, Chunxia Wang, Baoan Chen, Yongyuan Dai, Renyun Zhang, Min Song, Gang Lv, and Degang Fu. "The enhancement effect of gold nanoparticles in drug delivery and as biomarkers of drug-resistant cancer cells." *ChemMedChem: Chemistry Enabling Drug Discovery* 2, no. 3 (2007): 374–78.

195. Thambiraj, S., S. Hema, and D. Ravi Shankaran. "Functionalized gold nanoparticles for drug delivery applications." *Materials Today: Proceedings* 5, no. 8 (2018): 16763–73.

196. Sivaraj, Mehnath, Arjama Mukherjee, Rajan Mariappan, Arokia Vijayaanand Mariadoss, and Murugaraj Jeyaraj. "Polyorganophosphazene stabilized gold nanoparticles for intracellular drug delivery in breast carcinoma cells." *Process Biochemistry* 72 (2018): 152–61.

197. Lee, Hyun Uk, So Young Park, Eun Sik Park, Byoungchul Son, Soon Chang Lee, Jae Won Lee, Young-Chul Lee, et al. "Photoluminescent carbon nanotags from harmful cyanobacteria for drug delivery and imaging in cancer cells." *Scientific Reports* 4, no. 1 (2014): 1–7.

198. Olerile, Livesey David, Yongjun Liu, Shengjun Mu, Jing Zhang, Lesego Selotlegeng, and Na Zhang. "Near-infrared mediated quantum dots and paclitaxel co-loaded nanostructured lipid carriers for cancer theragnostic." *Colloids and Surfaces B: Biointerfaces* 150 (2017): 121–30.

199. Khodadadei, Fatemeh, Shahrokh Safarian, and Narges Ghanbari. "Methotrexate-loaded nitrogen-doped graphene quantum dots nanocarriers as an efficient anticancer drug delivery system." *Materials Science and Engineering: C* 79 (2017): 280–85.

200. Fakhri, Ali, Shiva Tahami, and Pedram Afshar Nejad. "Preparation and characterization of Fe_3O_4-Ag_2O quantum dots decorated cellulose nanofibers as a carrier of anticancer drugs for skin cancer." *Journal of Photochemistry and Photobiology B: Biology* 175 (2017): 83–88.

201. Thakur, Mukeshchand, Ashmi Mewada, Sunil Pandey, Mustansir Bhori, Kanchanlata Singh, Maheshwar Sharon, and Madhuri Sharon. "Milk-derived multi-fluorescent graphene quantum dot-based cancer theranostic system." *Materials Science and Engineering: C* 67 (2016): 468–77.

202. Esmaeili, Akbar, and Sepideh Ghobadianpour. "Vancomycin loaded superparamagnetic $MnFe_2O_4$ nanoparticles coated with pegylated chitosan to enhance antibacterial activity." *International Journal of Pharmaceutics* 501, no. 1–2 (2016): 326–30.

203. Aryal, Santosh, Che-Ming Jack Hu, and Liangfang Zhang. "Polymer-cisplatin conjugate nanoparticles for acid-responsive drug delivery." *ACS Nano* 4, no. 1 (2010): 251–58.

204. Roque, Luís, Pedro Castro, Jesús Molpeceres, Ana S. Viana, Amílcar Roberto, Cláudia Reis, Patrícia Rijo, et al. "Bioadhesive polymeric nanoparticles as strategy to improve the treatment of yeast infections in oral cavity: *In-Vitro* and *Ex-Vivo* studies." *European Polymer Journal* 104 (2018): 19–31.

205. Niwa, T., H. Takeuchi, T. Hino, N. Kunou, and Y. Kawashima. "Preparations of biodegradable nanospheres of water-soluble and insoluble drugs with D, L-Lactide/Glycolide copolymer by a novel spontaneous emulsification solvent diffusion method, and the drug release behavior." *Journal of Controlled Release* 25, no. 1–2 (1993): 89–98.

206. Sun, Weitong, Xiaoyan Chen, Chao Xie, Yong Wang, Liteng Lin, Kangshun Zhu, and Xintao Shuai. "Co-delivery of doxorubicin and anti-Bcl-2 siRNA by pH-responsive polymeric vector to overcome drug resistance *In Vitro* and *In Vivo* HEPg2 hepatoma model." *Biomacromolecules* 19, no. 6 (2018): 2248–56.

207. Szewczyk, Adrian, and Magdalena Prokopowicz. "Amino-modified mesoporous silica SBA-15 as bifunctional drug delivery system for cefazolin: Release profile and mineralization potential." *Materials Letters* 227 (2018): 136–40.

208. Wang, Fei, Xiaopan Cai, Yunzhang Su, Jingjing Hu, Qinglin Wu, Hongfeng Zhang, Jianru Xiao, and Yiyun Cheng. "Reducing cytotoxicity while improving anti-cancer drug loading capacity of polypropylenimine dendrimers by surface acetylation." *Acta Biomaterialia* 8, no. 12 (2012): 4304–13.

209. Yang, Hu, and Stephanie T. Lopina. "Penicillin V-conjugated PEG-PAMAM star polymers." *Journal of Biomaterials Science, Polymer Edition* 14, no. 10 (2003): 1043–56.

210. Du, Xiao, Shaoping Yin, Yang Wang, Xiaochen Gu, Guangji Wang, and Juan Li. "Hyaluronic acid-functionalized half-generation of sectorial dendrimers for anticancer drug delivery and enhanced biocompatibility." *Carbohydrate Polymers* 202 (2018): 513–22.

211. Baek, Jong-Suep, Jae-Woo So, Sang-Chul Shin, and Cheong-Weon Cho. "Solid lipid nanoparticles of paclitaxel strengthened by hydroxypropyl-B-cyclodextrin as an oral delivery system." *International Journal of Molecular Medicine* 30, no. 4 (2012): 953–59.

212. Pooja, Deep, Hitesh Kulhari, Madhusudana Kuncha, Shyam S. Rachamalla, David J. Adams, Vipul Bansal, and Ramakrishna Sistla. "Improving efficacy, oral bioavailability, and delivery of paclitaxel using protein-grafted solid lipid nanoparticles." *Molecular Pharmaceutics* 13, no. 11 (2016): 3903–12.

213. Rosiere, Remi, Matthias Van Woensel, Michel Gelbcke, Veronique Mathieu, Julien Hecq, Thomas Mathivet, Marjorie Vermeersch, et al. "New folate-grafted chitosan derivative to improve delivery of paclitaxel-loaded solid lipid nanoparticles for lung tumor therapy by inhalation." *Molecular Pharmaceutics* 15, no. 3 (2018): 899–910.

214. Kanwar, Rohini, Michael Gradzielski, and S. K. Mehta. "Biomimetic solid lipid nanoparticles of sophorolipids designed for antileprosy drugs." *The Journal of Physical Chemistry B* 122, no. 26 (2018): 6837–45.

215. Beloqui, Ana, María Ángeles Solinís, Anne des Rieux, Véronique Préat, and Alicia Rodríguez-Gascón. "Dextran–protamine coated nanostructured lipid carriers as mucus-penetrating nanoparticles for lipophilic drugs." *International Journal of Pharmaceutics* 468, no. 1–2 (2014): 105–11.

216. Lakhani, Prit, Akash Patil, Pranjal Taskar, Eman Ashour, and Soumyajit Majumdar. "Curcumin-loaded nanostructured lipid carriers for ocular drug delivery: Design optimization and characterization." *Journal of Drug Delivery Science and Technology* 47 (2018): 159–66.

217. Rocha, Kamilla Amaral David, Anna Paula Krawczyk-Santos, Lígia Marquez Andrade, Luana Clara de Souza, Ricardo Neves Marreto, Tais Gratieri, and Stephânia Fleury Taveira. "Voriconazole-loaded nanostructured lipid carriers (NLC) for drug delivery in deeper regions of the nail plate." *International Journal of Pharmaceutics* 531, no. 1 (2017): 292–98.

218. Kelidari, Hamid Reza, Maryam Moazeni, Roghayeh Babaei, Majid Saeedi, Jafar Akbari, Parisa Islami Parkoohi, Mojtaba Nabili, et al. "Improved yeast delivery of fluconazole with a nanostructured lipid carrier system." *Biomedicine & Pharmacotherapy* 89 (2017): 83–8.

219. De Jong, Wim H., and Paul J. A. Borm. "Drug delivery and nanoparticles: Applications and hazards." *International Journal of Nanomedicine* 3, no. 2 (2008): 133.

220. Magee, Donal Francis, and A. F. Dalley II. *Digestion and the Structure and Function of the Gut.* Karger Publishers, Berlin, 1986.

221. Nokhodchi, Ali, Shaista Raja, Pryia Patel, and Kofi Asare-Addo. "The role of oral controlled release matrix tablets in drug delivery systems." *BioImpacts: BI* 2, no. 4 (2012): 175.

222. Ramesan, Rekha M., and Chandra P. Sharma. "Challenges and advances in nanoparticle-based oral insulin delivery." *Expert Review of Medical Devices* 6, no. 6 (2009): 665–76.

223. Patton, Ohn S., C. Simone Fishburn, and Jeffry G. Weers. "The lungs as a portal of entry for systemic drug delivery." *Proceedings of the American Thoracic Society* 1, no. 4 (2004): 338–44.

224. Paranjpe, Mukta, and Christel C. Müller-Goymann. "Nanoparticle-mediated pulmonary drug delivery: A review." *International Journal of Molecular Sciences* 15, no. 4 (2014): 5852–73.

225. Patton, John S., and Peter R. Byron. "Inhaling medicines: Delivering drugs to the body through the lungs." *Nature reviews Drug Discovery* 6, no. 1 (2007): 67–74.

226. Mangal, Sharad, Wei Gao, Tonglei Li, and Qi Tony Zhou. "Pulmonary delivery of nanoparticle chemotherapy for the treatment of lung cancers: Challenges and opportunities." *Acta Pharmacologica Sinica* 38, no. 6 (2017): 782–97.

227. Xu, Caina, Yanbing Wang, Zhaopei Guo, Jie Chen, Lin Lin, Jiayan Wu, Huayu Tian, and Xuesi Chen. "Pulmonary delivery by exploiting doxorubicin and cisplatin co-loaded nanoparticles for metastatic lung cancer therapy." *Journal of Controlled Release* 295 (2019): 153–63.

228. Park, Chun Woong, Heidi M. Mansour, and Don Hayes. "Pulmonary inhalation aerosols for targeted antibiotics drug delivery." *European Pharmaceutical Review* 16 (2011): 32–36.

229. LiCalsi, Cynthia, Troy Christensen, John V. Bennett, Elaine Phillips, and Clyde Witham. "Dry powder inhalation as a potential delivery method for vaccines." *Vaccine* 17, no. 13–14 (1999): 1796–803.

230. Lu, Dongmei, and Anthony J. Hickey. "Pulmonary vaccine delivery." *Expert Review of Vaccines* 6, no. 2 (2007): 213–26.

231. Lam, P. L., and R. Gambari. "Advanced progress of microencapsulation technologies: *In Vivo* and *In Vitro* models for studying oral and transdermal drug deliveries." *Journal of Controlled Release* 178 (2014): 25–45.

232. Goyal, Ritu, Lauren K. Macri, Hilton M. Kaplan, and Joachim Kohn. "Nanoparticles and nanofibers for topical drug delivery." *Journal of Controlled Release* 240 (2016): 77–92.

233. Vaghasiya, Harshad, Abhinesh Kumar, and Krutika Sawant. "Development of solid lipid nanoparticles based controlled release system for topical delivery of terbinafine hydrochloride." *European Journal of Pharmaceutical Sciences* 49, no. 2 (2013): 311–22.

234. Liu, Jie, Wen Hu, Huabing Chen, Qian Ni, Huibi Xu, and Xiangliang Yang. "Isotretinoin-loaded solid lipid nanoparticles with skin targeting for topical delivery." *International Journal of Pharmaceutics* 328, no. 2 (2007): 191–95.

235. Nikam, K. R., M. G. Pawar, S. P. Jadhav, and V. A. Bairagi. "Novel trends in parenteral drug delivery system." *International Journal of Pharmacy and Technology* 5, no. 2 (2013): 2549–77.

236. Zavaleta, Cristina, Dean Ho, and Eun Ji Chung. "Theranostic nanoparticles for tracking and monitoring disease state." *SLAS Technology: Translating Life Sciences Innovation* 23, no. 3 (2018): 281–93.

237. Itani, Rasha, Mansour Tobaiqy, and Achraf Al Faraj. "Optimizing use of theranostic nanoparticles as a life-saving strategy for treating COVID-19 patients." *Theranostics* 10, no. 13 (2020): 5932.

238. Madamsetty, Vijay Sagar, Anubhab Mukherjee, and Sudip Mukherjee. "Recent trends of the bio-inspired nanoparticles in cancer theranostics." *Frontiers in Pharmacology* 10 (2019): 1264.

239. Zhang, Qize, Stephen O'Brien, and Jan Grimm. "Biomedical applications of lanthanide nanomaterials, for imaging, sensing and therapy." *Nanotheranostics* 6, no. 2 (2022): 184.

240. Zhao, Zhiwei, Rong Yan, Xuan Yi, Jingling Li, Jiaming Rao, Zhengqing Guo, Yanmei Yang, et al. "Bacteria-activated theranostic nanoprobes against methicillin-resistant *Staphylococcus Aureus* infection." *ACS Nano* 11, no. 5 (2017): 4428–38.

241. Zhang, Lu, Xiaoyuan Ji, Yuanyuan Su, Xia Zhai, Hua Xu, Bin Song, Airui Jiang, Daoxia Guo, and Yao He. "Fluorescent silicon nanoparticles-based nanotheranostic agents for rapid diagnosis and treatment of bacteria-induced keratitis." *Nano Research* 14, no. 1 (2021): 52–58.

242. Anwar, Ayaz, Ruqaiyyah Siddiqui, and Naveed A. Khan. "Importance of theranostics in rare brain-eating amoebae infections." *ACS Chemical Neuroscience* 10, no. 1 (2018): 6–12.

243. Moitra, Parikshit, Maha Alafeef, Ketan Dighe, Zach Sheffield, Dipendra Dahal, and Dipanjan Pan. "Synthesis and characterisation of N-gene targeted NIR-II fluorescent probe for selective localisation of SARS-COV-2." *Chemical Communications* 57, no. 51 (2021): 6229–32.

244. Delany, Isabel, Rino Rappuoli, and Ennio De Gregorio. "Vaccines for the 21st century." *EMBO Molecular Medicine* 6, no. 6 (2014): 708–20.

245. Ahmad, Mohammad Zaki, Javed Ahmad, Anzarul Haque, Mohammed Yahia Alasmary, Basel A. Abdel-Wahab, and Sohail Akhter. "Emerging advances in synthetic cancer nano-vaccines: Opportunities and challenges." *Expert Review of Vaccines* 19, no. 11 (2020): 1053–71.

246. Chaudhary, Jitendra Kumar, Rohitash Yadav, Pankaj Kumar Chaudhary, Anurag Maurya, Nimita Kant, Osamah AI Rugaie, Hoineiting Rebecca Haokip, et al. "Insights into COVID-19 vaccine development based on immunogenic structural proteins of SARS-COV-2, host immune responses, and herd immunity." *Cells* 10, no. 11 (2021): 2949.

247. Kubackova, Jana, Jarmila Zbytovska, and Ondrej Holas. "Nanomaterials for direct and indirect immunomodulation: A review of applications." *European Journal of Pharmaceutical Sciences* 142 (2020): 105139.

248. Yenkoidiok-Douti, Lampouguin, and Christopher M. Jewell. "Integrating biomaterials and immunology to improve vaccines against infectious diseases." *ACS Biomaterials Science & Engineering* 6, no. 2 (2020): 759–78.

249. Iwasaki, Akiko, and Saad B. Omer. "Why and how vaccines work." *Cell* 183, no. 2 (2020): 290–95.

250. Campos, Estefânia V. R., Anderson E. S. Pereira, Jhones Luiz De Oliveira, Lucas Bragança Carvalho, Mariana Guilger-Casagrande, Renata De Lima, and Leonardo Fernandes Fraceto. "How can nanotechnology help to combat COVID-19? Opportunities and urgent need." *Journal of Nanobiotechnology* 18, no. 1 (2020): 1–23.

251. Kheirollahpour, Mehdi, Mohsen Mehrabi, Naser M. Dounighi, Mohsen Mohammadi, and Alireza Masoudi. "Nanoparticles and vaccine development." *Pharmaceutical Nanotechnology* 8, no. 1 (2020): 6–21.

252. Mufamadi, Maluta Steven. "Nanotechnology shows promise for next-generation vaccines in the fight against COVID-19." *MRS Bulletin* 45 (2020): 981–2.

253. Krishnamachari, Yogita, Sean M. Geary, Caitlin D. Lemke, and Aliasger K. Salem. "Nanoparticle delivery systems in cancer vaccines." *Pharmaceutical Research* 28, no. 2 (2011): 215–36.

254. Schuster, Benjamin S., Jung Soo Suk, Graeme F. Woodworth, and Justin Hanes. "Nanoparticle diffusion in respiratory mucus from humans without lung disease." *Biomaterials* 34, no. 13 (2013): 3439–46.

255. Malabadi, Ravindra B., Neelambika T. Meti, and Raju K. Chalannavar. "Applications of nanotechnology in vaccine development for coronavirus (SARS-COV-2) disease (COVID-19)." *International Journal of Research and Scientific Innovations* 8, no. 2 (2021): 191–98.

256. Pati, Rashmirekha, Maxim Shevtsov, and Avinash Sonawane. "Nanoparticle vaccines against infectious diseases." *Frontiers in Immunology* 9 (2018): 2224.

257. Han, Jinyu, Dandan Zhao, Dan Li, Xiaohua Wang, Zheng Jin, and Kai Zhao. "Polymer-based nanomaterials and applications for vaccines and drugs." *Polymers* 10, no. 1 (2018): 31.

258. Bolhassani, Azam, Shabnam Javanzad, Tayebeh Saleh, Mehrdad Hashemi, Mohammad Reza Aghasadeghi, and Seyed Mehdi Sadat. "Polymeric nanoparticles: Potent vectors for vaccine delivery targeting cancer and infectious diseases." *Human Vaccines & Immunotherapeutics* 10, no. 2 (2014): 321–32.

259. Kim, Dongyoon, Yina Wu, Young Bong Kim, and Yu-Kyoung Oh. "Advances in vaccine delivery systems against viral infectious diseases." *Drug Delivery and Translational Research* 11, no. 4 (2021): 1401–19.

260. Seyfoori, Amir, Mahdieh Shokrollahi Barough, Pooneh Mokarram, Mazaher Ahmadi, Parvaneh Mehrbod, Alireza Sheidary, Tayyebeh Madrakian, et al. "Emerging advances of nanotechnology in drug and vaccine delivery against viral associated respiratory infectious diseases (VARID)." *International Journal of Molecular Sciences* 22, no. 13 (2021): 6937.

261. Poon, Christopher, and Amish A. Patel. "Organic and inorganic nanoparticle vaccines for prevention of infectious diseases." *Nano Express* 1, no. 1 (2020): 012001.

262. Ruiz-Hitzky, Eduardo, Margarita Darder, Bernd Wicklein, Cristina Ruiz-Garcia, Raquel Martín-Sampedro, Gustavo Del Real, and Pilar Aranda. "Nanotechnology responses to COVID-19." *Advanced Healthcare Materials* 9, no. 19 (2020): 2000979.

263. Verbeke, Rein, Ine Lentacker, Stefaan C. De Smedt, and Heleen Dewitte. "The dawn of mRNA vaccines: The COVID-19 case." *Journal of Controlled Release* 333 (2021): 511–20.

264. Thi, Thai Thanh Hoang, Estelle J. A. Suys, Jung Seok Lee, Dai Hai Nguyen, Ki Dong Park, and Nghia P. Truong. "Lipid-based nanoparticles in the clinic and clinical trials: From cancer nanomedicine to COVID-19 vaccines." *Vaccines* 9, no. 4 (2021): 359.

265. Nooraei, Saghi, Howra Bahrulolum, Zakieh Sadat Hoseini, Camellia Katalani, Abbas Hajizade, Andrew J. Easton, and Gholamreza Ahmadian. "Virus-like particles: Preparation, immunogenicity and their roles as nanovaccines and drug nanocarriers." *Journal of Nanobiotechnology* 19, no. 1 (2021): 1–27.

266. Al-Halifa, Soultan, Laurie Gauthier, Dominic Arpin, Steve Bourgault, and Denis Archambault. "Nanoparticle-based vaccines against respiratory viruses." *Frontiers in Immunology* 10 (2019): 22.

267. Khurana, Amit, Prince Allawadhi, Isha Khurana, Sachin Allwadhi, Ralf Weiskirchen, Anil Kumar Banothu, Deepak Chhabra, Kamaldeep Joshi, and Kala Kumar Bharani. "Role of nanotechnology behind the success of mRNA vaccines for COVID-19." *Nano Today* 38 (2021): 101142.

268. Tariq, Hasnat, Sannia Batool, Saaim Asif, Mohammad Ali, and Bilal Haider Abbasi. "Virus-like particles: Revolutionary platforms for developing vaccines against emerging infectious diseases." *Frontiers in Microbiology* 12 (2021): 4137.

269. Wang, Joshua W., and Richard B. S. Roden. "Virus-like particles for the prevention of human papillomavirus-associated malignancies." *Expert Review of Vaccines* 12, no. 2 (2013): 129–41.

270. Monie, Archana, Chien-Fu Hung, Richard Roden, and T. Cervarix Wu. "Cervarix™: A vaccine for the prevention of HPV 16, 18-associated cervical cancer." *Biologics: Targets & Therapy* 2, no. 1 (2008): 107.

271. Reed, Steven G., Sylvie Bertholet, Rhea N. Coler, and Martin Friede. "New horizons in adjuvants for vaccine development." *Trends in Immunology* 30, no. 1 (2009): 23–32.

272. Aguilar, J. C. and E. G. Rodriguez. "Vaccine adjuvants revisited." *Vaccine* 25, no. 19 (2007): 3752–62.

273. Reed, Steven G., Mark T. Orr, and Christopher B. Fox. "Key roles of adjuvants in modern vaccines." *Nature Medicine* 19, no. 12 (2013): 1597–608.

274. Singh, Manmohan, and Derek O'Hagan. "Advances in vaccine adjuvants." *Nature Biotechnology* 17, no. 11 (1999): 1075–81.

275. Petkar, Kailash C., Suyash M. Patil, Sandip S. Chavhan, Kan Kaneko, Krutika K. Sawant, Nitesh K. Kunda, and Imran Y. Saleem. "An overview of nanocarrier-based adjuvants for vaccine delivery." *Pharmaceutics* 13, no. 4 (2021): 455.

276. Zhu, Motao, Rongfu Wang, and Guangjun Nie. "Applications of nanomaterials as vaccine adjuvants." *Human Vaccines & Immunotherapeutics* 10, no. 9 (2014): 2761–74.

277. Sun, Bingbing, and Tian Xia. "Nanomaterial-based vaccine adjuvants." *Journal of Materials Chemistry B* 4, no. 33 (2016): 5496–509.

278. Shafaati, Maryam, Massoud Saidijam, Meysam Soleimani, Fereshte Hazrati, Rasoul Mirzaei, Bagher Amirheidari, Hamid Tanzadehpanah, et al. "A brief review on DNA vaccines in the era of COVID-19." *Future Virology* 17, no. 1 (2022): 49–66.

279. Moon, James J., Heikyung Suh, Mark E. Polhemus, Christian F. Ockenhouse, Anjali Yadava, and Darrell J. Irvine. "Antigen-displaying lipid-enveloped PLGA nanoparticles as delivery agents for a *Plasmodium Vivax* malaria vaccine." *PloS One* 7, no. 2 (2012): e31472.

280. Zhao, Kai, Xingming Shi, Yan Zhao, Haixia Wei, Qingshen Sun, Tingting Huang, Xiaoyan Zhang, and Yunfeng Wang. "Preparation and immunological effectiveness of a swine influenza DNA vaccine encapsulated in chitosan nanoparticles." *Vaccine* 29, no. 47 (2011): 8549–56.

281. Li, Xiaomin, Ronge Xing, Chaojie Xu, Song Liu, Yukun Qin, Kecheng Li, Huahua Yu, and Pengcheng Li. "Immunostimulatory effect of chitosan and quaternary chitosan: A review of potential vaccine adjuvants." *Carbohydrate Polymers* 264 (2021): 118050.

282. Reimer, Jenny M., Karin H. Karlsson, Karin Lövgren-Bengtsson, Sofia E. Magnusson, Alexis Fuentes, and Linda Stertman. "Matrix-M™ adjuvant induces local recruitment, activation and maturation of central immune cells in absence of antigen." *PloS One* 7, no. 7 (2012): e41451.

283. Ye, Tingting, Zifu Zhong, Adolfo García-Sastre, Michael Schotsaert, and Bruno G. De Geest. "Current status of COVID-19 (Pre) clinical vaccine development." *Angewandte Chemie International Edition* 59, no. 43 (2020): 18885–97.

284. Niikura, Kenichi, Tatsuya Matsunaga, Tadaki Suzuki, Shintaro Kobayashi, Hiroki Yamaguchi, Yasuko Orba, Akira Kawaguchi, et al. "Gold nanoparticles as a vaccine platform: Influence of size and shape on immunological responses *In Vitro* and *In Vivo*." *ACS Nano* 7, no. 5 (2013): 3926–38.

285. Mao, Lichun, Ziwei Chen, Yaling Wang, and Chunying Chen. "Design and application of nanoparticles as vaccine adjuvants against human corona virus infection." *Journal of Inorganic Biochemistry* 219 (2021): 111454.

286. Desai, Neil. "Challenges in development of nanoparticle-based therapeutics." *The AAPS Journal* 14, no. 2 (2012): 282–95.

287. Lu, Xuefei, Tao Zhu, Chunying Chen, and Ying Liu. "Right or left: The role of nanoparticles in pulmonary diseases." *International Journal of Molecular Sciences* 15, no. 10 (2014): 17577–600.

288. Hua, Susan, Maria B. C. de Matos, Josbert M. Metselaar, and Gert Storm. "Current trends and challenges in the clinical translation of nanoparticulate nanomedicines: Pathways for translational development and commercialization." *Frontiers in Pharmacology* 9 (2018): 790.

289. Naseri Neda, Hadi Valizadeh, and Parvin Zakeri. "Solid lipid nanoparticles and nanostructured lipid carriers: Structure, preparation and application." *Advanced Pharmaceutical Bulletin* 5, no. 3 (2015): 305–13.

290. Gottardo, Stefania, Agnieszka Mech, Jana Drbohlavová, Aleksandra Małyska, Søren Bøwadt, Juan Riego Sintes, and Hubert Rauscher. "Towards safe and sustainable innovation in nanotechnology: State-of-play for smart nanomaterials." *NanoImpact* 21 (2021): 100297.

10 The Era of Green Nanomaterials for Sensing

Shashwati Wankar, Swapnali Walake,
Rutuja Gumathannavar, Nidhi Sapre, and Atul Kulkarni
Symbiosis International University

CONTENTS

10.1 INTRODUCTION

10.1.1 IMPORTANCE AND NEED FOR GREEN NANOMATERIALS

To acknowledge modern science as a rational story, it is necessary to concede the achievements of the ancient Hindus. Ancient Hindu science has been tested for dependability, as well as remarkable objectivity, proper citations, and a substantial bibliography. It emphasizes this civilization's achievements through careful research of historical and scientific sources. With their holistic approach, they have already developed a world-acknowledged medical system, now known as Ayurveda, "an original source for medical discipline for human civilization". In addition, ancient Hindu science mastered metallurgical methods of metal extraction and purification, as well as the science of self-reformation, now known as Yoga [1].

Although Norio Taniguchi devised the term "nanotechnology" in 1974, the origins of nanotechnology can be traced back to the prehistoric Indian medical system. Ancient people had the knowledge regarding the beneficial properties of nanomaterials and synthesis processes long before the term "nanoera" came into use. Ancient Ayurvedic systems, such as Rasa-Shastra (Vedic chemistry), approach herbo-mineral/metal/nonmetal mixtures and Bhasmas and Vibhutis (metals, nonmetals, and herbals, as well as horns, shells, and feathers) [2]. Along with Rasa-Shastra, Rasayana-Shastra (immunomodulation and anti-aging properties), while yogavahi is an abbreviation for the ability to target drugs to the specific site, Rasibhava (absorbable, adaptive, assimilable, and nontoxic), Shigravyapi (spreads swiftly and is harmless, and additionally acts as a catalyst), and

DOI: 10.1201/9781003319153-12

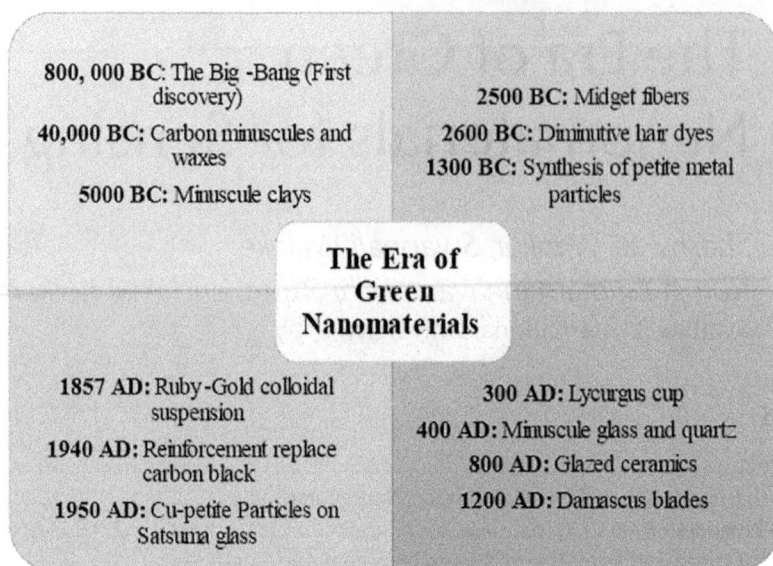

800, 000 BC: The Big -Bang (First discovery)

40,000 BC: Carbon minuscules and waxes

5000 BC: Minuscule clays

2500 BC: Midget fibers

2600 BC: Diminutive hair dyes

1300 BC: Synthesis of petite metal particles

The Era of Green Nanomaterials

1857 AD: Ruby-Gold colloidal suspension

1940 AD: Reinforcement replace carbon black

1950 AD: Cu-petite Particles on Satsuma glass

300 AD: Lycurgus cup

400 AD: Minuscule glass and quartz

800 AD: Glazed ceramics

1200 AD: Damascus blades

FIGURE 10.1   The evolution of nanomaterials over different eras.

Deepana Agni (Agni Deepana) (increases metabolism at the cellular level and acts as a catalyst) are herbo-mineral/metal/nonmetal mixture-based preparations that are also nontoxic, easily absorbable, adaptable, and easily assimilated in the body [3–5].

The origin of nanomaterials is attributed to a meteorite created during the Big Bang event about 800,000 BC, according to the ancient literature [6]. Figure 10.1 chronicles the progress of nanotechnology over time and across numerous ancient civilizations and ages (prehistoric, ancient, medieval, modern, and contemporary). The development of modern nanomaterials began in the 1960s. Carbon nanomaterials (fullerenes, graphene, and carbon nanotubes) formed from fire smoke and soot, coupled with fat, charcoal, and plant pigments, were utilized for cave paintings in 40,000 BC after the origin of nanomaterials [7,8]. During the year 5,000 BC, ancient Egypt utilized clay minerals and nano-platelets with a thickness of 1 nm as natural materials. Humanistic approaches utilized earthenware matrix materials strengthened with natural asbestos nanofibers (50–200 nm in diameter), and especially, approximately 2,500 and 2,600 BC, ancient Egyptians used PdS_2 (lead sulfide) nanomaterials in the manufacture of hair colorants. Egyptian and Mesopotamian scientists recorded the first chemical production of metal nanoparticles in 1,300 and 1,400 BC. This was followed by research into the effect of surface plasmon excitation from 1,200 to 1,000 BC [8–10]. Enamels made between 400 BC and 100 BC were also high in Cu NPs (copper nanoparticles) and cuprous oxide. In 800 AD, an azure pigment resistant to corrosion was created by chemically combining indigo dye with clay nanopores to create an environmentally steady color. In 700 AD, Swarna Bhasma (gold ash) and Lauha Bhasma (iron ash) were documented to be good therapeutic agents in the Indian medical literature. They were known to be nontoxic metallo-medicines that helped strengthen bones, improve iron absorption in the body, and maintain body alkalinity [9,10]. The widespread use of nanomaterials (especially metallic NPs) and nanostructures by Romans is one of the most prominent examples of historical nanomaterials usage and nanotechnology expertise. The colors used in the cave paintings at Bhimbetka region in India have definite traces of extremely fine white clay and iron-based red pigments in nanoscale. The Lycurgus Cup is one of the most stunning specimens of Roman glass manufacture, followed by Damascus blades, ruby-gold colloids, and reinforcement materials [10].

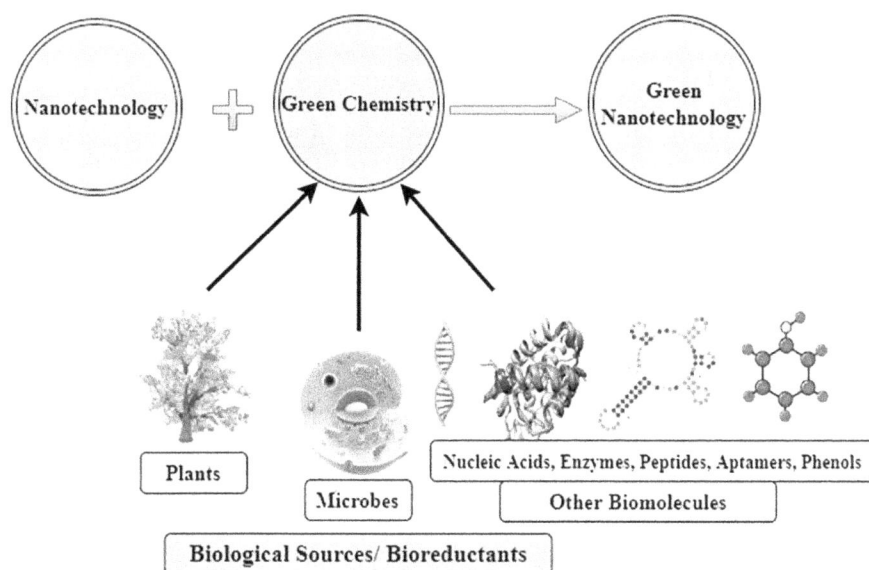

FIGURE 10.2 Technology integration depicting formation of green nanomaterials.

Nanoscience investigates molecules and structures on the nanometer scale, whereas nanotechnology employs these molecules and structures in practical applications such as devices, machines, and sensors [11]. Nanotechnology provides materials that have the potential to benefit society. The term "nanoform" refers to the altered physicochemical properties of the materials when compared to the same material in bulk, and it raises concerns about environmental and health issues of their production and use. To address such concerns, the EPA's Office of Pollution Prevention and Toxics in the USA introduced green chemistry-based nanomaterial synthesis in 1991 in order to minimize or totally abolish the use and generation of hazardous substances [12]. The key elements of green chemistry are energy efficiency, minimization of waste, safer product protocols and by-products, renewable resources, feasible atom economy, real-time monitoring solutions, and environmentally benign designs [12,13] (Figure 10.2).

The sustainability and persistent future of green nanomaterials is derived by the seamless integration of techniques of green chemistry with nanotechnology. This combination leads to the production of nanomaterials that are sustainable, innocuous, economic, and environmentally friendly [13,14]. The production of green nanomaterials eliminates the use of non-verdant solvents, substitutes biological materials (biomass, phytochemicals, and microbial) for chemically based materials, and employs clean energy storage materials (energy-efficient fuel/solar cells, green catalysis, etc.). Some breakthroughs and achievements, particularly in the field of nanomaterials, have resulted in significant improvements and the development of new medical solutions [14].

The production of green nanomaterials primarily incorporates biological precursors with nanofabrication design techniques, which can lead to the generation of sustainable green nanomaterials and can be carried out in the following ways:

- Scheming for minimum environmental impact
- Scheming safe, non-hazardous nanomaterials
- Scheming for the assurance of process safety
- Minimizing the production of waste
- Enhancing material efficiency and improvement in energy efficiency.

10.2   GREEN NANOMATERIALS

10.2.1   Origin and Evolution of Green Nanomaterial Synthesis

In ancient tradition, the metallic nanoparticles were found as reported in the literature. The Lycurgus Cup dating back to the 4th century AD found in Roman Empire shows dichroism when light was reflected back from it (yellowish-green color changes to translucent ruby red), depicting the mythological event of the death of the king of Edoni. This is a well-known example of the change in physicochemical properties of nanoparticles. Later, in British museum Dr. G. F. Claringbull stated that it was made of glass, and in 1959, it was confirmed by X-ray diffraction [15]. Further studies in the research laboratory of General Electrical Company were carried out for spectrographic analysis to determine the colorant. B. S. Cooper noted the presence of metallic elements (gold, silver, and copper) at ppm level, which were responsible for the dichroism [15,16]. The synthesis process required for this was based on dissolution of metallic salts in silicates and then partially reduction using heat treatment. Since the discovery of these metallic nanoparticles received great attention from researchers for developing new materials, the Damascus steel Dabres in the 17th century was discovered, which reveals the presence of carbon nanotubes and cementite [17]. The brief study of nanomaterials in ancient times also highlighted the importance of nature for developing nanostructures. Nacre is one of the examples of organic-inorganic nanocomposite found in mollusk shells with bio-macromolecular glue, which then inspired chemists to synthesize strong and stiff nanoceramics [17]. As the time passed, researchers used ceramic method, which included crushing or fine grinding and mixing powders of the integral carbonates, oxides, and other compounds to specific proportions, which required more time and effort. Therefore, identifying the ample biodiversity and availability of natural entities such as plants and animals, researchers worked a lot on the synthesis of nanomaterials from 1990 till date. The number of increasing scientific publications validates the "green era" of nanomaterials. In the middle of the 19th century, after the famous lecture by Richard. P. Feynman titled "There is plenty of room at the bottom", the boom of nanotechnology research increased after the discovery of the structure of carbon nanotubes by "Sumio Iijima" in 1991 [18]. Thereafter, the nanosized carbonaceous materials gained attention in the silicon technology with the invention of room-temperature graphene by Andre Geim and Novoselov bagging them the Nobel Prize in 2010 [19]. Although the discovery of carbon nanotubes was made in 1990s, recent excavations carried out in Keezhadi village of Tamil Nadu in India have uncovered oldest nanomaterials in the pottery relics from 600 BC that bear distinct traces of "unique black coatings" [20]. These black coatings were later found to be single-walled carbon nanotubes. The robust and ultra-stable structure of these carbon nanotube coatings on the pots has reportedly endured all adverse conditions for more than 2,600 years and still keep the pots in a sustained condition (Figure 10.3).

In the beginning of the 20th century, biomolecules, natural polymers, and waste materials gained great importance for capping, functionalization, and stabilization of the chemically synthesized nanomaterials that made them applicable in biomedical field. Various methods of biological synthesis using "bio-laboratory" evolved in the past two decades understanding the mechanisms and interactions involved in it [11,21]. Green synthesis techniques of nanomaterials encouraged green compatibility and protocols that deduct the use of toxic/unsafe chemicals, which in turn results in minimized environmental impacts and preservation of resources and human health. The research community is continuously focusing on the development of novel and eco-friendly nanomaterials which could result in a more reliable and sustainable solution to the different technological and environmental challenges in the diverse fields [22]. Therefore, identifying their need and availability of sources, top-down and bottom-up approaches are being used for the synthesis of nanomaterials that include physical, chemical, and biological synthesis routes for developing 0D, 1D, 2D, and 3D nanomaterials [23]. Figure 10.4 elicits the various physical, chemical, and biological approaches practiced for the synthesis of nanomaterials. To minimize the toxicity of

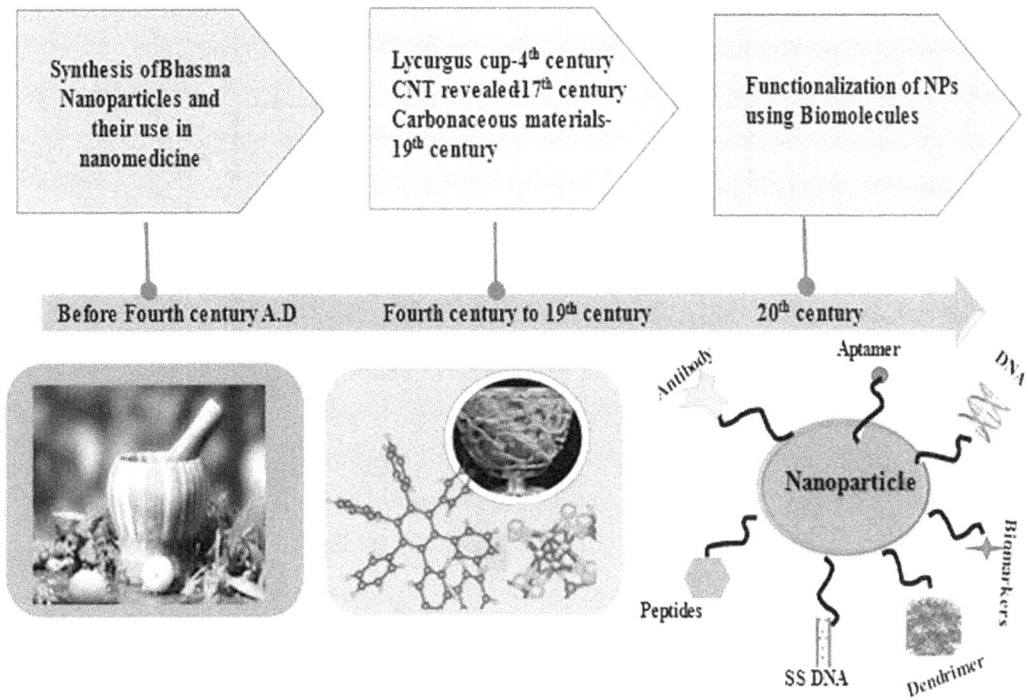

FIGURE 10.3 Evolution of biological and green synthesis methods.

nanomaterials, a green approach of synthesis (biological) is developed that uses intracellular and extracellular contents of plants and microorganisms. Easy availability and the low cost of biomolecules helped in several methods of biosynthesis of nanomaterials and are explained in the subsequent sections.

10.2.1.1 Plant-Mediated Synthesis

Plant-mediated synthesis of metallic nanomaterials has gained attention in the current market and research areas due to its rapid production, low cost, and non-hazardous nature. The production of various metal, metal oxide, and alloy nanomaterials is feasible using extracts derived from plant as a primary source with the fundamental method of "bio-reduction", which is performed at ambient environmental conditions. The phytochemicals extracted from various parts of plants (roots, stem, leaves, flowers, fruits, etc.) such as flavonoids, phenolic acids, isoflavones, curcumin, isothiocyanates, carotenoids, proteins, and enzymes favor the reduction of metal salts and thereby easy formation of metal nanoparticles. Callus is also widely utilized as a bio-reductant for the synthesis of metal nanoparticles [24]. The synthesis of silver nanoparticles (Ag NPs) is possible using neem leaf extract, where diterpenoids are integral to the synthesis mechanism [25]. The room-temperature synthesis of Ag NPs using guava (*Psidium guajava*) with tamarind (*Tamarindus indica*) leaves [26], aloe vera [27], and neem leaves (*Azadirachta indica*) [28] under hydrothermal condition has also been developed. The use of these phytochemicals as reducing agents provides unique surface functionality to the nanomaterials that make them much more effective in their respective applications. Gold nanoparticles (Au NPs) with SPR phenomenon also showed good colloidal stability using *Tilia sp.* linden bract aqueous leachate [29]. The optimized concentration of bio-reductant and the interaction time between biomass and precursors prevent the instability of colloidal gold nanoparticles. Metal oxide nanoparticles such as aluminum oxide [30], cerium oxide [31], cobalt oxide [32], iron oxide [33], nickel oxide [34], and zinc oxide [35] have been successively synthesized using plant extracts.

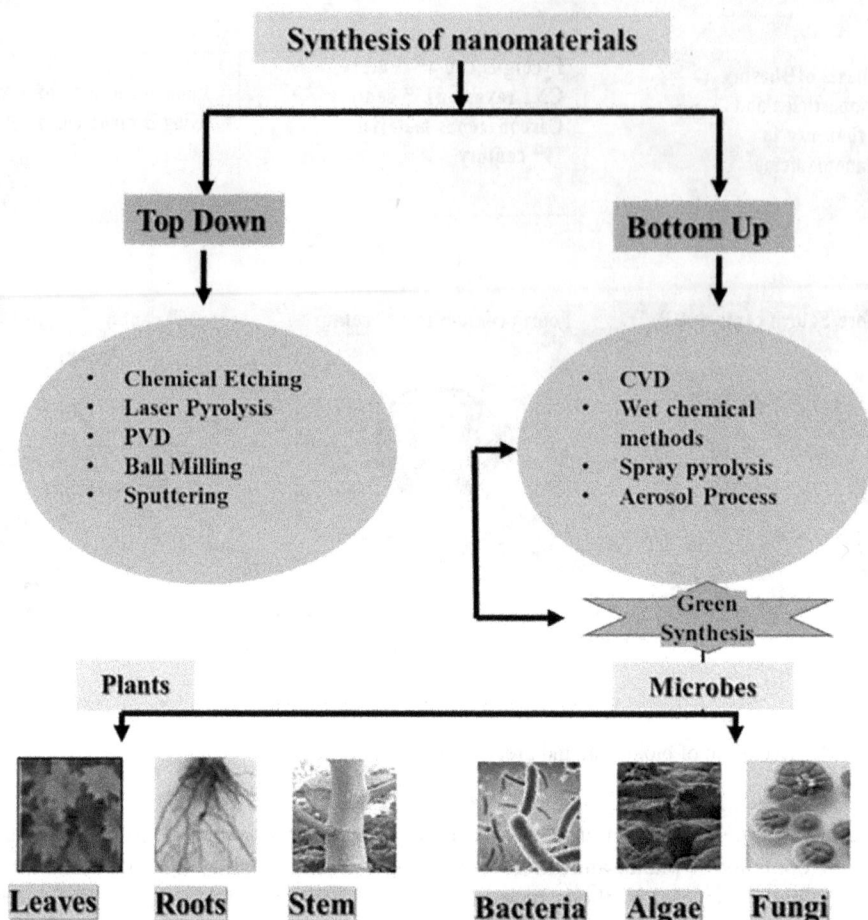

FIGURE 10.4 Conventional and green routes for the synthesis of nanomaterials.

10.2.1.2 Microbe-Mediated Synthesis

Green synthesis of nanomaterials also involves the use of intracellular and extracellular microbial biomass where mechanisms of synthesis vary from microbe to microbe. It comprises microbes as the reducing agents, such as algae, fungi, bacteria, actinomycetes, and yeasts [36]. In the intracellular microbial synthesis, the biomass is obtained by culturing microorganisms at optimal conditions with appropriate supply of nutrients and then washed with sterile water. The resulting supernatant is further mixed with solutions of metal ions, and nanomaterials are obtained by centrifugation, sonication, and washing. In the extracellular synthesis, the microbial culture is directly centrifuged and mixed with metallic salt solutions [37]. The synthesized nanomaterials are visually confirmed by color change of the solution. The synthesis using microbes is cheap and able to absorb and reduce the metal ions, but requires a tedious synthesis process, more time for growing microbes, and a suitable environment. Several bacterial strains are used for the synthesis of metal, metal oxide, and magnetic nanoparticles. Till date, all types of bacterial synthesis have been carried out using NADH, i.e. reductase enzyme mechanism. *Lactobacillus sp.* [38], *Bacillus sp.* [39], *Shewanella algae* [40], and *Rhodobacter sphaeroides* [41] are some of the strains utilized for the synthesis of silver, selenium, platinum, and zinc nanoparticles. Interestingly, the magnetotactic bacteria found in oxic-anoxic transition zone are capable of producing magnetite (Fe_3O_4) nanoparticles reducing Fe(III), where iron taken by the cell is re-oxidized as hydrous oxide and then to high-density ferrihydrite [42].

Algae and fungi are also easily available in the environment, are affordable, and can be used for developing durable nanomaterials from waste and hence called "bio-nanofactories". Algae are photoautotrophic, eukaryotic, aquatic, and aerobic microorganisms, and among all types of species, marine brown algae have the ability to accumulate heavy metals [43]. Fucoidans (brown seaweed), ulvans (green seaweeds), and agaroids/carrageenans (red seaweeds and Irish moss) are commercially important subclasses of sulfated polysaccharides that contain sulfated L-fucose [44], a signature chemical in brown seaweeds (40% of dry weight) that strengthens the cell wall during desiccation. The fucoidan polysaccharides present in the cell wall of algae also provide many applications including anti-coagulant, anti-inflammatory, and also anti-cancer [45]. For the reason of getting these properties in nanoparticles, *Spatoglossum asperum,* an algal genus, has been efficiently used for the synthesis of silver nanoparticles [46].

Fungi generally produce copious amounts of enzymes as compared to other microorganisms that can act as precursors for the reduction of gold and silver to the nanoparticle state [45,47]. Filamentous fungi have higher metal tolerance and bioaccumulation as compared to bacteria and algae [48]. Hence, they are capable of mass-producing nanomaterials, thereby cutting the production cost marginally and increasing the bioavailability of sources. The energy metabolic cycles of fungal cells are largely dependent on components such as sugars (glucose and fructose), proteins (ATPase and 3-glucan-binding proteins), and enzymes (glyceraldehyde-3-phosphate dehydrogenase). The common mechanism for synthesis using bacteria and fungi is based on enzymatic reduction process by electron shuttle using cofactors such as NADH- and NADPH-based nitrate reductase enzymes secreted by fungi [49].

10.2.2 Applications of Green Nanomaterials

In the past few decades, nanotechnology has revolutionized multiple fields by crossing over scientific barriers from electronics to medicine, to cosmetics, to point-of-care diagnostics, to environmental monitoring, to food fortification to biosensing applications, and so on. The use of natural, green ingredients to advance the nanomaterials synthesis process has rendered the process much more benign and eco-friendlier, thereby reducing the hazardous by-products generated during the chemical synthesis of nanomaterials [50]. Green nanotechnology is the breakthrough technology platform producing biocompatible, nontoxic, easily tunable and cost-effective nanomaterials that have found their way from laboratory to commercial applications in a multitude of areas [51]. Since 1900s, sensor technologies are making massive strides and are increasingly used in various divisions. The integration of plant phytochemicals and biological components with traditional facile synthesis methods results in the formulation of harmless and cost-effective nanomaterials that display high surface activity, specific size, and purity that make them suitable for versatile uses in industrial as well as research applications [52]. Biosensing platforms and healthcare modules together lead to the development of early diagnostic and effective treatment approaches for theranostics practices. Nanotechnology has enabled the effective management of materials in the nanometer scale, thus aiding enhanced biosensing and improved theranostics systems [53]. However, using chemical precursors and hazardous methods used to synthesize nanomaterials have adverse effects and hence act as threats to health and the environment. An eco-friendly and non-hazardous solution can be provided by green technology approach that integrates the principles of green chemistry with engineering to counter complications affecting the environment and human health [52–54].

Green nanotechnology has etched a new cognizance into material sciences and metal and metal oxide nanoparticles and bio-nanomaterials are used extensively in the field of healthcare monitoring, agri-food industry, and environmental remediation applications. Green methods can produce vital, diversified morphologies and characteristics that attribute better waste management and tunable physicochemical properties. The unique and variable optical properties of green nanomaterials make them classic entities to be used in the niche areas of is bioimaging, biosensing, and traceable drug delivery. The inhibitory activity of these green nanomaterials against microbes, cancer, and inflammation is also well established [55]. The agri-food sector is a major influence regulating the ecosystem, food cycle, and human health. Excessive use of pesticides, herbicides, and other chemicals is an alarming factor that cannot be neglected as the produce is up for direct consumption in most instances without any processing, causing acute toxicity and bioaccumulation [56,57]. The early detection and remediation of harmful chemicals and adulterants is of utmost importance to ensure healthy human vitality [58]. The development of ultrasensitive and accurate sensing platforms for the detection of contaminants and analytes in trace concentrations is the need of the hour.

10.3   SENSING REALMS OF GREEN NANOMATERIALS

Sensors are, in reality, anything that reacts predictably to natural stimuli. Simply put, they enable us to investigate phenomena that we cannot see, touch, or hear. Humankind is blessed with the senses such as touch, smell, sight, taste, and hearing. Sensors and biosensors are nothing but bio-mimetic sensing tools that enable us to measure signals arising from stimuli. The use of green nanomaterials in sensing technologies in various fields holds a unique appeal even before the term "nanotechnology" was coined. The ancient Indian Vedas and other traditional scriptures available to humankind chronicle the fascinating journey of nanomaterials through the eras. The earliest civilizations made use of sensing tools such as tracking celestial bodies, tracking climatic changes and seasonal cycles, and tracking the movements of birds and animals [59]. All these mechanisms point toward the advancement of our ancient civilizations to an extent that we can only imagine. This section accounts for the journey of sensor systems through the years and the current applications of nanomaterials-based sensors (Figure 10.5).

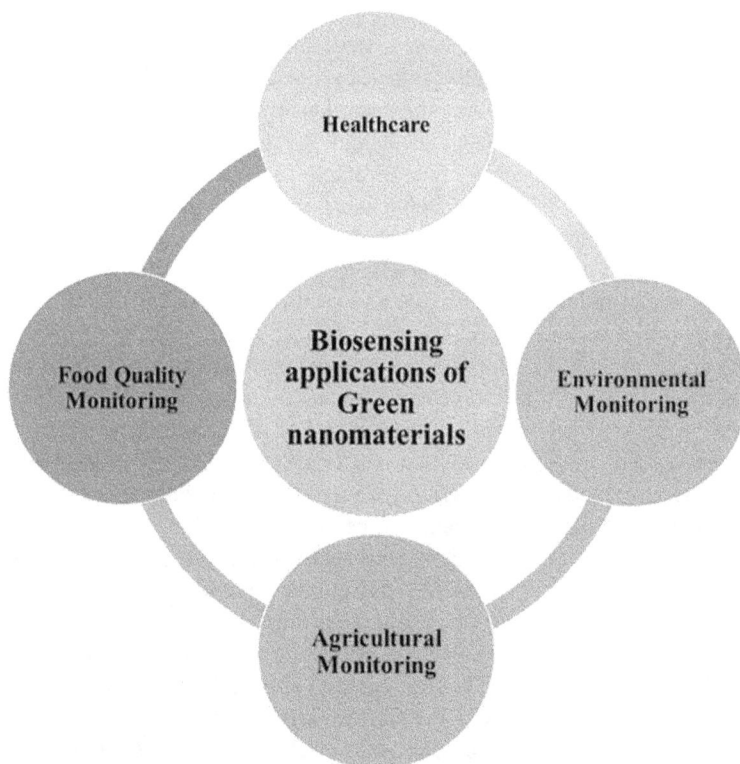

FIGURE 10.5   Biosensing applications of green nanomaterials.

10.3.1 Biomedical and Healthcare

Green nanomaterials are currently being developed to be used in physics, medicine, biomedicine, as well as chemistry, with the goal of developing subnormal devices. The breadth of biomedical applications of green nanomaterials domains is among the significant areas that are gaining traction, as all biological systems exhibit nanotechnology principles. Green nanomaterials are expected to have a massive influence on biology, biotechnology, and medicine as a result of their equivalent size to biological materials such as enzymes, antibodies, proteins, and nucleotides, allowing them to be used in medicinal applications [58]. The remarkable features of green nanomaterials, such as their high surface area, adjustable optical emission, and electrical and magnetic capabilities, can be employed in a variety of bioengineering applications, including drug delivery and biosensors [60]. Sensors form a broad category of materials and technologies used to capture physical, chemical, or biological stimuli and convert them to measurable output signals. Glucose sensors have been around since the first-generation enzymatic glucometer was developed in 1962. The primary goal of nanomaterial-based sensing mechanisms is to improve our understanding of biological processes and technology for medical diagnosis and treatment. Green nanomaterials have the potential to behave as promising molecules by increasing sensitivity and lowering detection limits for single molecules. The concept of "green" for sensor fabrication extends to much broader concepts encompassing their use in biomedical and healthcare [60,61].

Increasing demand for advanced medical technologies has revolutionized the green nanomaterials-based sensor industry in diagnosing, monitoring, and treating serious diseases. It is clear that the use of green nanomaterial-assisted techniques improves the sensitivity, affordability, and accessibility of diagnostics and treatment for a specific disease [62]. The chimeric performance of green nanomaterials researched for biomedical research is beneficial for the design and development of medicines that take individualized health management into account [63]. The biocompatibility, cost-effectiveness, and high sensitivity of nanomaterials have enabled point-of-care diagnostics to achieve the ASSURED (Affordable, Sensitive, Selective, User-friendly, Rapid & Robust, Equipment-free and Deliverable) criteria, which is a benchmark assessment standard set by WHO [64]. In addition to the conspicuous qualities of green nanomaterial-aided approaches, the introduction of statistical approaches, such as artificial intelligence (AI) and machine learning, and bioinformatics, has emerged as a highly important tool in comprehending forecasts and trends [65,66]. These prediction tools are useful for understanding epidemic variations, optimizing therapy, and assessing risk.

Green nanomaterials are also being studied and can be functionalized so that they can be used as a sensing tool in a variety of fields, including nucleic acid detection, hormone detection, nucleic acid modeling, biomedicines, tumor therapy, and anticancer drugs, distinguished by the sequestration of target analyte from other similar biomolecules like glucose, nucleic acid, H_2O_2, laccase, and heavy metals such as Cd^{2+} and Pb^{2+} [67,68]. Green nanomaterials are also being researched for sensing applications in HIV therapies, cardiovascular disease, and MRI scanning. Cell tracking, hyperthermia, bio-separation, nucleic acid and medication delivery, toxicity monitoring, and other techniques are available [68,69].

Along with these, printed sensors, implantable sensors, wearable sensors, ocular-intraocular pressure sensors, stress monitoring patches, and drug effect monitoring sensors hold great promise in revolutionizing the biomedical and healthcare industry [69]. The utilization of ecological and lucrative green nanomaterials for sensor components as well as reusability and multiplexed detection capability of green nanomaterials-based sensors come within the scope of green chemistry and holds a great future for green nanomaterials from future perspectives [68,69].

10.3.2 Agri-Food Industry

Nowadays, the increased population has created many problems to the mankind and environment such as problems in security of food and nutrition for global population, climate change, and management of natural resources. Identifying the need and solution to these problems by using green

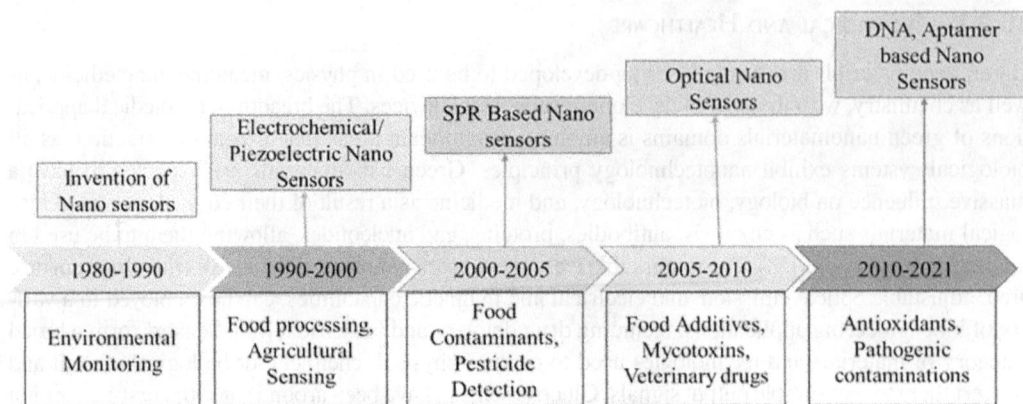

FIGURE 10.6 Timeline of green nanomaterials based sensors in food, agricultural, and environmental sensing.

and environment-friendly nanomaterials impacted positive in the 21st century. The gradual evolution in the use of nanomaterials in food and agriculture over the last 30–40 years has been magnificent in itself. Researchers have focused on developing systems based on nanomaterials for sustainable intensification in agriculture where increased yield of crops in small cultivated area is aimed. Crop productivity can also be targeted for increase using green nanopesticides and nanofertilizers, which also stimulate plant growth and improve the quality of soil [70]. The challenges in agriculture sector such as the detection of soil condition and diseases caused by infection to crops can be overcome by using nanosensors based on green nanomaterials. The sensing mechanism of those biosensors is commonly based on optical or electrochemical sensing. A rapid increase in population has led to the increased demand of agri-food, which resulted in more use of pesticides. Therefore, one can easily get poisoned by pesticides and numerous health issues can be caused which is needed to be detected by biosensors. The use of green nanomaterials in agri-food sensing applications over the years is depicted in Figure 10.6.

Green silver nanoparticles having strong inhibitory action against gram-negative, gram-positive and even multi-drug resistant bacteria are helpful in coating of crops for their protection from infection. Silver nanoparticles are extensively used for their antimicrobial property against a myriad of phytopathogens. Scientists have also reported that silver nanoparticles synthesized using *Tagetes erecta* (marigold) leaf and flower extracts enhance plant growth [70]. Indeed, the development of analytical devices dedicated to quality control and biosecurity have helped in the detection of agri-food toxins.

Food safety, analysis, and fortification are the most important steps throughout the production and supply chain of food. Presently, only centralized labs and a few private organizations are equipped with the conventional laboratory equipment and facilities to carry out the qualitative and quantitative detection and measurements of food additives and adulterants. Inexpensive portable assays with rapid detection abilities help in the identification of discrepancies in food producers, storage facilities, food quality and sterile transportation, stowage, and usage.

Green materials-based sensors can selectively detect pathogens such as *E. coli, Salmonella,* and *S. aureus,* contaminants such as heavy metals, illegal additives, residual veterinary antibiotics, pesticides, mycotoxins, allergens and nutrients, antioxidants. Green nanomaterials commonly used in these platforms include (i) green synthesized metal and metal oxide nanoparticles such as gold and silver that exhibit surface plasmon resonance (SPR) properties that promote their use in optical and electrochemical sensors, (ii) magnetic nanoparticles such as iron oxide for separation and augmentation of the targeted analyte, (iii) graphene- and carbon-based nanoparticles to amplify electrochemical signals and high electrical conductivity [19,70,76].

Antioxidants in green nanomaterials play an important role by fortifying the systems against excessive free radical production and accumulation, associated with pulmonary and neurodegenerative diseases, cancer, rheumatoid arthritis as well as aging. Therefore, there is a requirement for robust methods for the quantification of antioxidants in food and food constituents. Several assays dependent on gold and silver NPs, nanorods, nanocomposites, carbon nanomaterials are interfaced with electrochemical or spectroscopic transduction and give heightened electrochemical signals for the detection of three antioxidants: butylated hydroxytoluene (BHT), butylated hydroxyanisole (BHA), and butylated hydroquinone (TBHQ) [71]. Glycemic index (GI) of sugars in food is measured by electrochemical and biological sensors. Glucose and fructose were successfully measured by a conductive ink containing Cu NPs, graphite, and polystyrene by paper-based biosensors [71,72].

The oxidase-like activity of nanoceria enables the enzymatic and catalytic conversion of phytochemicals such as gallic acid and caffeic acid to their corresponding quinones, which helps in the detection mechanism; further, these were used in the detection of antioxidants in wine [70,71]. Dyes are exogenic compounds added to alter the properties and appearance of food. Food adulterants such as dopamine, melamine, and Sudan I are the common examples. Gold nanomaterials and graphene oxide are widely used in the detection of these dyes. UV-visible spectrophotometry with gold nanoparticles is best for the detection of the same [73]. Acid numbers and moisture content along with some toxic chemicals in edible oils were determined by water-soluble QDs [73–75]. Green synthesized R-GO was used for the detection of sunset yellow levels in the food items [76].

Fungal toxins that pose certain challenges due to their enormously high toxicity at low exposure levels. Researchers are developing the nanosensors based on oxide nanoparticles for the early detection of the same. The direct oxidation at electrode surfaces can be modified by iron oxide NPs, Au NPs, CNTs, and carbon materials to enhance detection capabilities of BPA [77]. These materials also functionalized with tyrosine to convert BPA into quinine for the detection mechanisms [77–78]. SERS-based silver and gold nanomaterials with DNA aptamer are used for the detection of kanamycin in milk [79].

10.3.3 ENVIRONMENTAL SENSING

Monitoring, protection, and remediation are the three major application areas of nanotechnology in the domain of environmental sustainability. Recent years have seen an increasing trend in the use of nanomaterials on all three fronts [80]. Since ancient times, nanomaterials have been utilized for various environmental applications, typically for improving the quality of the natural resources available to humans. Harsh weather conditions and harmful pollutants have always been a cause of concern for scientists all over the world. They are responsible for corroding and damaging various surfaces. However, a few outstanding examples of the use of nanomaterials in ancient culture for the preservation of historic monuments date back to centuries. The impressive, rust-less Iron Pillar of Mehrauli, Madhya Pradesh (now in Delhi), constructed between 375 and 415 BC, is one such example of prototypical use of crystalline nano-iron to avoid corrosion of metal from pollution and weather conditions [81]. Despite the presence of rampant pollution and extreme weather conditions in Delhi, the surface of the Iron Pillar is still withstanding all and standing rust-less and undamaged even today. The development of nanotechnology and green nanomaterials has thus made a strong contribution to the society and mitigation of the damage caused by environmental pollution [81].

The development of a diverse array of safe and eco-friendly advanced functionalized nanomaterials has opened an exciting new window in the areas of environmental biosensing, pollution monitoring, and environmental remediation and clean-up [11,82]. Pollutants produce harmful effects globally. Every year, tonnes of heavy metals, polycarbonated biphenyls (PCBs), volatile organic compounds (VOCs), dioxins and other toxins, phenolic and organophosphorous compounds are released in different strata of the environment [82,83]. The need to develop portable, economic, and rapid monitoring tools for trace level detection of environmental pollutants creates a niche area in the field of biosensor development using green nanomaterials. Biosensors function with the precision of ultrasensitive tools for the environment impact assessment (EIA) of ecological, biological, and chemical monitoring of the inorganic as well as organic pollutants. To detect trace pollutants from aqueous environment, Mishra et al. developed a portable electrochemical sensor device with DNA coupled gold core-shell nanoparticles [83]. Electrochemical sensing is an ultrasensitive technique in itself. However, when coupled with a bio-nanosensor, the limit of detection could be lowered up to picogram levels. The results obtained were fast enough, and the portable nature of the EC device was found ideal for field trials. Aptasensors are also an emerging class of biosensors that are being extensively coupled with nanomaterials for detection of pollutants. An electrochemical aptasensor based on thin film composed of chitosan-iron oxide nanocomposite (CHIT-IO) was developed for the detection of malathion, an organophosphate pesticide that is highly toxic and a known teratogenic agent [84,85]. The co-precipitation method was adopted for the preparation of iron oxide nanoparticles, and detection was carried out from fruit and vegetable samples as well as aqueous environmental samples [33,84]. The sensitivity of detection was found to be extremely high. Colloidal semiconducting nanomaterials are an emerging class of green nanomaterials that provide low-cost photovoltaic solutions due to their highly tunable surface charge properties and wide optical absorption range from visible to near-infrared region. Green colloidal nanomaterials have the additional functionalities of biogenic components, which enhance their surface plasmon properties and enable for ultra-trace detection of a variety of environmental pollutants such as pesticides, organic compounds, biomolecules, and heavy metals. Their remarkable optical properties make them ideal nanomaterial to be used in biosensors [86,87]. Recent times are seeing an increase in research carried out in biosensing by utilizing biological approaches and functional moieties to create safe, biocompatible, and environmentally benign nanomaterial-based sensors.

10.4 ADVANTAGES OF GREEN NANOMATERIALS IN SENSING

Green nanotechnology involves the principles of green chemistry whose purpose is to restrict and reinvent the use of harmful chemicals and generate highly safe and efficient products. Process and products are the two goals of green nanotechnology. Green nanotechnology is developing as one of the most effective branches of nanotechnology as it utilizes and applies the principles of green chemistry to the existing nanotechnology processes to develop nontoxic, highly efficacious, and most importantly, green functionalized nanomaterials. The design and potential of these nanomaterials can be easily applied to various fields without much management to promote sustainability and occupational health and process safety (Figure 10.7).

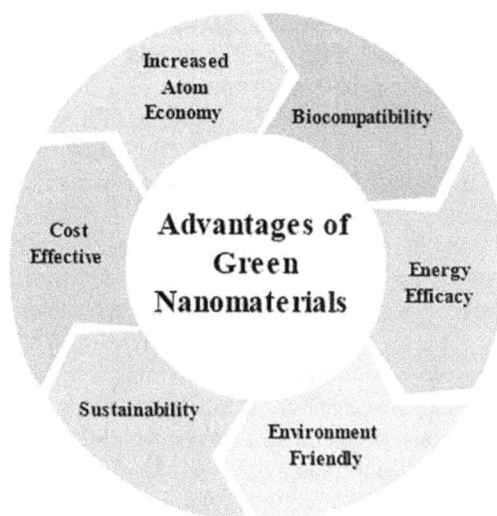

FIGURE 10.7 Advantages of green nanomaterials.

TABLE 10.1
Principles and Advantages of the Green Nanomaterials

Principle	Advantage
Atom economy	Maximization of material incorporation in the synthesis process
Natural reductive agents	Superior stoichiometry of plants and microbes as reagents
Less hazardous chemicals	No/less toxicity in the nanomaterials synthesis methods to human and environment
Energy efficacy	Minimized energy requirements of the synthesis process; minimized environmental and economic impacts
Biocompatibility	Increased biocompatibility; efficient for healthcare applications
Sustainability	Enhanced environmental sustainability of processes producing negative externalities

Physicochemical methods of nanomaterials synthesis are widely used, but these methods cause damage to the environment, use toxic and expensive chemicals, have high energy consumption and less productivity, and generate hazardous waste. Therefore, there is a pressing need for adopting green technology route for the synthesis of nanomaterials which is eco-friendly, biocompatible, and safe. Merging green chemistry and green engineering is accepted as "green nano". It is developed to minimize the environmental and human health risks allied to the fabrication and use of nanomaterials and products. Green nanotechnology refers to the biogenic synthesis of nanomaterials. The advantages of green nanomaterials that surface to the forefront are simplicity, ease of procuring the bio-sources, less time-consuming process, high yield, lower toxicity, and biocompatibility. Another major advantage is the easy, time-temperature-pH controllable size and morphological characteristics of the nanomaterials. As discussed previously, various natural products such as plants, microorganisms, and actinomycetes are used for the synthesis of green nanomaterials. The advantages of green nanomaterials are listed in Table 10.1. The advantages of green nanomaterials involve cleaner production and process such as synthesis of nanomaterials using green methods and further recycling of industrial surplus materials into nanomaterials.

10.5 CONCLUSIONS AND FUTURE PROSPECTS

The study of development of novel green nanomaterials over the prehistoric and ancient eras may help in driving both engineering and technology fields to greater heights. The field of biomimetics is gaining superior trends by mimicking the natural and structural arrangements and functions of biological entities. Composite nanomaterials and nanopolymers are also showing rapid progression in the areas of identification and theranostics, environmental monitoring, and agri-food sensing. In the sensing field, green nanomaterials have a significant ability to improve transduction mechanisms and augmented diagnosis capability for a wide range of pathogens, allergens, toxicants, antibiotics, nutraceuticals, freshness and ripening indicators (ethylene and CO_2), cancer biomarkers, and pesticides. Green synthesized magnetic nanomaterials have extensive applications as MRI contrasting agents for heightened signal ratios and improved sensitivity. Size- and shape-dependent opto-electrical properties of green metal nanomaterials such as gold, silver, platinum, and selenium will play an important role in the no-contact detection of DNA, immunoassays, cancer biomarkers, pathogenic infections, virus detection (HIV and coronavirus), and drug and hormone detection. Green metal oxide nanomaterials such as copper oxide, zinc oxide, cesium oxide, and titanium oxide will be promising analytes for the detection of H_2O_2, pesticides, insecticides, sugars, nucleic acids, laccase, and heavy metals. The modern-day nanotechnology is evolving their subfields; one of the fields is DNA nanotechnology that proposes simplistic and competent design pathways with unique targeting and sensitivity applications. A lot of wide-spectrum studies need to be undertaken, and a thorough life cycle examination of nanomaterials in food, healthcare, and environment needs to be carried out to understand the underlying risks and limitations. The precision in guidance and decision-making about the perils, risks, and control of nanomaterials is required with scientific clarity to identify and mitigate the health effects of nanoparticle exposure. The future of sustainable nanomaterials can be fortified by carefully studying and identifying the nuances of ancient nano-practices.

ACKNOWLEDGEMENT

The authors wish to thank *Symbiosis International University (SIU)* profusely for their support and providing the facilities to carry out this study.

REFERENCES

1. Kumar, A. (2019). Ancient hindu science: Its transmission and impact on world cultures. *Synthesis Lectures on Engineering, 13*(1), 1–211.
2. Kumar Pal, S. (2015). The ayurvedic bhasma: The ancient science of nanomedicine. *Recent patents on Nanomedicine, 5*(1), 12–18.
3. Sharma, R., & Prajapati, P. K. (2016). Nanotechnology in medicine: Leads from Ayurveda. *J Pharm Bioallied Sci, 8*(1), 80–1.
4. https://sdlindia.com/blog/ayurveda-utilizes-basic-principles-of-nanotech.html.
5. Pal, D., Sahu, C. K., & Haldar, A. (2014). Bhasma: The ancient Indian nanomedicine. *Journal of Advanced Pharmaceutical Technology & Research, 5*(1), 4.
6. Dai, Z. R., Bradley, J. P., Joswiak, D. J., Brownlee, D. E., Hill, H. G., & Genge, M. J. (2002). Possible in situ formation of meteoritic nanodiamonds in the early Solar System. *Nature, 418*(6894), 157–159. https://doi.org/10.1038/nature00897.
7. Barhoum, A., García-Betancourt, M. L., Jeevanandam, J., Hussien, E. A., Mekkawy, S. A., Mostafa, M., ...& Bechelany, M. (2022). Review on natural, incidental, bioinspired, and engineered nanomaterials: History, definitions, classifications, synthesis, properties, market, toxicities, risks, and regulations. *Nanomaterials, 12*(2), 177.
8. Bayda, S., Adeel, M., Tuccinardi, T., Cordani, M., & Rizzolio, F. (2019). The history of nanoscience and nanotechnology: From chemical–physical applications to nanomedicine. *Molecules, 25*(1), 112.
9. Kumar, A., Nair, A. G. C., Reddy, A. V. R., & Garg, A. N. (2006). Unique ayurvedic metallic-herbal preparations, chemical characterization. *Biological Trace Element Research, 109*(3), 231–254.
10. Joshi, N., Dash, M. K., Dwivedi, L., & Khilnani, G. D. (2016). Toxicity study of Lauha Bhasma (calcined iron) in albino rats. *Ancient Science of Life, 35*(3), 159.
11. Hutchison, J. E. (2008). Greener nanoscience: A proactive approach to advancing applications and reducing implications of nanotechnology. *ACS Nano, 2*(3), 395–402.
12. Bai, R. G., Sabouni, R., & Husseini, G. (2018). Green nanotechnology—a road map to safer nanomaterials. In: *Applications of Nanomaterials* (pp. 133–159). Woodhead Publishing, Sawston. https://doi.org/10.1016/B978-0-08-101971-9.00006-5.
13. Adawy, A., El-Bassyouni, G., Ibrahim, M., & Abdel-Fattah, Wafa. (2012). Bio Nano Material: The Third Alternative. doi: 10.13140/2.1.4974.9284.
14. Nath, D., & Banerjee, P. (2013). Green nanotechnology - a new hope for medical biology. *Environmental Toxicology and Pharmacology, 36*(3), 997–1014. https://doi.org/10.1016/j.etap.2013.09.002.
15. Drozdov, A., Andreev, M., Kozlov, M., Petukhov, D., Klimonsky, S., & Pettinari, C. (2021). Lycurgus cup: The nature of dichroism in a replica glass having similar composition. *Journal of Cultural Heritage, 51*, 71–78.
16. Reibold, M., Paufler, P., Levin, A. A., Kochmann, W., Pätzke, N., & Meyer, D. C. (2006). Materials: Carbon nanotubes in an ancient Damascus sabre. *Nature, 444*(7117), 286. https://doi.org/10.1038/444286a.
17. Bouville, F., Maire, E., Meille, S. (2014). Strong, tough and stiff bioinspired ceramics from brittle constituents. *Nature Materials, 13*, 508–514. https://doi.org/10.1038/nmat3915.
18. Aqel, A., Abou El-Nour, K. M., Ammar, R. A., & Al-Warthan, A. (2012). Carbon nanotubes, science and technology part (I) structure, synthesis and characterisation. *Arabian Journal of Chemistry, 5*(1), 1–23.
19. Geim, A. K., & Novoselov, K. S. (2010). The rise of graphene. In *Nanoscience and Technology: A Collection of Reviews from Nature Journals* (pp. 11–19). https://doi.org/10.1142/9789814287005_0002.

20. Kokarneswaran, M., Selvaraj, P., Ashokan, T., Perumal, S., Sellappan, P., Murugan, K. D., ... & Chandrasekaran, V. (2020). Discovery of carbon nanotubes in sixth century BC potteries from Keeladi, India. *Scientific Reports*, *10*(1), 1–6.

21. Aboul-Nasr, M. B., Mohamed, S. S., & Yasien, A. A. (2021). Fungus-Mediated synthesis of silver nanoparticles using Aspergillosis causing fungi. *Journal of Environmental Studies*, *24*(1), 42–52.

22. Lu, Y., & Ozcan, S. (2015). Green nanomaterials: On track for a sustainable future. *Nano Today*, *10*(4), 417–420.

23. Paramasivam, G., Palem, V. V., Sundaram, T., Sundaram, V., Kishore, S. C., & Bellucci, S. (2021). Nanomaterials: Synthesis and applications in theranostics. *Nanomaterials*, *11*(12), 3228.

24. Khajuria, A. K., Bisht, N. S., Manhas, R. K., & Kumar, G. (2019). Callus mediated biosynthesis and antibacterial activities of zinc oxide nanoparticles from Viola canescens: An important Himalayan medicinal herb. *SN Applied Sciences*, 1(5), 1–13.

25. Varghese Alex, K., Tamil Pavai, P., Rugmini, R., Shiva Prasad, M., Kamakshi, K., & Sekhar, K. C. (2020). Green synthesized Ag nanoparticles for bio-sensing and photocatalytic applications. *ACS Omega*, *5*(22), 13123–13129.

26. Le, N. T. T., Trinh, B. T., Nguyen, D. H., Tran, L. D., Luu, C. H., & Hoang Thi, T. T. (2021). The physico-chemical and antifungal properties of eco-friendly silver nanoparticles synthesized by Psidium guajava leaf extract in the comparison with Tamarindus indica. *Journal of Cluster Science*, *32*(3), 601–611.

27. Tippayawat, P., Phromviyo, N., Boueroy, P., & Chompoosor, A. (2016). Green synthesis of silver nanoparticles in aloe vera plant extract prepared by a hydrothermal method and their synergistic anti-bacterial activity. *PeerJ*, *4*, e2589. https://doi.org/10.7717/peerj.2589.

28. Mukherjee, P., Ahmad, A., Mandal, D., Senapati, S., Sainkar, S. R., Khan, M. I., ... & Sastry, M. (2001). Fungus-mediated synthesis of silver nanoparticles and their immobilization in the mycelial matrix: A novel biological approach to nanoparticle synthesis. *Nano Letters*, *1*(10), 515–519.

29. Holišová, V., Urban, M., Konvičková, Z., Kolenčík, M., Mančík, P., Slabotinský, J., ... & Plachá, D. (2021). Colloidal stability of phytosynthesised gold nanoparticles and their catalytic effects for nerve agent degradation. *Scientific Reports*, *11*(1), 1–9.

30. Chen, H., Huang, D., Su, X., Huang, J., Jing, X., Du, M., ... & Li, Q. (2015). Fabrication of Pd/γ-Al$_2$O$_3$ catalysts for hydrogenation of 2-ethyl-9,10-anthraquinone assisted by plant-mediated strategy. *Chemical Engineering Journal*, *262*, 356–363. doi: 10.1016/j.cej.2014.09.117.

31. Singh, K. R., Nayak, V., Sarkar, T., & Singh, R. P. (2020). Cerium oxide nanoparticles: properties, bio-synthesis and biomedical application. *RSC advances*, *10*(45), 27194–27214.

32. Khalil, A. T., Ovais, M., Ullah, I., Ali, M., Shinwari, Z. K., & Maaza, M. (2020). Physical proper-ties, biological applications and biocompatibility studies on biosynthesized single phase cobalt oxide (Co3O4) nanoparticles via Sageretia thea (Osbeck.). *Arabian Journal of Chemistry*, *13*(1), 606–619.

33. Makarov, V. V., Makarova, S. S., Love, A. J., Sinitsyna, O. V., Dudnik, A. O., Yaminsky, I. V., ... & Kalinina, N. O. (2014). Biosynthesis of stable iron oxide nanoparticles in aqueous extracts of Hordeum vulgare and Rumex acetosa plants. *Langmuir: The ACS Journal of Surfaces and Colloids*, *30*(20), 5982–5988. https://doi.org/10.1021/la5011924.

34. Sabouri, Z., Akbari, A., Hosseini, H. A., Hashemzadeh, A., & Darroudi, M. (2019). Eco-friendly biosyn-thesis of nickel oxide nanoparticles mediated by okra plant extract and investigation of their photocata-lytic, magnetic, cytotoxicity, and antibacterial properties. *Journal of Cluster Science*, *30*(6), 1425–1434.

35. Vanathi, P., Rajiv, P., Narendhran, S., Rajeshwari, S., Rahman, P. K., & Venckatesh, R. (2014). Biosynthesis and characterization of phyto mediated zinc oxide nanoparticles: A green chemistry approach. *Materials Letters*, *134*, 13–15.

36. Arshad, A. (2017). Bacterial synthesis and applications of nanoparticles. *Nano Science & Nano Technology: An Indian Journal*, *11*(2), 1–30.

37. Vetchinkina, E., Loshchinina, E., Kupryashina, M., Burov, A., Pylaev, T., & Nikitina, V. (2018). Green synthesis of nanoparticles with extracellular and intracellular extracts of basidiomycetes. *PeerJ*, *6*, e5237.

38. Dakhil, A. S. (2017). Biosynthesis of silver nanoparticle (AgNPs) using Lactobacillus and their effects on oxidative stress biomarkers in rats. *Journal of King Saud University-Science*, *29*(4), 462–467.

39. Singh, N., Saha, P., Rajkumar, K., & Abraham, J. (2014). Biosynthesis of silver and selenium nanopar-ticles by Bacillus sp. JAPSK2 and evaluation of antimicrobial activity. *Der Pharm Lett*, *6*(6), 175–181.

40. Konishi, Y., Ohno, K., Saitoh, N., Nomura, T., Nagamine, S., Hishida, H., ... & Uruga, T. (2007). Bioreductive deposition of platinum nanoparticles on the bacterium Shewanella algae. *Journal of Biotechnology*, *128*(3), 648–653.

41. Bai, H. J., Zhang, Z. M., & Gong, J. (2006). Biological synthesis of semiconductor zinc sulfide nanoparticles by immobilized Rhodobacter sphaeroides. *Biotechnology Letters*, *28*(14), 1135–1139.

42. Li, S. N., Wang, R., & Ho, S. H. (2021). Algae-mediated biosystems for metallic nanoparticle production: From synthetic mechanisms to aquatic environmental applications. *Journal of Hazardous Materials*, *420*, 126625.

43. Menon, S., Rajeshkumar, S., & Kumar, V. (2017). A review on biogenic synthesis of gold nanoparticles, characterization, and its applications. *Resource-Efficient Technologies*, *3*(4), 516–527.

44. Ravichandran, A., Subramanian, P., Manoharan, V., Muthu, T., Periyannan, R., Thangapandi, M., ... & Marimuthu, P. N. (2018). Phyto-mediated synthesis of silver nanoparticles using fucoidan isolated from Spatoglossum asperum and assessment of antibacterial activities. *Journal of Photochemistry and Photobiology B: Biology*, *185*, 117–125.

45. Oladipo, O. G., Awotoye, O. O., Olayinka, A., Bezuidenhout, C. C., & Maboeta, M. S. (2018). Heavy metal tolerance traits of filamentous fungi isolated from gold and gemstone mining sites. *Brazilian Journal of Microbiology*, *49*, 29–37.

46. Lu, Y., & Ozcan, S. (2015). Green nanomaterials: On track for a sustainable future. *Nano Today*, *10*(4), 417–420.

47. Rani, M., Yadav, J., Chaudhary, S., & Shanker, U. (2021). An updated review on synthetic approaches of green nanomaterials and their application for removal of water pollutants: Current challenges, assessment and future perspectives. *Journal of Environmental Chemical Engineering*, *9*(6), 106763.

48. Noah, N., & Ndangili, P. (2021). Green synthesis of nanomaterials from sustainable materials for biosensors and drug delivery. *arXiv preprint arXiv:2112.04740*.

49. Yadav, A., Kon, K., Kratosova, G., Duran, N., Ingle, A. P., & Rai, M. (2015). Fungi as an efficient mycosystem for the synthesis of metal nanoparticles: Progress and key aspects of research. *Biotechnology Letters*, *37*(11), 2099–2120.

50. Parveen, K., Banse, V., & Ledwani, L. (2016, April). Green synthesis of nanoparticles: their advantages and disadvantages. In *AIP Conference Proceedings* (Vol. 1724, No. 1, p. 020048). AIP Publishing LLC.

51. Verma, R., Pathak, S., Srivastava, A. K., Prawer, S., & Tomljenovic-Hanic, S. (2021). ZnO nanomaterials: Green synthesis, toxicity evaluation and new insights in biomedical applications. *Journal of Alloys and Compounds*, *876*, 160175.

52. Bartolucci, C., Antonacci, A., Arduini, F., Moscone, D., Fraceto, L., Campos, E., ... & Scognamiglio, V. (2020). Green nanomaterials fostering agrifood sustainability. *TrAC Trends in Analytical Chemistry*, *125*, 115840.

53. García-Guzmán, J. J., López-Iglesias, D., Bellido-Milla, D., Palacios-Santander, J. M., & Cubillana-Aguilera, L. (2020). Green synthesis of nanomaterials for biosensing. In: Inamuddin, Asiri, A. (eds.) *Nanosensor Technologies for Environmental Monitoring* (pp. 135–217). Springer, Cham.

54. Singh, J., Dutta, T., Kim, K. H., Rawat, M., Samddar, P., & Kumar, P. (2018). 'Green' synthesis of metals and their oxide nanoparticles: Applications for environmental remediation. *Journal of Nanobiotechnology*, *16*(1), 1–24.

55. Malhotra, B. D., & Ali, M. A. (2018). Nanomaterials in biosensors: Fundamentals and applications. In: *Nanomaterials for Biosensors* (pp. 1–74). https://doi.org/10.1016/B978-0-323-44923-6.00001-7.

56. Holzinger, M., Le Goff, A., & Cosnier, S. (2014). Nanomaterials for biosensing applications: A review. *Frontiers in Chemistry*, *2*, 63.

57. Kaushik, A. K., & Dixit, C. K. (eds.). (2016). *Nanobiotechnology for Sensing Applications: from Lab to Field*. CRC Press, Boca Raton, FL.

58. Kaushik, A., Jayant, R. D., & Nair, M. (eds.). (2017). *Advances in Personalized Nanotherapeutics*. Springer International Publishing, New York.

59. Crane, T. (2000). The origins of qualia. In: Tim Crane & Sarah Patterson (eds.), *The History of the Mind-Body Problem*. London: Routledge.

60. Kaushik, A., & Mujawar, M. A. (2018). Point of care sensing devices: Better care for everyone. *Sensors*, *18*(12), 4303.

61. Nair, M., Jayant, R. D., Kaushik, A., & Sagar, V. (2016). Getting into the brain: potential of nanotechnology in the management of NeuroAIDS. *Advanced Drug Delivery Reviews*, *103*, 202–217.

62. Kaushik, A., Jayant, R. D., & Nair, M. (2018). Nanomedicine for neuroHIV/AIDS management. *Nanomedicine*, *13*(7), 669–673.

63. Kaushik, A. (2019). Biomedical nanotechnology related grand challenges and perspectives. *Frontiers in Nanotechnology*, *1*, 1.

64. Naseri, M., Ziora, Z. M., Simon, G. P., & Batchelor, W. (2022). ASSURED-compliant point-of-care diagnostics for the detection of human viral infections. https://doi.org/10.1002/rmv.2263.

65. Yu, K. H., Beam, A. L., & Kohane, I. S. (2018). Artificial intelligence in healthcare. *Nature Biomedical Engineering, 2*(10), 719–731.
66. Zhu, L., & Zheng, W. J. (2018). Informatics, data science, and artificial intelligence. *JAMA, 320*(11), 1103–1104.
67. Chou, K. C. (2004). Structural bioinformatics and its impact to biomedical science. *Current Medicinal Chemistry, 11*(16), 2105–2134.
68. Saratale, R. G., Karuppusamy, I., Saratale, G. D., Pugazhendhi, A., Kumar, G., Park, Y., ... & Shin, H. S. (2018). A comprehensive review on green nanomaterials using biological systems: Recent perception and their future applications. *Colloids and Surfaces B: Biointerfaces, 170*, 20–35.
69. Kamarudin, S. F., Mustapha, M., & Kim, J. K. (2021). Green strategies to printed sensors for healthcare applications. *Polymer Reviews, 61*(1), 116–156.
70. Kumar, P., Pahal, V., Gupta, A., Vadhan, R., Chandra, H., & Dubey, R. C. (2020). Effect of silver nanoparticles and Bacillus cereus LPR2 on the growth of Zea mays. *Scientific Reports, 10*(1), 1–10.
71. Sharpe, E., Frasco, T., Andreescu, D., & Andreescu, S. (2013). Portable ceria nanoparticle-based assay for rapid detection of food antioxidants (NanoCerac). *Analyst, 138*(1), 249–262.
72. Alkasir, R. S., Ganesana, M., Won, Y. H., Stanciu, L., & Andreescu, S. (2010). Enzyme functionalized nanoparticles for electrochemical biosensors: A comparative study with applications for the detection of bisphenol A. *Biosensors and Bioelectronics, 26*(1), 43–49.
73. Manikandan, V. S., Adhikari, B., & Chen, A. (2018). Nanomaterial based electrochemical sensors for the safety and quality control of food and beverages. *Analyst, 143*(19), 4537–4554.
74. Njagi, J., Chernov, M. M., Leiter, J. C., & Andreescu, S. (2010). Amperometric detection of dopamine in vivo with an enzyme-based carbon fiber microbiosensor. *Analytical Chemistry, 82*(3), 989–996.
75. Zhao, Y., Xu, Y., Shi, L., & Fan, Y. (2021). Perovskite nanomaterial-engineered multiplex-mode fluorescence sensing of edible oil quality. *Analytical Chemistry, 93*(31), 11033–11042.
76. Chen, Q., Zhang, L., & Chen, G. (2012). Facile preparation of graphene-copper nanoparticle composite by in situ chemical reduction for electrochemical sensing of carbohydrates. *Analytical Chemistry, 84*(1), 171–178.
77. Chen, Y., Liu, L., Xu, L., Song, S., Kuang, H., Cui, G., & Xu, C. (2017). Gold immunochromatographic sensor for the rapid detection of twenty-six sulfonamides in foods. *Nano Research, 10*(8), 2833–2844.
78. Hayat, A., Paniel, N., Rhouati, A., Marty, J. L., & Barthelmebs, L. (2012). Recent advances in ochratoxin A-producing fungi detection based on PCR methods and ochratoxin A analysis in food matrices. *Food Control, 26*(2), 401–415.
79. Jiang, Y., Sun, D. W., Pu, H., & Wei, Q. (2019). Ultrasensitive analysis of kanamycin residue in milk by SERS-based aptasensor. *Talanta, 197*, 151–158.
80. Ciambelli, P., La Guardia, G., & Vitale, L. (2020). Nanotechnology for green materials and processes. In: *Studies in Surface Science and Catalysis*, vol. 179, pp. 97–116. https://doi.org/10.1016/B978-0-444-64337-7.00007-0.
81. Srinivasan, S., & Ranganathan, S. (2004). *India's Legendary Wootz Steel: An Advanced Material of the Ancient World* (pp. 1–153). National Institute of Advanced Studies, Bangalore.
82. Verma, M. L., & Rani, V. (2021). Biosensors for toxic metals, polychlorinated biphenyls, biological oxygen demand, endocrine disruptors, hormones, dioxin, phenolic and organophosphorus compounds: A review. *Environmental Chemistry Letters, 19*(2), 1657–1666.
83. Mishra, G. K., Sharma, V., & Mishra, R. K. (2018). Electrochemical aptasensors for food and environmental safeguarding: A review. *Biosensors, 8*(2), 28.
84. Prabhakar, N., Thakur, H., Bharti, A., & Kaur, N. (2016). Chitosan-iron oxide nanocomposite based electrochemical aptasensor for determination of malathion. *Analytica Chimica Acta, 939*, 108–116.
85. Cook, L. W., Paradise, C. J., & Lom, B. (2005). The pesticide malathion reduces survival and growth in developing zebrafish. *Environmental Toxicology and Chemistry: An International Journal, 24*(7), 1745–1750.
86. Priyanka, U., Akshay Gowda, K. M., Elisha, M. G., & Nitish, N. (2017). Biologically synthesized PbS nanoparticles for the detection of arsenic in water. *International Biodeterioration & Biodegradation, 119*, 78–86.

11 Agro-Industrial Wastes-Derived Carbon Nanomaterials

Synthesis and Multi-faceted Applications

Sakshi Kabra Malpani
Save the Water™

Rajesh Kumar Meena
Kalindi College

Stuti Katara
University of Kota

Deepti Goyal
Gautam Buddha University

CONTENTS

DOI: 10.1201/9781003319153-13

11.1   INTRODUCTION

Onset of the 21st century marked the tremendous revolution in agricultural and industrial sectors, population explosion, advances in urbanization, etc., which has resulted in the generation of huge amounts of agro-industrial wastes. If these wastes are not treated judiciously and get discharged into environment as such, then they can cause serious effects on our health and eco-system. In the last 50 years, agricultural production has been tripled worldwide.[1] According to a review, in 2011, the global industrial waste production was estimated to be more than 9000 million tons, out of which more than 50% industrial waste was produced in China, India, Japan, South Korea, and Australia (Asia-Pacific region).[2] In developing countries where waste management is still an unplanned sector, the proper disposal of these wastes is a big challenge. Usually, they are burnt or dumped in landfills and water bodies, pollute our environment, and are responsible for more than 25% of greenhouse gas emissions. These so-called wastes are treasure of numerous valuable components like carbon (in form of starch, cellulose, lignin, etc.), metals, metal oxides (silica, alumina, iron oxide, etc.), thus, could be of great economic, social, and industrial importance and produce various value-added products, if they are re-utilized in appropriate manner. Agro-industrial wastes such as rice and wheat husk ash, bamboo leaves, sugarcane bagasse, corn cob ash, fly ash, and plastic and tire wastes are abundant sources of carbon and hence can be used to produce carbon nanomaterials (CNMs) via cost-effective, green routes, having potential applications in diverse fields, as shown in Scheme 11.1.

The unique physical, chemical, thermal, and mechanical properties of nanomaterials showcase their special position in the category of sustainable materials that play a strong role in the agro-industrial revolution. Next to oxygen, carbon holds the place of second most abundant element

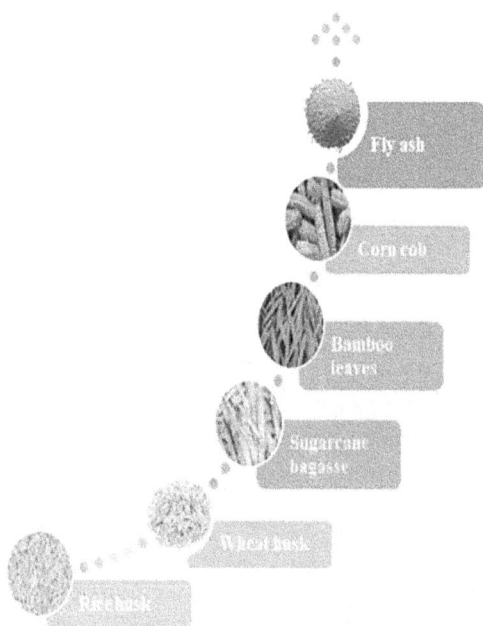

SCHEME 11.1   Different agro-industrial wastes utilized for CNMs production.

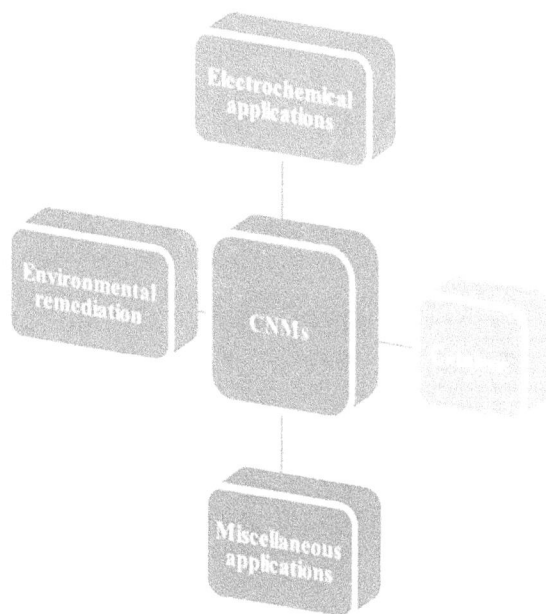

SCHEME 11.2 Various applications of CNMs.

on this planet and has extensively been used in the synthesis of CNMs. CNMs become popular in 1980s after the discovery of fullerene nanotubes, followed by the synthesis of CNTs by Iijima in 1991. The Nobel Prize winning discovery of graphene in 2D form in 2010 adds glory in the research of CNMs. In the past decade, CNMs has gained attraction in various sectors, viz. energy storage, heterogeneous catalysis, environmental remediation, technologies related to renewable energy conversion, biofuels, gas separation, and biomedical applications.[3-8] CNMs are known for their high porosity, large specific surface area, lightweight, great flexibility and elasticity, good sorbent capacity, etc., making them suitable candidates for new-age applications as shown in Scheme 11.2. Nano-activated carbon, graphene and graphene oxide, carbon nanotubes (CNTs), fullerenes, and nanodiamonds are widely used CNMs. The selection of carbon source is the main driving force toward sustainable and greener aspects of formation of CNMs, and the use of waste materials as carbon sources solves the purpose; thus, it has become a topic of significant research interest these days. CNTs are extensively studied CNMs. Highly pure single-walled CNTs and multi-walled CNTs are preferred to be produced from renewable sources containing less amount of hydrogen, to reduce the amount of by-product (amorphous carbon) generation. Similarly, if nano-activated carbon has been produced utilizing conventional precursors such as coal, wood, and petroleum residues, the whole process becomes expensive and hazardous; thus, agro-industrial wastes that cause pollution in the ecosystem can be used for the production of CNMs in an effective, greener manner. This holistic approach to utilize waste materials to generate beneficial CNMs is quite easier, cost-effective, and convenient and thus could be projected for mass-scale greener production of CNMs.

This chapter presents a collection of various reports and research work that gained momentum in the past decade regarding production and derivation of green, environmentally friendly CNMs from renewable sources, agro-industrial wastes. It gives an overview of several types of agro-industrial wastes prevailing in the synthesis of CNMs, followed by different types of CNMs produced from these methods. Further, the general methods used to generate these CNMs were discussed. Then various characterization techniques used to study the analytical features of the as-synthesized CNMs were figured out with the help of reported results. We have also summarized different applications of these CNMs, such as catalysis, electrochemical and biomedical applications, environmental remediation, adsorption, and sensing. At the end, conclusions drawn from this study are presented with some flaws of existing in the procedures and prospects of this work in coming years.

11.2 AGRICULTURAL AND INDUSTRIAL WASTES AS SOURCES

11.2.1 BAMBOO LEAVES

The bamboo plant is mostly found in tropical, subtropical, and temperate zones of Asian countries.[9] Every year, approximately 10 million tons of dried bamboo leaves are burnt in the environment, which causes lots of environmental issues. East Java has an abundance of bamboo leaves, which are underutilized and are of low economic worth. Therefore, the proper disposal of this natural resource is mandatory. The main components of bamboo leaves are hydrocarbon compounds, such as cellulose, which may be used to synthesize carbonaceous materials. Due to the less toxic nature, high solubility in water, desirable chemical inertness, excessive biocompatibility, and smooth functionalization processes, bamboo leaves can be utilized as a biomedical agent.[10] As an amorphous carbon material, it has a high surface area and is suitable for numerous applications, including anode materials for high energy density storage batteries, water absorbents, and many others.[11–15] Various studies have confirmed the antioxidant, antibacterial and antimicrobial activities of bamboo leaves. Waste bamboo leaves are reported to be utilized in the synthesis of carbon dots.[16]

11.2.2 CORNCOBS

Corn cob is mostly found in the tropical or subtropical regions of Indonesia. It is an agricultural waste of corn production, having high carbon content (cellulose (41%), hemicellulose (36%), lignin (6%)) and low trace element content.[17] Every year a huge amount of this waste is generated and disposed to the environment without any safety measures. This can cause severe environmental issues. To overcome these problems corn cob can be utilized to produce several value-added chemicals such as catalysts, fuels, and adsorbents.[18] As a carbon source, corncobs are excellent precursors for making porous CNMs with high specific surface areas as well as the special biogenetic textures of corn cob are responsible for the production of carbon nanomaterials.[19,20] The potential applications of corn cob include its use as a polishing tool, which has a smooth texture, as well as a pipe for absorbing tobacco cigarette smoke. Furthermore, corncobs can be used to synthesize CNMs, such as CNTs and graphite when carbonized at a specific temperature.[17,21,22] The high surface area of corn cob ash makes it a suitable candidate to act as catalyst or catalyst support materials.[23] Also corn cob-derived activated carbon has been utilized in energy storage applications due to its high surface area and low cost. In a recent study, corn cob is utilized as a cost-effective carbon precursor to produce biochar nanomaterial.[24]

11.2.3 RICE HUSK

Rice is one of the most important plants in the world, serving as a carbohydrate source for humans. Rice husk is a by-product of the rice milling industry and mainly contains almost 40% of carbon as an organic component.[25,26] The global production of rice till 2018 was reported approximately 996 million tons, which consequently responsible to produce approximately 199 million tons of rice husk as waste. Thus-produced rice husk disposed to landfill and cause environmental pollution. In addition, the generation of rice husk ash on the burning of rice husk can also cause several respiratory diseases in human beings.[27] Consequently, conversion of rice husk into value-added materials is not only worthwhile in terms of environmental protection also it is very helpful in the generation of industrially important materials. Several studies reported the utilization of rice husk ash for the synthesis of CNMs such as biochar, graphene, and graphene oxide, CNTs and many more.[28–30] In a recent study, the synthesis of porous silica-carbon composite has been reported using rice husk as carbon precursor. Thus, synthesized nanocomposite consists of high surface area and increased strength.[31]

11.2.4 SUGARCANE BAGASSE

Sugarcane bagasse is a by-product of sugar industry, mainly composed of lingo cellulosic content with more than 70% of the total content along with some trace amount of ash.[32] Approximately 270 kg of dry sugarcane are produced from every one ton of sugarcane used in the process. Due to high production of sugarcane bagasse, its proper utilization and safe disposal is of environmental concern. A research has demonstrated the use of this abundant waste in the production of carbon-based nanomaterials in a novel way. The process involved the use of gases released during the pyrolysis of sugarcane bagasse and other wastes (Boston, USA).[33] Another study demonstrated the synthesis of heterogeneous catalyst using pyrolysis gases of sugarcane bagasse. The synthesis of carbon-based nanomaterials from sugarcane bagasse can also help to reduce the cost of the production of nanomaterials.[34]

11.2.5 WHEAT HUSK

Wheat is the commonly used cereal, which is produced about 2.6 Mtons globally.[35] India, Canada, Europe, Russia, and USA are the major wheat-producing countries. Wheat husk is generated as waste during milling of wheat. Wheat husk makes up approximately 15%–20% of wheat grain. An average of 125 million tonnes of wheat husk are produced worldwide.[36] Wheat husk, an important lignocellulosic agricultural waste, is mostly utilized in making sheets for animals, as food for animals, and as a fuel.[37,38] Recently, several researchers have been working on its utilization in the production of value-added chemicals. Due to high carbon content (more than 70%), wheat husk can be taken as an efficient precursor for the synthesis of CNMs.[39]

The conversion of wheat husk ash into nanoporous activated carbon is well reported in the literature. In this study, three samples of activated carbon, calcined at three different temperatures were synthesized and successfully employed as an energy storage device showing maximum capacitance (271.5 F/g).[40,41]

11.2.6 INDUSTRIAL WASTES

Increasing population has led to increase the global consumption of goods, which is responsible for increased industrial activities. Such industrial activities, increases the amount of industrial waste. Therefore, it has focused increasing attention of chemists and researchers to find out some alternative pathway to minimize the production of such industrial wastes and their conversion into value added chemicals. Industrial wastes are mainly classified into three categories, namely gaseous industrial wastes (hydrocarbons such as methane, ethane, acetylene, and ethylene), liquid industrial wastes (liquid hydrocarbons from petroleum industries), and solid industrial wastes (plastic wastes, fly ash, etc.) containing high percent of carbon content. The conversion of these industrial wastes into CNMs is gaining popularity due to the cost-effective and sustainable nature of the process.[42]

The gases released in petrochemical down streaming such as methane, carbon monoxide, carbon dioxide, propane, and ethylene, can be utilized to produce CNMs in the presence of a catalyst.[43,44] Similarly, some liquid by-products of petroleum industries such as heavy hydrocarbon residue can also be used as a carbon precursor. Solid wastes such as plastic create huge environmental issues. The conversion of such plastic wastes into valuable CNMs can be an important and eco-friendly alternate for industries as well as for the environment. Recently various researchers have been adopted various techniques for conversion of plastic wastes such as PVC, PP, PE, LDPE, and HDPE into CNTs.[45] Fly ash, a solid waste of thermal power plants, has also been utilized in synthesis of CNTs under optimized reaction conditions.[46] The presence of metal oxides in fly ash, makes it a suitable catalyst used for production of CNTs.[47]

11.3   DIFFERENT CARBON-BASED NANOMATERIALS

Nanomaterials are the materials with particle size ranging from 1 to 100 nm. Among various types of nanomaterials, CNMs found important applications in various fields such as in batteries, nanocomposites, textile industries, catalysis, and environmental remediation.[48,49] CNMs are classified based on their geometrical structure; for example, if the nanoparticles are tube-shaped, they are named as CNTs, and if the nanoparticles are spherical, they are named as fullerenes. Naturally occurring CNMs are very rare, so most of the CNMs are synthetic and their physical, chemical, and electrical properties can be modified easily and are related to their hybridization. Based on the bonding between carbon atom and its neighboring atoms, most of the CNMs show sp, sp^2, or sp^3 configuration.

CNMs mainly include CNTs, fullerenes, graphene, nanoporous activated carbon and their derivatives such as graphene oxide (GO) and nanodiamonds (Figure 11.1). Fullerene was discovered in 1985, containing spherical shape nanoparticles, while CNTs were discovered in 1991, containing graphite sheet rolled in the form of a tube in the nanometer range. In the recent years, graphene is discovered with a single carbon atom layer in the form of a sheet. CNMs show potential applications in computer and electronics,[50] data processing and storage units,[51] sensors[52] and composite materials,[53] and most recently biomedicine.[54]

11.3.1   CNTs

The most widely used carbon nanomaterials are CNTs. They are 1D analogue of 0D fullerene molecules. Theoretically, CNTs are made-up of a cylindrical graphene sheet and capped with a spherical fullerene. The combined properties of fullerene and graphene are responsible for exceptional behavior of CNTs. Based on the number of cylindrical graphene sheets, CNTs can be categorized into two parts: single-walled CNTs (SWCNTs) and multi-walled CNTs (MWCNTs). In some studies, double-walled CNTs have also been reported. In general, SWCNTs have a diameter in the range 0.5–3 nm and a length of a few μm, while MWCNTs have 5–50 nm diameters and around 10 μm

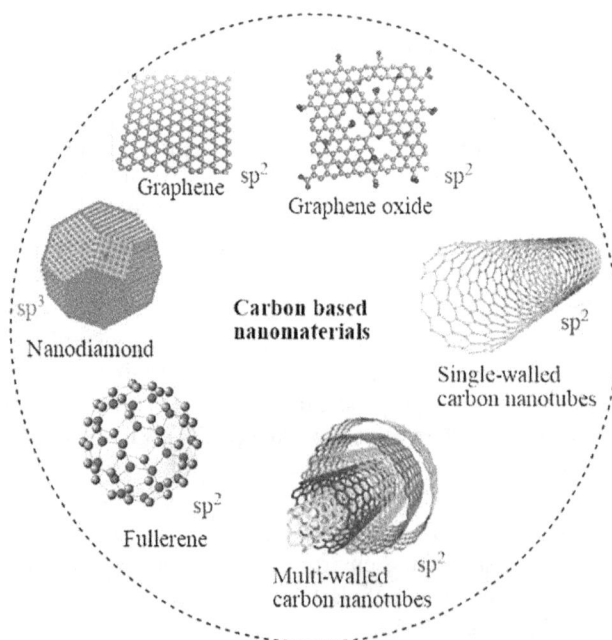

FIGURE 11.1   Carbon-based nanomaterials.

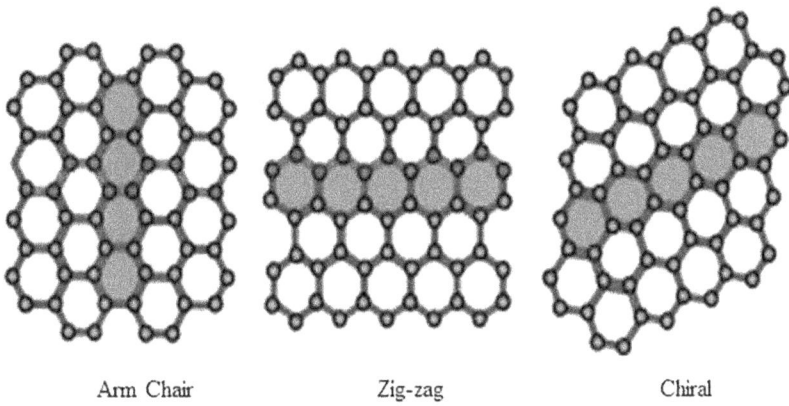

Arm Chair   Zig-zag   Chiral

FIGURE 11.2   Different types of SWCNTs.[55]

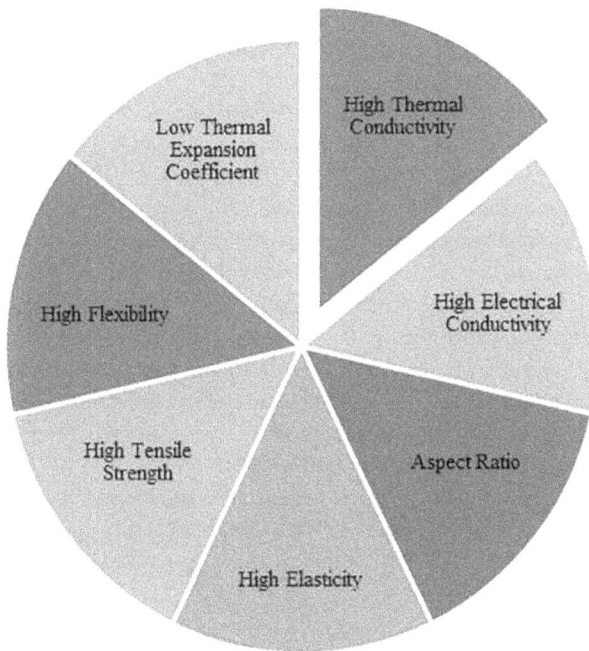

FIGURE 11.3   Properties of CNTs.

lengths. Depending upon the way of wrapping into a cylindrical shape, SWCNTs are further classi-fied into three types, namely armchair, zig-zag, and chiral (Figure 11.2).[55]

SWCNTs can be compared by MWCNTs in the following manner:

CNTs show exceptional properties such as rigidity, strength, and elasticity due to their unique structure as compared to other materials (Figure 11.3).

On comparing the properties of SWCNTs and MWCNTs, we observed that SWCNTs display high length to diameter ratio and exhibit extraordinary electrical properties which depend on their chirality, on the other hand MWCNTs exhibit extraordinary mechanical properties due to multiple carbon layers. CNTs composed of graphite sheets (one carbon atom bonded with three neighboring atoms via strong bonds) are also considered as high strength fibers. SWCNTs are much stiffer than the steel because of one plane graphitic C-C bonds and are very resistant to damage.[56]

Conventional methods for the synthesis of CNTs are arc discharge, laser ablation and chemical vapor deposition (CVD) method. Among these three methods, chemical vapor deposition is the most utilized method due to its low temperature requirement. CNTs have been utilized in many industrial applications due to their exceptionally high mechanical strength and flexibility. Similarly, due to their high electrical conductance, CNTs have also been applied in the production of nano-electronic circuits and in highly efficient e⁻ emission devices. Due to its luminescent properties, it is also being utilized in several illuminating substances.[57]

11.3.2 BUCKMINSTERFULLERENE

A fullerene may be defined as an allotrope of carbon consisting single and double bonded carbon atoms, arranged in a closed or partially closed network. The most used fullerene is Buckminster fullerene, also known as buckyball is denoted by C60. The structure of C60 consists of 60 carbon atoms (sp^2 hybridized) arranged in a cage-like structure, which is made up of 20 hexagons and 12 pentagons and is stabilized due to resonance (Figure 11.4). The physicochemical properties of C60 are an intermediate of an aromatic compound and a straight chain aliphatic compound. The resonating structure of C60 is supported by its electronic configuration and bonding geometry for each carbon atom.[58]

The physicochemical properties of fullerenes are summarized below.

The unusual stability of C60 makes it a suitable starting material for chemical transformations. Owing to the unique physical, chemical, and physicochemical properties, C60 has been utilized in synthesis of several nanomaterials such as nanorods, nanotubes, and nanosheets. Moreover, C60 can form stable crystalline nanoparticles with the particles size ranges from 25 to 500 nm in various solvents.[59,60] Furthermore, due to the high ionization energy and electron affinity, it can easily donate or accept the electrons. Owing to this behavior, C60 can be used in synthesis of various batteries and electronic devices. In addition, only a specific surface site can react with another molecule, so that the remaining surface gets unaffected. Different fullerene derivatives can be prepared by replacing the surface atoms of C60 by other functional groups such as phenyl. C60 molecules can also be act as superconductor only at very high temperature. This superconductivity can be produced by reaction of C60 with alkali and alkaline earth metals.

The most common method for the synthesis of fullerene is arc discharge method, in which carbon is heated in a furnace at very high temperature. Thus, obtained carbon clusters are then annealed at

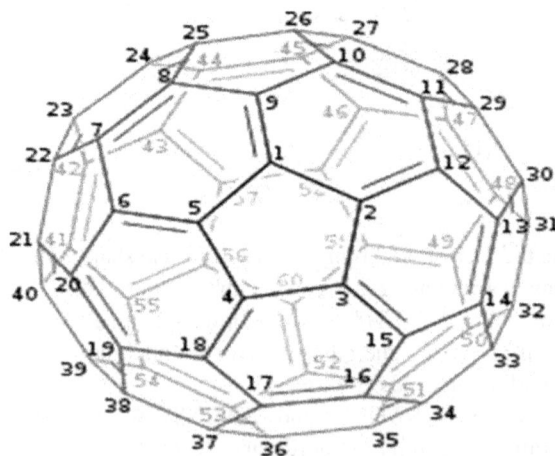

FIGURE 11.4   Chemical structure of Buckminsterfullerene.

very high temperature around 1000°C in a quartz tube. The condensed fullerene then collected in a water cooler trap.[61]

11.3.3 GRAPHENE AND GRAPHENE OXIDE

Graphene consists of a honeycomb lattice structure, formed by sharing sp^2 electrons with their three neighboring carbon atoms. It is also known as planer monolayer graphite in which each carbon atom makes bonds with the three other sp^2 hybridized carbon atoms and forms a benzene layer, leaving one free electron on each atom (Figure 11.5a). Each benzene layer in graphene exhibit 2 delocalized pi electrons, which are responsible for electrical conductivity of graphene. The unit cell of graphene contains two carbon atoms with C-C bond length 0.142 nm and lattice constant of 2.46 Å (Figure 11.5b). Graphene is considered as highly stable compound consisting of combined structure of graphite, CNTs and fullerenes. Its thickness is very less (0.35), i.e., 1/200,000th the diameter of a human hair.

Graphene is considered as a most thrilling material due to its unique properties such as high thermal and electrical conductivity, stability, and high specific surface area.[62] These properties are responsible for its potential applications in various fields such as transparent conductive electrodes. Though, CNTs and graphene show almost similar electrical, thermal, and optical properties, but graphene shows different electronic features than CNTs. Moreover, the elasticity of graphene is much more as compared to other CNMs, which makes it suitable for generating nanoengineered materials.[63]

There are two types of graphene, namely graphene oxide (containing C, H, and O atoms) and reduced graphene oxide (containing more number of C atoms than O atoms), are reported in the literature. Graphene oxide (GO) can be synthesized by oxidation of graphene in the presence of some strong oxidizing agents such as H_2SO_4. The presence of oxygen containing functional groups

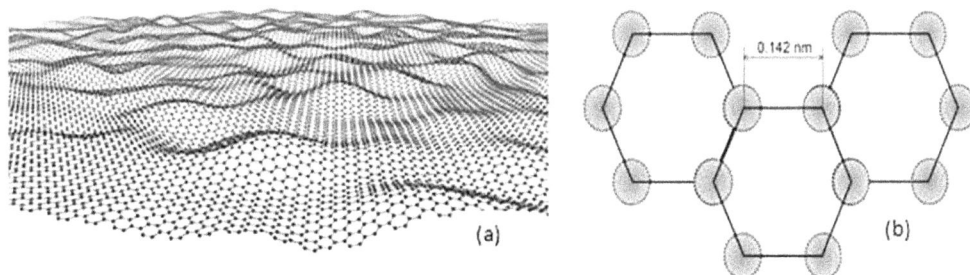

FIGURE 11.5 (a) Graphene monolayer; (b) unit cell structure of graphene.

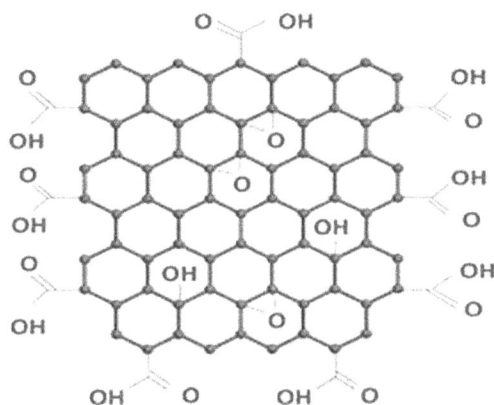

FIGURE 11.6 Chemical structure of graphene oxide.

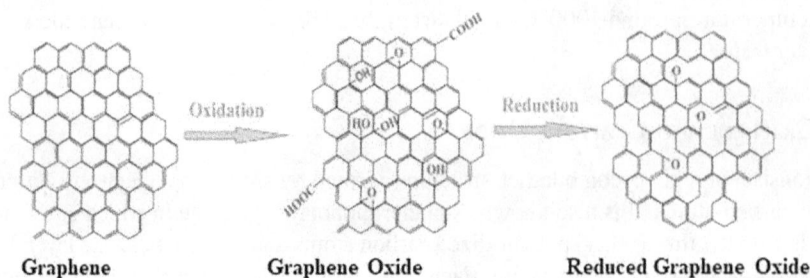

Graphene   Graphene Oxide   Reduced Graphene Oxide

FIGURE 11.7   Schematic representation of conversion of GO to rGO.[67]

TABLE 11.1
Physicochemical Properties of Graphene

Electronic properties	No band gap due to overlapping of π e$^-$
	High mobility of e$^-$ and thus high electrical conductivity
Thermal properties	Can conduct in-plane and inter-plane conductivity
	High in-plane conductivity
Mechanical properties	High tensile strength of 1.3×10^{11} Pa
	Light in weight
	Increases the strength of composites
Flexibility	High

on the surface of the graphene oxide is responsible for its increased hydrophilicity and greater layer separation (Figure 11.6). Thus, improved hydrophilicity, makes GO capable to react under ultrasonic conditions, which generate a few graphene layers with high hydrophilicity. GO show different physical and chemical properties including its high solubility in various solvents such as water, organic solvents, and various matrices. GO considered as the highly adjustable material for several applications owing to oxygen rich functional groups as well as the graphene backbone.

Despite this, some researchers have been observed that these oxygen containing functional groups are also responsible for reduced electrical conductivity and structural defects in GO.[64–66] This limitation can be eliminated by partial reduction of GO, which than convert into reduced graphene oxide (rGO). The most common methods for reduction of graphene oxide are chemical, thermal, and electrochemical reduction. The conversion of GO into rGO is shown in Figure 11.7. Apart from this, some other methods are also reported such as annealing, laser, and microwave reduction. Thus, the produced rGO shows improved properties such as electrical and thermal conductivity.[67]

The synthesis of graphene can be done in four different ways:

- Chemical vapor deposition
- Chemical reduction
- Mechanical exfoliation
- Thermal exfoliation.

The various properties of graphene are summarized in Table 11.1.

Owing to these physical and chemical properties, graphene shows potential applications in various fields, viz. in sensors, batteries, nanocomposites, polymer composites, catalyst support, medicine, and electron emission displays.

11.3.4 NANOPOROUS ACTIVATED CARBON

Nanoporous activated carbon possesses high surface area with highly porous structure, due to which it is considered as an important absorptive material and is being utilized as an air- and water-purifying agent in several industries.[68] It can be produced from lignocellulosic precursors at high temperature. Other potential applications of nanoporous activated carbon are electrochemical energy storage and gas storage. Such applications are due to the presence of porous network embedded with some heteroatoms such as O, N, and S (Figure 11.8), which are located at the ends of the basal panels. The presence of heteroatoms depends upon the synthesis method, precursor used, and the post-synthesis functionalization of activated carbon. The porosity of activated carbon is the most important property, and according to the size of the pores, nanopores can be classified as:[69]

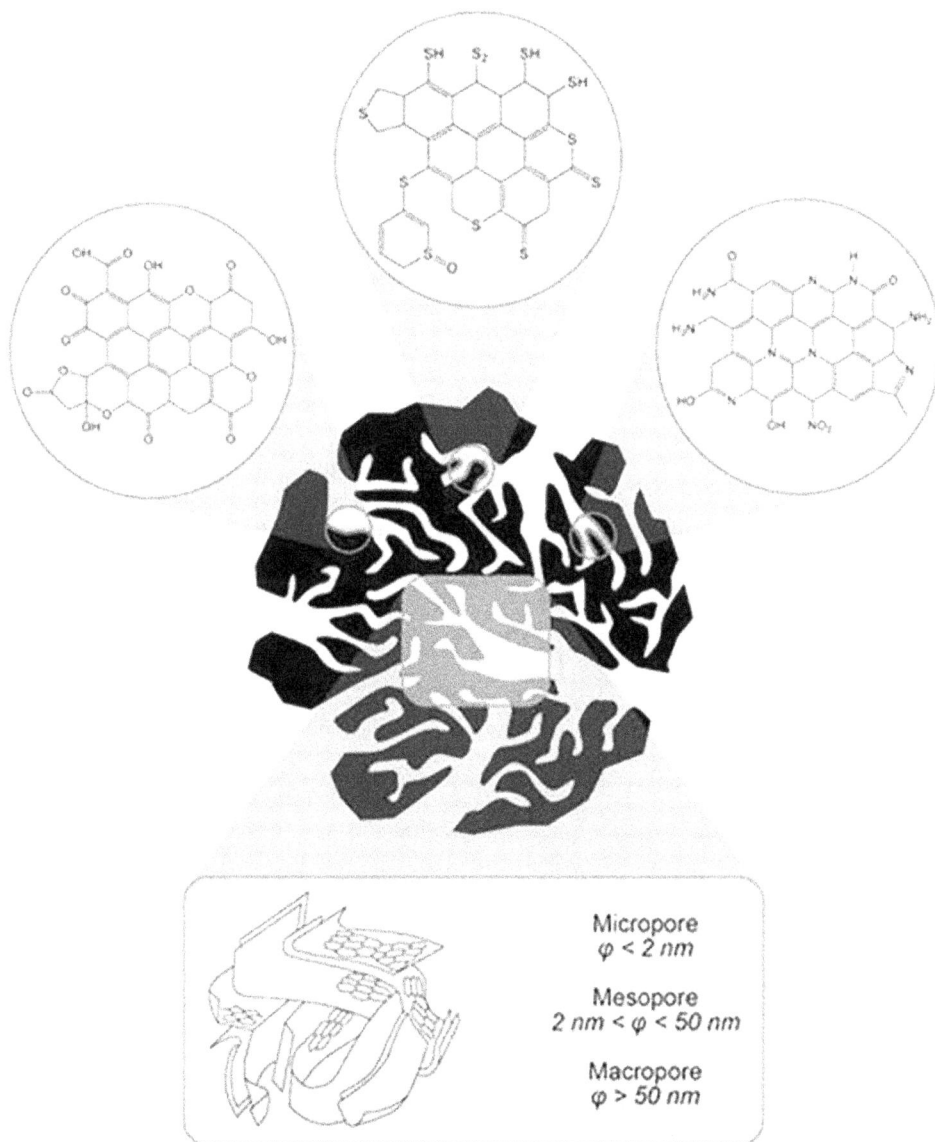

FIGURE 11.8 Schematic representation of the structure of nanoporous activated carbon.[69]

- Micropores (pore size < 2 nm)
- Mesopores (pore size from 2 to 50 nm)
- Macropores (pore size > 50 nm).

Agro-industrial waste-derived nanoporous activated carbon shows distinct properties. This may be due to the fact that these nanoporous activated carbon can be produced via different methods such as pyrolysis and activation of different plant-based materials such as wood chips, lignocellulosic materials, rice husk, wheat husk and rice straw etc. The most common application of nanoporous activated carbon is adsorption, by which various pollutants from wastewater can be removed efficiently.[70]

11.4 SYNTHESIS AND CHARACTERIZATION OF CARBON-BASED NANOMATERIALS

11.4.1 SYNTHESIS OF GRAPHENE

GO can be easily derived from agro-industrial wastes. After juice was extracted from sugarcane, the remaining fiber (sugarcane bagasse) was repeatedly crushed and ground well to produce fine powder. About 0.5 g of ground sugarcane bagasse powder mixed with 0.1 g of ferrocene was placed in a crucible and put directly into a muffle furnace at 300°C for 10 minutes under atmospheric conditions. The as-produced black solid is graphene oxide as shown in Figure 11.9.[61] Graphene microcrystals were reportedly prepared from lignin refined from sugarcane bagasse by following two technical schemes: the pyrolysis reaction of lignin in a tubular furnace at atmospheric pressure, and the hydrothermal carbonization of lignin at lower temperature, followed by pyrolysis at higher temperature.[71] Graphene synthesis from RHA is also reported, with simple KOH treatment, consisted of nano-sized crystalline graphene and corrugated graphene with atomically smooth surfaces and edges. The nano-sized crystalline graphene exhibited a monolayer or multilayer structure with clean edges, whereas the corrugated grapheme consisted of domains a few nanometers in size (200–300 carbon atoms), showing clear grain boundaries as well as non-hexagonal carbon rings (heptagons and pentagons). It was clearly demonstrated that RHA could be converted to high-value-added graphene in a rapid, reliable, scalable, and cost-effective manner.[72]

11.4.1.1 Characterization

The XRD image of graphene as synthesized from sugarcane bagasse (Figure 11.10), shows a peak at $2\theta = 11.6°$ indicating that the agricultural sugarcane bagasse is fully oxidized into GO with the

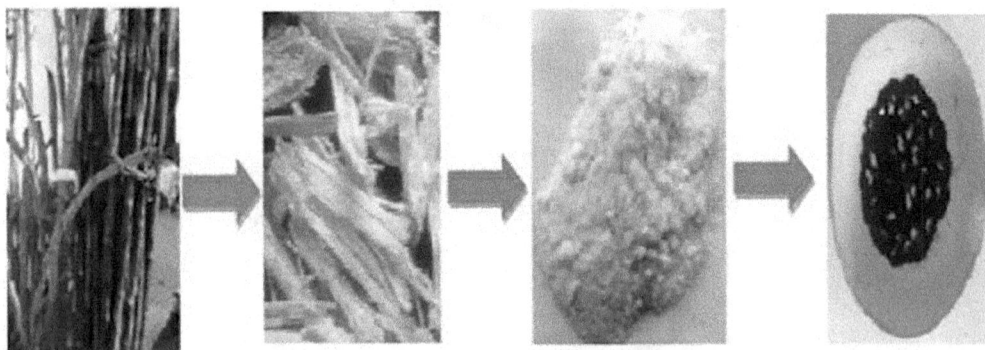

FIGURE 11.9   GO synthesis from sugarcane bagasse (Somanathan et al.).[61]

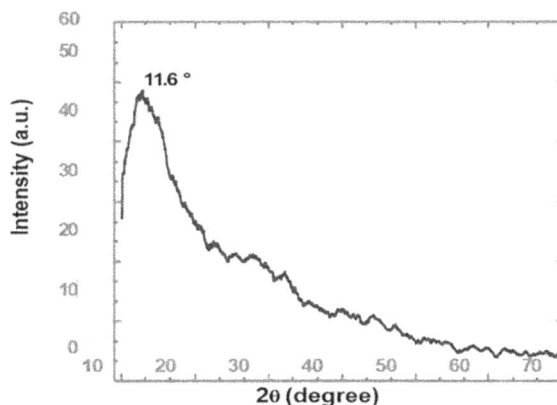

FIGURE 11.10 X-ray diffraction (XRD) pattern of GO obtained from sugarcane bagasse.[61]

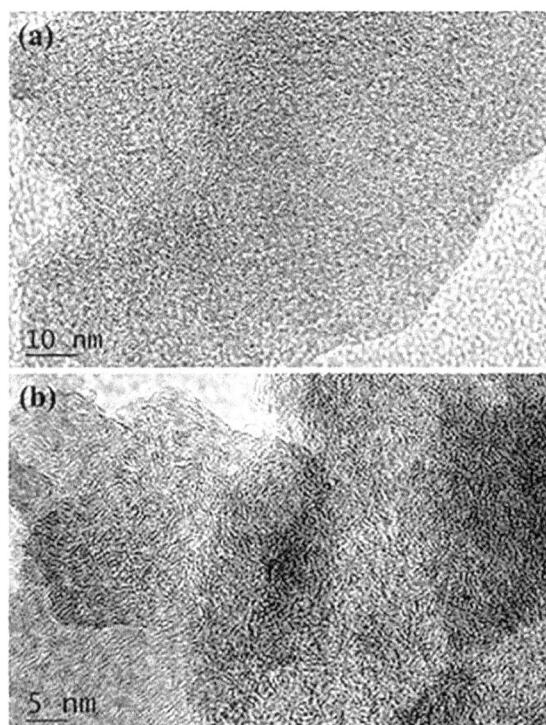

FIGURE 11.11 The HRTEM images of the graphene microcrystals produced from sugarcane bagasse.[71]

interlayer distance of 0.79 nm. Single step oxidations of sugarcane bagasse yield GO whose 2θ peaks are quite similar to the peaks reported in some past research work. Also, this XRD pattern resembles with graphitized 2D-structures made of GO sheets.[61]

In another work, graphene microcrystals have been formed from sugarcane bagasse. In XRD pattern of these microcrystals, a peak around $2\theta = 25–26.5°$ appeared due to the reflection of its 002 planes. On increasing temperature, the intensity of diffraction peak of the 002 plane has increased and it moved from 25° to 26.5°. This peak value at 26.5° is attributed to the interlayer distance, 0.34 nm, present in the graphene microcrystals. It reveals the excellent graphitization of graphene microcrystals samples at elevated temperatures. The HRTEM images of the graphene microcrystals

(Figure 11.11) also confirm the formation of glass-kind of graphene microcrystals along with existence of some crystalline fragments on graphene sheets. Here, generally graphene fragments are composed of 3–5 parallel nanographene sheets in the size range of 3–5 nm.[71]

11.4.2 Synthesis of Activated Carbon

Nanoporous activated carbon can be derived from rice husk by using KOH treatment. Rice husk was thoroughly washed, dried, powdered, and then carbonized at 500°C/1 h under nitrogen atmosphere. They were then mixed with KOH solution in different weight ratios (KOH:carbonized rice husk = 1:1, 2:1, 3:1, 4:1, 5:1). The mixture was then dried at 110°C for 24 hours followed by heating between 600°C and 900°C for 0.5–2 hours. Resultant activated carbon was filtered, washed with water up to pH =7 and again dried at 110°C for 24 hours (Figure 11.12).[73]

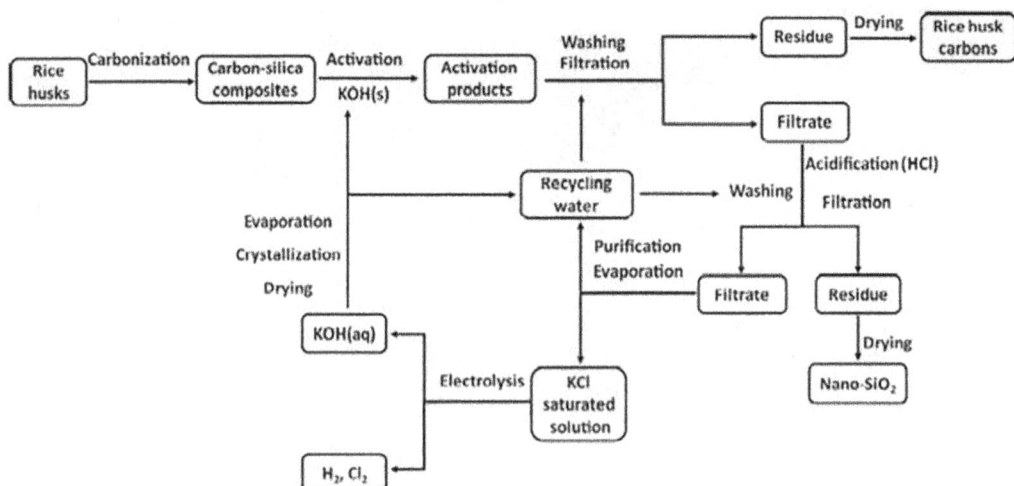

FIGURE 11.12   Activated carbon synthesis from rice husk. (Reproduced with permission.)[73]

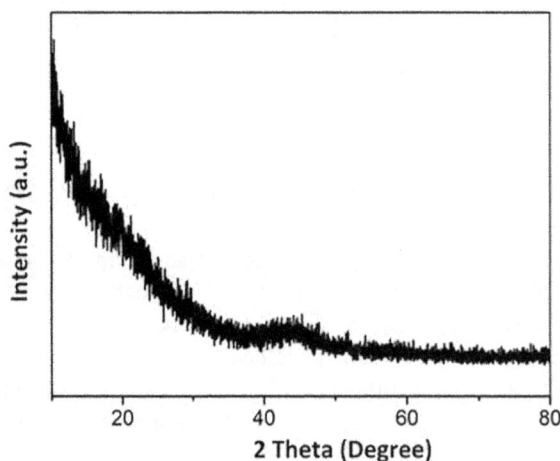

FIGURE 11.13   XRD pattern of activated carbon derived from rice husk. (Reproduced with permission.)[73]

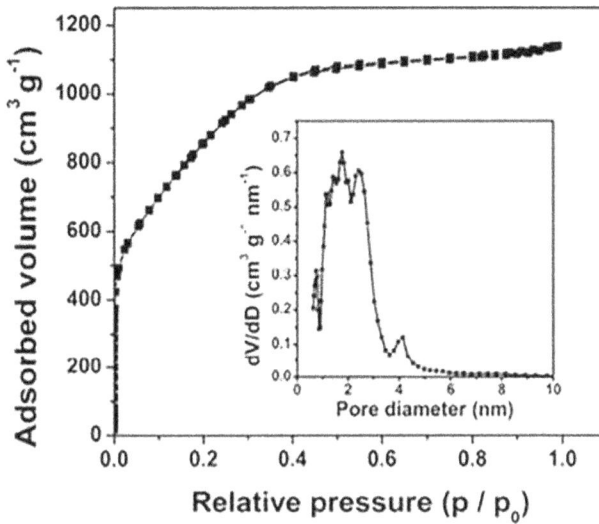

FIGURE 11.14 N_2 adsorption-desorption isotherms and pore size distribution of activated carbon.

11.4.2.1 Characterization

XRD pattern of activated carbon derived from rice husk reported above,[72] showed two broad peaks, between $2\theta = 15°$ and $30°$ and $44°$ attributed to [002] and [100] planes of graphite plane, thus showing amorphous nature of the as-produced activated carbon (Figure 11.13). Sometimes disappearance of [002] diffraction peak suggests vanishing of crystalline fragments due to intercalation of potassium during KOH treatment and generation of porosity in activated carbon formed here.

In Figure 11.14, type I isotherm can be seen for activated carbon derived from rice husk. It also reveals high microporosity, narrow pore size distribution in range of 0.5–4 nm and larger specific surface area of activated carbon.[73]

11.4.3 SYNTHESIS OF CNTs

To obtain CNTs from rice husk, they are properly cleaned with water under sonication, dried at 65°C for 24 hours, then grinded to form powder.[30] This dried powder (120 mg) was mixed with ethanolic solution of ferrocene, which acted as catalyst and placed as a substrate on an aluminum sheet, covered with aluminum casing and kept at center of quartz tube placed in a microwave oven for 38 minutes deposition time. This process is known as plasma processing. After this, the sample was cooled down to room temperature slowly and a black soot-like material appeared on the top surface of aluminum case, as shown in Figure 11.15.

CNTs can also be synthesized by using fly ash as carbon and acetylene as co-carbon precursors. About 1 g of dried, sonicated, ultrafine grey fly ash was heated in quartz boat, up to 650°C maintaining the growth time and chamber pressure (Salah et al. 2016).[74] After attaining desired temperature, a mixture of acetylene and argon gas is flowed inside in an optimal flow rate. After completion of process, black-colored product was obtained. Hintsho et al. utilized South Asian Coal fly ash to form CNTs by using CVD method. Coal fly ash used in this work was heated between 400°C and 700°C and then acetylene gas was purged at 100 mL/min for 30 minutes. After completing this reaction time, acetylene flow was stopped, and reactor was cooled at room temperature under hydrogen atmosphere.[75]

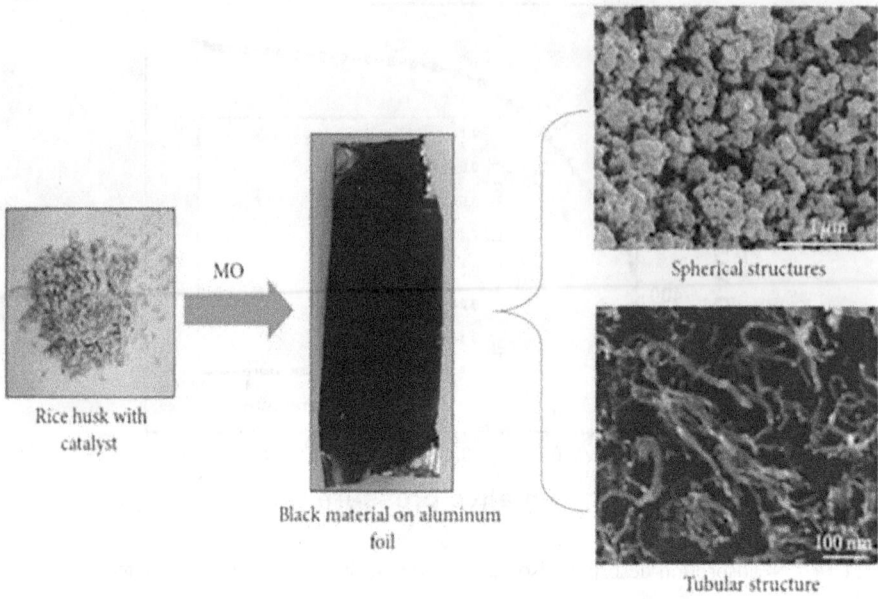

FIGURE 11.15   Synthesis of CNTs from rice husk[30] (Reproduced with permission.)[73]

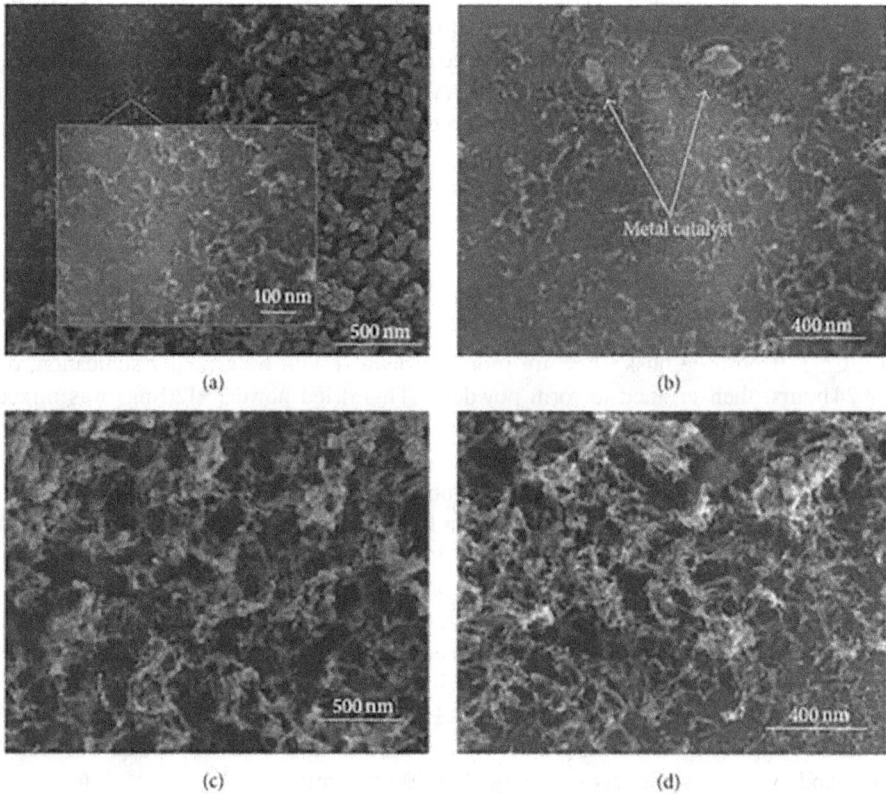

FIGURE 11.16   FESEM images of rice husk-derived CNTs.[30]

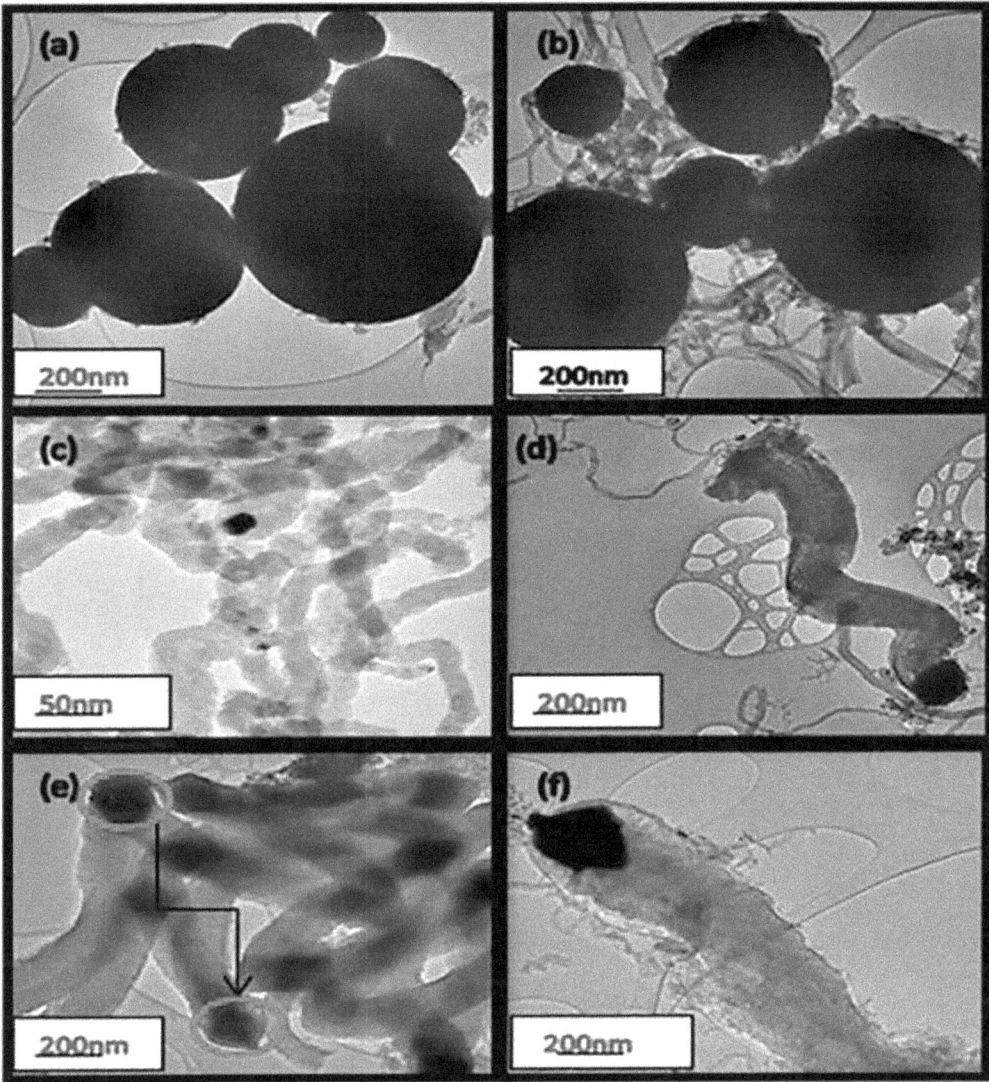

FIGURE 11.17 TEM images of (a) coal fly ash and synthesized carbon nanofibers/CNTs at (b) 400°C, (c) 500°C, (d) 600°C and (e, f) 700°C.[75]

11.4.3.1 Characterization of CNTs

FESEM images of CNTs derived from rice husk showed twisted, long, web-like network for most of CNTs.[30] While some onion-like carbon nanostructures can also be seen in this image. They also suggest that surface morphology of CNTs is greatly impacted by nature and amount of catalyst used in the process (Figure 11.16).

TEM images of fly ash and CNTs (Figure 11.17) subsequently derived from the fly ash in the work stated above[75] shows how spherical, glassy, smooth fly ash agglomerates converted into irregular, tubular CNTs on activation with acetylene. Later, due to higher temperature treatment, large carbon nanofibers got transformed into small CNTs.

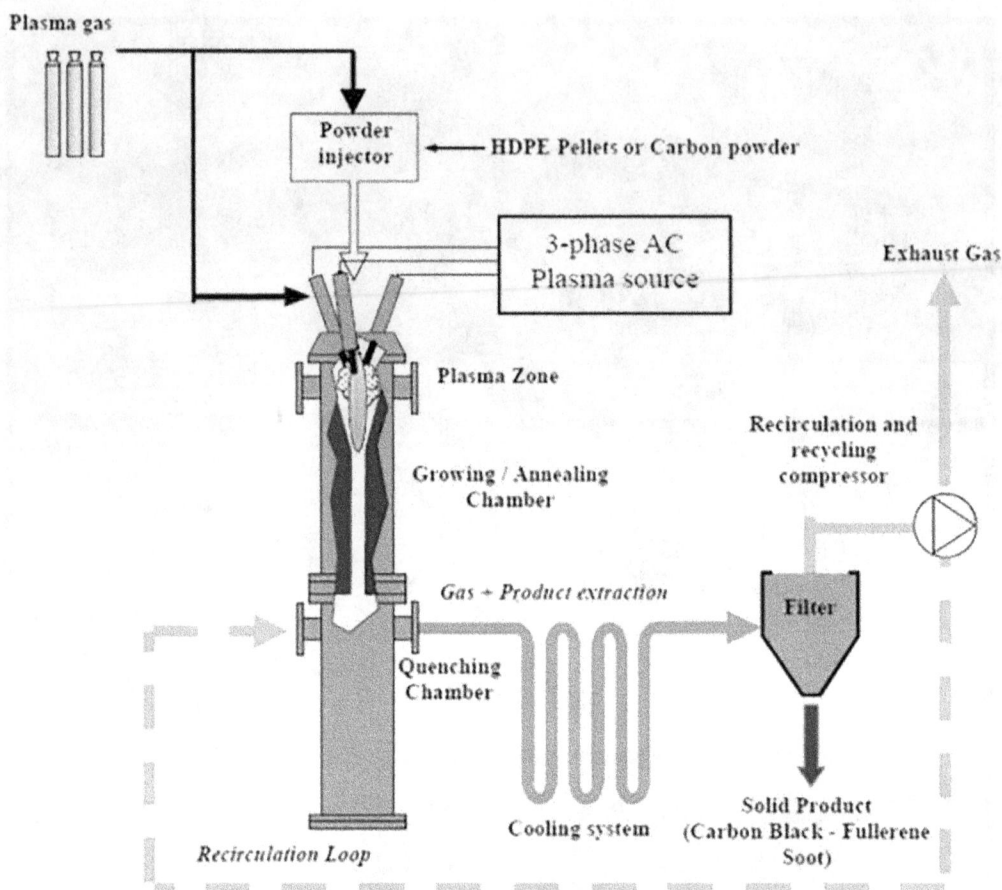

FIGURE 11.18 Synthesis of fullerenes in plasma reactor using plastic waste as carbon precursor.[77]

11.4.4 SYNTHESIS OF FULLERENE

Traditionally, fullerenes can be synthesized from simple carbon precursors such as CH_4 and C_2H_4 on metallic nanoparticles templates using various techniques such as laser ablation, arc discharge, arc jet plasma, and combustion[76] and has versatile applications like other CNMs. Although fewer reports are available on derivation of fullerenes from agro-industrial wastes, we have tried to sum up them here. Fabry and Fulchery investigated the production of fullerenes from industrial wastes such as plastic waste and tire wastes by using three-phase AC thermal plasma technique during thermochemical processing.[77] This technique facilitates synthesis of fullerenes from low thermal value organic, biomass and industrial wastes. In this process, waste carbon precursors were mixed with N_2 to form aerosols and injected inside plasma reactor (Figure 11.18). When they passed hottest temperature zone, plasma zone (>3,500°C), they were partially vaporized, which were then re-condensed on non-vaporized part during annealing at around 1700°C. The mixture of fullerene and plasma gas is quenched from plasma region and rapidly cooled to room temperature by using gas recirculation loop and then filtered to get purified product.

Biomass is also economical, carbon-rich, non-toxic waste source, which is easily available in huge amounts and have been extensively reported for CNMs synthesis.[28] For better understanding about analytical features of fullerenes, we are giving synthesis and characterization of fullerene-like CNMs derived from a biomass source, brown algae. Marriott and team used porous carbon materials derivatized from biomass sources – alginic acid-derived Starbon and calcium alginate-derived mesoporous

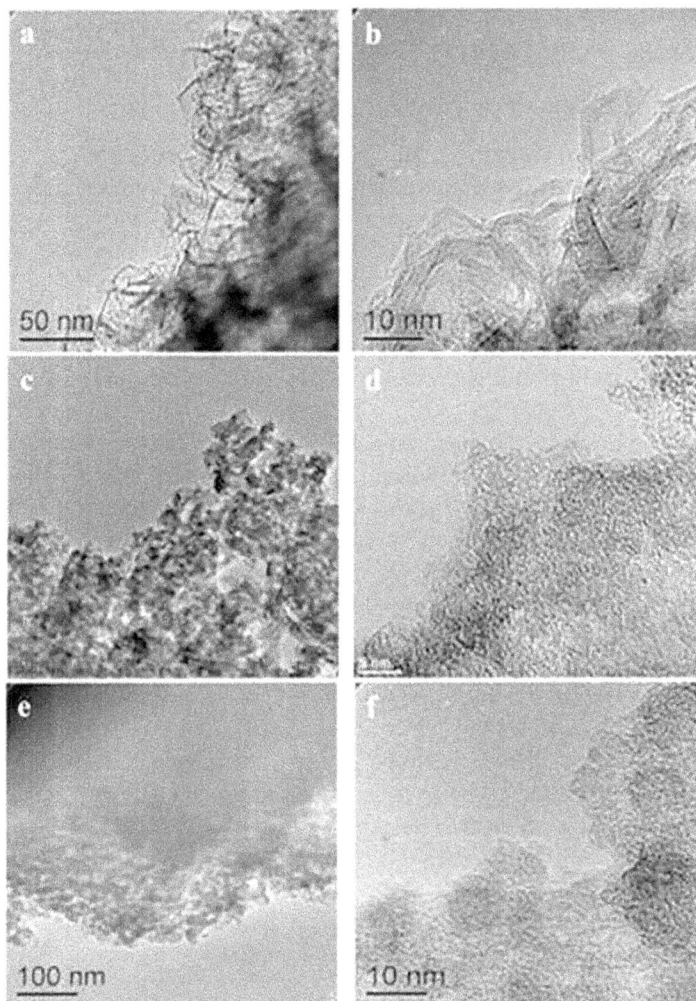

FIGURE 11.19 High-resolution TEM images of (a) and (b) commercial carbon precursor; (c) and (d) fullerene-type CNMs at 800°C; (e) and (f) fullerene-type CNMs at 1,000°C. (Reproduced with permission.)[78]

carbon spheres, which were pyrolyzed in ceramic furnace at 800°C–1,000°C for 20 minutes in the presence of N_2.[78]

11.4.4.1 Characterization of Fullerenes

UV-Vis analysis of fullerenes extracted from industrial wastes by 3-phase AC thermal plasma technique showed the presence of C60 and C70 fullerene molecules attributed to two absorption peaks at 330 and 470 nm, respectively.[77] High-resolution TEM images of commercial carbon precursor show the presence of closed loop, curved, graphitic stacks (Figure 11.19a and b), while Figure 11.19c-f shows the transformation of biomass into fullerene-like CNMs at different temperatures.[78] Here ordered graphitic planes were minimized to 2–3 layers. In addition to graphitic hexagonal structures, the presence of carbonaceous pentagons and heptagons to form very closed-loop structures revealed the presence of greater fullerene-like CNMs in the sample which increases with increasing pyrolysis temperature to 1000°C during synthesis. TEM images (Figure 11.19e and f), show distinct fullerene-like loop structures with almost absolute disappearance of graphitic carbon.

11.5 APPLICATIONS OF AGRO-INDUSTRIAL WASTE-DERIVED CNMS IN CATALYSIS

11.5.1 Environmental Remediation

CNMs exhibit excellent porosity, surface area, size dimensions, electronic properties, and presence of ample surficial oxygen functional groups (-OH, -COOH, etc.); thus, they can be widely utilized as adsorbents in certain environmental remediation applications such as wastewater treatment and adsorption of hazardous dyes, pollutant gases, harmful insecticides and metal ions. Gupta et al. studied the efficient adsorption of harmful organic dyes (methylene blue and methyl orange) from industrial wastewater, using porous graphene-like CNMs derived from the husk of Bengal gram beans. Dye molecules were trapped by these adsorbents due to outstanding Π-Π interaction, hydrogen bonding, electrostatic interaction between dye molecules and CNMs.[7] Pol and Thiyagrajan utilized polymer waste and converted low- and high-density polyethylene into MWCNTs through thermal dissociation method which can efficiently adsorb gold nanoparticles from water and used in water purification.[79] A detailed report on production of sustainable CNMs from renewable resources, wastes also suggest the use of these materials in adsorption of phenolic and organic dyes from industrial wastewater.[80] Sugarcane bagasse, rice straw, pinewood saw dust were used to produce nano-GO which can remove Ni(II) ions from agricultural wastewater. Adsorption studies claimed that GO derived from sugarcane bagasse showed excellent adsorptive properties up to 85.6% Ni(II) ions removal.[81] Various CNMs obtained from rice husk such as CNTs, fullerenes, GO and carbon nanofibers by utilizing both conventional and sustainable approaches found suitable applications in water remediation including detection of heavy metal ions, antibiotics in wastewater, and desalination membrane.[82] Nano-sized graphene obtained from rice husk via KOH activation method possess clean and stable edges thus could be utilized in producing water filters.[72] Another example of porous nano-activated carbons derived from sawdust (biomass) was analyzed by using KOH activation of hydrothermally carbonized precursor. Existence of nanoporous structure helps in CO_2 storage and adsorption and can be applied in CO_2-N_2 separation with easier regeneration and recycling properties.[83] Waste plastics can give highly pure, porous carbon nanosheets upon KOH activation which had been used in rapid adsorption of methylene blue from wastewater sample up to ten cycles.[84]

11.5.2 Energy-Based Applications

Porosity, high surface area, good specific conductance, clean edges, outstanding mechanical strength, variations in intrinsic atomic structure, carrier mobility, thermal conductivity are some of the peculiar features of CNMs which enable them to act as anodic material in electrochemical applications. Polypropylene wastes were utilized to form MWCNTs through chemical vapor deposition method which exhibit excellent transmittance properties, comparable with commercial films used in touchscreen devices, thus, having potential scope in manufacturing optoelectronic devices.[85] Porous graphene-like nanosheets formed from waste coconut shell through simultaneous activation-graphitization technique demonstrated high surface area, cycle durability, excellent specific conductance and over 99.5% Coulombic efficiency and thus could be used in forming low-cost electrodes for high-power supercapacitors.[86] Nano-sized graphene obtained from rice husk via KOH activation method possesses clean and stable edges and thus could be utilized in producing high-performance energy storage and conversion devices based on carbon, such as supercapacitors and H_2 storage systems.[72] Mangrove charcoal, a by-product of power plants, was carbonized at high temperatures to form fullerene-like CNMs having very good discharge capacity, cyclic batteries.[87] Waste coffee husk can also be converted to porous carbon nanosheets via in-situ carbonization and KOH activation methods and used as supercapacitors giving brilliant cyclic performance of around 5,000 cycles.[88]

Nanocarbons formed from leather industry wastes via simple thermal treatment possess partially graphite-like structure, was used as anode material in Li-ion batteries, exhibited about 50 charge-discharge cycles.[89] Empty fruit bunches and palm oil wastes have been utilized in generation of nanoparticles of activated carbon, CNTs, graphene by traditional hydrothermal and pyrolysis methods and their mixture was used to design supercapacitor cell. Carbonization and chemical activation of brown rice husk yielded graphene nanosheets with high energy and power density, specific conductance of 115 F/g at 0.5 mA/cm^2, 88% cyclic stability over 2,000 cycles, used in electrochemical energy storage (Sankar et al. 2017).[90] Lemon peels can also form porous CNMs with 100% cyclic stability over 1,000 cycles, thus have promising applications in fabrication of economical supercapacitors.[91]

11.5.3 CATALYSIS

Extraordinary porosity and high surface area of CNMs play a vital role in heterogeneous catalysis. The use of several carbonaceous materials in different organic transformations is well reported in the literature. A group of researchers have been prepared palladium nanoparticles supported silica carbon catalyst by acid treatment of rice husk ash followed by calcination. The catalytic activity of this catalyst was checked on Suzuki-coupling reaction of aryl iodides (Scheme 11.3).[92]

To improve the recovery of spent catalyst, some research have been done to synthesize magnetic nanocatalysts. The use of such magnetic nanocatalysts can eliminate the tedious centrifugation and filtration step for recovery of the catalyst. In this way, magnetic carbon supported Pd catalyst has been synthesized and utilized in hydrogenation reactions, animation reactions and arylation of halides and oxidation of alcohols.[93]

In a recent study, nanocomposites of nanocellulose and NiFe$_2$O$_4$ was synthesized from pine needles and was successfully utilized as an efficient catalyst for degradation of Remazol Black 5 dye (RB5) and reduction of nitro-phenols (Scheme 11.4). The regeneration study showed that the catalyst was equally efficient till three reaction cycles (Gupta et al. 2017).[94] In a similar study, TiO$_2$ nanoparticles supported activated carbon derived from olive residue, was utilized a photocatalytic degradation of methylene blue dye.[95]

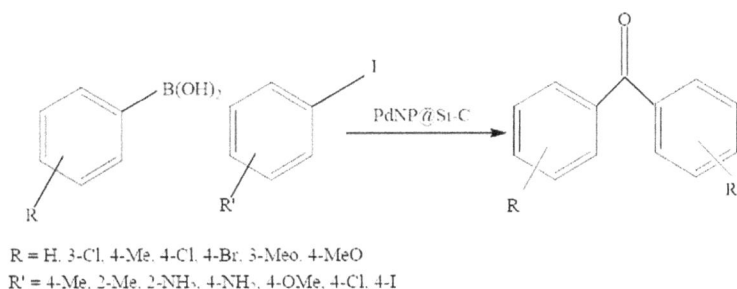

R = H, 3-Cl, 4-Me, 4-Cl, 4-Br, 3-Meo, 4-MeO
R' = 4-Me, 2-Me, 2-NH$_2$, 4-NH$_2$, 4-OMe, 4-Cl, 4-I

SCHEME 11.3 Suzuki coupling reaction over PdNP@Si-C catalyst.

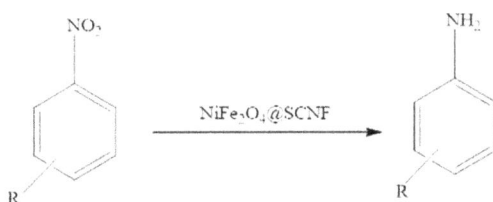

SCHEME 11.4 Reduction of nitrophenols over NiFe$_2$O$_4$@SCNF catalyst.

Furthermore, another study demonstrated the methanation of syngas over rice husk-derived activated carbon catalyst synthesized via pyrolysis method. The activity of this catalyst was compared with commercial Ru catalyst and the results showed that with the activated carbon conversion % and selectivity % was 99% and 85%, respectively, while with Ru catalyst it was 100% and 98%, respectively.[96] In a similar way, N-doped biochar (derived from *Pinus sylvestris*) modified with Ru nanoparticles via impregnation method, was utilized in CO_2 methanation reaction. The best results were obtained with the catalyst pyrolyzed at 600°C. In addition, biochar catalyst without N doping was also employed in the reaction, which showed worst catalytic performance.[97]

11.5.4 MISCELLANEOUS APPLICATIONS

Owing to their excellent magnetic, thermal, low cytotoxic and optical features, CNMs have been widely used in biomedical applications, tumor imaging, cancer diagnosis and therapy, nanocarriers in biomedicines, etc. Carbon nanoparticles derived from industrial waste proved their potency as antimicrobial and bactericidal agent by causing bacterial cell damage and formation of complexes with bacterial plasmid.[98] Owing to their hemocompatibility, they were also utilized as encapsulation vehicle in protein protection. Muramatsu et al. synthesized nanographene from rice husk having potential applications in manufacturing nanocomposites, energy storage and conversion devices.[72] Sha and group derived graphene-like mesoporous nanosheets from okara agro-waste, doped them with nitrogen and then used as non-enzymatic biosensor in amperometric detection of Vitamin C concentration in different beverages with good sensitivity and detection limits (as shown in Figure 11.20).[99] This work has future perspectives in human diagnostics, foodstuff safety etc.

Somanathan et al. derived graphene oxide nanosheets from sugarcane bagasse and its detailed characterization study concluded that these nanosheets could serve in economical fabrication of gas sensors and energy storage devices.[61] Mubarik et al. studied various CNMs from rice husk including graphene oxide, CNTs, carbon dots, etc. and reported a review on their various applications in field of bioimaging, methane gas sensing, fuel cells, solar cells, bacterial treatment etc.[82] In another work, carbon nanoparticles were derived from agro-waste and doped with nitrogen and potassium to form nanocomposite fertilizer which significantly improved growth parameters, height of plant, number of leaves and flowers, yield, mineral content, flavonoid and phenolic content, etc. in common bean crop (*Phaseolus vulgaris* L., Figure 11.21).[100]

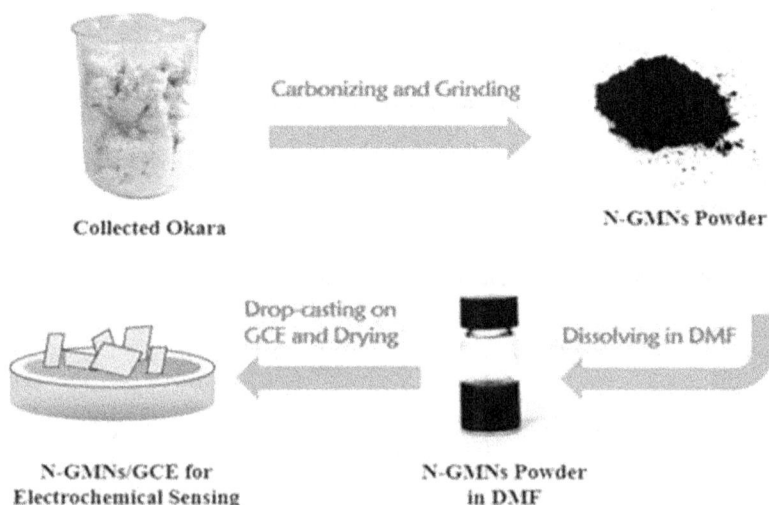

Collected Okara

Carbonizing and Grinding

N-GMNs Powder

Drop-casting on
GCE and Drying

Dissolving in DMF

N-GMNs/GCE for
Electrochemical Sensing

N-GMNs Powder
in DMF

FIGURE 11.20 Synthesis of N-doped graphene mesoporous nanosheets from okara waste for electroanalysis. (Reprinted with permission.)[99]

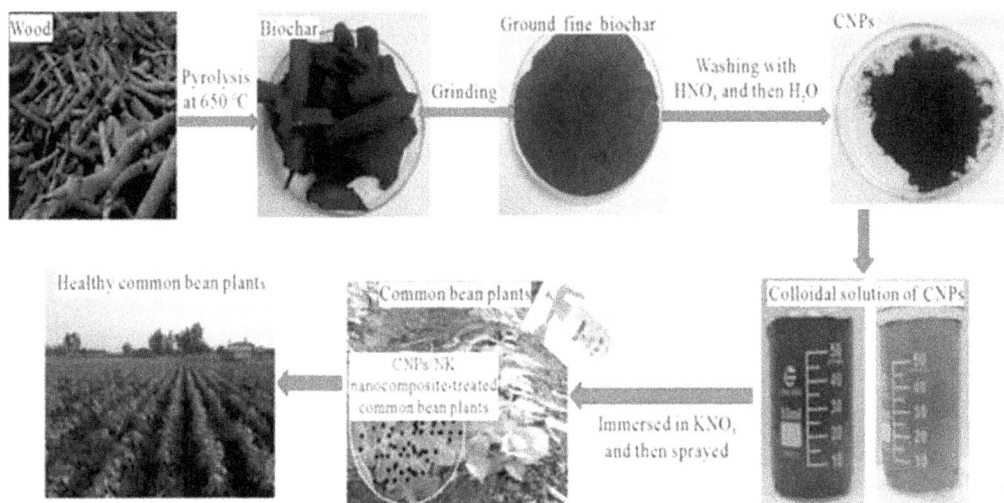

FIGURE 11.21 Steps for the preparation of carbon nanoparticles/nanocomposites from agro-wastes and their use as fertilizer for common bean crop. (Reprinted with permission.)[100]

Another review accounts for versatile applications of CNMs, particularly CNTs derived from industrial wastes in pollutant gas detection, handling of spilled oil, removal of antibiotics, etc.[42] Rice husk-derived carbon nanoparticles have also been used in the optical/fluorescence detection of Sn(-II) ions with good repeatability and better sensitivity.[101] Rhee and co-workers form cement mortar by incorporating rice husk-derived graphene nanosheets having higher surface area and analyzed their bond strength, compressive strength, freeze-and-thaw resistance, thermal performance and compared with them other CNMs.[102,103] They found increased reinforcing effect and mechanical properties in their synthesized materials and thus could be utilized as a conductive filler in cementitious materials, mortar, and cement mixes.

11.6 CONCLUSIONS AND FUTURE PERSPECTIVES

After the adoption of 3R (Reduce, Reuse, Recycle) policy in different nations, the trend of recycling and utilization of agro-industrial wastes has increased in the past few years. Since inappropriate disposal of such wastes can cause serious detrimental effects on human health and our ecosystem, there is a pressing need for feasible re-exploitation of these wastes at larger scale. The global annual production of CNMs had been projected to increase by 31% in the beginning of 2010, so researchers have also been inspired to develop CNMs in a cost-effective manner with these low-cost wastes. In this series, we have tried to accumulate and report on the work of several researchers carried out in the past decade, to produce different kinds of CNMs by recycling various agro-industrial wastes with high carbon content. This chapter highlights various types of agro-industrial wastes utilized for the production of different kinds of CNMs such as fullerene, GO and rGO, CNTs, and nano-activated carbon. Several conventional and eco-friendly approaches utilized in the production of CNMs have been discussed with analytical studies of the as-produced CNMs from agro-industrial wastes. CNMs are lightweight, chemically stable, electrical, thermal conductive, and mechanically strong, with higher specific surface areas. These reports concluded that low-cost, green CNMs can be easily derived from these wastes, which have versatile applications in different fields such as catalysis, electrochemical, biomedical applications, environmental remediation, adsorption, and sensing. This chapter aims at providing an overview of different sustainable CNMs derived from cost-effective agro-industrial wastes such as sugarcane bagasse, rice husk, wheat husk, and electroplating residues, to the readers. It also provides insights into various potential applications of the as-derived CNMs and projects their large-scale production in future. It also reveals the facts that a lot more interdisciplinary research work is still required to develop greener, sustainable approaches for recycling of wastes in the production of value-added products.

REFERENCES

1. Duque-Acevedo, M., Belmonte-Ureña, L.J., Cortés-García, F.J. and Camacho-Ferre, F. 2020. Agricultural waste: Review of the evolution, approaches and perspectives on alternative uses. *Glob. Ecol. Conserv.* 22: e00902.
2. Millati, R., Cahyono, R.B., Ariyanto, T., Azzahrani, I.N., Putri, R.U. and Taherzadeh, M.J. 2019. Agricultural, industrial, municipal, and forest wastes: An overview. *Sustainable Resource Recovery and Zero Waste Approaches*, pp. 1–22. doi: 10.1016/B978-0-444-64200-4.00001-3.
3. Titirici, M.M., White, R.J., Brun, N., Budarin, V.L., Su, D.S., Del Monte, F., Clark, J.H. and MacLachlan, M.J. 2015. Sustainable carbon materials. *Chemical Society Reviews* 44: 250–290.
4. Nasir, S., Hussein, M.Z., Zainal, Z. and Yusof, N.A. 2019. Development of new carbon-based electrode material from oil palm waste-derived reduced graphene oxide and its capacitive performance evaluation. *J. Nanomater.* 2019: 1–13.
5. Deng, J., You, Y., Sahajwalla, V. and Joshi, R.K. 2016. Transforming waste into carbon-based nanomaterials. *Carbon N. Y.* 96: 105–115.
6. Nasir, S., Hussein, M.Z., Yusof, N.A. and Zainal, Z. 2017. Oil palm waste-based precursors as a renewable and economical carbon sources for the preparation of reduced graphene oxide from graphene oxide. *Nanomaterials* 7: 1–18.
7. Gupta, K., Gupta, D. and Khatri, O.P. 2019. Graphene-like porous carbon nanostructure from Bengal gram bean husk and its application for fast and efficient adsorption of organic dyes. *Appl. Surf. Sci.* 476: 647–657.
8. Bhagat, R., Panakkal, H., Gupta, I. and Ingle, A.P. 2021. Carbon-based nanocatalysts in biodiesel production. In: *Nano- and Biocatalysts for Biodiesel Production*, pp. 157–181. doi: 10.1002/9781119729969. ch7.
9. Umemura, M. and Takenaka, C. 2014. Biological cycle of silicon in moso bamboo (Phyllostachys Pubescens) forests in central Japan. *Ecol. Res.* 29: 501–510.
10. Rangaraj, S. and Venkatachalam, R. 2017. Lucrative chemical processing of bamboo leaf biomass to synthesize biocompatible amorphous silica nanoparticles of biomedical importance. *Appl. Nanosci.* 7: 145–153.
11. Ghosh, S., Santhosh, R., Jeniffer, S., Raghavan, V., Jacob, G., Nanaji, K., Kollu, P., Jeong, S.K. and Grace, A.N. 2019. Natural biomass derived hard carbon and activated carbons as electrochemical supercapacitor electrodes. *Sci. Rep.* 9(1): 16315.
12. Isahak, W.N.R.W., Hisham, M.W.M. and Yarmo, M.A. 2013. Highly porous carbon materials from biomass by chemical and carbonization method: A comparison study. *J. Chem.* 2013: 620346.
13. Liu, Y., Zhao, Y. and Zhang, Y. 2014. One-step green synthesized fluorescent carbon nanodots from bamboo leaves for copper (II) ion detection. *Sens. Actuators B: Chem.* 196: 647–652.
14. Khalil, H.P.S. Abdul, Jawaid, M., Firoozian, P., Rashid, U., Islam, A. and Akil, H. Md. 2013. Activated carbon from various agricultural wastes by chemical activation with KOH: Preparation and characterization. *J. Biobased Mater. Bioenergy* 7: 1–7.
15. Deng, J., Li, M. and Wang, Y. 2016. Biomass-derived carbon: Synthesis and applications in energy storage and conversion. *Green Chem.* 18(18): 4824–4854.
16. Fahmi, M.Z., Haris, A., Permana, A.J., Wibowo, D.L.N., Purwanto, B., Nikmah, Y.L. and Idris, A. 2018. Bamboo leaf-based carbon dots for efficient tumor imaging and therapy. *RSC Adv.* 8: 38376–38383.
17. Zhou, Q., Cai, W., Zhang, Y., Liu, J., Yuan, L., Yu, F., Wang, X. and Liu, M. 2016. Electricity generation from corn cob char though a direct carbon solid oxide fuel cell. *Biomass Bioenerg.* 91: 250–258.
18. Rafiee, E., Shahebrahimi, S., Feyzi, M. and Shaterzadeh, M. 2012. Optimization of synthesis and characterization of nanosilica produced from rice husk (a common waste material). *Int. Nano Lett.* 29(2): 1–8.
19. Mulyana, C., Wulandari, A.P. and Hidayat, D. 2016. Development of Indonesia corncob and rice husk biobriquette as alternative energy source. *AIP Conf. Proc.* 1712: 050014.

20. Permatasari, N., Sucahya, T.N., Permatasari, N., Sucahya, T.N., Bayu, A. and Nandiyanto, D. 2016. Review: Agricultural wastes as a source of silica material. *Indonas. J. Sci. Technol.* 1: 82.

21. Sun, Y. and Webley, P.A. 2010. Preparation of activated carbons from corncob with large specific surface area by a variety of chemical activators and their application in gas storage. *Chem. Eng. J.* 162: 883.

22. Liu, X., Pan, L., Lv, T., Zhu, G., Sun, Z. and Sun, C. 2011. Microwave-assisted synthesis of CdS–reduced graphene oxide composites for photocatalytic reduction of Cr(vi). *Chem. Commun.* 47: 11984.

23. Okoronkwo, E.A., Imoisili, P.E. and Olusunle, S.O.O. 2013. Extraction and characterization of amorphous silica from corn cob ash by sol-gel method. *Chem. Mater. Res.* 3: 68–72.

24. Genovese, M., Jiang, J., Lian K. and Holm, N. 2015. High capacitive performance of exfoliated biochar nanosheets from biomass waste corn cob. *J. Mater. Chem. A* 3: 2903.

25. Seyfferth, A.L., Morris, A.H., Gill, R., Kearns, K.A., Mann, J.N., Paukett, M. and Leskanic, C. 2016. Soil incorporation of silica-rich rice husk decreases inorganic arsenic in rice grain. *J. Agric. Food Chem.* 64: 3760–3766.

26. Vargas, C., Simarro, R., Reina, J.A., Bautista, L.F. and Molina, M.C. N., 2019. New approach for biological synthesis of reduced graphene oxide. *Biochem. Eng. J.* 151: 107331.

27. Vaibhav, V., Vijayalakshmi, U. and Roopan, S.M. 2015. Agricultural waste as a source for the production of silica nanoparticles. *Spectrochim. Acta A: Mol. Biomol. Spectrosc.* 139: 515–520.

28. Wang, Z., Shen, D., Wu, C. and Gu, S. 2018. State-of-the-art on the production and application of carbon nanomaterials from biomass. *Green Chem.* 20: 5031–5057.

29. Guan, L., Pan, L., Peng, T., Gao, C., Zhao, W., Yang, Z., Hu, H. and Wu, M. 2019. Synthesis of biomass-derived nitrogen-doped porous carbon nanosheets for high-performance supercapacitors. *ACS Sustain. Chem. Eng.* 7: 8405–8412.

30. Asnawi, M., Azhari, S., Hamidon, M.N., Ismail, I. and Helina, I. 2018. Synthesis of carbon nanomaterials from rice husk via microwave oven. *J. Nanomater.* 2018: 2898326.

31. Bakar, R.A., Yahya, R. and Gan, S.N. 2016. Production of high purity amorphous silica from rice husk. *Proc. Chem.* 19: 189–195.

32. Islam, M.R., Haniu, H., Islam, M.N. and Uddin, M.S. 2010. Thermochemical conversion of sugarcane bagasse into bio-crude oils by fluidized-bed pyrolysis technology. *J. Therm. Sci. Technol.* 5: 11–23.

33. Zhuo, C., Alves, J.O., Tenório, J.A.S. and Levendis, Y.A. 2012. Synthesis of carbon nanomaterials through up-cycling agricultural and municipal solid wastes. *Ind. Eng. Chem. Res.* 51: 2922–30.

34. Alves, J.O., Tenório, J.A.S., Zhuo, C. and Levendis, Y.A. 2012. Characterization of nanomaterials produced from sugarcane bagasse. *J. Mater. Res. Technol.* 1(1): 31–34.

35. Terzioglu, P., Yucel, S., Rababah, T.M. and Özçimen, D. 2013. Characterization of wheat hull and wheat hull ash as a potential source of SiO$_2$. *Bio Resources.* 8: 4406–4420.

36. Irfan, M., Zhao, Z.Y., Panjwani, M.K., Mangi, F.H., Li, H., Jan, A., Ahmad, M. and Rehman, A. 2020. Assessing the energy dynamics of Pakistan: Prospects of biomass energy. *Energy Rep.* 6: 80–93.

37. Goodman, B.A. 2020. Utilization of waste straw and husks from rice production: A review. *J. Bioresour. Bioprod.* 5(3): 143–162.

38. Bledzki, A. K., Mamun, A.A. and Volk, J. 2010. Physical, chemical and surface properties of wheat husk, rye husk and soft wood and their polypropylene composites. *Compos. A: Appl. Sci. Manuf.* 41: 480–488.

39. Gou, G., Meng, F., Wang, H., Jiang, M., Wei, W. and Zhou, Z. 2019. Wheat straw-derived magnetic carbon foams: In-situ preparation and tunable high-performance microwave absorption. *Nano Res.* 12(6): 1423–1429.

40. Terzioglu, P., Yücel, S. and Kus, C. 2019. Review on a novel biosilica source for production of advanced silica-based materials: Wheat husk. *Asia Pac. J. Chem. Eng.* 14: 2262.

41. Zhang, Y., Song, X., Xu, Y., Shen, H., Kong, X. and Xu, H. 2019. Utilization of wheat bran for producing activated carbon with high specific surface area via NaOH activation using industrial furnace. *J. Clean. Prod.* 210: 366–375.

42. Kerdnawee, K., Termvidchakorn, C., Yaisanga, P., Pakchamsai, J., Chookiat, C., Eiad-ua, A., Wongwiriyapan, W., Chaiwat, W., Ratchahat, S., Faungnawakij, K., Suttiponparnit, K. and Charinpanitkul, T. 2017. Present advancement in production of CNTs and their derivatives from industrial waste with promising applications. *KONA Powder Part. J.* 34: 24–43.

43. Morales, N.J., Goyanes, S., Chiliotte, C., Bekeris, V., Candal, R.J. and Rubiolo, G.H. 2013. One-step chemical vapor deposition synthesis of magnetic CNT–hercynite (FeAl$_2$O$_4$) hybrids with good aqueous colloidal stability. *Carbon* 61: 515–524.

44. Acomb, J.C., Wu, C. and Williams, P.T. 2014. Control of steam input to the pyrolysis-gasification of waste plastics for improved production of hydrogen or CNTs. *Appl. Catal. B: Environ.* 147: 571–584.

45. Nahil, M.A., Wu, C. and Williams, P.T. 2015. Influence of metal addition to Ni-based catalysts for the co-production of CNTs and hydrogen from the thermal processing of waste polypropylene. *Fuel Proc. Technol.* 130: 46–53.

46. Salah, N., Habib, S.S., Khan, Z.H., Memic, A. and Nahas, M.N. 2012. Growth of CNTs on catalysts obtained from carbon rich fly ash. *Dig. J. Nanomater. Bios.* 7: 1279–1288.

47. Nath, D.C. and Sahajwalla, V. 2011. Application of fly ash as a catalyst for synthesis of carbon nanotube ribbons. *J. Hazard. Mater.* 192: 691–697.

48. Yin, F., Yue, W., Li, Y., Gao, S., Zhang, C., Kan, H., Niu. H., Wang, W. and Guo, Y. 2021. Carbon-based nanomaterials for the detection of volatile organic compounds: A review. *Carbon* 180: 274–297.

49. Maiti, D., Tong, X., Mou, X. and Yang, K. 2019. Carbon-based nanomaterials for biomedical applications: A recent study. *Front. Pharmacol.* 9: 1401.

50. Peng, L.M., Zhang, Z. and Wang, S. 2014. Carbon nanotube electronics: Recent advances. *Mater. Today* 17(9): 433–442.

51. Lu, Z., Raad, R., Safaei, F., Xi, J., Liu, Z. and Foroughi, J. 2019. Carbon nanotube based fiber supercapacitor as wearable energy storage. *Front. Mater.* 6:138.

52. Vinícius, D.N.B., Thaís, L.A.M., de Menezes, B.R.C., Renata, G.R., Victor, A.N.R., Karla, F.R. and Gilmar, P.T. 2019. Carbon nanostructure-based sensors: A brief review on recent advances. *Adv. Mater. Sci. Eng.* 2019: 21.

53. Muralidharan, N., Teblum, E., Westover, A.S., Schauben, D., Itzhak, A., Muallem, M., Nessim, G.D. and Pint, C.L. 2018. Carbon nanotube reinforced structural composite supercapacitor. *Sci. Rep.* 8: 17662.

54. Zaytseva, O. and Neumann, G. 2016. Carbon nanomaterials: Production, impact on plant development, agricultural and environmental applications. *Chem. Biol. Technol. Agric.* 3:17.

55. Eatemadi, A., Daraee, H., Karimkhanloo, H., Kouhi, M., Zarghami, N., Akbarzadeh, A., et al. 2014. CNTs: Properties, synthesis, purification, and medical applications. *Nanoscale Res. Lett.* 9: 393–405.

56. Cha, C., Shin, S.R., Annabi, N., Dokmeci, M.R. and Khademhosseini, A. 2013. Carbon-based nanomaterials: Multi-functional materials for biomedical engineering. *ACS Nano.* 7(4): 2891–2897.

57. Babu, J.S., Prasanna, H.B.N., Babu, J.S., Rao, Y.N. and Beyan, S.M. 2022. Environmental applications of sorbents, high-flux membranes of carbon-based nanomaterials. *Adsorp Sci. Technol.* 2022: 13.

58. Deguchi, S., Mukai, S., Sakaguchi, H. and Nonomura, Y. 2013. Non-engineered nanoparticles of C60. *Sci. Rep.* 3: 2094.

59. Macovez, R. 2018. Physical properties of organic fullerene cocrystals. *Front. Mater.* 4: 46.

60. Yadav, J. 2018. Fullerene: Properties, synthesis and application. *Res. Rev. J. Phys.* 6(3): 1–6.

61. Somanathan, T., Prasad, K., Ostrikov, K., Saravanan, A. and Krishna, V.M. 2015. Graphene oxide synthesis from agro waste. *J. Nanomater.* 5: 826–834.

62. Song, H., Zhang, L., He, C., Qu, Y., Tian, Y.F. and Lv, Y. 2011. Graphene sheets decorated with SnO_2 nanoparticles: In situ synthesis and highly efficient materials for cataluminescence gas sensors. *J. Mater. Chem.* 21: 5972–5977.

63. Bo, Z., Shuai, X., Mao, S., Yang, H., Qian, J., Chen, J., Yan, J. and Cen, K. 2014. Green preparation of reduced graphene oxide for sensing and energy storage applications. *Sci. Rep.* 4: 1–5.

64. Mao, S., Pu, H. and Chen, J. 2012. Graphene oxide and its reduction: Modeling and experimental progress. *RSC Adv.* 2: 2643–2662.

65. Chen, J. 2011. A new reducing agent to prepare single-layer, high-quality reduced graphene oxide for device applications. *Nanoscale* 3: 2849–2853.

66. Lu, G., Park, S., Yu, K., Ruoff, R.S., Ocola, L.E., Rosenmann, D. and Chen, J. 2011. Toward practical gas sensing with highly reduced graphene oxide: A new signal processing method to circumvent run-to-run and device-to-device variations. *ACS Nano* 5(2): 1154–1164.

67. Priyadarsini, S., Mohanty, S., Mukherjee, S., Basu, S. and Mishra, M. 2018. Graphene and graphene oxide as nanomaterials for medicine and biology application. *J. Nanostructure Chem.* 8: 123–137.

68. Rwayhah, Y.M.N., Hassan, M.L. and Shehata, M.R. 2017. Nanoporous activated carbon from olive stones wastes. *J. Sci. Ind. Res. (JSIR).* 76(11): 725–732.

69. Mestre, A.S. and Carvalho, A.P. 2017. Nanoporous carbon synthesis: An old story with exciting new chapters. In: T. H. Ghrib (ed.) *Porosity - Process, Technologies and Applications.* doi:10.5772/intechopen.72476.

70. Rajbhandari, R., Shrestha, L.K. and Pradhananga, R.R. 2012. Nanoporous activated carbon derived from Lapsi (Choerospondias axillaris) seed stone for the removal of arsenic from water. *J. Nanosci. Nanotechnol.* 12(9): 7002–7009.

71. Tang, P.-D., Du, Q.-S., Li, D.-P., Dai, J., Li, Y.-M., Du, F.-L., Long, S.-Y., Xie, N.-Z., Wang, Q.-Y. and Huang, R.-B. 2018. Fabrication and characterization of graphene microcrystal prepared from lignin refined from sugarcane bagasse. *Nanomaterials* 8: 565.

72. Muramatsu, H., Kim, Y.A., Yang, K.S., Cruz-Silva, R., Toda, I., Yamada, T., Terrones, M., Endo, M., Hayashi, T. and Saitoh, H. 2014. Rice husk-derived graphene with nano-sized domains and clean edges. *Small* 10: 2766–70.

73. Liu, D., Zhang, W., Lin, H., Li, Y., Lu, H. and Wang, Y. 2016. A green technology for the preparation of high capacitance rice husk-based activated carbon. *J. Cleaner Prod.* 112(1): 1190–98.

74. Salah, N., Al-ghamdi, A.A., Memic, A., Habib, S.S. and Khan, Z.H. 2016. Formation of CNTs from carbon-rich fly ash: Growth parameters and mechanism. *Mater. Manuf. Process.* 31(2): 146–156.

75. Hintsho, N., Shaikjee, A., Masenda, H., Naidoo, D., Billing, D., Franklyn, P. and Durbach, S. 2014. Direct synthesis of carbon nanofibers from South African coal fly ash. *Nanoscale Res. Lett.* 9: 387.

76. Mojica, M., Alonso, J.A. and Mendez, F. 2013. Synthesis of fullerenes. *J. Phys. Org. Chem.* 26(7): 526–39.

77. Fabry, F. and Fulcheri, L. 2016. Synthesis of carbon blacks and fullerenes from carbonaceous wastes by 3-phase AC thermal plasma. In *6th International Conférence on Engineering for Waste and Biomass valorisation – Waste Eng 2016*, May 2016, Albi, France.

78. Marriott, A.S., Hunt, A.J., Bergström, E., Wilson, K., Budarin, V.L., Thomas-Oates, J., Clark, J.H. and Brydson, R. 2014. Investigating the structure of biomass-derived non-graphitizing mesoporous carbons by electron energy loss spectroscopy in the transmission electron microscope and X-ray photoelectron spectroscopy. *Carbon* 67: 514–24.

79. Pol, V.G. and Thiyagarajan, P. 2010. Remediating plastic waste into CNTs. *J. Environ. Monit.* 12: 455–59.

80. Ravi, S. and Vadukumpully, S. 2016. Sustainable carbon nanomaterials: Recent advances and its applications in energy and environmental remediation. *J. Env. Chem. Eng.* 4: 835–56.

81. Tohamy, H-A.S., Anis, B., Youssef, M.A., Abdallah, A.E.M., El-Sakhawy, M. and Kamel, S. 2020. Preparation of eco-friendly graphene oxide from agricultural wastes for water treatment. *Desalin. Water Treat.* 191: 250–62.

82. Mubarik, S., Qureshi, N., Sattar, Z., Shaheen, A., Kalsoom, A., Imran, M. and Hanif, F. 2021. Synthetic approach to rice waste-derived carbon-based nanomaterials and their applications. *Nanomanufacturing* 1: 109–159.

83. Sevilla, M. and Fuertes, A.B. 2011. Sustainable porous carbons with a superior performance for CO_2 capture. *Energy Environ. Sci.* 4: 1765–71.

84. Gong, J., Liu, J., Chen, X., Jiag, Z., Wen, X., Mijowska, E. and Tang, T. 2015. Converting "real-world" mixed waste plastics into porous carbon nanosheet with excellent performance in adsorption of organic dye from wastewater. *J. Mater. Chem. A* 3: 341–51.

85. Mishra, N., Das, G., Ansaldo, A., Genovese, A., Malerba, M., Povia, M., Ricci, D., Fabrizio, E., Zitii, E.D., Sharon, M. and Sharon, M. 2012. Pyrolysis of waste polypropylene for the synthesis of CNTs. *J. Anal. Appl. Pyrolysis* 94: 91–98.

86. Sun, L., Tian, C., Li, M., Meng, X., Wang, L., Wang, R., Yin, J. and Fu, H. 2013. From coconut shell to porous graphene-like nanosheets for high-power supercapacitors. *J. Mater. Chem. A* 1: 6462–70.

87. Liu, T., Luo, R., Qia, W., Yoon, S.-H. and Mochida, I. 2010. Microstructure of carbon derived from mangrove charcoal and its application in Li-ion batteries. *Electrochim. Acta* 55: 1696–1700.

88. Yun, Y.S., Park, M.H., Hong, S.J., Lee, M.E., Park, Y.W. and Jin, H.-J. 2015. Hierarchically porous carbon nanosheets from waste coffee grounds for supercapacitors. *ACS Appl. Mater. Interfaces* 7(6): 3684–90.

89. Ashokkumar, M., Narayanan, N.T., Mohana Reddy, A.L., Gupta, B.K., Chandrasekaran, B., Talapatra, S., Ajayan, P.M. and Thanikaivelan, P. 2012. Transforming collagen wastes into doped nanocarbons for sustainable energy applications. *Green Chem.* 14: 1689–95.

90. Sankar, S., Lee, H., Jung, H., Kim, A., Ahmed, A.T.A., Inamdar, A.I., Kim, H., Lee, S., Im, H. and Kim, D.Y. 2017. Ultrathin graphene nanosheets derived from rice husks for sustainable supercapacitor electrodes. *New J. Chem.* 41: 13792–97.

91. Mehare, M.D., Deshmukh, A.D. and Dhoble, S.J. 2021. Bio-waste lemon peel derived carbon based electrode in perspect of supercapacitor. *J. Mater. Sci: Mater. Electron.* 32: 14057–71.

92. Ketike, T., Velpula, V.R.K., Madduluri, V.R., Kamaraju, S.R.R. and Burri, D.R. 2018. Carbonylative Suzuki-Miyaura cross-coupling over Pd NPs/ Rice-Husk carbon-silica solid catalyst: Effect of 1, 4-dioxane solvent. *Chem. Select.* 3, 7164–7169.

93. Nasir Baig, R.B. and Varma, R.S. 2014. Magnetic carbon-supported palladium nanoparticles: An efficient and sustainable catalyst for hydrogenation reactions. *ACS Sustain. Chem. Eng.* 2(9): 2155–2158.

94. Gupta, K., Kaushik, A., Tikoo, K.B., Kumar, V. and Singhal, S. 2017. Enhanced catalytic activity of composites of $NiFe_2O_4$ and nano cellulose derived from waste biomass for the mitigation of organic pollutants. *Arab. J. Chem.* 13: 783–798.

95. Donar, Y.O., Bilge, S., Sınağ, A. and Pliekhov, O. 2018. TiO_2 /Carbon materials derived from hydrothermal carbonization of waste biomass: A highly efficient, low-cost visible-light-driven photocatalyst. *Chem. Cat. Chem.* 10: 1134–1139.

96. Zhu, L., Yin, S., Yin, Q., Wang, H. and Wang, S. 2015. Biochar: A new promising catalyst support using methanation as a probe reaction. *Energy Sci. Eng.* 3: 126–134.

97. Wang, X., Liu, Y., Zhu, L., Li, Y., Wang, K., Qiu, K., Tippayawong, N., Aggarangsi, P., Reubroycharoen, P. and Wang, S. 2019. Biomass derived N-doped biochar as efficient catalyst support for CO_2 methanation. *J. CO_2 Util.* 34: 733–741.

98. Das Purkayastha, M., Manhar, A.K., Mandal, M. and Mahanta, C.L. 2014. Industrial waste-derived nanoparticles and microspheres can be potent antimicrobial and functional ingredients. *J. Appl. Chem.* 2014: 1–12.

99. Sha, T., Liu, J., Sun, M., Li, L., Bai, J., Hu, Z. and Zhou, M. 2019. Green and low-cost synthesis of nitrogen-doped graphene-like mesoporous nanosheets from the biomass waste of okara for the amperometric detection of vitamin C in real samples. *Talanta* 200: 300–306.

100. Salama, D.M., Abd El-Aziz, M.E., El-Naggar, M.E., Shaaban, E.A. and Abd El-Wahed, M.S. 2021. Synthesis of an eco-friendly nanocomposite fertilizer for common bean based on carbon nanoparticles from agricultural waste biochar. *Pedosphere* 31: 923–933.

101. Ngu, P.Z.Z., Chia, S.P.P., Fong, J.F.Y. and Ng, S.M. 2016. Synthesis of carbon nanoparticles from waste rice husk used for the optical sensing of metal ions. *New Carbon Mater.* 31(2): 135–143.

102. Rhee, I., Kim, Y.A., Shin, G.O., Kim, J.H. and Muramatsu, H. 2015. Compressive strength sensitivity of cement mortar using rice husk-derived graphene with a high specific surface area. *Constr. Build. Mater.* 96: 189–197.

103. Rhee, I., Lee, J.S., Kim, J.H. and Kim, Y.A. 2017. Thermal performance, freeze-and-thaw resistance, and bond strength of cement mortar using rice husk-derived graphene. *Constr. Build. Mater.* 146: 350–359.

Section III

Green Nanomaterials

Case Studies

12 Case Studies on Applications of Green Nanotechnology

Nidhi Jain
Bharati Vidyapeeth's College of Engineering

Mona Kejariwal
R.D. & S.H. National College of Arts and
Commerce and S.W.A. Science College

CONTENTS

12.1 INTRODUCTION

Sustainable technologies available currently are modified by the use of nanotechnology to provide promising therapeutic solutions. Applications of the nanotechnology are achieving heights by providing solution to the monotonous problems. It was possible to provide improved nanotechnology in the field of agriculture, electronics, cosmetics, food, materials, medicine, energy, etc. [1]. Five different case studies of the nanotechnology are discussed at length in this chapter.

DOI: 10.1201/9781003319153-15

12.2 CASE STUDY I: RELEVANCE OF NANOTECHNOLOGY IN CANCER THERAPEUTICS AND IMAGING – USE OF HYPERTHERMIA FOR CANCER TREATMENT

In the contemporary age, still the cancer disease is considered synonymous to death. Worldwide, the fear of cancer disease is increasing continuously for the reasons of high mortality rate, painful therapies, time-consuming treatments such as chemotherapy, surgery, side effects of the cancer drugs, etc. The limitation encountered with the contemporary treatment is non-targeted drug delivery to the tumor cells, unequal distribution of the drugs, and insufficient monitoring of the drug responses. Inadequate treatment at a specific time causes complications leading to the multidrug resistance of the disease.

The two major milestones achieved by the nanotechnology in the cancer treatment are superior targeting discrimination and enhanced delivery possible due to cancer therapeutics and imaging. These therapies will reduce the side effects to the ordinary tissues. Monoclonal antibody (mAb) drugs/ligands that specifically combine to antigens/receptors help in treating the tumor cell completely. Conservative chemotherapy drugs have preclinical and clinical side effects, which are overcome by the ligand-targeted therapeutics. It was found in the literature that 1–10 parts could be able to achieve target out of intravenously administered 100,000 drugs [2–4].

With the advent of the nanotechnology, new tools have been developed for cancer treatment and imaging. It not only provided different methods of imaging, but also helped in developing new devices, which will syndicate with different molecules such as ligands, drugs, antibodies, and imaging probes. The size of these nanodevices is smaller than cancer cells, sometimes 100- to 1,000-fold smaller compared to that of cancer cells. The smaller size of the nanodevices helps in the easy penetration through the effected blood vessels and targeting the tumor protein both inside and outside the cell [5].

Nanoparticles used for the cancer treatment are of different types such as metals and ceramics. Their shape and properties are governed by the raw materials used and the process of manufacturing. The nanoparticles used in the drug delivery systems are liposomes, polymeric nanoparticles, lipid-based nanoparticles, quantum dot nanoparticles, magnetic iron oxide nanoparticles, etc. [6,7].

The delivery of nanoparticles on the tumor tissues is possible by active or passive targeting. There are several reasons for the easy passive targeting of nanoparticles in the tumor cells such as inherent size small size and unique properties of the nanoparticles, tumor vasculature including permeability and retention (EPR) effect and microenvironment. The EPR effect is enhanced in the cancer cells because of the defective vascular and pitiable lymphatic drainage [8–10].

12.2.1 ACTIVE TARGETING:

Sometimes, scientists are not able to achieve the desired results on therapeutic delivery of drugs by nanoparticles on tumor cells because of the absence of the targeting moiety, over-dependency on the EPR effect, limitations of favorable microenvironment, and tumor angiogenesis. These are certain limitations in passive targeting through EPR. The poor lymphatic drainage one side it increases the drug flow and other side leads to the drug outflow from the cells because of the advanced osmotic pressure in the interstitium [11].

However, another strategy to surmount the limitations of passive targeting is by incorporating nanoparticles with the targeting ligand or an antibody. The combination of both targeted ligand and nanoparticles is able to selectively deliver drugs to tumor cell with superior competence. Several more studies are going on for increasing the effective drug delivery [12].

12.2.2 EXPLOITATION OF HYPERTHERMIA FOR CANCER TREATMENT

Hyperthermia is a treatment given to the patient by maintaining temperature 42°C–45°C. In case of cancer patients, it helps in killing cancer cells without harming normal tissues as depicted in

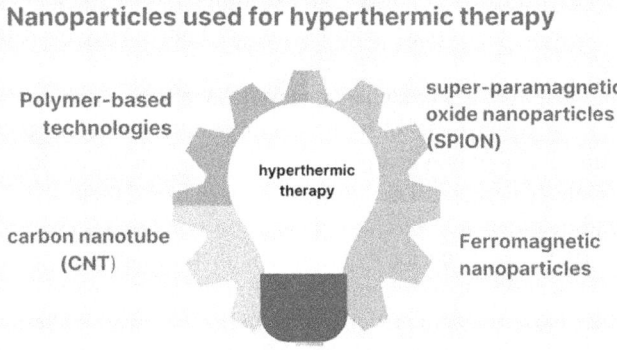

FIGURE 12.1 The use of hyperthermia for cancer treatment.

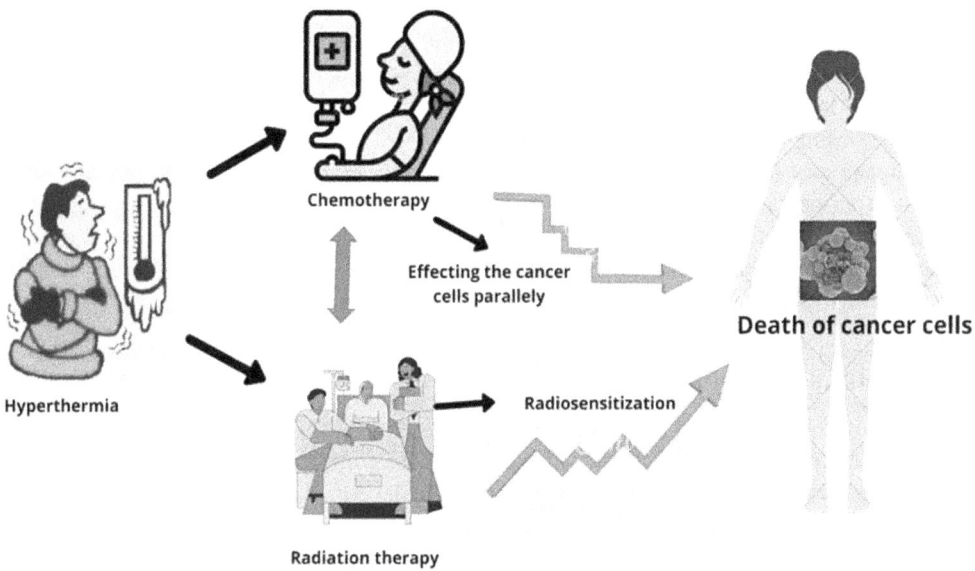

FIGURE 12.2 Hyperthermia is also helpful in sensitizing chemotherapeutics agents and also enhances radiation therapy for cancer treatment.

Figure 12.1. This study was initiated in 1893 by Coley. Practically, it is quite difficult to heat specific area to treat the cancer tumor area. So, hyperthermia is backed with chemotherapy and radiation to get the desired results, as depicted in Figure 12.2. The thermal stress helps in radiosensitizing the cancer tissue, which enhances the radiotherapy in a more profound manner. Hyperthermia is also helpful in sensitizing chemotherapeutics agents for cancer treatment. Induction of ER-mediated intrinsic or extrinsic apoptosis could be possible for malignant melanoma when combined with hyperthermia in the temperature range from 43°C to 45°C, respectively [13–15].

Various preclinical finding have proved that the available cancer treatment such radio- and chemotherapy have become more effective if they are backed by hyperthermic therapy [16]. The traditional use of hyperthermic therapy was not as effective because of the application to full body rather than specific application to one body part. The use of hyperthermic therapy on full body is not able to provide the desired result as well as is prone to many side effects. It was given in the literature that the whole-body hyperthermia may lead to cardiovascular and gastrointestinal diseases. More promising treatment of cancer requires targeted localized hyperthermia added with nanoparticles mediation. Photothermal therapy (PTT) nowadays is gaining lots of attention, in which subsequent cellular necrosis or apoptosis takes place with the help of light energy (near-infrared region) converted into heat energy [17].

As an external stimulus, light is an excellent material that can be easily controlled, concentrated, and distantly measured. A better focus of light on a targeted organ leads to minimum damage to the tissue. The inadequate penetration less usefulness for tumor of the photothermic light was seen because of the inadequate penetration. The wavelength range from 800 to 1,200 nm and temperature from 55°C to 95°C and in vivo conditions have more impact on tumor cells [18,19].

A range of nanoparticles are used for hyperthermic therapy, such as polymer-based technologies, carbon nanotubes (CNTs), super-paramagnetic iron oxide nanoparticles (SPION), and ferromagnetic nanoparticles. Ferromagnetic nanoparticles in the presence of alternating magnetic fields (AMFs) can produce heat for thermal therapy. The drawback of the magnetic nanoparticles is that it is hard to generate well-tuned and precise treatment of tumors required for effective photothermal once the magnetic field is removed from outside [20].

CNTs are nanomaterials are used for cancer therapy as they can accomplish the passing of light both in NIR spectrum and in visible spectrum. Both the types of CNTs, SWCNTs and MWCNTs, are successfully utilized in cancer treatment of different types like in mice by means of NIR illumination of tumor xenograft squamous cell carcinoma and of renal cancer xenografts by short pulses of small-power laser illumination, respectively. However, the use of CNT may cause certain limitations also regarding long-term biocompatibility.

In contemporary methods of treatment (photothermal therapy), polymeric materials are also in use, such as polyaniline nanoparticles, polypyrrole, and dopamine-melanin (polydopamine). There are certain limitations associated with the polymers, such as poor photothermal efficiency (polydopamine) and vulnerability to degradation in polymers making unsuitable for their long-term biocompatibility [21].

12.2.3 MEDICINAL USE OF GOLD THROUGH HYPERTHERMIC CANCER THERAPEUTICS

Gold is known from the ancient times for its medicinal use. Gold (Au) is a noble metal known for its non-reactive, corrosion resistance nature with other exclusive properties. The evidence for healing and diagnosis has been reported in the literature of the colloidal Au. $K[Au(CN)^2]$ bacteriostatic properties also impacted eventually the modern medical treatments. In nanomedicine drugs, the therapeutic delivery of Au has been recognized. In case of selective cancer targeting, colloidal Au is linked with adenoviral vectors covalently, which induces hyperthermia as soon as near-infrared (NIR) light is applied. Multi-functional design of gold nanoparticles enhanced NIR light-absorbing capacity which leads to better multiple desired drugs delivery at guarded and targeted manner.

Due to numerous benefits gold nanoparticles, it is suitable for photothermal healing of cancer. Some of the benefits are as follows: (i) Localized treatment of cancer is possible with the gold nanoparticles; (ii) activation of gold nanoparticles is possible by near-infrared (NIR) laser light, which makes it penetrate more into biological tissues; (iii) gold nanoparticles are tailored, which makes it suitable for multilayered cancer photothermal treatment; (iv) they help in drug delivery systems; (v) they help in the removal of toxins; (vi) they are used as a dye for target discrimination and enhanced delivery with the help of cancer therapeutics and imaging, depicted in Figure 12.3 [22,23].

12.3 CASE STUDY II: UTILIZATION OF RADIOACTIVE GOLD NANOPARTICLES (GA-198AUNP) IN CANCER THERAPIES AND ITS STUDY ON MICE BEARING PROSTATE TUMOR

In vivo studies and non-hazardous features of GA-198AuNPs (β-emitting gold (198Au; approximately βmax = 0.97 MeV; half-life of approximately 2.7 days) have been studied for cancer

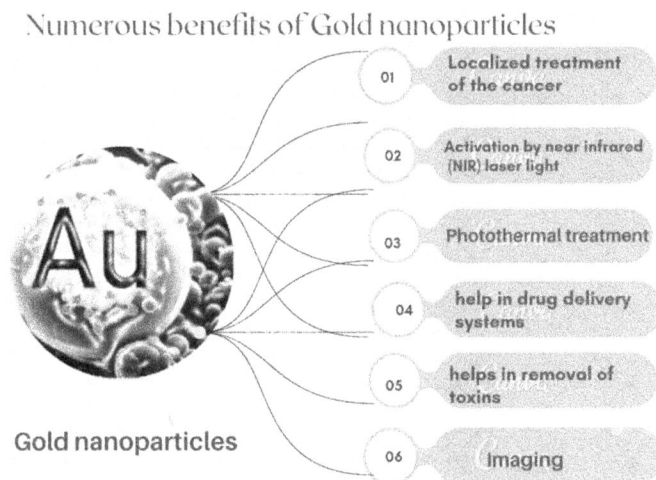

Numerous benefits of Gold nanoparticles

Gold nanoparticles

01 Localized treatment of the cancer

02 Activation by near infrared (NIR) laser light

03 Photothermal treatment

04 help in drug delivery systems

05 helps in removal of toxins

06 Imaging

FIGURE 12.3 The properties of the gold were successfully documented in the literature with preclinical and clinical data, and it is found that gold is best suitable for the cancer therapeutics.

treatment. by Hydrodynamic diameter of gum arabic glycoprotein and AuNPs with size ranging from approximately 12 to 19 nm center diameter and approximately 86 nm is best suitable for cancer tumor cells targeting. The result obtained was found very promising in tumor cell and growth of prostate tumors by means of the help of β-emitting GA-198AuNPs over 30 days. A significant reduction in the tumor cells was seen, that is 82% smaller within 3 weeks. Pharmacokinetic studies proved that there was no or very leakage of radioactivity GA-198AuNPs during the treatment. The count of the various components of the blood such as RBC, WBC, and platelets are simpler as normal SCID mice [24].

12.3.1 Synthesized GA-AuNPs and GA-198AuNPs and Subsequent Characterization

GA-conjugated AuNPs were synthesized at University of Missouri Research Reactor irradiation facilities. The characterization of the solution was done by means of ultraviolet-visible spectrophotometry, and as absorption band around 541 nm was found for 198AuNPs. In vivo function of the GA-198AuNPs for therapy was done after studying the stability and biocompatibility characteristic of the GA-198AuNPs in vivo function of the GA-198AuNPs is possible. Transmission electron Microscopy (TEM) was used for the determination of morphology of GA-AuNPs within the metallic gold core with size ranging between 12-17 nm.

Dynamic light scattering (DLS) was used for the determination of the hydrodynamic diameter of GA-AuNPs and is found to be 85 nm with the zeta potential of approximately –24±1.5 mV, as seen in Figure 12.1. This information proves that the dispersion of nanoparticles obtained is stable in an aqueous solution. The biological stability of nanoparticles was also studied using different solutions such as 10% NaCl and 0.2 M histidine with buffers of different pH values. See Figure 12.4.

Figure 12.5 shows the reduction in the palpable tumors in SCID mice. The treatment was continued for 30 days. Figure 12.5b depicts the TEM image intake of GA-AuNPs in prostate cancer cells. The prostate tumor-bearing SCID mice models were treated with the help of GA-198AuNPs. The significant decrease in the dimension of the tumor was found after 30 days of the treatment. After the use of the single-dose intratumoral administration of GA-198AuNPs (408 μCi), the overall reduction in size was found to be 82%. The absorption of GA-198AuNPs by the other non-target organs is minimum, up to 2%–5%. The excess dose was cleared by renal pathway.

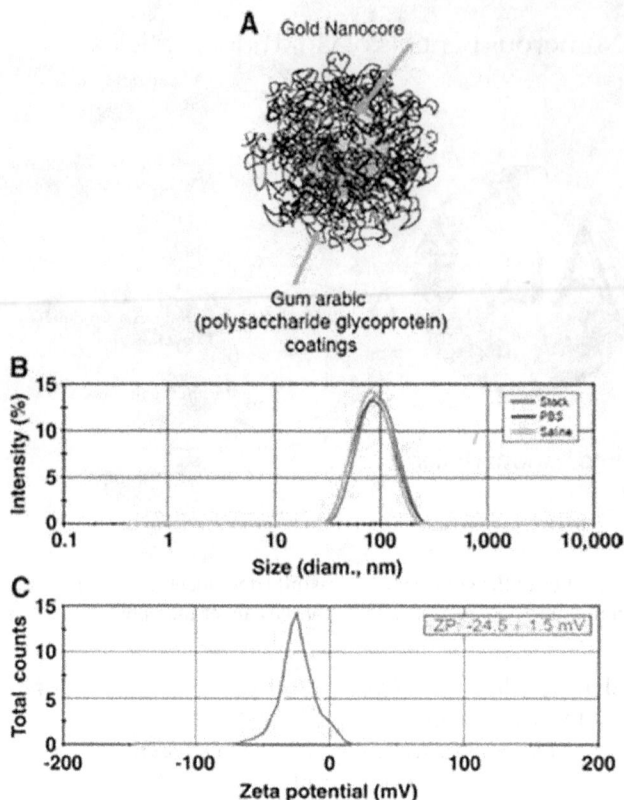

FIGURE 12.4 The characterization details of gum arabic glycoprotein-gold nanoparticles (GA-AuNPs). (A) The depiction of GA-198AuNPs. (B) Dynamic light scattering spectra representing hydrodynamic sizes of GA-AuNPs in deionized water, in addition to normal saline and buffers of different pH values. (C) Zeta potential (ZP) of GA-AuNPs. (Reprinted with permission from Ref. [24]. Copyright 2010, Elsevier.)

The reduction in the size may have ruled out the possibility of surgical resection in certain circumstances. To know the treatment effect, the body weight loss graph was also monitored with blood parameters. The recovery group showed only transitional weight loss, which was regained after the treatment.

In case of the untreated control group, the death of the two animals with major weight loss proves the decline in the health status. The components of the blood of the treated mice, such as white and red blood cells, lymphocytes, and platelets, are found normal compared to the untreated mice. The hormone therapy is most widely used, but the GA-AuNPs radioactive therapy has left a good impact on the treatment of metastatic lesions. All these studies clearly prove that gold nanoparticulate radioactive therapy shows significant improvement of the SCID mice without minimum or no outflow of the radioactive substance [25].

12.4   CASE STUDY III: GREEN NANOTECHNOLOGY HELPFUL IN THE CHEMOTHERAPY OF OVARIAN CANCER

As per gynecological studies, in females, ovarian cancer is considered to be the major cause of death, which is due to the lack of early diagnosis and subsequent treatment. Due to late detection of cervical cancer where the infection spreads in abdomen in three to four stages, only 30% of patients survive in the last 5 years. Not only the detection of cervical cancer is a challenge, but its treatment also poses many challenges with few options. Also, it is resistant to regular cancer

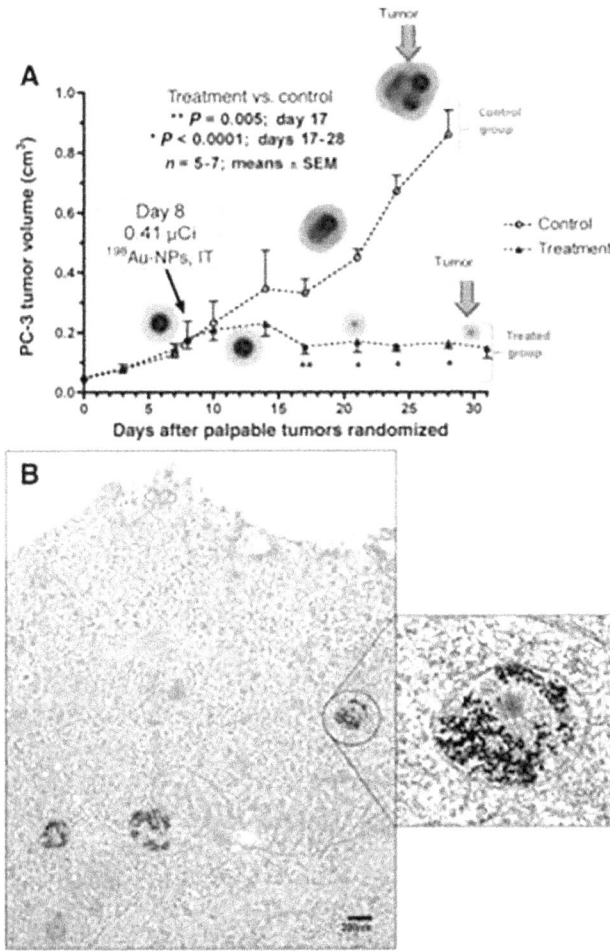

FIGURE 12.5 The reduction of the palpable tumors in SCID mice. (Reprinted with permission from Ref. [24]. Copyright 2010, Elsevier.)

treatments, and specifically due to cytotoxic side effects. In this case study, we will highlight the role of nanotechnology for the prognosis of ovarian carcinoma and the technology involved in the process. It will lead to prospective improvement in detection and curing of cervical carcinoma. It helps in the early diagnosis (testing and scanning) of cervical cancer. It also helps to overcome drug resistance and recurrence of cervical cancer. The biggest challenge in the implementation of such technology is the lack of training for biomedical fraternity for the advancement and handling of nano-based expertise [26].

Nanotechnology is a new field that is an application-based technology where a nanoscale particle contributes to increasing the surface area of a reaction tremendously changing the entire process altogether. Here, nanoscale parts of at least one size of 1–100 nm are used. The application of nano-based technology to the field of biology is called nanobiotechnology, and the use of nano-based technology in the medical field is called nanomedicine. The focus of nano-based technology is to diagnose, treat or both treatment and diagnosis which is also called theranostics for target specific diseases as well as cancer. In this case study, we will focus on the nanomaterials used in therapies, diagnoses, and theranostics. Nanocarriers have many functions as they are loaded with many therapeutic molecules with the help of visible adsorption or chemical compounding drugs, imaging agents, identifying components such as ligands or antibodies and polyethylene glycol (PEG);

this increases the half-life of therapeutic agents and promotes a stable and active tumor directing [27,28–36].

12.4.1 NANOTECHNOLOGY METHODS FOR OVCA DIAGNOSIS

Current OVCA tests: Currently, there is no standard test for healthy OVCA women; however, based on the serum blood levels for CA125 biomarker, routine examination, and sonography, the diagnosis is performed. These methods are not for early detection of disease and therefore not very helpful for detecting OVCA precisely. The current study gives a solution for OVCA prognosis by detecting OVCA biomarkers by creating a profile of improperly expressed molecules through better imaging skills. This will lead to better prediction, which ultimately increases the survival rate. For OVCA biomarker testing, the sensitivity should be 75% with a higher specificity rate of 96.6%. OVA1TM from QuestDiagnostics Madison, NJ, has been approved by FDA for determining ovarian tumor malignancy for prognosis as well as for post-surgery diagnosis [36].

In the study, the first method for biomarker testing by harvesting low molecular weight compounds from sample biological fluids as reviewed by Geho and colleagues [37] and Luchini [38]. In this method, nanoparticles compete with the functional proteins for their affinity in the form of change, mass, acid base reaction, etc. The best example of such nanocarriers are hydrogen and carbon nanoparticles [38–40]. The second method uses MALDI-TOF mass spectroscopy for the detection of low molecular weight compounds based on optical and thermal properties of carbon nanotubes, where nanotubes behave as a matrix in spectroscopy giving slow laser power limit for desorption and low ionization molecules, which leads to the development of signal-to-back ratio as well improve the detection of low MW molecules [40,41]. The third method is a chip-based microfluidic lab device to read OVCA cell features [42–45].

Despite the large data in cancer research and application of nanotechnology to mitigate it, the analysis is required at large scale. This requires proper classification based on types of nanoparticles, their modification, and their applicability in the cancer research. Nanotechnology can be helpful in reversing the ontology, prognosis of OVCA and in improvement of treatment methods which include drug delivery systems. Here, training of medical professional in handling these nanocarriers and related instruments is must with the amalgamation of engineering sciences to improve better delivery systems [46].

12.5 CASE STUDY IV: THE ROLE OF BIOACTIVE PHYTOCHEMICAL COMPOUNDS (BPC) NANO-FORMULATION IN BIOMEDICAL DRUG DEVELOPMENT AND DELIVERY SYSTEMS

Natural biological phytochemical compounds (BPCs) from plants and other natural resources are being used since the existence of human civilization. They are extensively used all over the world for treatment and improving human health issues. They are important resources for manufacturing and development of allopathic as well as traditional drugs. These phytochemicals have efficacy to show potent and promising biological functions. However, they also show some limitations for their effective use as drugs for therapeutic applications for their low melting point, instable structures and uneven distribution to the various parts of plant. These chemical phytochemicals are now being converted into nano-formulations (NF) to overall solubility, higher stability and pharmacokinetics and pharmacodynamics due to their improved potency. The case study outlines techniques to improve nano-formation from BPCs to address the relevant problems BPCs are used to have in drug making and their recent advances in prognosis and therapeutic use of imaging and treatment. This review presents a summary of technologies involved in processing of the next generation of nano-biophytochemicals for establishing and finding advanced therapeutic drugs [47].

Among them, nanoparticles, nanoemulsion, or lipid-based nanocarrier-based NFs account for about one-third, as well as liposomal BPCNF studies. There are liposome-based bioactive phytochemical nano-formulations which are called solid lipid nanoparticles (SLNs) and nano-lipid carriers (NLCs). Nanoparticle-based BPCNF has been evaluated since 1989. The number of publications in the NF has amplified exponentially in recent years, surpassing that of the liposome-based BPCNF publication since 2007, bringing the total to 5,713 by 2021. Other data on BPCNF in 2021 are about 1,200 articles. Publishing trends show a growing interest in BPCNF research and changes in BPC NF technology [47].

The next section shows the recent biomedical use of BPCNF, from prognosis to treatment in various applications. In particular, this case study focuses on various types of nano-formulations to improve drug delivery. It also discusses the role of new nano-formulation-based methods for discovering and developing next-generation phytochemical drugs based on advanced medical systems.

12.5.1 CONSTRUCTION OF NANO-FORMULATION

A key goal in nano-formulation development is to control particle size and surface properties, and to release the profile of bioactive phytochemical compounds to achieve site-specific accumulation and efficiency of the drug with appropriate treatment doses. The enhancement of overall property of bioactive phytochemical-based nano-formulation is achieved typically by encapsulation, which is designed typically of 10–200 nm in diameter. Another approach is the subtraction of drugs to nanocarriers, or combinations, the combined binding of drug molecules to nanocarriers suitable for delivery system. These carriers are used in drug delivery research to improve bioavailability, to enhance solubility, to increase blood circulation, and to target drug delivery while minimizing side effects produced in other drug delivery systems. NF can enter the body by inhaling, inhaling, entering the skin, or injecting and engaging in biological processes. The NF design increases the chances of access to the post-management environment and helps protect the property from important factors such as pH, enzymes and biological degradation. Additionally, NF can reduce doses by improving targeted delivery, site-specific delivery and the therapeutic effect of the drug. Nano-formulations can also be programmed to respond to extrinsic stimuli such as temperature, light, magnetic field, and ultrasound or internal stimuli, for example, pH, redox energy, and change in enzyme concentration [48] (Figure 12.6).

12.6 CASE STUDY V: GREEN NANOTECHNOLOGY FOR BRAIN TUMOR DIAGNOSIS AND SPECIFIC TREATMENT

As described in the previous case study, phytochemical drugs can be targeted in various forms and applications of nanocarriers. In this case study, we focus on how these carriers or nanomolecules contribute to brain tumor diagnosis and its specific treatment [49].

Statistics released in 2019 show that in some countries, about 2 million cases of cancer have been diagnosed. From these, around 23,000 cases are associated with brain tumors, which are abnormal masses of brain cells [50]. Brain tumors are known to be the most deadly type of cancer, and they occur in people of all ages; statistics show that it is more common in children and adults [51]. There are more than 20 different types of brain tumors, and they can be classified as just and cruel. The most common benign tumors are meningiomas (arising from the meninges and associated with the spine cord injury), schwannomas (affecting the nerves around the cranium, spine, and organs by isolation and, consequently, resulting in loss of function), as well as the pituitary adenomas (found in the pituitary causing its dysfunction) [52,53]. On the other hand, the most common malignant tumors are gliomas. Gliomas are divided into different subtypes, depending on the related cells. For example, astrocytoma starts in the astrocytes and can change the tissues that support the brain, while oligodendrogliomas can affect the mycelium and jeopardize brain differentiation. However, one of the most studied types of malignant brain tumors is glioblastoma multiforme (GBM), which

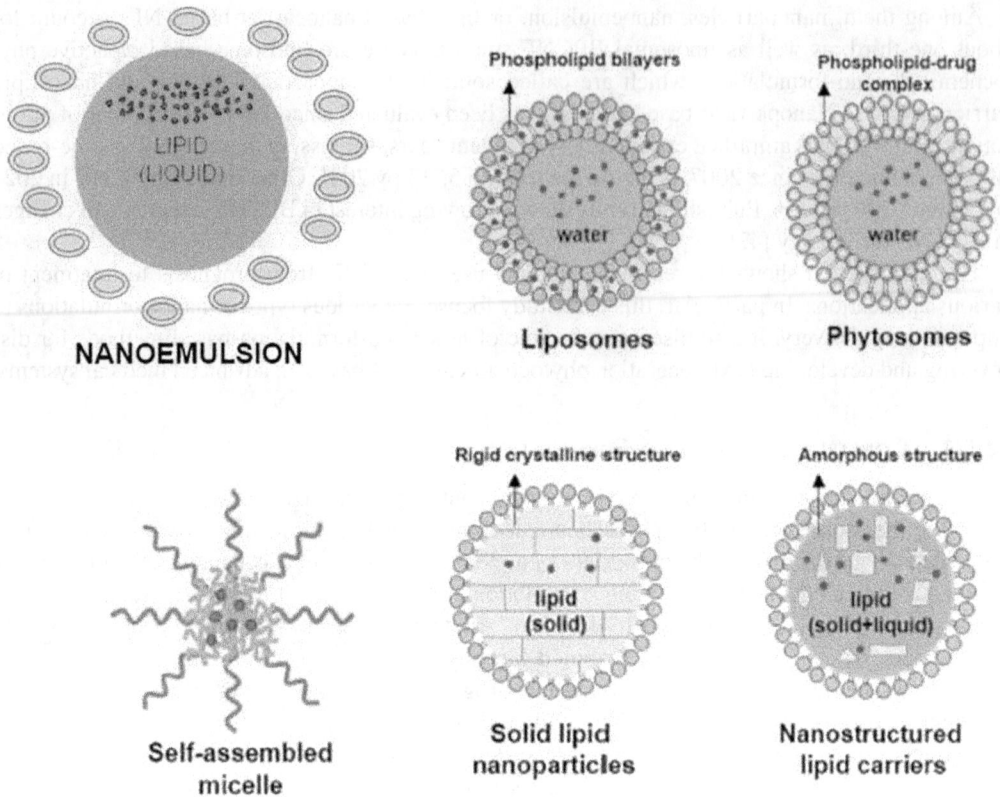

FIGURE 12.6 Commonly occurring bioactive phytochemical nano-formulations.

can convert both astrocytes and oligodendrocytes [54–56]. Lastly, damage to each of these glands can be graded on a 4-point scale, with 4 representing a highly invasive and progressive cancer [57].

12.6.1 Example of Diagnostic Performance of Nano-Based Technology

Nanodiagnostics, as the use of nanotechnology in medical diagnostics, is promising due to its high sensitivity, low cost, and effective diagnostic applications as well as high power of intracellular detection [58]. With the help of malignant images brain tumors using NPs, Yan et al. designed and integrated optical/paramagnetica dendrimer-based nanoprobe with many labels. This nanoprobe has shown a combination of BBB infiltration and improved penetration and retention effect (EPR) for non-invasive detection of glioblastomas, high sensitivity and specificity identification [58]. Kircher et al. used multimodal NPs based on magnetic resonance imaging (MRI) and found iron oxide cores mixed with material near-infrared fluorescent (NIRF) fluorochrome for preoperative imaging and intra-operative optical definition of brain tumor, respectively [59]. Using a different approach, Zhou et al. developed targeted micelles for optical/MRI operation with encapsulating inserts superparamagnetic iron oxide NPs to determine the location of glioma before and during performance. In addition, in vitro cytotoxicity studies with micelles have shown cytocompatibility with respect to mouse embryo fibroblast cell line, NIH-3T3 cells [60]. Similarly, Lu et al. used a single nanostructure that included the target foam gold nanoparticles (HAuNPs) in the xenograft mouse model of glioma in both photoacoustic tomography and photothermal ablation [61].

The use of raw nanomedicine in the treatment of brain cancer is a growing field [55,56]. However, while the ideas are good, there are not many examples available books. The case study reflects some light on the available literature on recent developments in methods based on raw nanotechnology from which all raw materials are extracted from biological or natural resources and environmentally friendly methods. One or more commonly used raw materials have been replaced by natural or biological means, or using one or more biological molecules aims to provide a highly efficient and customized treatment and management solution in the brain tumor research. The methods are described as follows:

- Microbiologically derived nanotechnologies for the treatment of brain tumors
- *Nanotechnologies are found in bacteria*: Bacteria are considered among the first green or biogenic candidates for the production of NMs due to the potential and widespread availability of these organisms on the planet. They offer a wide variety of biochemical chemical structures for the production of elemental nanostructures [62].

When referring to brain tumors, many examples are related to the use of iron oxide (Fe_3O_4) nanoparticles (IONPs) or super-paramagnetic IONPs (SPIONs). The feasibility of their production by conventional means, these NPs are generally prepared by polymeric or hydrophilic materials to reduce cytotoxicity, mutation hydrophilicity, and loading of drugs/genetics magnetosomes, membranous structures produced by magnetotactic bacteria (MTB). Magnetosomes containing NP-rich magnetic NPs are trapped inside the lipid bilayer membrane, which provides biocompatibility in a system [63,64]. These NPs are easily produced by using microorganisms designed as regulatory bio-factory with integrated method for the production of mineral core and shell, and impregnated with selected ligands [65,66].

- *Mammalian cell-derived nanotechnologies*: Using cell power as a biofactory for NMs, researchers have successfully removed the NM insertion step inside plant cell as they are produced within the cell membrane itself. Therefore, the cell-mediated NM integration has been reported as one of the latest raw nanotechnology ways to get into cells and tumors. In existing ways in studies, cells are able to quickly produce NMs efficiently and effectively through an environmentally friendly approach [67].
- *Virus-derived nanotechnologies*: Sceintists have designed NMs that can mimic viral features provided different structures with better input into cells and more formal accommodation than NMs. Therefore, viral-mimetic particles are regarded as attractive candidates as a platform for medical delivery, with a lot of applications in brain tumor research. For example, Lee et al. successfully developed gold nanorods. They mimic the rabies virus in terms of size, shape, surface glycoprotein structure, and in vivo behavior. Using rabies virus glycoprotein (RVG) which carries rabies virus to the CNS through neuronal pathways and passes through the blood brain barrier, this group produced rabies virus-mimetic silica-coated nanorods gold to treat gliomas in the brain. The virus-like nanosystem was transplanted into neuronal in vitro cells and caused severe heat effect in response to NIR laser irradiation. RVG peptide modified and improved the in vivo distribution of nanorods in the CNS [68].

As already presented, the use of organic matter as nano-biofactories and the use of different biological molecules are presented as two powerful mechanisms for the production of NMs with different pharmacokinetics, physicochemical properties, and drug delivery skills that may support significant developments in the sector. As an important consideration, the use of raw nanotechnology suggests that the production of NMs is a cost-effective, environmentally friendly, and straightforward manner. The methods of raw nanotechnology are based on existing principles of green chemistry, which means that all processes are directed by a few uses of consumables and renewable inputs, where possible, and hence energy savings. As a result, when you apply such principles, it is easy to gain an increase in energy efficiency, reducing waste and greenhouse gas emissions and reducing the use of non-renewable energy sources [69]. Most of the active components are used as endless renewable natural resources. Therefore finding efficient conversion of waste disposal and production patterns into a method that will not significantly affect the planet is the need of hour. [69–72].

12.7 CONCLUSIONS

The expansion of eco-friendly technologies in material synthesis is of substantial significance to get bigger opportunities in biological applications. In contemporary time, a numerous types of green nanoparticles with distinct chemical composition, shape, and structure are available. This chapter highlights five different case studies and future prospects of the same. There are numerous products and applications that are widely known in biomedical and medicine. Some of the commonly known applications known are drug delivery, medical treatment, etc., which are discussed in this chapter. The exceptional properties of nanoparticles are also pronounced as nanoscale studies. The two major milestones achieved by the nanotechnology in the cancer treatment are superior target discrimination and enhanced delivery possible due to cancer therapeutics and imaging. Hyperthermia is backed with chemotherapy and radiation to get the desired results as discussed in the first case study. In the second study, the use of gold nanoparticulate in radioactive therapy for the improvement of SCID mice without minimum or no outflow of the radioactive substance is discussed. In the third and fourth case studies, a brief discussion about the use of green nanotechnology in chemotherapy of ovarian cancer, the role of bioactive phytochemical compounds (BPCs), and nano-formulation in biomedical drug development and delivery system is done. The last case focuses on brain tumor diagnosis and its specific treatment. These case studies help to give a wide spectrum of potential uses of green nanotechnology and its applications.

REFERENCES

1. Virkutyte J.; Varma R.S. Green synthesis of metal nanoparticles: Biodegradable polymers and enzymes in stabilization and surface functionalization. *Chem Sci*, **2011**, 2 (5), 837–846.
2. Sapra P.; Tyagi P.; Allen T.M. Ligand-targeted liposomes for cancer treatment. *Curr Drug Deliv*, **2005**; 2, 369–381.
3. Ferrari M. Cancer nanotechnology: Opportunities and challenges. *Nat Rev Cancer*, **2005**; 5, 161–171.
4. Li K.C.; Pandit S.D.; Guccione S.; Bednarski M.D. Molecular imaging applications in nanomedicine. *Biomed Microdevices*, **2004**; 6: 113–116.
5. Wang X.; Yang L.; Chen Z.G.; Shin D.M. Application of nanotechnology in cancer therapy and imaging. *CA Cancer J Clin*, **2008**; 58: 97–110.
6. LaVan D.A; McGuire T.; Langer R. Small-scale systems for in vivo drug delivery. *Nat Biotechnol* **2003**; 21: 1184–1191.
7. Allen T.M. Ligand-targeted therapeutics in anticancer therapy. *Nat Rev Cancer* **2002**; 2: 750–763.
8. Gao X.; Cui Y.; Levenson R.M., et al. In vivo cancer targeting and imaging with semiconductor quantum dots. *Nat Biotechnol* **2004**; 22: 969–976.
9. Deryugina E.I.; Quigley J.P. Matrix metalloproteinases and tumor metastasis. *Cancer Metastasis Rev* **2006**; 25: 9–34.
10. Brannon-Peppas L.; Blanchette J.O. Nanoparticle and targeted systems for cancer therapy. *Adv Drug Deliv Rev* **2004**; 56: 1649–1659.
11. Stohrer M.; Boucher Y.; Stangassinger M.; Jain R.K. Oncotic pressure in solid tumors is elevated. *Cancer Res* **2000**; 60: 4251–4255.
12. Vicent M.J.; Duncan R. Polymer conjugates: Nanosized medicines for treating cancer. *Trends Biotechnol* **2006**; 24: 39–47.
13. Mantso T.; Vasileiadis S.; Anestopoulos I.; Voulgaridou G.P.; Lampri E.; Botaitis S.; Kontomanolis E.N.; Simopoulos C.; Goussetis G.; Franco R., et al. Hyperthermia induces therapeutic effectiveness and potentiates adjuvant therapy with non-targeted and targeted drugs in an in vitro model of human malignant melanoma. *Sci Rep* **2018**; 8: 10724.
14. Luk K.H.; Hulse R.M.; Phillips T.L. Hyperthermia in cancer therapy. *West J Med* **1980**; 603(132): 179–185.
15. Kampinga H.H. Cell biological effects of hyperthermia alone or combined with radiation or drugs: A short introduction to newcomers in the field. *Int J Hyperther* **2006**; 22: 191–196.

16. Peeken J.C.; Vaupel P.; Combs S.E. Integrating hyperthermia into modern radiation oncology: What evidence is necessary? *Front Oncol* **2017**; 7(132): 616.

17. Kaur P.; Aliru M.L.; Chadha A.S.; Asea A.; Krishnan S. Hyperthermia using 618 nanoparticles–Promises and pitfalls. *Int J Hyperther* **2016**; 32: 76–88.

18. Wilson B.C.; Patterson M.S. The physics, biophysics and technology of photodynamic therapy. *Phys Med Biol* **2008**; 53: R61–109. doi: 10.1088/0031-9155/53/9/R01.

19. Thomsen S. Pathologic analysis of photothermal and photomechanical effects of laser-tissue interactions. *Photochem Photobiol* **1991**; 53: 825–835.

20. Huang N.; Wang H.; Zhao J.; Lui H.; Korbelik M.; Zeng H. Single-wall carbon nanotubes assisted photothermal cancer therapy: Animal study with a murine model of squamous cell carcinoma. *Lasers Surg Med* **2010**; 42: 638–648. doi:10.1002/lsm.20968.

21. Dong Z.; Gong H.; Gao M.; Zhu W.; Sun X.; Feng L.; Fu T.; Li Y.; Liu Z. Polydopamine nanoparticles as a versatile molecular loading platform to enable imaging guided cancer combination therapy. *Theranostics* **2016**; 6: 1031–1042. doi:10.7150/thno.14431.

22. Sapra P.; Tyagi P.; Allen T.M. Ligand-targeted liposomes for cancer treatment. *Curr Drug Deliv* **2005**; 2: 369–381.

23. Kennedy L.C.; Bickford L.R.; Lewinski N.A.; Coughlin A.J.; Hu Y.; Day E.S.; West J.L.; Drezek R.A. A new era for cancer treatment: Gold-nanoparticle-mediated thermal therapies. *Small* **2011**; 7: 169–183. doi:10.1002/smll.201000134.

24. Chanda N., et al. Radioactive gold nanoparticles in cancer therapy: Therapeutic efficacy studies of GA-[198]AuNP nanoconstruct in prostate tumor–bearing mice. *Nanomed Nanotechnol Biol Med* **2010**; 6: 201–209.

25. Kannan R.; Rahing V.; Cutler C.; Pandrapragada R.; Katti K.K.; Kattumuri V., et al. Nanocompatible chemistry toward fabrication of target-specific gold nanoparticles. *J Am ChemSoc* **2006**; 128: 11342–11343.

26. Kim P.S.; Djazayeri, S.; Zeineldin, R. Novel nanotechnology approaches to diagnosis and therapy of ovarian cancer. *Gynecol Oncol* **2011**; 120 (3): 393–403.

27. Van Vlerken L.E.; Vyas T.K.; Amiji M.M. Poly(ethylene glycol)-modified nanocarriers for tumor-targeted and intracellular delivery. *Pharm Res* **2007**; 24: 1405–1414.

28. Cho K.; Wang X.; Nie S.; Chen Z.G.; Shin D.M. Therapeutic nanoparticles for drug delivery in cancer. *Clin Cancer Res* **2008**; 14: 1310–1316.

29. Baker Jr. J.R. Dendrimer-based nanoparticles for cancer therapy. *Hematol Am Soc Hematol Educ Program* **2009**; 12: 708–719.

30. Prato M.; Kostarelos K.; Bianco A. Functionalized carbon nanotubes in drug design and discovery. *AccChem Res* **2008**; 41: 60–68.

31. Liu Z.; Tabakman S.; Welsher K.; Dai H. Carbon nanotubes in biology and medicine: In vitro and in vivo detection, imaging and drug delivery. *Nano Res* **2009**; 2: 85–120.

32. Veiseh O.; Gunn J.W.; Zhang M. Design and fabrication of magnetic nanoparticles for targeted drug delivery and imaging. *Adv Drug Deliv Rev* **2010**; 62: 284–304.

33. Arvizo R.; Bhattacharya R.; Mukherjee P. Gold nanoparticles: Opportunities and challenges in nanomedicine. *Expert Opin Drug Deliv* **2010**; 7: 753–763.

34. Agasti S.S.; Rana S.; Park M.H.; Kim C.K.; You C.C.; Rotello V.M. Nanoparticles for detection and diagnosis. *Adv Drug Deliv Rev* **2010**; 62: 316–328.

35. Bharali D.J.; Khalil M.; Gurbuz M.; Simone T.M.; Mousa S.A. Nanoparticles and cancer therapy: A concise review with emphasis on dendrimers. *Int J Nanomed* **2009**; 4: 1–7.

36. Das P.M.; Bast Jr. R.C. Early detection of ovarian cancer. *Biomark Med* **2008**; 2: 291–303.

37. Geho D.H.; Jones C.D.; Petricoin E.F.; Liotta L.A. Nanoparticles: Potential biomarker harvesters. *Curr Opin Chem Biol* **2006**; 10: 56–61.

38. Luchini A.; Fredolini C.; Espina B.H.; Meani F.; Reeder A.; Rucker S.; et al. Nanoparticle technology: Addressing the fundamental roadblocks to protein biomarker discovery. *Curr Mol Med* **2010**; 10: 133–141.

39. Longo C.; Patanarut A.; George T.; Bishop B.; Zhou W.; Fredolini C.; et al. Core–shell hydrogel particles harvest, concentrate and preserve labile low abundance biomarkers. *PLoS ONE* **2009**; 4: e4763.

40. Gaspari M.; Ming-Cheng Cheng M.; Terracciano R.; Liu X.; Nijdam A.J.; Vaccari L.; et al. Nanoporous surfaces as harvesting agents for mass spectrometric analysis of peptides in human plasma. *J Proteome Res* **2006**; 5: 1261–1266.

41. Terracciano R.; Gaspari M.; Testa F.; Pasqua L.; Tagliaferri P.; Cheng M.M.; et al. Selective binding and enrichment for low-molecular weight biomarker molecules in human plasma after exposure to nanoporous silica particles. *Proteomics* **2006**; 6: 3243–50.

42. Najam-ul-Haq M.; Rainer M.; Szabo Z.; Vallant R.; Huck C.W.; Bonn G.K. Role of carbon nano-materials in the analysis of biological materials by laser desorption/ionization-mass spectrometry. *J Biochem Biophys Meth* **2007**; 70: 319–328.

43. Xu S.; Li Y.; Zou H.; Qiu J.; Guo Z.; Guo B. Carbon nanotubes as assisted matrix for laser desorption/ionization time-of-flight mass spectrometry. *Anal Chem* **2003**; 75: 6191–6195.

44. Wang C.H.; Li J.; Yao S.J.; Guo Y.L.; Xia X.H. High-sensitivity matrix-assisted laser desorption/ionization Fourier transform mass spectrometry analyses of small carbohydrates and amino acids using oxidized carbon nanotubes prepared by chemical vapor deposition as matrix. *Anal Chim Acta* **2007**; 604: 158–164.

45. Jokerst J.V.; Raamanathan A.; Christodoulides N.; Floriano P.N.; Pollard A.A.; Simmons G.W., et al. Nano-bio-chips for high performance multiplexed protein detection: Determinations of cancer biomarkers in serum and saliva using quantum dot bioconjugate labels. *Biosens Bioelectron* **2009**; 24: 3622–3629.

46. Thomas D.G.; Pappu R.V.; Baker N.A. Nano particle ontology for cancer nanotechnology research. *J Biomed Inform* **2010**; 44: 59–74.

47. Han H.S.; Koo S.Y.; Choi K.Y. Emerging nanoformulation strategies for phytocompounds and applications from drug delivery to phototherapy to imaging. *Bioactive Materials* **2021**; 13: 121–128.

48. Patra J.K.; Das G.; Fraceto L.F.; Campos E.V.R.; Rodriguez-Torres M.d.P.; Acosta-Torres L.S.; Diaz-Torres L.A.; Grillo R.; Swamy M.K.; Sharma S.; Habtemariam S.; Shin H.-S. Nano based drug delivery systems: Recent developments and future prospects. *J Nanobiotechnol* **2018**; 16: 71.

49. Mostafavi E.; Medina-Cruz D.; Vernet-Crua A.; Chen J.; Cholula-Díaz J.L.; Guisbiers G.; Webster T.J. Green nanomedicine: The path to the next generation of nanomaterials for diagnosing brain tumors and therapeutics? *Expert Opin Drug Deliv* **2021**; 18(6): 715–736.

50. Siegel R.L.; Miller K.D.; Jemal A. Cancer statistics. *CA Cancer J Clin* **2019**; 69(1), 7–34. doi: 10.3322/caac.21551.

51. Deepa A.R.; Sam Emmanuel W.R. An efficient detection of brain tumor using fused feature adaptive firefly backpropagation neural network. *Multimed Tools Appl* **2018**; 1(3): 12–16.

52. Aghi M.; Barker II, F.G. Benign adult brain tumors: An evidence-based medicine review. In Pollock B.E. (ed.) *Guiding Neurosurgery by Evidence*, **2006**, vol. 19, Basel: Karger, 80–96.

53. Black P.M. Benign brain tumors. Meningiomas, pituitary tumors, and acoustic neuromas. *Neurol Clin* **1995**; 13(4): 927–52.

54. Lapointe S.; Perry A.; Butowski N.A. Primary brain tumours in adults. *Lancet* **2018**, 392: 432–446. doi: 10.1016/S0140-6736(18)30990-5.

55. Kavin Kumar K.; Meera Devi T.; Maheswaran S. An efficient method for brain tumor detection using texture features and SVM classifier in MR images. *Asian Pac J Cancer Prev* **2018**; 19(10): 2789–2794.

56. Collins V.P. Brain tumours: Classification and genes. *J Neurol Neurosurg Psychiatry* **2004**; 75 (2), 2. doi: 10.1136/jnnp.2004.040337.

57. Zacharaki E.I.; et al. Classification of brain tumor type and grade using MRI texture and shape in a machine learning scheme. *Magn Reson Med* **2009**; 62 (6): 1609–1618. doi: 10.1002/mrm.22147.

58. Yan H.; et al. Imaging brain tumor by dendrimer-based optical/paramagnetic nanoprobe across the blood-brain barrier. *Chem Commun* **2011**; 47 (28): 8130. doi: 10.1039/c1cc12007g.

59. Kircher M.F.; Mahmood U.; King R.S.; Weissleder R.; Josephson L. A multimodal nanoparticle for preoperative magnetic resonance imaging and intraoperative optical brain tumordelineation. *Cancer Res* **2003**; 63 (23): 8122–8125.

60. Zhu Y.; et al. Glioma-targeting micelles for optical/magnetic resonance dualmode imaging. *Int J Nanomedicine* **2015**; 10: 1805. doi: 10.2147/IJN.S72910.

61. Lu W.; et al. Effects of photoacoustic imaging and photothermal ablation therapy mediated by targeted hollow gold nanospheres in an orthotopic mouse xenograft model of glioma. *Cancer Res* **2011**; 71 (19): 6116–6121.

62. Horváth I.T. Introduction: Sustainable chemistry. *Chem Rev* **2018**; 118 (2): 369–371. doi: 10.1021/acs.chemrev.7b00721.

63. Varshney R.; Bhadauria S.; Gaur M.S.; Pasricha R. Characterization of copper nanoparticles synthesized by a novel microbiological method. *JOM* **2010**; 62(12): 102–104. doi: 10.1007/s11837-010-0171-y.

64. Thakkar K.N.; Mhatre S.S.; Parikh R.Y. Biological synthesis of metallic nanoparticles. *Nanomed Nanotechnol Biol Med* **2010**; 6 (2): 257–262. doi: 10.1016/j.nano.2009.07.002.

65. Fayaz A.M.; Balaji K.; Girilal M.; Yadav R.; Kalaichelvan P.T.; Venketesan R. Biogenic synthesis of silver nanoparticles and their synergistic effect with antibiotics: A study against gram-positive and gram-negative bacteria. *Nanomed Nanotechnol Biol Med* **2009**; 6 (1): 103–109. doi: 10.1016/j.nano.2009.04.006.

66. Plan Sangnier A.; et al. Targeted thermal therapy with genetically engineered magnetite magneto-somes@RGD: Photothermia is far more efficient than magnetic hyperthermia. *J Control Release* **2018**; 279: 271–281. doi: 10.1016/J.JCONREL.2018.04.036.

67. Anshup A.; et al. Growth of gold nanoparticles in human cells. *Langmuir* **2005**; 21(25): 11562–11567. doi: 10.1021/LA0519249.

68. Lee C.; et al. Rabies virus-inspired silica-coated gold nanorods as a photothermal therapeutic platform for treating brain tumors. *Adv Mater* **2017**; 29 (13): 1605563. doi: 10.1002/adma.201605563.

69. Maksimović M.; Omanović-Mikličanin E. Towards green nanotechnology: Maximizing benefits and minimizing harm. In: Badnjevic, A. (eds.) *CMBEBIH*, **2017**. Singapore: Springer, pp. 164–170.

70. Jain N.; Gupta E.; KanuNand J. Plethora of carbon nanotubes applications in various fields-A state-of-the-art-review. *Smart Sci* **2022**; 10(1): 1–24.

71. Jain N.; KanuNand J. The potential application of carbon nanotubes in water treatment: A state-of-the-art-review. *Mater Today Proc* **2021**; 43(5): 2998–3005.

72. Jain N.; Tiwari S. Biomedical application of carbon nanotubes (CNTs) in vulnerable parts of the body and its toxicity study: A state-of-the-art-review. *Mater Today Proc* **2021**; 46(17): 7608–7617.

13 Case Studies on Multifunctional Green Quantum Dots – From Lab Bench to Commercialization

Sayoni Sarkar and Ajit R. Kulkarni
Indian Institute of Technology Bombay

CONTENTS

13.1 INTRODUCTION

Quantum dots (QDs) have become increasingly eminent because of their fascinating size-dependent optical, biological, physiochemical, and electrical properties. In addition to this, the quantum confinement effect arising due to the size of the particles being smaller than or close enough to that of the electron's wavelength enhances the surface chemistry of the QDs. These exclusive traits have opened up newer possibilities for the QDs in energy storage devices, smart healthcare, catalysis, agriculture, and preserving the environment. Thus, this stupendous growth in the demand for QDs has resulted in the implementation of advanced synthesis techniques that often threaten the environment by producing harmful by-products and inefficient energy consumption. The researchers are now focusing on innovating state-of-the-art fabrication strategies for green QDs (GQDs), which

DOI: 10.1201/9781003319153-16

overcome these concerns and ensure that the QDs are sustainable, eco-friendly, and harmoniously fit within the principles of "green chemistry".

Endeavors including stringent surface functionalization of QDs by environmentally benign sources and encouraging the development of "green" fabrication protocols have paved the way for GQDs as the best substitute for quantum dots that risk the environment. Humungous efforts around the globe in recent years have gradually culminated in the flourishing of GQDs that have necessitated the formulation and development of robust synthesis methods that are easily scalable and ensure the fabrication of superior-quality GQDs. The conventional batch processes adopted for synthesizing GQDs are riddled with several cons – poor yield, non-uniform heat distribution, high cost, and consumption of high energy. Rapid technological advancements have successfully endorsed facile scalable eco-friendly synthesis routes based on the crux of flow chemistry. Unlike the batch reactors, continuous flow reactors can effectively tune the processing conditions for pilot-scale production of the most preferred GQDs, setting the dice rolling for newer innovations. For instance, in a pioneering study, surface-functionalized, biocompatible nanoparticulate iron oxide was synthesized in a continuous flow setup for applications in biology and served as a proof-of-concept for economic pilot-scale production of revolutionary nanomaterials [1]. These continuous flow synthesis assemblies additionally ensure efficient mixing of the precursors and fast nucleation-growth kinetics for which the critical reaction parameters namely temperature, pressure, pH, residence time, and flow rate governing these phenomena require appropriate optimization. In another work, green silver nano-architectures, which have tremendous potential as antimicrobial agents, were synthesized by employing biosurfactants via a continuous flow method. To obtain monodispersed particles, the process intensification was carried out in traditional batch reactors, then extrapolated to continuous flow synthesis to enhance their commercial potential [2]. The main objective of ascertaining good control over the synthesis process is to bridge the gap between lab-scale synthesis and large-scale production of GQDs by eco-friendly means as stepping stones to a sustainable future. Thus, the production of GQDs by continuous flow chemistry from biomass-derived sources as initial precursors is a one-step solution for minimizing the environmental risks that typically originate from the fabrication of quantum dots, increasing energy efficiency, bringing down waste generation, and operating costs. In the life cycle assessment studies of the green carbon QDs by the continuous flow routes, the primary point of focus revolved around the process efficiency and pre-eminence of the QDs' unique traits for sensing toxic and beneficial metal ions [3].

Several breakthroughs in the conceptualization and designing methodology of attractive continuous flow platforms for GQDs synthesis have brought about drastic changes by adopting simulation-based approaches through applying fundamental aspects of computational fluid dynamics (CFD). CFD suffices as a promising tool to understand the heat and mass transfer profiles for evaluating the performance of continuous flow reactors by solving the associated equations for particular flow rate, temperature, and pressure conditions. It is essential to examine the reactor's competence to boost the desirable properties of GQDs as there is a strong interlink between the nucleation-growth kinetics and fluid dynamics inside the reactor. These analyses are driven mainly by some user-friendly standard software packages such as ANSYS Fluent, COSMOL Multiphysics, and OpenFOAM. All these cumulative attempts have led to significant progress in the seamless transition of fabrication of high-quality GQDs from lab bench to commercial scale. Herein, we discuss the sustainable and affordable continuous flow synthesis of zinc oxide (ZnO) and cerium oxide (CeO_2) GQDs, their path toward successful commercialization facilitated by rigorous process intensification relying on numerical simulations and distinctive case studies on the application of these GQDs in biomedicine, agriculture, and environmental remediation.

13.2 CONTINUOUS FLOW SYNTHESIS OF ZNO AND CEO₂ GQDS – BRIDGING THE GAP BETWEEN LAB BENCH TO PILOT-SCALE SYNTHESIS

13.2.1 OVERVIEW

13.2.1.1 Nucleation-Growth Mechanisms Governing the Formation of Green QDs Inside the Continuous Flow Reactor

Several factors govern the synthesis of monodispersed superior-quality QDs. The backbone of the formation of colloidal nanoparticles can be elucidated by the LaMer mechanism that suggested the "burst of nucleation" phenomenon as best suited for narrow particle size distribution. In the continuous flow synthesis of green QDs, the kinetic and thermodynamic factors of nucleation and growth processes determine the size- and morphology-dependent properties of the QDs [4]. Under conditions of rapid heat transfer in the reaction chamber, when the temperature is higher than the crystallization temperature, a burst of nucleation ensues, resulting in the generation of seed QDs; thereupon, the growth of the seeds occurs. Recurrently, in the traditional batch reactors, the development of spatial heat transfer gradient is responsible for the undesirable nucleation that results in a broad particle size distribution of GQDs. However, the flow reactors are designed to corroborate uniform heat distribution and precisely control the other deciding factors of the quality of GQDs being produced involving flowrate, mixing, residence time, and pressure [5,6]. These parameters tailor the kinetics of nucleation-growth mechanisms for obtaining monodispersed particles exhibiting essential characteristics for advanced applications. The uniqueness of continuous flow reactors also lies in their potential to generate dean vortices causing ultrafast rapid mixing and optimized residence time, significant for uniform particle size [7]. It is important to note that the theory of nucleation-growth reached from the studies done by Lamer and his co-workers is limited by their assumptions and modifications. Over the years, the evolution of these theories has led to many recent developments wherein numerical simulations gave an insight into the dynamics of nucleation-growth of the nanocrystals, thereby opening broader possibilities of improvising the contemporary synthesis protocols [8]. Powerful *in situ* tools for rapid real-time information on the nucleation-growth kinetics of GQDs at short length scales during continuous flow synthesis enables the quantitative estimation of the crucial parameters. For instance, in one of the studies, the nucleation and growth of the colloidal nanoparticles were investigated by *in situ* small-angle X-ray scattering (SAXS). To understand the kinetics of the Ostwald ripening growth phenomenon of gold NPs, real-time SAXS measurements accurately provided a quantitative estimation of the distinctive phases of nucleation and growth [9]. Such robust operando characterization methods are critical toward devising, developing, and validating the theories on the nucleation-growth mechanism of GQDs in a continuous flow platform (Figure 13.1).

13.2.1.2 Conventional Synthesis Routes versus Continuous Flow Reactors

The several benefits of continuous flow synthesis of GQDs include an eco-friendly approach, high throughput, superior quality, safety, and easy down-streaming. Under steady-state conditions, flow reactors are capable of achieving uniform distribution of heat and mass throughout their cross section such that there is no inhomogeneity in the end product, thereby enriching the quality [11]. On the other hand, traditional batch reactors have the disadvantage of non-uniform heat transport resulting in gradients that influence the efficiency of the GQDs [12,13]. Furthermore, batch processing lacks feasible and convenient scale-up options. This is why continuous flow synthesis for GQDs has gained immense popularity in the recent past. The small channel dimensions of the flow reactors facilitate significant improvement in the exchange of heat between the reactor wall to the reaction chamber (interior of the reactor) [14]. Unlike the batch reaction vials, this reduced size of the flow reactors also serves as the one-stop solution in ameliorating the problems associated with

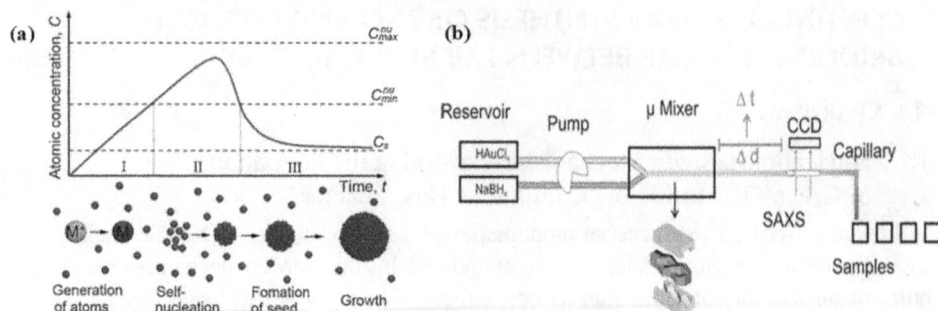

FIGURE 13.1 (a) Nucleation and growth phases of seed QDs by LaMer mechanism. The first and second phases lead to nucleation, and the third one depicts the growth of the seed QDs. (b) Continuous flow synthesis platform for gold nanoparticles, which consists of a static micromixer connected to the SAXS assembly for in-line monitoring [4,10].

space crunch. Besides, the in-line monitoring of the flow process of GQDs assures maximum safety in operations. This added advantage of real-time analysis expedites the processing time and minimizes wastage of energy and raw materials drastically, which is in stark contrast to a batch method. This is extremely useful in understanding why and how the events of nucleation and growth occur and the changes that take place inside the reactor thereof [9,15].

Synthesis along the lines of batch chemistry is usually affected by poor mixing of the precursor solutions, which causes nucleation to occur during the growth of other nuclei. Hence, GQDs with variations in sizes are formed. These limitations can be very efficiently overcome by the design and optimization of flow reactors that can exhibit enhanced hydrodynamic performance. Depending on the morphology of the reactor, the dynamics of the fluid flow result in the formation of dean vortices which causes ultrafast rapid mixing. Another benefit of the flow process for GQDs synthesis is tailoring the flow rate as desired to meet the required residence time for complete nucleation and growth [16–18]. This is of paramount importance during process modeling, where the interlink between the synthesis conditions and the favorable characteristics of the GQDs becomes crucial to ensure high reproducibility. Despite the numerous advantages of a continuous flow process over batch synthesis, certain difficulties often go unnoticed. The first and foremost is that each continuous flow reactor is designed uniquely to suit the process conditions of the fabrication of a particular type of GQDs. Although making the flow reactors versatile is not a tedious job if rigorous numerical simulations for conceptualizing the fundamental fluid dynamics are employed. Additionally, rapid changes in heat and flow chemistry during the reaction in the flow reactor are notably different from that of a batch process. Therefore, a transition from the batch environment to a continuous mode of processing needs thorough validation of the critical synthesis factors of the batch. These parameters are refined further and tuned according to the requirements of the continuous flow synthesis assembly to use their full potential. Hence, it is indispensable to ultimately test the steady-state working condition of the flow reactors before initiation of any reaction; else, it might prove to be catastrophic. Sometimes in flow synthesis of GQDs involving multiple steps, generation and removal of undesired intermediate by-products are troublesome, and in such cases, additional purification is required. In light of the pros and cons of the conventional batch process and the continuous method of flow synthesis, the scope of adopting the latter synthesis route for scaling up the production is of enormous significance [14,19,20] (Figure 13.2).

13.2.1.3 Leading Flow Processes toward Sustainability through Design and Engineering

Continuous flow synthesis of advanced green quantum dots (GQDs) has accelerated the innovations in major domains of biomedicine, conservation of the energy and environment, and agriculture. Out of the many advantages, superior transport of heat, precise control over reaction parameters,

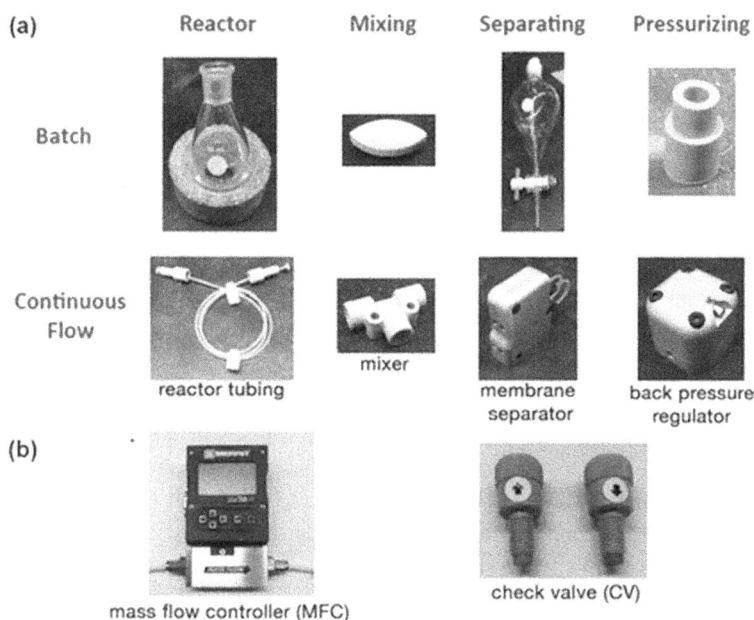

FIGURE 13.2 (a) Operating units of batch synthesis compared with continuous flow synthesis. (b) Mass flow controller and check valves used for precise regulation of the flow conditions [13].

maximization of safe operations, reproducibility and throughput and reducing energy usage and cost, are some benefits that give this process an edge over the others. Incorporating the key ideologies of green engineering and chemistry in a holistic way into the design and development of the continuous flow setup facilitates effortless translation from lab scale to pilot scale. This also makes it an environmentally benign process with a thrust toward sustainability. Focusing on the "how" of the design process for continuous flow routes for fabricating GQDs, divulge the main drivers that make this unconventional synthesis method so appealing. In-depth analysis and stressing the need to adopt greener footprints, requires formulating reaction-kinetics-based protocols and modifying the reactor configuration for uniform temperature distribution and mass transfer [21]. This makes it apparent that the formation of GQDs via rapid reactions and extreme processing conditions are best governed by accurate residence time and efficient mixing. Mixing of the reactants plays an integral role in the nucleation and growth by maintaining homogeneity in mass transfer. The phenomenon of mixing differs in a batch process and continuous flow synthesis. In batch mixing, the solution is agitated by a magnetic stirrer with a tunable rate of stirring, whereas, in the flow reactors, the mixing efficiency is largely determined by the structure of the reactor and the hydrodynamics of the fluid flow [18,22]. For scaling up, higher flow rates associated with increased Reynolds (Re) number are desirable to enhance the productivity of the reactor. Typically, in a microstructured channel, the fluid flow is laminar, which does not have much influence on the mixing conditions. With advancements in technology, the geometries of flow reactors have undergone many changes, involving square-coiled reactors, serpentine, and helical-shaped reactors, that have enhanced the flow dynamics remarkably owing to the impact of the centrifugal force that generates the dean vortices in a direction perpendicular to the primary flow. This yields GQDs with narrow particle size distribution and enriched characteristics. Another common design methodology adopted to conveniently mix the precursor solutions for synthesizing GQDs is to periodically alter the direction of fluid flow. All these strategies align with the objectives of green chemistry and the industrial scale of production [23–25]. Besides, rapid molecular diffusion is time and again attained by reduction of the reactor size that further guarantees an increment in mixing efficiency and shortens the residence time. One of the well-known criteria of immense significance to increasing the flow process efficiency is heat transfer between the reactor's interior and the external boundary in an accurately controlled mode. Thus, improvising the transfer characteristics by detailed investigation of the kinetics associated enables boosting the relation between yield-conversion rate-reaction time. The large surface-area-to-volume ratio of the designed flow reactor expedites the heating rate. It results in uniform temperature distribution throughout the entire reactor volume that prevents any build-up of localized hot spots, further facilitating GQDs formation. For example, Lin et al. determined the heat distribution profile based on the assumption that the flow inside the helical reactor was laminar to estimate the time required to reach the steady-state conditions and the effect on size distribution. Nusselt number for convective heat transfer was related to Prandtl number and helical number for optimization of the time taken to complete the nucleation and growth for the fixed reactor length. A close look into the thermodynamics of the synthesis manifested that the coupled effect of increased heating rate and high reaction temperature assisted in effectuating monodispersed particles [26]. In an interesting work, a green chemistry approach for synthesizing bismuth nanoparticles was implemented (Figure 13.3). Initially, an attempt was made to synthesize the nanoparticles in batch reactors. However, a meticulous inspection of the fabrication process delineated the flaws that made the batch reactor setup inadequate to fit into the ideologies of green chemistry because the polydispersed nanoparticles were limited by strikingly low yield, prolonged operation time, and high temperature, which caused huge consumption of energy. Due to these limitations, a different mode of heating that homogenously heated the reactor volume was employed that escalated the

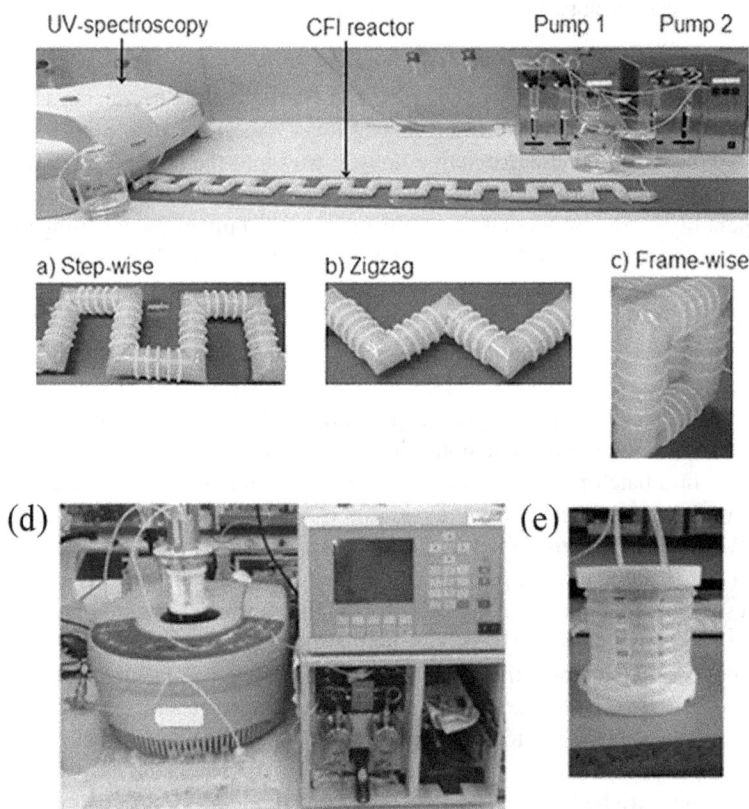

FIGURE 13.3 Coiled flow inverter assembly made up of PTFE in a "step" design with coiled silicon tubing for residence time distribution measurements. (a–c) PTFE-silicon tubing coiled flow inverter of varied geometry – step, zigzag, and square-frame. (d) Digital image of the millifluidic reactor with a microwave heating source. (e) An image of Teflon tubing held inside the cylinder made of Teflon [25,27].

reaction kinetics. This green continuous flow synthesis protocol enhanced productivity massively by efficient transport of heat and mass [27].

The trade-off between the efficiency of continuous flow processes by experimental validation after the reaction completion and the interrelated kinetics is understood by real-time in-line reaction telescoping. To gain detailed insight, the reaction conditions are varied over a wide range and precisely regulated by a variety of experimental techniques such as SAXS, UV-Vis, or fluorescence spectroscopy and parallel reactions. The specific parameters that attribute to the kinetically controlled non-equilibrium reaction dynamics in the flow reactors for GQDs are fixed. Based on these, a set of experiments is repeated to gather noticeable reproducibility. The information collected is subsequently analyzed and refined further based on kinetic models and simulations on fluid flow before being implemented for the next set of experiments. This methodology ensures minimizing the variation in experimental results and incorporates high process repeatability and accuracy, which not only holds promise in the reduction of operation cost for large-scale synthesis, but also makes it an environmentally benign process. The ease in rational design, optimization, and automation of the continuous flow processes for GQDs is made possible by the integration of Design of Experiments (DoE) and other relevant statistical algorithms. In the study by Kitson et al., a 3D printed versatile microfluidic flow reactor was designed and quickly fabricated along with the scope of reaction-centric in-line detection by UV-visible absorption spectrofluorometer to record and analyze the data of reaction kinetics. Combined with the LabVIEW program, the real-time data allowed controlling the key parameters involving residence time, the composition of the product, and hence, the reactor's performance [28]. Recent advancements in the frontier of statistical models employed to determine the interlink between the various process parameters have outlined the importance of deep learning. Deep learning aided experimental planning streamlines simultaneous optimization of the reaction conditions. Neural network models in a riveting way predict the influence of the precursor ratio, stirring rate, flow rate on the particle size after being trained with the key attributes of the particular synthesis under consideration. In one such machine learning-based model developed for the "green" synthesis of carbon nanodots with natural precursor sources such as tapioca flour, the model agreed well with the experimentally validated data. It helped enhance the photoluminescence quantum yield and additional specifications of the reaction [29].

13.2.2 IMPORTANCE OF COMPUTATION-BASED TOOLS FOR BETTER CONTROL OVER REACTION PARAMETERS

The transition of simulated continuous flow reactor design, process optimization and feedback system from being a novel approach to being the most preferred strategy to enable automation in synthesis routes by researchers, engineers, and industrialists have escalated over a short time. With advancements in the domains of statistics and computation, the power to rapidly identify the experimental conditions and improve them simultaneously is no longer an existing "black box". Albeit the combinatorial strategies of quick screening of the reaction parameters, comprehensive knowledge of kinetics and thermodynamics of the reaction is crucial as it helps identify the degree of efficient heat and mass transfer in the flow reactors. This assists in smoothening the scale-up by maximizing the reactor's overall performance through numerical simulations via developing mathematical models and machine learning (ML)-based algorithms (Figure 13.4). Problems associated with exhaustive experimentation for investigating the transport phenomena occurring inside the flow reactors with small channel dimensions requiring sophisticated, time-consuming, and costly measurements, are overcome by smart practice of merging simulations into the continuous flow platform. The computational fluid dynamic (CFD) simulations are employed to solve the mathematical models and equations that monitor the heat and mass transport, particle size distribution and residence time distribution profiles. This helps to carefully regulate the favorable reaction environment while also reducing the cost and time. Leybros et al. integrated the population balance model with CFD simulations for probing the hydrodynamic performance of the reactor, transfer characteristics taking

(a)

(b)

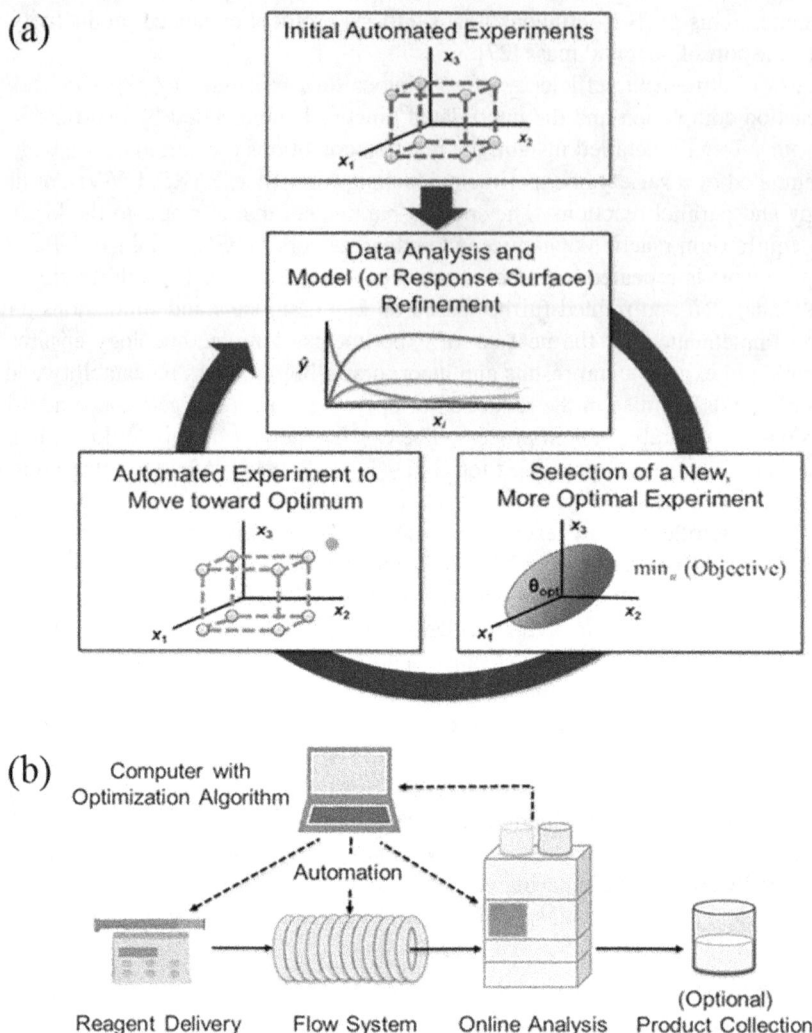

FIGURE 13.4 (a) Protocol for automated process intensification via a feedback loop. (b) Schematic of the automated continuous flow operation and optimization [34].

place inside it and their role on the particle size variation of nano-ZnO synthesized by a "green" continuous flow process using supercritical water. The simulation was performed with Fluent© software to understand the hydrodynamics, nucleation, and growth kinetics of the nanoparticles. SIMPLE algorithm was employed to obtain the temperature distribution profile and the optimal flow conditions. The population balance module validated and agreed with the particle size distribution data estimated from transmission electron microscopy (TEM). Thus, these findings substantiated that typically numerical simulations serve as smart and attractive tools for the effective design of continuous flow reactors [30]. In another work, the continuous flow process for the production of cerium oxide nanoparticles in a tubular-shaped reactor with supercritical water was carried out. The reactors involved had various configurations. Through numerical simulation and advanced experiments such as neutron radiography, the influence of different flow rates on the fabricated nanoparticle size and correlation with the reactor geometry were explored. ANSYS FLUENT 12.1 was used for the three-dimensional numerical simulations on unsteady conditions. For the verification of the simulated temperature and flow profiles, the neutron radiography technique was employed for imaging and analyzing the thermal characteristics. The observations depicted that the formation of the vortices was caused due to the varying density of the precursor solution kept at room temperature and that of the supercritical water was responsible for the buoyancy force. As a consequence, an increment in the average size of the nanoparticles was noted. Therefore, the computational power of solving mathematical equations and specific models is beneficial for ensuring eco-friendly routes of synthesis by predominantly diminishing the energy consumption, wastage of raw materials for optimization, and multiplying the process safety [31].

Over the years, the focus has been on efficient design and prompt optimization methodologies of reliable continuous flow synthesis platforms for ascertaining the intrinsic dynamics of the reactions to precisely control the desirable properties of the GQDs being fabricated. The increasing popularity of a continuous workflow at the laboratory and pilot scale has ushered in numerous changes in the experimental data analysis and process intensification. This led to the application of machine learning to predict the flow reactor's performance. Tao et al. utilized a novel ML algorithm for developing an autonomous microfluidic assembly. This new protocol was designed to understand the interlink between the structure, targeted property, and process condition. An environment-friendly synthesis route for the fabrication of gold nanoparticles was employed. For varying nanoparticle sizes, the reaction parameters were manipulated by real-time monitoring through in-line spectroscopic measurements. During the process enhancement, three different ML algorithms were incorporated for examining the data received from the characterization techniques, which guided in comprehending the relation between the desired properties of nanoparticulate gold and the reaction environment. From the predictions made by the models, it was clear that the results matched well with the data procured from experimentation. The gold nanoparticles synthesized by this autonomous platform exhibited superior optical properties. However, it was revealed that if the need arises, then the model can be tuned to get the best morphology and particle size distribution at the expense of the properties under consideration. Besides, the number of targets could be increased to accomplish higher precision with increased selectivity. Such versatility and flexibility of this state-of-the-art ML-driven continuous flow synthesis of GQDs hold tremendous importance in the years to come [32]. The implementation of machine learning algorithms for rational design and development of sustainable continuous flow synthesis has recently experienced a boom. The ubiquitous impact of machine learning to augment the new GQDs fabrication is now often realized. The recent era of "Industry 4.0" is proving time and again that machine learning and continuous flow synthesis of GQDs are two sides of the same coin as has been perfectly revealed in the studies by Mekki-Berada et al. where a deep neural network was utilized for achieving the favored absorbance profile of silver nanoparticles synthesized in a microfluidic setup. They used the Bayesian optimization algorithm that proved to be highly efficient and was in agreement with the experimental results. Information on the reason behind varying spectra for different parameters and the influence of the nanoparticle shape was derived. These examples further stress the necessity to couple continuous flow approach with simulations and ML to fasten the data acquisition, analysis, and experimentation [33]. Table 13.1 summarizes the significance of computational tools for a meticulous understanding of the various process parameters and how they influence the targeted characteristics and nucleation-growth kinetics.

13.3 CASE STUDIES ON GREEN QUANTUM DOTS, THEIR ADVANTAGES, DISADVANTAGES, AND STRUCTURE-PROPERTY CORRELATION

13.3.1 CONTINUOUS FLOW SYNTHESIS OF ZnO GQDs AND THEIR BIOLOGICAL APPLICATIONS

ZnO quantum dots (ZnO QDs) have become the center of attention among researchers and engineers for the myriad of opportunities that they hold in the vital areas of energy, environment, optoelectronics, agriculture, and biology. In the past few years, ZnO QDs with newer architectures and novel physiochemical, optical, biological, and electrical properties have been designed and fabricated. Thanks to the flexible reaction kinetics of ZnO seed QD formation, several well-established synthesis routes

TABLE 13.1

Summary on the Significance of Computational Methods for Varying Reactor Designs to Synthesize Nanomaterial or Quantum Dots with/without Real-Time Monitoring

Nanoparticles/Quantum Dots	Reactor Design	Simulation/Computation-Based Approaches	Importance	Real-Time Monitoring Techniques
Ag nanoparticles	Flow reactor with heart-shaped cells [35]	CFD simulation	Mixing and flow characteristics were studied. It revealed the flow pattern was affected by the variation in flow rate. The mixing efficiency increased because of the unique reactor geometry	CCD camera mounted on top of a stereo microscope
CeO$_2$ nanoparticles	Static microstructured channel T-mixers [36]	CFD simulation	Better mixing efficiency was observed due to flow engulfment. This influenced the key properties and particle size distribution of the nanoparticles	
ZnO nanoparticles	Microfluidic reactor [37]	Numerical simulation	Optimization of time pulsing for enhancing the degree of mixing	
Au nanoparticles	Microfluidic reactor [38]	Numerical simulation	Size-controlled time-resolved formation of Au nanoparticles	Transmission electron microscopy
ZnO nanoparticles	Microfluidic reactor [39]		Narrow particle size distribution obtained by close monitoring of the nucleation and growth of particles	In situ SAXS/WAXS and UV-Vis experiments
Ag nanoparticles	Tubular micro-reactor [40]	CFD simulation	Enhance the mixing efficiency and correlate with the residence time, and flow rate. The surface-area-to-volume ratio of the micro-reactors, and their influence on the particle size were studied.	
Carbon dots	Micro-reactor [41]		Viable rapid screening of a large number of reaction conditions. The targeted PL property was achieved	PL spectroscopy
Au nanoparticles	Microfluidic reactor [32]	Machine learning algorithms	The synthesis is made fully automated, saving cost and time, and prevents wastage – a greener route. Desired optical properties are enriched	Spectroscopic measurements
Ag nanoparticles	Microfluidic reactor [33]	Deep learning	Relationship between varying shapes of nanoparticles and the experimental conditions that influenced the recorded spectra. Fast process optimization	

have been molded by the principles of green chemistry for obtaining high-performing ZnO green quantum dots (ZnO GQDs). However, these batch syntheses have poor reproducibility, complexity in scaling up, and inadequate process control. Hence, addressing these stumbling blocks and devising continuous flow chemistry-based protocols for easing the scalability, ensuring safety, better heat and mass transfer, and high yield, has become an absolute necessity. The reduced dimensions of the flow reactors having a high aspect ratio or surface-area-to-volume ratio enable improved fluid flow dynamics and homogenous distribution of heat. These attributes of flow reactor technology radically diminish energy consumption and material wastage by reaction telescoping, making it one of the best suited green chemistry approaches. This method of fabrication has become increasingly recognized for ZnO GQDs for its utility in biology. ZnO nanowires synthesized by hydrothermal route in a microfluidic reactor by Kim et al. emphasized the feasibility of economical, environment-friendly synthetic chemistry for obtaining the most appropriate geometry to be employed for mechanical cell lysis. The synthesis involved a polydimethylsiloxane microstructured channel that was oriented to bond to the substrate made of glass which consisted of a layer of ZnO seed. The precursor solutions were fed at a constant flow rate into the micro-channel through syringe pumps. The substrate was heated at 95°C, and the synthesis time was controlled. After the subsequent synthesis, DI water was used to rinse the channel thoroughly, followed by annealing the sample at 95°C by allowing airflow into the channel. It was compared with a conventional hydrothermal synthesis of ZnO nanowires under different arrays of microposts inside the channel. This device was utilized for extracting intracellular material. Besides, it showed a boosted performance in cell lysis for different cell types. Thus, this process not only saved time and resources, but also paved the way for inexpensive lysis [42]. The possibilities of ZnO QDs in bio-imaging and as fluorescent probes or labels have grown by leaps and bounds owing to its wide bandgap, reduced toxicity, and enriched surface defect chemistry [43,44]. The visible emission that they exhibit is attributed to the presence of intermediate defect states, of which singly ionized oxygen vacancies are believed to be the dominant ones. The transition of electrons from the band edge to the defect levels results in the aberrant photoluminescent (PL) properties of ZnO QDs, which are enhanced with diminishing particle sizes. These PL features have unfathomable implications in the fields of biology. This intrinsic defect chemistry has established ZnO QDs as one of the most stable nanomaterial systems against harmful UV radiations with superior PL quantum efficiency achieved through the design of new synthesis processes. Efficient surface functionalization via appropriate synthesis approaches minimizes the potential of ZnO QDs to undergo dissolution to generate Zn^{2+}, thereby augmenting its biocompatibility. These merits affirm the potential of ZnO QDs as a biological fluorescent probe, and a green continuous flow fabrication method reported by Aleksandra et al. represented the formation of monodispersed, colloidally stable GQDs. The micro-reactor assembly consisted of two syringe pumps that fed the precursor solutions into the inlet of the cylindrical pipes, which then led to the micromixer. The temperature of the reaction, residence times, and flow rates were adjusted to obtain GQDs with PL quantum efficiency of 30%, superior UV absorbing capabilities, and altered surface chemistry assisted by befitting capping agents. Additionally, the influence of flow conditions on the particle size regulated by the nucleation and growth kinetics was noted. It was evident that continuous flow micro-reactors enable subtle and accommodating reaction microenvironment for betterment of the characteristics. In the age of leading-edge innovations, this "green" synthesis strategy provided an effortless translation of lab-bench synthesis to industrial-scale production while guaranteeing higher precision and safety. This further strengthened the reputation of ZnO GQDs fabricated by continuous flow synthesis in biological domains [45]. The fluorescence-enhancing potential of nanostructured ZnO was integrated with a microfluidic platform to systematically utilize the device as detectors for cancer biomarkers. Rigorous optimization strategies for fluorescence immunoassays revealing remarkable performance are underway, and with time it has been realized that coupling microfluidic chemistry will eventually boost the throughput. The microfluidic device was designed such that the pumps could inject the precursor fluids into the glass capillaries of specified dimensions (Figure 13.5). The kinetically controlled reaction conditions confirmed the formation of ZnO

(a)

(b)

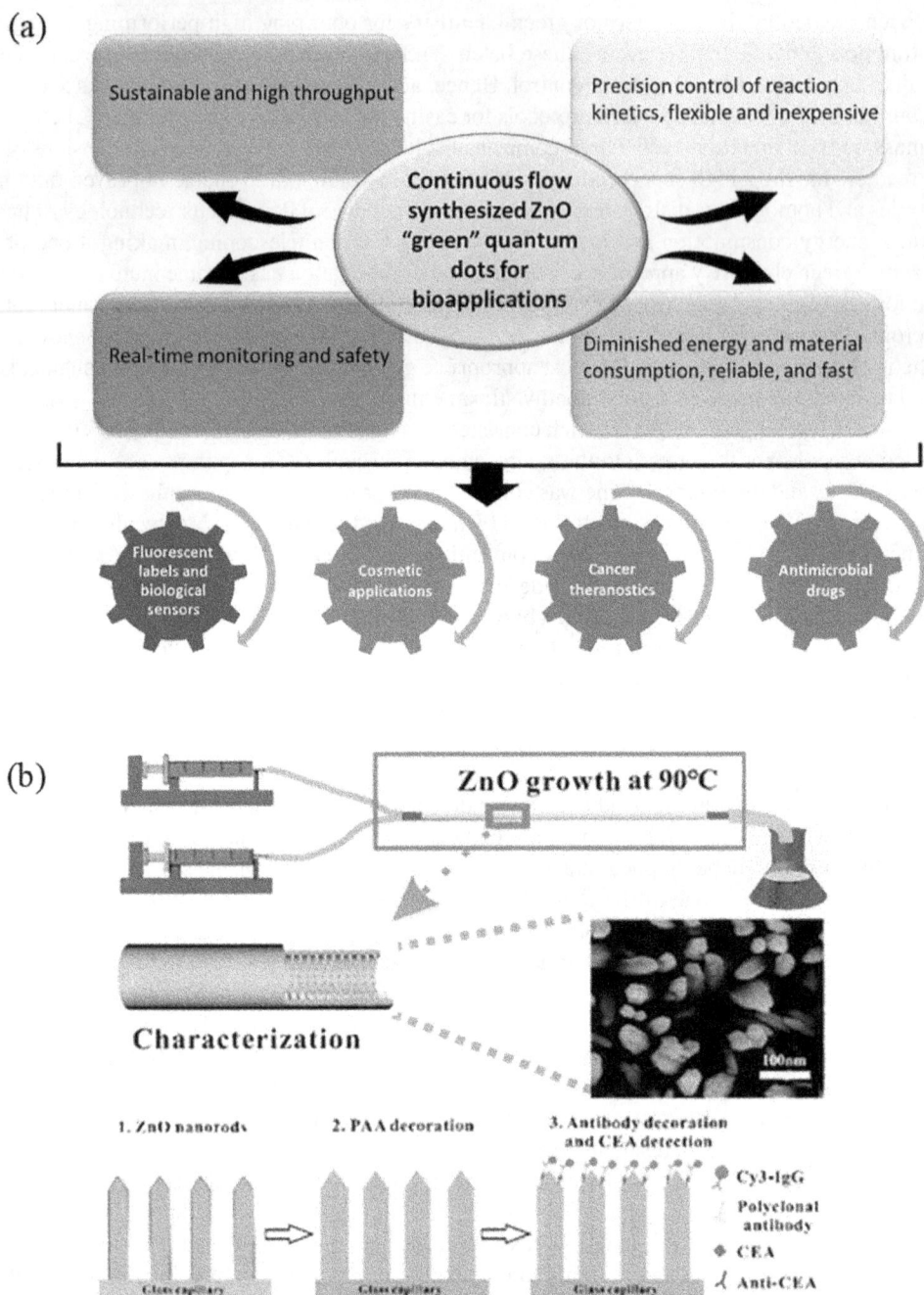

FIGURE 13.5   (a) Illustration of the promising opportunities of ZnO "green" quantum dots synthesized in continuous flow reactors in biology. (b) Schematic of the microfluidic platform designed for the construction of ZnO nanorods inside the glass capillaries and diagram of the optimized fluorescence immunoassay exhibited by the PAA surface-modified ZnO nanorods grown in the glass capillary's inner surface [46].

nanorods with the coveted aspect ratio critical for intensifying the fluorescence characteristics. It was discerned that those flow parameters could appreciably modify the dimensions of the nanorods; for increasing flowrate, the aspect ratio was hampered. Furthermore, the microfluidic process chemistry simplifies the surface passivation methods of ZnO by applying PAA. Therefore, continuous flow synthesis renders plenty of opportunities for rapid point-of-care tests as a solution for smart healthcare services [46].

13.3.2 At the Crossroads of Flow Chemistry and Bioactive CeO_2 GQDs for Treatment of ROS-Mediated Diseases and Improvement in Crop Health

Ceria (CeO_2) quantum dots have gained considerable attention in the fields of catalysis, remediation of the environment, biomedicine, and agriculture. In cerium, owing to their fundamental electronic configuration, Ce^{3+} and Ce^{4+} coexist as a redox couple. At the nano-regime, with a higher surface area, the self-regenerative cycle of the +3 and +4 valence states of Ce acts as the key player for tailoring surface chemistry by generating oxygen vacancies. In nanostructured CeO_2, this inherent behavior becomes predominant, and for that reason, CeO_2 is often referred to as a reservoir for oxygen. Therefore, in CeO_2 QDs, many of the surface traits, particularly their redox behavior, are notably modified. Regulating the Ce^{3+}/Ce^{4+} ratio judiciously renders these QDs with phenomenal antioxidant potential, which is enhanced when the particle sizes are further reduced and are essential for preventing the build-up of intracellular oxidative stress. This makes CeO_2 QDs a perfect candidate for being widely exploited in treating reactive oxygen species (ROS)-mediated ailments and improving crop health. These intrinsic size-tailored physiochemical properties that make CeO_2 QDs so novel and primarily influence their utility in biology and agriculture are governed by the synthesis methods adopted. Thus, advanced fabrication methodologies which decisively control the reaction kinetics and parameters such as temperature, pressure, and pH are of utmost priority to produce QDs with narrow particle size distribution. These prerequisites have led to the emergence of continuous flow synthesis for industrial-scale production of CeO_2 QDs with improved properties. Efforts of the competing technologies in agriculture and biomedicine have eliminated the environmentally hazardous chemical processes and substituted these with synthesis approaches for CeO_2 green quantum dots (CeO_2 GQDs) that follow the footprints of green chemistry.

Exposito and his group employed a continuous flow micro-reactor setup involving a non-toxic deep eutectic solvent system (reline consisting of water). The syringe pump fed the precursor with the solvent into the coiled PFA reactor placed inside a hot oil bath. The end product was collected, at the output end in a vial immersed in a water bath kept at room temperature, as portrayed in Figure 13.6. Under such conducive environment, the reaction was rapid, with enhanced yield and simultaneously saved time. The reaction of nanostructured CeO_2, which otherwise took hours to complete in well-known batch reactors, was completed in a few minutes by means of flow synthesis with absolutely no compromise on the nanoparticles' quality. This was possible because of the exceptional heat and mass transfer that quickens the nucleation-growth mechanism. Under varying flow rates and residence times, the growth phenomena were governed by the Ostwald ripening and oriented attachment crystallizing into particles with favored shapes. Estimations based on the pseudo-first-order kinetics and considering the reactor as isothermal revealed that the flow was laminar inside the reactor. The transition of morphology is attributed to the hydrodynamics of the reactor, thereby indicating the significant implications this has on the structural changes and surface functionalization of the nanoparticles. Above all, this affordable, unconventional, continuous flow process involving the use of biodegradable and eco-friendly solvents have paved the path of "green" manufacturing protocols not only for CeO_2 GQDs but other green quantum dots too [47].

In another work, Yao et al. explored alternative solutions to combat the poor mixing efficiency and uncontrolled reaction parameters leading to the spatial temperature gradients, thereby impacting

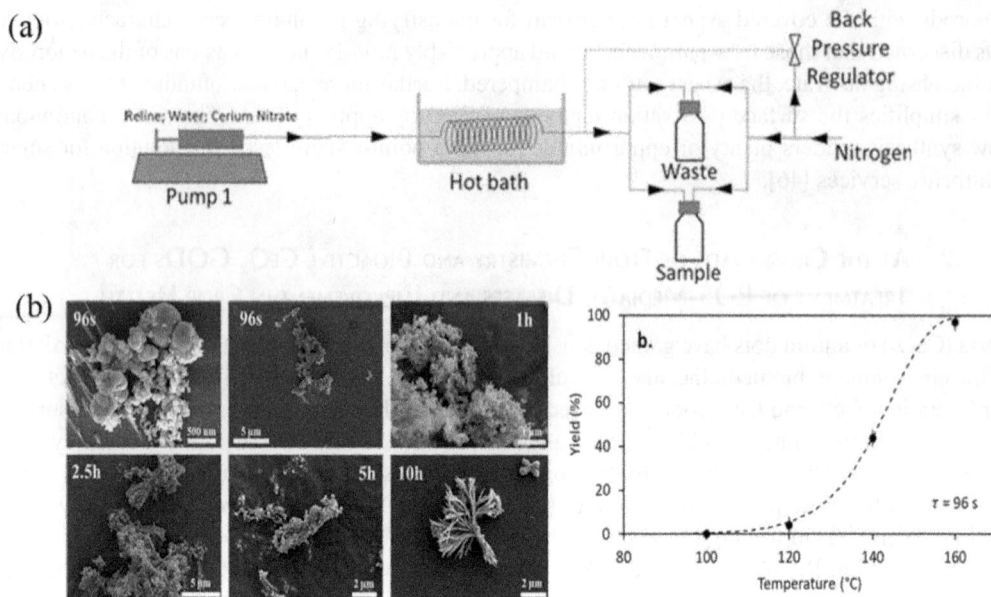

FIGURE 13.6 (a) Continuous flow micro-reactor assembly. (b) Scanning electron microscopy images showing the morphology of CeO_2 nanoparticles formed at different residence times at constant temperature conditions. The plot reveals the influence of temperature on the yield for a residence time calculated to be 96 seconds [47].

the nanoparticle sizes fabricated in the traditional stirring tank reactors. Hence, they developed a one-of-a-kind micro-reactor integrated with membrane dispersion technology for nano-CeO_2 synthesis and analyzed the consequence of the degree of mixing, temperature, supersaturation, and pH of the environment. Evaluation of the particle quality by X-ray diffraction patterns and transmission electron microscopy signified that a regulated reaction environment led to non-agglomerated, equi-sized nanocrystals formation by the micro-reactor synthesis route, and therefore, the larger surface-area-to-volume ratio of the particles drastically improved the catalytic activity checked in the presence of H_2O_2 [48].

A similar catalytic response at the nano-biointerface has been reported previously too for nano-CeO_2 with very small particle sizes [49]. Thus, it obviates the need to produce a greater number of monodispersed CeO_2 quantum dots assisted by specialized continuous flow reactors for superior functionality with no potential threat to the environment. Besides, the surface defect chemistry is enriched when exposed to the biological media as understood from the first-principles investigations of nano-CeO_2 [50]. Moreover, several preclinical studies have revealed that the excellent biocompatibility is attributed to the enzyme mimicking and antioxidant or ROS scavenging potential as a result of the high concentration of oxygen vacancy defects. Hence, its multifunctional radical scavenging enzyme mimetic response has fostered the potential for treatment of ROS-related disorders such as chronic wound healing, neurodegenerative diseases and even cancer [51].

Similarly, the abiotic stress-related ROS in plants is scavenged by defect-rich CeO_2 GQDs, thus improving the plant growth conditions. However, the activity of CeO_2 GQDs varies as a function of the soil moisture content. In addition, CeO_2 GQDs can effectively regulate the rate of photosynthesis in plants, subjected to surface modification and size [52]. Although it still remains an unexplored regime, the rapidly expanding "green" quantum dots industry calls for an immediate paradigm shift in developing synthesis routes along the lines of microfluidics and green chemistry for CeO_2 GQDs, which is the crux for heralding newer technologies and innovations in biomedicine and agriculture (Figure 13.7).

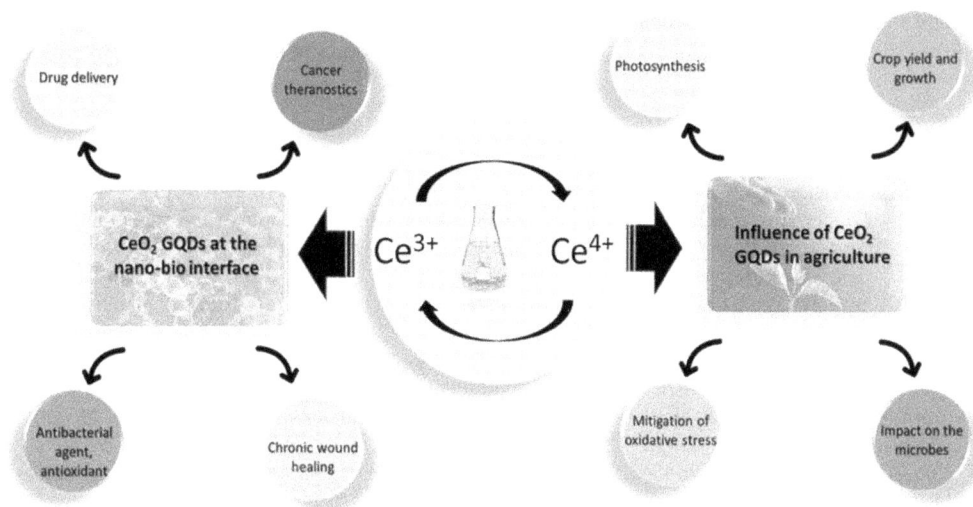

FIGURE 13.7 Schematic diagram for depicting the importance of CeO_2 GQDs in agriculture and treatment of ROS-mediated diseases.

13.4 CONCLUSIONS AND OUTLOOK

The formulation of continuous flow synthesis strategies for "green" quantum dots fabrication has become almost indispensable in all walks of leading-edge innovations in biological, agricultural, and energy domains. Since its inception, the conjugation of green chemistry and flow synthesis protocols has broken the stereotypical ideas revolving around GQDs production. The uniformity in heat and mass transfer accomplished through miniaturized reactor dimensions (larger aspect ratio) accelerated the GQDs reaction kinetics under the preferred microenvironment inside the reactor chamber. The scope of flow chemistry is no longer confined within the narrow boundaries of lab-bench microfluidic novel nanomaterial synthesis. Nonetheless, it has dramatically advanced to industrial-scale fabrication with assured safety, tremendous process intensification, and a high degree of reproducibility. Some promising implications of this have culminated in the adoption of real-time reaction monitoring and analytical methods, simulation-based approaches, and machine learning algorithms to optimize synthesis conditions. This enables to elevate product profile, boost yield, reduce cost, and eliminate toxic wastes and intermediates, and it is in accordance with the philosophy of environmentally benign quantum dots and their corresponding sustainable synthetic platforms.

Despite several technological breakthroughs in flow chemistry, several bottlenecks exist in the detailed understanding of GQDs nucleation-growth complexities and reaction engineering "know-how". Hence, emphasis must be laid on maximizing the full potential of the flow reactors by building a smooth interface for the execution of mathematical simulations, AI-based feedback mechanisms, mechanistic studies, in-line characterization techniques, and simultaneous process scale-up. Even though attaining these objectives can often be puzzling at one's disposal, however, the explicit recognition of continuous flow methods has expedited the interest of researchers to come up with elegant platforms that holistically demonstrate the advantages of efficient reactor format in combination with parallel automated optimization protocols.

The impactful case studies on multifunctional continuous flow synthesized ZnO and CeO_2 GQDs in biology and agriculture demystified many of the blind alleys regarding the industry-centric relevance of these GQDs and a way forward, and how the credibility of being inevitably "green" is attributed to the high selectivity of reaction pathway, minimum energy consumption, and safe experimental conditions. Therefore, beyond any doubt, continuous flow reactor assemblies have proved revolutionary with their intrinsic environmental benefits and hold immense promise in unique green quantum dots discovery, the significance of which is expected to grow in leaps and bounds in the years to come.

REFERENCES

1. Mahin, J., Franck, C.O., Fanslau, L., Patra, H.K., Mantle, M.D., Fruk, L. and Torrente-Murciano, L., Green, scalable, low cost and reproducible flow synthesis of biocompatible PEG-functionalized iron oxide nanoparticles. *Reaction Chemistry & Engineering*, 6(10) (2021), 1961–1973.
2. Kumar, D.R., Kasture, M., Prabhune, A.A., Ramana, C.V., Prasad, B.L.V. and Kulkarni, A.A., Continuous flow synthesis of functionalized silver nanoparticles using bifunctional biosurfactants. *Green Chemistry*, 12(4) (2010), 609–615.
3. Baragau, I.A., Power, N.P., Morgan, D.J., Lobo, R.A., Roberts, C.S., Titirici, M.M., Middelkoop, V., Diaz, A., Dunn, S. and Kellici, S., Efficient continuous hydrothermal flow synthesis of carbon quantum dots from a targeted biomass precursor for on–off metal ions nanosensing. *ACS Sustainable Chemistry & Engineering*, 9(6) (2021), 2559–2569.
4. You, H. and Fang, J., Particle-mediated nucleation and growth of solution-synthesized metal nanocrystals: A new story beyond the LaMer curve. *Nano Today*, 11(2) (2016), 145–167.
5. Sui, J., Yan, J., Liu, D., Wang, K. and Luo, G., Continuous synthesis of nanocrystals via flow chemistry technology. *Small*, 16(15) (2020), 1902828.
6. Długosz, O. and Banach, M., Inorganic nanoparticle synthesis in flow reactors–applications and future directions. *Reaction Chemistry & Engineering*, 5(9) (2020), 1619–1641.
7. Choi, C.-H., Su, Y.-W., Chang, C.-H. Effects of fluid flow on the growth and assembly of ZnO nanocrystals in a continuous flow microreactor, *CrystEngComm* 15 (17) (2013), 3326–3333.
8. Talapin, D.V., Rogach, A.L., Haase, M. and Weller, H., Evolution of an ensemble of nanoparticles in a colloidal solution: Theoretical study. *The Journal of Physical Chemistry B*, 105(49) (2001), 12278–12285.
9. Harada, M. and Kizaki, S., Formation mechanism of gold nanoparticles synthesized by photoreduction in aqueous ethanol solutions of polymers using in situ quick scanning x-ray absorption fine structure and small-angle X-ray scattering. *Crystal Growth & Design*, 16(3) (2016), 1200–1212.
10. Polte, J., Erler, R., Thunemann, A.F., Sokolov, S., Ahner, T.T., Rademann, K., Emmerling, F. and Kraehnert, R., Nucleation and growth of gold nanoparticles studied via in situ small angle X-ray scattering at millisecond time resolution. *ACS Nano*, 4(2) (2010), 1076–1082.
11. Jolliffe, H.G. and Gerogiorgis, D.I., Process modelling and simulation for continuous pharmaceutical manufacturing of ibuprofen. *Chemical Engineering Research and Design*, 97 (2015), 175–191.
12. Powell, M.J., Marchand, P., Denis, C.J., Bear, J.C., Darr, J.A. and Parkin, I.P., Direct and continuous synthesis of VO_2 nanoparticles. *Nanoscale*, 7(44) (2015), 18686–18693.
13. Lummiss, J.A., Morse, P.D., Beingessner, R.L. and Jamison, T.F., Towards more efficient, greener syntheses through flow chemistry. *The Chemical Record*, 17(7) (2017), 667–680.
14. Wegner, J., Ceylan, S. and Kirschning, A., Ten key issues in modern flow chemistry. *Chemical Communications*, 47(16) (2011), 4583–4592.
15. Maceiczyk, R.M., Lignos, I.G. and Andrew, J.D., Online detection and automation methods in microfluidic nanomaterial synthesis. *Current Opinion in Chemical Engineering*, 8 (2015), 29–35.
16. Hessel, V., Hardt, S., Löwe, H. and Schönfeld, F., Laminar mixing in different interdigital micromixers: I. Experimental characterization. *AIChE Journal*, 49(3) (2003), 566–577.
17. Kováts, P., Velten, C., Mansour, M., Thévenin, D. and Zähringer, K., Mixing characterization in different helically coiled configurations by laser-induced fluorescence. *Experiments in Fluids*, 61(9) (2020), 1–17.
18. Gürsel, I.V., Kurt, S.K., Aalders, J., Wang, Q., Noël, T., Nigam, K.D., Kockmann, N. and Hessel, V., Utilization of milli-scale coiled flow inverter in combination with phase separator for continuous flow liquid–liquid extraction processes. *Chemical Engineering Journal*, 283 (2016), 855–868.
19. Uson, L., Sebastian, V., Arruebo, M. and Santamaria, J., Continuous microfluidic synthesis and functionalization of gold nanorods. *Chemical Engineering Journal*, 285 (2016), 286–292.
20. Newman, S.G. and Jensen, K.F., The role of flow in green chemistry and engineering. *Green Chemistry*, 15(6) (2013), 1456–1472.
21. Yoshida, J.I., Kim, H. and Nagaki, A., Green and sustainable chemical synthesis using flow microreactors. *ChemSusChem*, 4(3) (2011), 331–340.
22. Bourne, J.R., Mixing and the selectivity of chemical reactions. *Organic Process Research & Development*, 7(4) (2003), 471–508.

23. Kováts, P., Velten, C., Mansour, M., Thévenin, D. and Zähringer, K., Mixing characterization in different helically coiled configurations by laser-induced fluorescence. *Experiments in Fluids*, 61(9) (2020), 1–17.

24. Pal, S.K., Dhasmana, P., Nigam, K.D.P. and Singh, V., Tuning of particle size in a helical coil reactor. *Industrial & Engineering Chemistry Research*, 59(9) (2019), 3962–3971.

25. Kurt, S.K., Gelhausen, M.G. and Kockmann, N., Axial dispersion and heat transfer in a milli/microstructured coiled flow inverter for narrow residence time distribution at laminar flow. *Chemical Engineering & Technology*, 38(7) (2015), 1122–1130.

26. Lin, X.Z., Terepka, A.D. and Yang, H., Synthesis of silver nanoparticles in a continuous flow tubular microreactor. *Nano Letters*, 4(11) (2004), 2227–2232.

27. Hallot, G., Cagan, V., Laurent, S., Gomez, C. and Port, M., A greener chemistry process using microwaves in continuous flow to synthesize metallic bismuth nanoparticles. *ACS Sustainable Chemistry & Engineering*, 9(28) (2021), 9177–9187.

28. Kitson, P.J., Rosnes, M.H., Sans, V., Dragone, V. and Cronin, L., Configurable 3D-Printed millifluidic and microfluidic 'lab on a chip' reactionware devices. *Lab on a Chip*, 12(18) (2012), 3267–3271.

29. Pudza, M. Y., et al. Sustainable synthesis processes for carbon dots through response surface methodology and artificial neural network. *Processes* 7, 704 (2019)

30. Leybros, A., Piolet, R., Ariane, M., Muhr, H., Bernard, F. and Demoisson, F., CFD simulation of ZnO nanoparticle precipitation in a supercritical water synthesis reactor. *The Journal of Supercritical Fluids*, 70 (2012), 17–26.

31. Sugioka, K.I., Ozawa, K., Kubo, M., Tsukada, T., Takami, S., Adschiri, T., Sugimoto, K., Takenaka, N. and Saito, Y., Relationship between size distribution of synthesized nanoparticles and flow and thermal fields in a flow-type reactor for supercritical hydrothermal synthesis. *The Journal of Supercritical Fluids*, 109 (2016), 43–50.

32. Tao, H., Wu, T., Kheiri, S., Aldeghi, M., Aspuru-Guzik, A. and Kumacheva, E., Self-driving platform for metal nanoparticle synthesis: Combining microfluidics and machine learning. *Advanced Functional Materials*, 31(51) (2021), 2106725.

33. Mekki-Berrada, F., Ren, Z., Huang, T., Wong, W.K., Zheng, F., Xie, J., Tian, I.P.S., Jayavelu, S., Mahfoud, Z., Bash, D. and Hippalgaonkar, K., Two-step machine learning enables optimized nanoparticle synthesis. *NPJ Computational Materials*, 7(1) (2021), 1–10.

34. Reizman, B.J. and Jensen, K.F., Feedback in flow for accelerated reaction development. *Accounts of Chemical Research*, 49(9) (2016), 1786–1796.

35. Yang, M., Yang, L., Zheng, J., Hondow, N., Bourne, R.A., Bailey, T., Irons, G., Sutherland, E., Lavric, D. and Wu, K.J., Mixing performance and continuous production of nanomaterials in an advanced-flow reactor. *Chemical Engineering Journal*, 412 (2021), 128565.

36. Palanisamy, B. and Paul, B., Continuous flow synthesis of ceria nanoparticles using static T-mixers. *Chemical Engineering Science*, 78 (2012), 46–52.

37. Kang, H.W., Leem, J., Yoon, S.Y. and Sung, H.J., Continuous synthesis of zinc oxide nanoparticles in a microfluidic system for photovoltaic application. *Nanoscale*, 6(5) (2014), 2840–2846.

38. Li, Y., Sanampudi, A., Raji Reddy, V., Biswas, S., Nandakumar, K., Yemane, D., Goettert, J. and Kumar, C.S., Size evolution of gold nanoparticles in a millifluidic reactor. *ChemPhysChem*, 13(1) (2012), 177–182.

39. Herbst, M., Hofmann, E. and Förster, S., Nucleation and growth kinetics of ZnO nanoparticles studied by in situ microfluidic SAXS/WAXS/UV–Vis experiments. *Langmuir*, 35(36) (2019), 11702–11709.

40. Liu, H., Huang, J., Sun, D., Lin, L., Lin, W., Li, J., Jiang, X., Wu, W. and Li, Q., Microfluidic biosynthesis of silver nanoparticles: Effect of process parameters on size distribution. *Chemical Engineering Journal*, 209 (2012), 568–576.

41. Lu, Y., Zhang, L. and Lin, H., The use of a microreactor for rapid screening of the reaction conditions and investigation of the photoluminescence mechanism of carbon dots. *Chemistry–A European Journal*, 20(15) (2014), 4246–4250.

42. Kim, J., Hong, J.W., Kim, D.P., Shin, J.H. and Park, I., Nanowire-integrated microfluidic devices for facile and reagent-free mechanical cell lysis. *Lab on a Chip*, 12(16) (2012), 2914–2921.

43. Ma, Y.Y., Ding, H. and Xiong, H.M., Folic acid functionalized ZnO quantum dots for targeted cancer cell imaging. *Nanotechnology*, 26(30) (2015), 305702.

44. Xiong, H.M., Xu, Y., Ren, Q.G. and Xia, Y.Y., Stable aqueous ZnO@ polymer core– shell nanoparticles with tunable photoluminescence and their application in cell imaging. *Journal of the American Chemical Society*, 130(24) (2008), 7522–7523.

45. Schejn, A., Frégnaux, M., Commenge, J.M., Balan, L., Falk, L., Schneider, R. Size-controlled synthesis of ZnO quantum dots in microreactors. *Nanotechnology*, 25(14) (2014), 145606.

46. Wu, Z., Zhao, D., Hou, C., Liu, L., Chen, J., Huang, H., Zhang, Q., Duan, Y., Li, Y. and Wang, H., Enhanced immunofluorescence detection of a protein marker using a PAA modified ZnO nanorod array-based microfluidic device. *Nanoscale*, 10(37) (2018), 17663–17670.

47. Exposito, A.J., Barrie, P.J. and Torrente-Murciano, L., Fast synthesis of CeO$_2$ nanoparticles in a continuous microreactor using deep eutectic reline as solvent. *ACS Sustainable Chemistry & Engineering*, 8(49) (2020), 18297–18302.

48. Yao, H., Wang, Y. and Luo, G., A size-controllable precipitation method to prepare CeO$_2$ nanoparticles in a membrane dispersion microreactor. *Industrial & Engineering Chemistry Research*, 56(17) (2017), 4993–4999.

49. Walkey, C., Das, S., Seal, S., Erlichman, J., Heckman, K., Ghibelli, L., Traversa, E., McGinnis, J.F. and Self, W.T., Catalytic properties and biomedical applications of cerium oxide nanoparticles. *Environmental Science: Nano*, 2(1) (2015), 33–53.

50. Fronzi, M., Piccinin, S., Delley, B., Traversa, E. and Stampfl, C., Water adsorption on the stoichiometric and reduced CeO$_2$ (111) surface: A first-principles investigation. *Physical Chemistry Chemical Physics*, 11(40) (2009), 9188–9199.

51. Xu, C. and Qu, X., Cerium oxide nanoparticle: A remarkably versatile rare earth nanomaterial for biological applications. *NPG Asia Materials*, 6(3) (2014), e90.

52. Prakash, V., Peralta-Videa, J., Tripathi, D.K., Ma, X. and Sharma, S., Recent insights into the impact, fate and transport of cerium oxide nanoparticles in the plant-soil continuum. *Ecotoxicology and Environmental Safety*, 221 (2021), 112403.

Index

Note: **Bold** page numbers refer to tables and *italic* page numbers refer to figures.

For Product Safety Concerns and Information please contact our EU
representative GPSR@taylorandfrancis.com
Taylor & Francis Verlag GmbH, Kaufingerstraße 24, 80331 München, Germany

www.ingramcontent.com/pod-product-compliance
Lightning Source LLC
Chambersburg PA
CBHW081054220326
41598CB00038B/7087